Backcountry Press Presents

THE KLAMATH MOUNTAINS

A NATURAL HISTORY

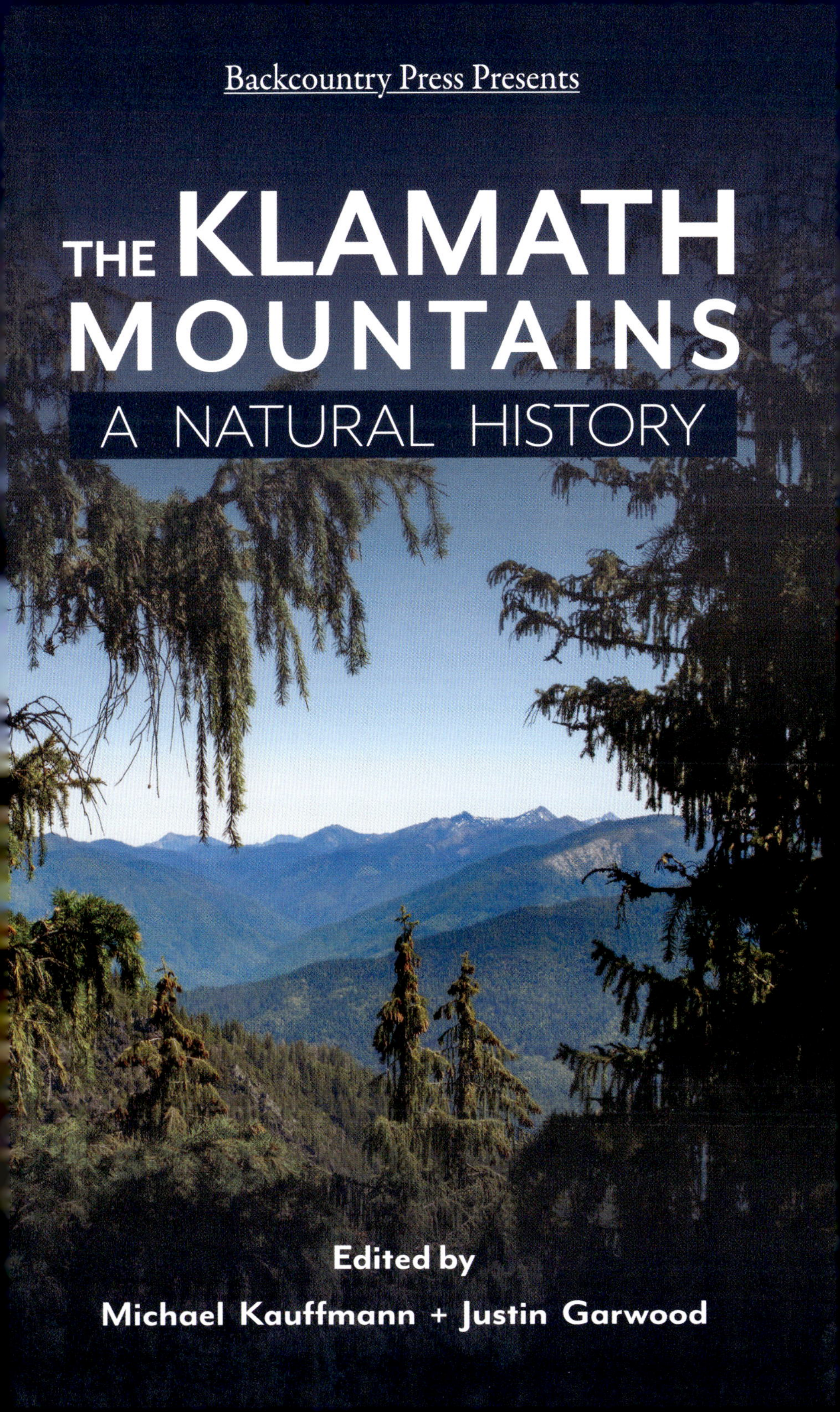

Edited by

Michael Kauffmann + Justin Garwood

Klamath Mountains Geomorphic Province

So overlooked are some of these mountain areas that they have escaped appellation and remain anonymous in the truest sense of the word.

—Bubba Suess

The Klamath Mountains

A Natural History

Plants – Animals – Geology – Fire
Water – Climate – Cultural History

Edited by
Michael Kauffmann and **Justin Garwood**

With
James K. Agee, Frank Lake, David Rains Wallace, Matt Johnson, Dawn Blake, Charley Reed, Keith Parker, Mark Bailey, Rosemary Sherriff, Tim Palmer, Jeffrey Kane, Jordan Mayor, Dana York, Tom Carlberg, Eric Peterson, Lucy Kerhoulas, Julie Evens, Julie Kierstead, Michael Murray, Chris Lee, Len Mazur, Brian Dykstra, Rod Crawford, Dana Ross, Paul Hammond, Chris Feldman, Andrew Kinziger, Greg Gray, Daniel Barton, Tiana Williams-Claussen, and Karen Reiss

Tracy's penstemon (*Penstemon tracyi*) in the Trinity Alps Wilderness. Photo by Christian Schwarz.

First Edition
9 8 7 6 5 4 3 2
www.backcountrypress.com

Cover painting of the Trinity Alps by Ken Jarvela
Book layout and cover design by Backcountry Press

Published by Backcountry Press • Kneeland, California

ISBN 978-1-941624-09-8
Library of Congress Control Number: 2022936955

BACKCOUNTRY PRESS
Humboldt County, CA

Created and Printed in the USA

FSC www.fsc.org
MIX
Paper from responsible sources
FSC® C005010

In Praise of

The Klamath Mountains: A Natural History

“*The Klamath Mountains: A Natural History* will be seen as one of the great natural history works of western North America.

—**REED NOSS**, Conservation Biologist and Author

“An absolutely necessary book for understanding the world we inhabit, and it's a model for how natural histories should be written. It's educational, entertaining, and thought-provoking, and will serve as the definitive reference for years to come.

—**AMY STEWART**, *New York Times* Bestselling Author

“A truly impressive collaborative effort, this book is a treasure trove of useful information for all who wish to learn more about the unique and fascinating Klamath Mountains bioregion.

—**STEPHEN C. SILLETT**, Kenneth L. Fisher Chair in Redwood Forest Ecology at Cal Poly Humboldt

“Hallelujah! Far too long ignored in favor of the Sierra Nevada, the richly diverse Klamath Mountains are at last the subject of a thorough and beautifully illustrated guide. This much-needed contribution will guide exploration and appreciation of the Klamaths for many years to come.

—**SUSAN HARRISON**, Professor of Ecology at the University of California, Davis

“This volume should be required reading for all visitors to the region, and for everyone interested in the natural history of this unique corner of North America.

—**PEPPER TRAIL**, Naturalist

“This book is a tour de force and will have a well-deserved and long-lasting impact on the region.

—**ERIK JULES**, Professor of Biology, Cal Poly Humboldt

“An invitation to a wider world of understanding California's unique Klamath Mountains and their role as an intrinsic, vital piece of this planet's biodiversity.

—**OBI KAUFMANN**, Author of *The California Field Atlas*

“Every map, every table, every figure and beautiful photograph guide you on an unforgettable journey through the natural history of the magical Klamath Mountains.

—**MATT RITTER**, Author of *A Californian's Guide to the Trees Among Us*

“Kauffmann and Garwood have assembled a detailed and thorough account of the Klamath Mountains that goes well beyond the flora and fauna, providing a solid introduction for any curious person visiting the area. I wish every part of California had a book like this!

—**KEN-ICHI UEDA**, Co-founder of iNaturalist

Table of Contents

Preface

by Michael Kauffmann & Justin Garwood

Organizing my (Michael's) world as a youngster was foundational in developing my relationship with the natural world. My approaches emerged with matchbox cars and baseball cards, but nature classification followed by way of birds and soon trees, rocks, bugs, and mammals. With a path toward connecting with the natural world, I graduated college having focused on macrobiology but my experiences were compartmentalized—as our American education systems tend to do. I have no early recollection of reading or discussing natural history by way of broad, regional interconnectedness. Upon graduation, I moved to California and began to learn a true natural history. On the way into Sequoia National Park one weekend, at the visitor's center bookstore in Three Rivers, I discovered Stephen Whitney's *A Sierra Club Naturalist's Guide to the Sierra Nevada* and carried a copy backpacking from the Mineral King to Eagle Lake (where I also saw my first foxtail pines). The Sierra Nevada is legendary for open vistas, stately trees, and gentle wanders across glacially polished mountainsides. Whitney's book allowed me to see how granite defines the Sierra Nevada.

Writing a natural history happens with definable landscapes. For it to be comprehensive, regional boundaries defined by geology, ecology, and climatic patterns—or realistically all three—create a space wherein a natural history emerges. The Sierra's granitic boundaries have produced numerous natural histories. Other regions of the West that have their own natural histories include Daniel Mathew's *Natural History of the Pacific Northwest Mountains* that weaves climate and ecology. A new tome, *Mountains of Nevada,* by David Charlet uses political boundaries to define an entire state's flora by way of 300+ mountain ranges. Lawrence R. Walker and Frederick H. Landau use climate to tell *A Natural History of the Mojave Desert*. The list goes on. These books are exciting because natural history is foundational in building and maintaining the human relationship with nature. The written relationship of natural history in the western world started with Charles Darwin and Alexander von Humboldt. Aldo Leopold, John Muir, Annie Dillard, Peter Matthiessen, Robin Wall Kimmerer, and others have continued these traditions—today, I believe we are experiencing a natural history renaissance.

Human survival is inherently linked to our ability to recognize natural assets, name them, and then to communicate that information with others. Without relationships with rocks, climates, fire, and other living species our ancestors would not have survived. Humans have never been without this relationship. By writing this book, I now understand that the relationship Native Peoples have with natural history is quite different than the relationship I have. I am now, thanks to this book and the connections it has allowed me to develop with others, beginning to develop an entirely new understanding of what a relationship to natural history means. It is common for western scientists, trained as I have been, to under-appreciate or altogether miss the connections Native Peoples have with a region's ecology. Missing that perspective leads to misconceptions and blind spots.

What I am learning is that much of what we see today in the Klamath Mountains was affected, to some degree, by long-term human habitation. Another major misconception taught in western science is the description of the assumed wild and wilderness being absent of human impact—when this is far from the truth. Much of what we have designated as wilderness in the United States were sculpted by Native People's stewardship. For example, numerous travel routes through our current wilderness were maintained for securing basketry, medicine, food resources, or reaching ceremonial sites (see First Peoples chapter). And today, biomass accumulation over the past 150 years is choking our forests. With fire suppression comes fuel accumulation compounded by higher fire intensity and subsequent forest loss. I'm learning that we can't just leave wilderness alone.

As we move deeper into our relationships with modern technologies, our direct relationship to the natural world fades and natural history is marginalized. Top academic institutions have removed or shrunk natural history curricula. Our youth can recognize hundreds of corporate logos, but few can identify more than a handful of local plant or animal species. How do we go about creating a society in which children of all ages can interpret and understand the natural world? The answer is, most likely, nature study. Place-based instruction should include basic natural history like plant identification and animal life histories. Nature study can take many forms, but it must include leaving the perceived safety of our homes or schoolrooms to encounter nature in our own learning landscape.

I am proud to say that many enlightened organizations across the Klamath Mountains are crafting connections to natural history. Groups are collaborating around place-based, experiential curriculum for K-12 students. Classroom lessons include iNaturalist Community Science projects, river ecology studies, game camera data collection, and beaver dam engineering design activities. Kids are outside, observing their worlds, and sharing it with other students across the region. The Klamath Mountains are the perfect learning laboratory and there is so much more to know, share, and relearn through the eyes of others. New scientists are engaging with this amazing landscape and becoming stewards—ensuring a better future for the Klamath Mountains.

Together, through this book and with each other, we are connected to the natural history of the Klamath Mountains. Climate, soils, fire, and geology connect all living things across space and time. From those connections to the land, interpreted by the First People in the beginning and built upon by western scientists who followed, the deep knowledge for this place is helping to reinvigorate relationships to the land and with each other. In this bond, we all have something to offer—and even more to the mountains and rivers and forests. We will continue to share our current knowledge and better understand what those before us have done and thought. Through these connections, we can only hope that some of our old approaches and understandings fade away and a better path for place-based connection and stewardship continue to grow.

The Natural History Institute defines natural history as the "...practice of intentional, focused attentiveness and receptivity to the more-than-human world, guided by honesty and accuracy." Justin and I present our honest and accurate work for the Klamath Mountains.

A region's natural history is available to anyone who attends to it, such as studying species and their biogeography, harvesting food and other natural resources, or collectively cultivating a deep connection to the landscape through practice and experience. I, along with countless other naturalists, have found the Klamath Mountains to be a profound setting by absorbing its splendor of ancient rocks, water, lifeforms, and the relationships among them.

I (Justin) grew up in the Klamath Mountains. In Lewiston, California I lived in a small valley of oaks with grasslands, gallery ponderosa pine forests, and chaparral—all bisected by the Trinity River and its narrow alluvial floodplain. The valley is shouldered by tall ridges draped with rigid, green stands of Douglas-fir. From a distance, the taller granite horns and the red-rock domes of the Trinity Alps seemed to float over the stacked green horizons—adding grandeur, depth, and enticing mystery. There are many small, river valley towns throughout the Klamath Mountains with similar layered landscapes and, because of their habitat mosaics, are spectacular hosts of regional biodiversity. To be clear, these hamlets are not intact landscapes of some former period. Most of these have been ravaged, over time, by unnatural landscape-scale disturbances such as the displacement of Native Americans, the gold rush, road building, dozer logging, dam building, the green rush, and wildfire. Like many other towns, Lewiston is a museum of such prospects—both in relics and scars—and this history is also woven into local cultures.

Looking back, I realize I became a naturalist as a child through innate curiosity and imprinting on annual phenological patterns. What I learned through a combination of wanderings, and through the responsibilities required of mountain living, generated a unique and continuous anticipation for the annual Klamath Mountain cycles. For me, dogwoods, shooting stars, and hound's tongue represent spring. Swimming, Mom's garden, and hiking means summer. Woodcutting with Dad along with smoked salmon, and the smell of rotting salmon on the river means fall. Winters are cold with steelhead running while snow brings silence and the magic of animal tracks. In stark contrast, my science training has fostered a sturdy discipline of learning and studying illuminating features of the Klamath Mountains. Books that chronicle intense observations of the natural world, such as Loren Eiseley's *The Immense Journey* and Annie Dillard's *Pilgrim at Tinker Creek*, helped me realize that natural history is both accessible and also infinitely complex. Based on my own experience, a naturalist's journey is one of anticipation, suspense, and surprise while also being predictably humbling.

The beauty in cultivating a deep understanding of a landscape is in the long, gratifying journey it promotes. With enough time, any Klamath naturalist will bear witness to phenomenal events. One October morning I was rafting down the Rogue River where its deep wilderness canyon cuts west through the Klamath Mountains. A flock of wood ducks flew by—low along the steep curved gorge—when a peregrine falcon dove from above into the flock, precisely striking one of the birds killing it on impact. As the wood duck fell to the bedrock, a plume of its downy feathers floated down the canyon. The falcon landed at its meal and, as if this event before us was some orchestrated natural history finale, a golden eagle dropped from a Douglas-fir, looming high above the canyon bedrock and displaced the falcon to take possession of the prey. Years later this spectacular observation still resonates—feeding my curiosity for what other naturalists have witnessed in the Klamath Mountains.

Other observations I have had are rich in personal discovery. Once, while conducting fish surveys in a headwater stream of the Smith River, I located a bleach-white skeleton of a mountain lion, the Klamath's most revered apex predator. Usually mammal remains are scattered, or many of the small bones are chewed by rodents. This skeleton laid in the footprint of a recent wildfire and was so complete and purely organized that even the sickle-shaped claws and small tarsal bones of each toe laid perfectly *in situ*. The spinytail fairy shrimp is a dazzling transparent glacier-ice blue, and their strange undulating upside-down locomotion in water makes them resemble tiny, deep-sea creatures. The first time I found them was in a drying pond in the Trinity Alps and the experience was so puzzling and spectacular I naively thought I had discovered something new to science—regardless, it was fun to celebrate at the time.

Some of my most enlightening Klamath Mountain experiences have been more planned adventures, such as descending over 5,000 feet down a drainage in a steep trailless canyon. This transect took me from the subalpine zone down to the valley floor where I ended in darkness—scratched and exhausted. In the early descent, the cliffs and dense vegetation decided my every move. Eventually, out of necessity, I discovered and traced a narrow deer trail for a few steep miles, revealing the only viable travel route for a large mammal such as myself. Several of these gravity-led descents over the years have provided great personal lessons in Klamath Mountain geology and ecological zonation with botanical discoveries along the way. Wow, that black oak is at an impressive elevation! Look—ponderosa pine—I must be close to the valley floor now! Finally, on a trail out, a salamander was illuminated in my dimming headlamp—the first ensatina I had ever seen in the Trinity Alps. Oh, the journey I took just to discover that salamander!

Some naturalist encounters result from spontaneous journeys through abrupt boundaries found in the Klamath Mountains. One fall afternoon I stood with Michael van Hattem and Ken Lindke, at the foot of Grizzly Glacier—the last remaining glacial ice in the range. It was after a good snow year, so the glacier was safer to experience up close than in drought years where it melts unsustainably and can shed massive ice boulders at any moment. Here a narrow crevasse invited us deep into the belly of the glacier where we eventually shimmied into a cavernous room below a hulking ice ceiling. I have never seen colors quite like those radiating throughout the fissured ceiling above. Some were neon glacier blue while others were chrome black. Another boundary I often cross is into the underwater world of the Klamath Mountains with a mask and snorkel. In addition to many amazing underwater critters, my kids and I have watched green sturgeon in their natural habitats, placidly circling in clear sun-dappled river pools lined with ancient Klamath bedrock. Those brilliant glacier colors—and sturgeon holding their ancient evolutionary secrets—are among the countless delicate products hosted and maintained by this mountain range.

I hope this book helps more of us make meaningful connections to the Klamath Mountains. The more we understand this beautifully complex landscape, the better stewards we will be in return.

Prologue

David Rains Wallace

When I started backpacking in the Klamath Mountains in the 1970s, I carried Storer and Usinger's *Sierra Nevada Natural History*, since no comprehensive field guide existed for the Klamath region. That suited me in a way, because I liked to think I was exploring a little-known place. Growing up back east, I'd thought of western mountains as "wide open spaces" like the Sierra or Cascades, but the region's deep, heavily forested river gorges seemed different, mysterious. As I learned more, it seemed a place where I might explore in time as well as in space, where the ancient past—deep time—could be unusually accessible to the imagination. I decided to write a kind of evolutionary creation story based on the region's diverse natural history, from the "rock bottom" of geology, to the "primal ooze" of lakes and wetlands, to the "salt water wealth" of anadromous river life, to the ancient forests and high meadows. I called the book *The Klamath Knot,* borrowing from a geologist's nickname for the region, and, if not universally acclaimed, it helped to describe the place and encourage conservation activism.

Of course, the fact that I had to drive past dams, mine tailings, and logging to reach the wilderness showed the wishful thinking in my attempts to see into deep time. More logging has gone on since the 1970s, and the Klamath Mountains remain less known and protected than the Sierra or Cascades, a neglect that began early with their exclusion from John Muir's classic natural history guide, *The Mountains of California.* A map in its 1894 first edition doesn't even show mountains in the northwest part of the State. Muir does cover "coast ranges" in his final chapter, but it is a cursory look at what he described elsewhere as "dense dark woods of the Douglas spruce." It was the Sierra and Cascades—"ranges of light"—that inspired him, and when logging and grazing diminished them, he moved on to the ice-carved vastness of Alaska.

Since big national parks often followed in Muir's footsteps, the Klamath Mountains might have one today if he had noticed them. Instead, most of the region passed into the realm of his political rival, Gifford Pinchot, an industrial forester. That largely shaped the twentieth-century history. It got little attention when the eminent botanist Willis Linn Jepson, in 1907, described the old-growth of "Madrona, Tan Oak, Douglas Fir, and Black Oak" as "the most beautiful woods in California ... more beautiful than the Sierra woods ... with a lightness, diversity, and color that appeal with the liveliest effect to the imagination..."

Industrial forestry came into full force by the 1960s, when Congress compensated the timber industry for Redwood National Park's creation by mandating increased logging inland, and the U.S. Forest Service planned the Gasquet-Orleans Road across the southern Siskiyou range as a timber transport route and scenic highway for tourism. One outcome of such developments was logging of the conifer-hardwood old growth that had inspired Jepson. The Forest Service justified its plans by saying that the region is not very scenic compared to places like Yosemite—"average mountain country." And some of it is not outstandingly scenic, for example (paradoxically, given the rationale for building the GO Road) much of the terrain that the highway was to cross.

Scenic grandeur was perhaps the main thing that early conservationists like Muir invoked when advocating parks, and it remains important, as I once found when advocating a Klamath Mountains park on a radio phone-in show. A caller said that the region isn't worthy of one because it lacks an outstanding scenic feature like the great depth that justified park status for Crater Lake National Park. I said that there are other reasons for protecting land beside scenic superlatives—biodiversity, for example—and that Crater Lake National Park is an unfortunate example of biodiversity loss because developments like logging have isolated it. But he didn't seem convinced.

This book is a big step toward convincing people that there is much more worth protecting in the Klamath Mountains than spectacular scenery. To be sure, they have plenty of scenery, as attested by Preston Peak and its companion summits in the Siskiyou Wilderness north of the GO Road area. But many of the superlatives here go deeper than early conservationists' focus on lakes, peaks, or giant trees. In fact, every chapter in this book cites such biodiversity superlatives.

The most important factor in historical terms is the region's diverse Native American cultures. They featured prominently in regard to the GO Road when the fact emerged that—if the terrain the road was to traverse is not outstandingly scenic, it is something very important to people who have lived here for thousands of years—sacred. That fact's influence on the controversy that eventually stopped the road's completion set a precedent for spiritual and cultural significance of land that will only grow in importance.

Euro-American explorers were quick to recognize the region's qualities. The Wilkes Expedition that traveled through the eastern Klamaths in 1841 found it the most botanically diverse area they crossed. Later scientists like Jepson and Alice Eastwood confirmed this, and, in the 1960s and 70s, the region acquired global significance when ecologist Robert Whittaker identified it, with the southern Appalachians, as North America's major locus of temperate forest diversity, and ecologists John Sawyer and Dale Thornburgh found the "miracle mile," the Russian Peak watershed that contains the world's record of temperate conifer diversity.

Much of what makes the region so diverse goes on "behind the scenes," through the processes that *make* scenery. The glaciers and volcanoes that Muir admired are important land forms, but recent ones in deep time. Klamath Mountain geology is an outstanding example of just how ancient and complicated deep time's processes are. Instead of the Sierra's wall of glaciated granite or the Cascades' row of lava cones, Klamath rocks lie in deep, arcuate belts that extend—from west to east—back to the beginnings of life. Roughly 600-million-year-old sedimentary rocks at the region's eastern edge contain fossils of "Ediacaran cyclomedusoids" that may have been among the first multicellular organisms to evolve. The fact that they resemble fossils found in Australia and China is another aspect of the complex past.

The geological features that most distinguish the region are the ultramafic rocks that tectonic plate movement has pushed up from the Earth's mantle. Vast expanses of these reddish and greenish rocks are exposed in the Kalmiopsis, Siskiyou, Red Buttes, and Trinity Alps wilderness areas, and their peculiar chemistry has generated a unique vegetation of Jeffrey pine and incense-cedar "savanna" that can seem strangely out of sync with adjacent areas' lush forest.

Climate provides another window into deep time, in this case, into the late dinosaur age, when a forest of unsurpassed diversity stretched across the northern hemisphere in weather so much milder than today's that the Arctic supported diverse trees. As Whittaker remarked, this "Arcto-Tertiary" forest has retreated in North America to two regions where such relatively mild climates remain—the southern Appalachians and the Klamath Mountains. In the Klamaths, where a Mediterranean-type regime of wet winters and dry summers prevails, climate combines with geology to generate a particularly wide diversity of weather and vegetation, from the lush western ranges, which get more recorded annual rainfall than anywhere else in North America, to the semi-arid eastern ones, which have vegetation much like the Great Basin's.

Wildfire has become a problematic aspect of climate as industrial forestry's suppression policies have obstructed the processes long prevalent in fluctuations between dry, hot summers and wet, stormy winters. This book gives an illuminating account of wildfire's complexity in the diverse landscapes, including its beneficial effects on many plant species, which native cultures long have understood. Traditional Ecological Knowledge around fire is now re-shaping a fire-suppressed landscape through Tribal partnerships. Where good fire has been returned, the land has become more resilient—but there is still a long way to go. We can only hope that these practices continue to grow for the health of the region and its people.

Precipitation and topography have interacted to create some of North America's most unspoiled watersheds. The fact that the greatest rivers—the Rogue and Klamath—have their sources to the north and east of the region shows just how old those rivers are, since they must have carved their gorges as the mountains rose. Other main streams—from the Illinois, Chetco, and Smith in the north and west to the Scott, Salmon, and Trinity in the south and east—constitute the lower forty-eight states' greatest concentration of wild rivers, and thus some of the healthiest populations of riverine life, while the springs, wetlands, ponds, lakes, and creeks that feed them harbor many unique organisms.

While the region's rocks are spectacularly old, many of its living organisms have equally ancient antecedents. Although, as far as we know, life is no longer evolving from non-living matter, the processes that arose with the first organisms still operate here, from algal mats in warm ponds, to the untold diversity of bacteria, algae, and fungi of the soil, to the lichens, mosses, horsetails, ferns, and other "cryptogams" that occur from the forest floor to the treetops and cover even barren-seeming rocks.

The region's higher plants, gymnosperms and angiosperms, are world-class examples of how evolution continues today, in both "conservative" and "progressive" ways. Climate and terrain have made the region a treasure house of paleoendemics—plants that once had a wider distribution but now survive only here. These include species like Brewer spruce, Port Orford-cedar, and Siskiyou kalmiopsis, and herbaceous ones like the carnivorous cobra plants that rise majestically in fens. Equally, climate and terrain have made the region a treasure house of neoendemics, particularly herbaceous species like Siskiyou false-hellebore, Klamath Mountain buckwheat, and Scott Mountain fawn lily, all of which have branched off from relatives elsewhere because of isolation and a changing environment. Neoendemics are especially numerous in "ultramafic" landscapes, due to the unique environmental conditions present there.

Exotic plant diseases, particularly the *Phytophthora* species that parasitize Port Orford-cedars and high elevation pines, threaten this botanical wealth. But the region's high biodiversity also includes many *native* organisms that attack forest plants in various ways, some quite fascinating, as with dwarf mistletoes which parasitize various local conifer species. They are so small as to be unnoticeable, but can cause notable effects like the "witches' broom" that erupts on Douglas-firs. The forest displays both an ancient resilience to disease as well as recent vulnerabilities.

Many forest "pests" are insects, which, along with other invertebrates, are the most numerous group of animals. They certainly are the most diverse group in the Klamath Mountains, bewilderingly so, and biologists constantly discover new things about them. Some of the recent discoveries have been in the region's limestone and marble caves. Twenty-five species of invertebrates that live only in caves are known here, and since much cave exploration dates from the 1970s, more probably will be found. They include a scorpion species found only in one cave; a newly discovered family of spiders; and a species of grylloblattid or "ice crawler," a predacious insect resembling a large cricket that occurs only in caves and on snowfields around Marble Mountain.

Terrestrial animals like insects and arachnids are predictably diverse in forested mountains, and the region is a "global hotspot" for butterflies and moths. But it is also full of animals often thought of as marine: mollusks and crustaceans. Eighty known taxa of land snails and slugs are present, making them the most diverse mollusks here, but aquatic diversity teems in the Klamath Mountains' waters as well, which, like tide pools, the longer one looks in them the more one sees, from snails, limpets, mussels, and clams in streams; to amphipods, ostracods, and copepods that throng ponds and lakes; to fairy shrimps that appear in ephemeral pools.

Aquatic invertebrate abundance supports two other globally significant regional groups—fishes and amphibians. Anadromous fishes define biological communities of streams. Despite the impacts from dams, mines, and logging, salmonids are also cultural keystone species and provide major economic support to human cultures. As links between the Pacific's teeming life and the river systems that run throughout the region, the three lamprey, two sturgeon, and seven salmonid species here are key to the whole ecosystem's vitality.

As with the flora, the Klamath Mountains are classed with the Southern Appalachians as a global biodiversity hotspot for amphibians. They are the most abundant vertebrates in some ecosystems, and they parallel plants as exemplars of ancient and recent evolution. The coastal tailed frog, one of two sister species representing the oldest frog species on the planet, inhabits icy streams here—its nearest relatives live in New Zealand. The coastal giant salamander of streams and lakes probably isn't as ancient, but its size makes it seem like a relic of the Paleozoic era. On the other hand, some of the many little lungless salamander species that live from soil to treetops have evolved in response to local and regional climatic and geological changes of the past few million years.

The Klamath Mountains' latitude—at a complex crossroads of a variety of biotic communities—is and will continue to be a biodiversity refuge. Given the

Klamath Mountains' parallels with the Southern Appalachians, conservationists have long maintained that this region also should be a UNESCO Biosphere Reserve. The region retains significant populations of old-growth species: spotted owls, goshawks, martens, and fishers. Others such as California condors and beavers were once common, and could be again. Instead of being isolated and impoverished by mines, dams, and logging, the region's wilderness areas could be linked by corridors of restored mid- to low-elevation old-growth forest and riparian vegetation. The "most beautiful woods" that inspired Jepson could again cover the steep, unstable slopes between Redwood National Park and the Siskiyou Wilderness, and elsewhere.

The Klamath Mountains as a Teacher

Keith Parker, Matt Johnson, Dawn Blake, and Charley Reed

Among us we are fathers, a mother and grandmother, sons, a daughter, scientists, students, and stewards. Three of our families have been here since time immemorial, one of us is a comparatively new arrival. For each of us, the Klamath region is, and has been, a teacher. The mountains, forests, rivers, and the life therein have taught people for thousands of years, and they continue to do so today. Here, we each share a personal testimony of how the Klamath has been a teacher in our lives. We hope our stories, and by extension this entire book, help portray the wonder of the Klamath, and inspire others to learn the lessons it can offer.

We are all relatives, not resources

The Native Tribes of the Klamath River Basin have been stewards of their respective lands since time immemorial. The aboriginal land is especially important to regional Tribes as traditionally they believe they are living near the centers of their worlds respectively. Each Tribe has a specific and designated physical location for their center where the cultural, ceremonial, and subsistence use of lands are interwoven. The customs, laws, and rights of land and water use are all part of the weave that exemplifies the daily life of Tribal communities. In particular, the Klamath River is central to the identity of the Yurok, Karuk, Hupa, Shasta, and, outside the Klamath Mountains but in the Klamath River Basin, the Modoc, Klamath, and Yahooskin people. The river is the principal landscape feature of the reservations and provides the fundamental basis for culture, philosophy, and religion. The paradigm is exemplified by one elder who stated, "The river flows like our blood, it is our veins and arteries."

I (Keith Parker) live on the Yurok Reservation near the house my grandfather was born in, overlooking the mouth of the Klamath River as it flows into the Pacific Ocean. My life journey led me away for a number of years, but the River always whispered to me, like home whispers to a soldier fighting someone else's battle in a foreign land. The Klamath River provides me and my people with traditional foods—a foundational part of Klamath Mountain Tribal cultures that feeds much more than physical bodies...traditional foods feed Spirits. This is because these foods represent the living link with the land—the Klamath River is our grocery store. The indigenous worldview holds the belief living resources and culture are singular, with nature completely intertwined with humanity. Language, customs, ceremony, cultural development, and food sources evolved synchronously, not independently. This is the reason indigenous people of the Klamath Mountains regard all people, plants, and animals as relatives rather than resources. Many traditional laws dictate that people care for their relatives in responsible ways; saying "my mother is in pain" is very different from saying "the Earth is undergoing global warming." Salmon, sturgeon, lamprey, and other fish have been keystone cultural species for thousands of years. Such a relationship exemplifies Tribe's ability to maintain a sustainable fishery and illustrates the depth of knowledge Tribes possess regarding fish as well as ecosystem management.

My desire to become a Fisheries Biologist has been innate in me since I was young, learning to row a boat in Klamath River estuary before I even drove a car. The completion of graduate school was my opportunity to merge my Traditional Ecological Knowledge (TEK) with western scientific knowledge. I learned that science and social justice are intertwined—the two cannot be separated. From the perspective of Traditional Ecological Knowledge, the loss of fish and other species in the Klamath River Basin represents not only a loss of biodiversity but a loss of important cultural heritage as well. The place-based identities of Tribal peoples cannot coexist with continued over-extraction of trees, minerals, fishes, and water. All things are relatives, not resources.

Rethinking our relationship to the Klamath Mountains

As a professor of wildlife ecology at Cal Poly Humboldt, the Klamath has taught me (Matt Johnson) natural history including conifers, birds, and fishes. But more importantly, my time in the area and my experiences with people whose ancestors have lived here since time immemorial have taught me humility and urged me to re-think the relationships between people and the land. The region has also provided inspiration for how to engage college students in the interdisciplinary work of environmental conservation.

I grew up in California, and I've long had a passion for wildlife and the environment. Professionally, my passion is birds, though my experiences birding, hiking, and working in various places around the world also sparked a love of trees. So after moving to the North Coast in 1999, I eagerly began to explore the Klamath Mountains, eager to see its birds and famous conifer diversity. The mountains taught me patience and appreciation, as my first attempts to find Brewer spruce were both unsuccessful and glorious. Later, with my wife and two teenage boys (and the help of Kauffmann's conifer book), the Klamath taught my whole family the appreciation of diversity: the revelation of seeing the pendulous branches of Brewer spruce, the thrill of noticing the Pacific yew was *right next to us* as we snacked along a creek. The satisfying certainty offered by the tight bottle brushes of needles *revealed that yes, THAT is a foxtail* after examining so many whitebark pines before it. The high-elevation lakes taught my sons and me to fly fish, with brook trout rising to dry flies at dusk while the lilting flutes of hermit thrushes drifted through the shoreline firs. Those same lakes taught my colleagues and students—who then taught me—that non-native trout send unwelcome ripples through a trophic web ultimately affecting rare, endemic Cascades frogs.

All along the way, I've had the privilege of working with and befriending students and colleagues whose people have lived here for thousands of years. The people of the Klamath have taught me to re-consider our position with Nature, not as above it or below or beside it, but rather within it. Even as a young conservation biologist I always bristled at the notion of "protecting nature *from* people," with its undertones of exclusion and elitism. Instead, my interests in human welfare and rural livelihoods led me to research so-called ecosystem services—the processes and values of Nature that can help both sustain and fulfill us. In other words, why we might conserve nature *for* people. It wasn't until living here, and interacting with the people of the Klamath, that I fully appreciated the concept of conserving nature *with* people. I now try to pass some of these lessons on to students as part of my work on curricula for first-year natural resource science students at Cal Poly Humboldt in a program we call the Klamath Connection. Working with many other faculty and staff and Tribal partners, our hope is that the Klamath can teach the next generation of natural resource scientists about how environmental conservation and social justice are two sides of the same coin.

Exercising body, mind, and spirit in the Klamath Mountains

I (Dawn Blake) live on the Hoopa Valley Indian Reservation in the heart of my ancestral territory. While I have had the privilege of participating in the cultural heritage of my people, my heart has often yearned for what has been lost. I embarked on a career that complements that essential part of my life and have integrated into the natural sciences, where I have found comfort and purpose.

Astoundingly, the first contact with European settlers here in the Klamath Mountains occurred less than 200 years ago. My great-grandfather lived to be 104. One generation separated him from the original incursion. I was 12 when he died. Pondering how close that first contact is to me, especially though my relationship with him, is bewildering when considering my own level of assimilation into Western society. I have welcomed education and have found a certain symbiosis in reconciling my traditional life with my life as a scientist. I'm grateful for the understanding that I have obtained as I pursued each in tandem.

Science has facilitated my journey of awareness. My studies paired working as a biologist with developing my comprehension of the natural landscape. After the years that it took to be comfortable navigating the mountains by myself, I now find that it's my preference. I can tune into my senses—consciously and subconsciously—listening, seeing, smelling, and touching. All of these senses help me to recognize the subtleties in the landscape with an awareness that has kept me safe when danger has arisen. I have learned my limitations and on occasion exceeded what I had perceived them to be. In testing my own limits, I have pushed myself to exhaustion, and even to the bounds of my physical strength.

Though my graduate research on pileated woodpecker ecology, I have incessantly contemplated bridging spirit and science. Through numerous chance encounters, I determined that the pileated woodpecker effectively taught me what my curiosity couldn't formulate into a question. Now I appreciate a long-standing, reciprocated, and esteemed relationship between the woodpecker and my people. At the same time, I have also found that it's easy and natural to pray in the mountains, as my ancestors have for millennia. This enriches my life and grounds me, making me happy, well, and ultimately keeps me close to my ancestors. Often my troubles will dissipate, never to return. My time in the mountains is now melded into every aspect of my life and has made me a better person. The reverence that I have gained for ecological and natural systems has opened me to knowledge and allowed me to teach. I've learned that the ways of my people aren't lost when I quiet my mind and that my inherited features prime me for this work. In this landscape, I've learned to trust myself and something greater than me.

The woven fabric of our World

I (Charley Reed) am a Karuk, Hupa, and Yurok person of the Klamath-Trinity watershed that practices World Renewal Ideology. I was raised near the confluence of the Salmon and Klamath rivers at Ka'tim'îin, a Karuk ceremonial site we believe to be the center of the world. All my communities are rooted in this place-based religion, and our sacred ceremonies transfer knowledge across generations. The ceremonies reinforce how to manage an optimal ecological and cultural lifestyle based on harmony and diversity, which teaches us how to treat all things with respect and reciprocity. As a Karuk person, my worldview is shaped by the interconnectedness between people and place, which makes Mother Earth my most influential teacher/mentor I have encountered.

World Renewal Ideology provides a sense of wellness mentally, emotionally, and spiritually, while also preserving the environment we live in today for generations to come. As a traditional dipnet fisherman, I have learned many of my life lessons down at Ishi Pishi Falls. These include how to provide for family and community and how our traditional stories have guided my understanding and relationships in this world. Two stories in particular have provided me with teachings—the story of the fisherman who became a moss-eater and the story of the brawl between the coho salmon and the turkey vulture.

Every time I go fishing with my family down at Ishi Pishi Falls, I see the "moss-eater" bird. With each interaction, I am reminded how fortunate I am to be in a position to provide for my family and community. This reminder is associated with a story of "The Greedy Father." In this story, the father held out on his family in times of hunger. He lost his privilege as a fisherman by being turned into a "moss-eater" bird, who can only eat mud and moss. This story reminds me that as a human being and a fisherman, I have a responsibility to protect, manage, harvest, and provide for my people, Mother Earth, and our more-than-human relatives. Otherwise, I will lose my privileges as a caretaker of this world. The moss-eater bird is a reminder that we must be humble, respectful, and considerate of others—a lesson that shapes my experience as a person, a provider, and a father.

Through our oral history and reinforced through lived experience, I have been taught that there are indicator species that tell us when to expect the arrival of the coho salmon. Before the creation of humankind, coho and turkey vulture got into a fight. Turkey vulture punched coho in the nose, and in return he dunked the vulture in acorn soup. That's why coho has a hooked nose, turkey vulture is bald, and when coho returns to the river, turkey vulture leaves. So now when there are no more vultures around, we know that the coho are running. We also know that when they are together at the river, the system is out of balance. The cultural traditions and creation stories tied to this land have woven the fabric of my lived experiences and relationships to the world around me. A part of what makes these teachings powerful is the fact that not only have they been transferred across generations, but they have also survived calculated acts of colonial violence. As a river person, it is my responsibility to take care of this land as my ancestors intended, in order for my child and her children—and all generations to come—to be able to learn their own lessons.

Chapter 1: First Peoples

Frank Kanawha Lake

The chapter's author, Frank Lake, harvesting tanoak acorns on his property in Orleans.

Introduction

The history of the First Peoples in the Klamath Mountains is an interconnected story that weaves long-standing relationships with the natural world into complex cultures that tend the varied and biodiverse landscapes. While some understandings of the First People's history in the Klamath Mountains have emerged through the sharing of stories by Tribal members, others are slowly emerging through archeological and ethnographic studies. As you read this chapter, remember that there is still much to learn about the history of regional Tribes. What follows is an exploration of understandings based on both Traditional Ecological Knowledge and Western science. The discussion is based on a broad collective knowledge, with some aspects being more understood than others.

Various Native American groups have inhabited the Klamath Mountain region for at least 9,000 years—though across the various western science disciplines there is, as yet, no agreement regarding the dates of Native American occupation in North America. Therefore, 9,000 years is simply the best current estimate. The region has been home to at least 15 Tribes having a minimum of 10 different languages. These Indigenous people have survived by knowing and adapting to what the Klamath Mountains offer and what they, in turn, can provide and do to nurture healthy relationships between humans and the natural world. Within the parameters of these relationships, over long periods of time, First Peoples developed complex socio-ecological systems. Traditional Ecological Knowledge has played, and continues to play, a major role in affecting the diversity of habitats in the Klamath Mountains.

This chapter focuses on the evolution of Indigenous Agroforestry systems and fire-resource management practices in the Klamath region before 1850 CE, and the current efforts by Tribes to recover and revitalize these traditions including fire stewardship with culturally-informed Traditional Ecological Knowledge-based practices that integrate the management of trees, plants, fungi, and animals to serve the needs of Indigenous communities, Tribes, and the broader society. Interdisciplinary methods, drawing from archaeological, ethnographic, oral and fire histories, and regional paleo-climatic data, are used to examine the influences Indigenous people had on the ecology of the Klamath Mountains.

Indigenous peoples were given instructions for how to live by the Creator, First, or Spirit People. Their creation stories explain how, after the Great Transformation, these Spirit Peoples became the celestials: i.e., the Sun, Moon, Stars and Planets. Some of the Spirit Peoples became parts of Earth, such as mountain ranges, bodies of water, meadows, rocks, etc. Some became Plant People (i.e. trees, shrubs, forbs, etc.), Fish People (i.e. salmon/steelhead, sturgeon, sculpin, suckers, lamprey eels, etc.), and Bug People (i.e., mollusks and insects). Some of the Spirit People transformed into the elements: air, earth, fire, thunder, lightning, rain, mist, and snow. These ways of Tribal understandings—their world view, their role and responsibilities, their obligations and commitments—come from ancient times and are acknowledged among the Tribes today.

Cultural Traditions Emerge and Evolve

The Indigenous peoples of the region consider themselves having lived in their ancestral territory since time immemorial. Many have creation and origin stories rooting their people in place since a time when this world was void of life, dark, and covered with water. The beings we know today emerged as specific spirits or as the result of Creation heroes. Such accounts formalize and ground each Tribe's relationship with the land, places, and with each other. Many cultural practices, including burning, horticultural tending, and more complex Tribal agroforestry stewardship approaches are rooted in ritual, ceremonial, and other sacred acts that emphasize living responsibly, respectfully, and with reciprocity. This is a collective way of being for Tribal traditions and cultures, providing interconnected relationships and forming socio-ecological systems that influenced human communities and the environment.

> "At the upriver end of the world is where they first made the Deerskin Dance, Brush Dance and Gambling Game, Girls Flower Dance, and a few others. The Deerskin Dance was on one side (requiring 10 days), and all the short dances (1–5 days) on the other. Then they wanted to bring the Deerskin Dance downriver because they did not have it. So they arrived at Okonile'l [Yurok]/Inam [Karuk]/Clear Creek [English]. They were in a hurry to get down river, so they danced with [willow] brush, not deerskins for a short while, and kept going down river to Segwu' [Yurok]/Katimin [Karuk]/Somes Bar [English]. Arriving in the evening they danced on the water in a boat, then they went up to the town to see if they also wanted to dance. Ha' [Yes-Karuk], so they went back down to the river and brought up their regalia and danced. Some at Katimin didn't like it, but by the time they finished dancing it was like a war dance. It is like now—the dances are nearly the same but not quite alike. Then they continued down river and arrived at Olege'l [Yurok]/Tishunnik [Karuk]/Camp Creek-Orleans [English]. There they danced well, and used various kinds of skins, white deer skin, silver fox, and others. Now they came down to Weitspus [Yurok]/Weitchpec [Anglicized Yurok]. They stayed and danced longer (10 days), facing uphill and everything went well. As they went all the way down river, stopping at some villages and increasing the stature of the regalia and even more dancing in boats at times. They arrived at Rekwoi [Yurok]/Requa [Anglicized Yurok], they discussed the dances, and thought there were many White Deerskin dances up river..."
>
> **—Billy Werk of Weitspus, Origins of Deerskin Dances**

Cultural Adaptations Shaped by Ecology and Knowledge Systems

Western scientific knowledge and understanding of Tribal histories—including dates, migration timing, and cultural development—is far from a settled, undisputed fact. Therefore, the dates provided here are only general estimates. Though based on a variety of research, some Tribes consider them to be an opinion at best.

As people arrived in the Klamath Mountains, some cultural land-management practices developed *in situ*, while others were imported or adapted to meet the needs of various Tribal groups. It is believed that during the last 3,000 years, a diversity of cultural-linguistic groups arrived, established territories, and adapted to regional climatic and local environmental conditions (Connolly 1988). Indigenous traditions and customs evolved out of cultural knowledge and practices. These traditions established an exchange of environmental relationships within geographic settings to form the Tribal territories documented by western science circa 1850 CE, illustrated in Figure 1 (Lalande 1991, permission in Lake 2007).

Yet, western scientific understanding reveals that with the establishment and movements of Tribes comes the exchange of cultural traditions and knowledge systems. Indigenous Peoples believe that they emerged from a particular location at the beginning of time. A collection of archeological evidence points to at least four distinct archaeological traditions across the Klamath Mountains (Table 1). These patterns, as seen in a variety of tools documented at previous villages, camps, and hunting and gathering sites, appear to have been introduced to the landscape beginning at the end of the Pleistocene. These patterns emerged and evolved within the region due to the immigration of different Tribes developing and bringing new knowledge, technologies, ways of subsistence, and resource stewardship techniques, including ceremonies and rituals.

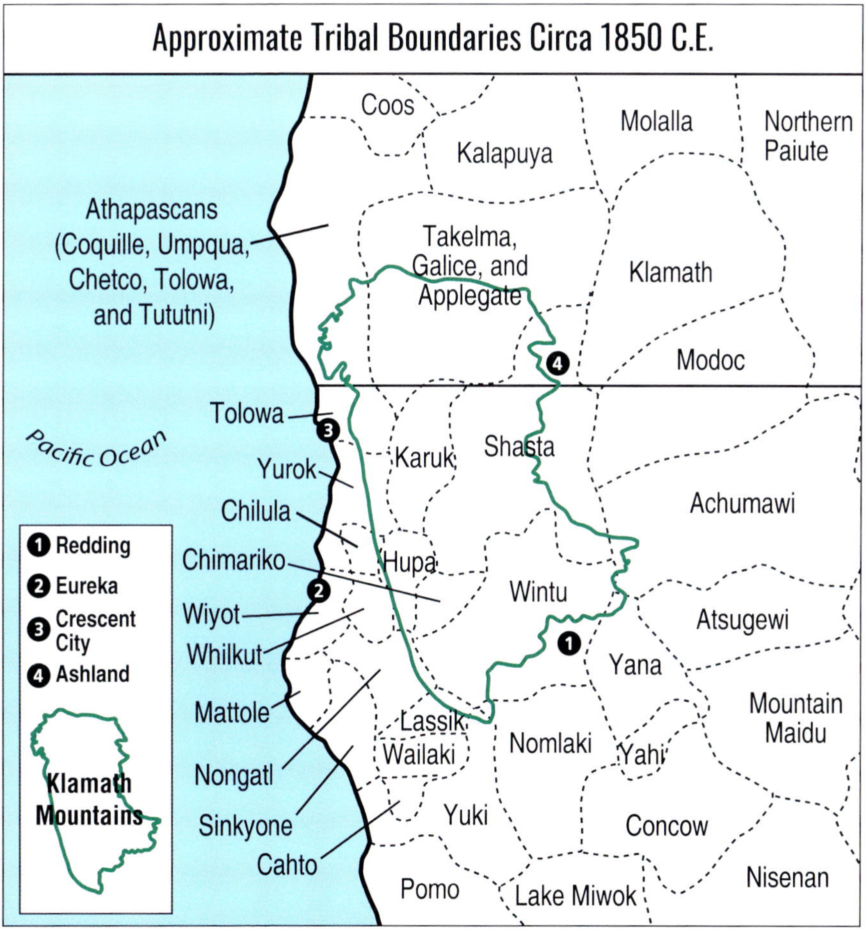

Figure 1: Approximate Tribal territories/boundaries map pre-contact. Adapted from Adapted from LaLande (1991).

The cultures of the earliest known Tribal migrants of the Klamath Mountains, and potential earliest inhabitants, are characterized by western scientific scholars as Paleo-Indians. These groups are believed to have migrated to the area by coastal marine and interior western North American travel routes. With them, they brought technologies related to the Clovis and Fluted types of stone weaponry. The Borax Lake Pattern is the earliest type recovered to date in the Klamath Mountains, with artifacts only found in the southern Klamath and not yet in Oregon. The temporal and geographical boundaries continue to be interpreted but it appears that some of the earliest objects could be between 10,000 and 12,000 years before present (ybp). This pattern is represented by crescentic objects, fluted and wide-stemmed projectile points, coarse single-flake blades, manos, and millingstones. These small, socially organized Tribes focused a lifestyle around seed collecting and hunting in drier plant communities (Sundahl and Henn 1993).

The Glade Tradition emerged regionally circa 9,000 ybp (Aikens 1993). These tools have attributes including foliate points and edge-faceted cobbles. These attributes suggest there is a technological link between the early Pacific Northwest regional Cascade Pattern and assemblages assigned to the Glade Tradition in southwest Oregon (Connolly 1988).

The Siskiyou Pattern was the next to emerge circa 1,700 ybp (Aikens 1993). With this cultural exchange, dramatic transformations occur from the Glade Tradition including projectile points that shifted from shouldered to barbed; milling equipment that shifted from stone bowl mortars to metates and hopper mortars; the emergence of nucleated pit house villages; and evidence of long-distance trade in marine shells from the Pacific Ocean and obsidian from east of the Cascades, Great Basin, and north central California. The significant differences between the Siskiyou Pattern and earlier Glade Traditions suggests fundamental cultural and related subsistence changes. Contact between Tribes—often from neighboring ecoregions (e.g., the Pacific Northwest, Columbia Plateau, Great Basin, and various California provinces)—increased as evidenced by artifacts brought in via long-distance travel and knowledge exchange. There were also shifts in the social climate as denoted by nucleated villages and the increased use of status and complex ceremonial objects (Connolly 1988).

The last major emergent traditions are found within the Tuluwat (formerly Gunther) Pattern, beginning circa 1,500 ybp. This pattern was sculpted through

Visits from Polynesia

The Yurok stories tell of across-the-ocean cultural knowledge and ceremonies exchange with a fine race of people—like and similar to themselves—with those who dance with full feather capes, elaborately crafted feathered headdresses, finely carved items of stone (jade and obsidian) and shell. It is believed that Polynesian cultural groups arrived via canoes from across the Pacific Ocean. These people visited the California coast at least two separate times in the past circa 1,700 years across several centuries. Other California coastal Tribes employed Polynesian techniques in fishing as well as in the creation of various crafts. See Kroeber 1976. Yurok myths. University of California Press.

Table 1: Overview of Cultural-Historical Models for the region. While not all Tribes and Tribal members agree with these demarcations, we present them here as a way to begin to build a framework for the movement of traditions and patterns into and out of the Klamath Mountains. Table adapted from Hildebrandt (2007).

Cultural - Historical Model for the Klamath Mountain Region

	Eco-region Origin	Cultural Models	Select Tools	
Borax Lake Pattern Emerged regionally between circa 10,000 and 12,000 ybp	California Floristic Province	Socially organized and small Tribes focused around seed collecting and hunting in drier plant communities	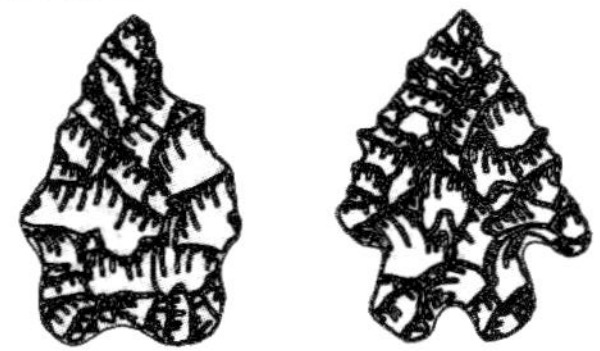	Represented by crescentic objects, fluted and wide-stemmed projectile points, coarse single-flake blades, manos, and millingstones
Glade Tradition Emerged regionally circa 9,000 ybp	Pacific Northwest	Seasonal camps of small groups of highly mobile hunter-gatherers		Tool attributes include foliate points and edge-faceted cobbles
Siskiyou Pattern Emerged regionally circa 1,700 ybp	Great Basin - Northern Sierra Nevada	Nucleated villages, increased use of status objects, and long-distance travel		Projectile points that shifted from shouldered to barbed; changes in milling equipment from stone bowl mortars to metates and hopper mortars
Tuluwat Pattern Emerged regionally circa 1,500 ybp	Coastal Pacific Northwest	Dramatic artistic elaboration and the establishment of long-distance trade	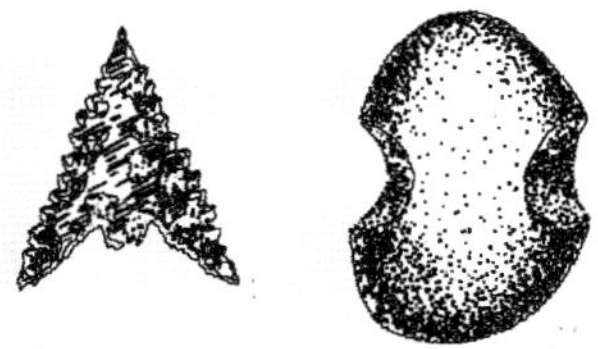	Predominance of triangular, concave-based projectile points, line sinkers (pictured), and harpoon tips used for hunting sea mammals

a relationship with the Pacific Ocean while paralleling the establishment of the Siskiyou Pattern. The Tuluwat Pattern differs from earlier coastal traditions in its increased use of artistic elaboration and, again, the establishment of long-distance trade relationships. Other key differences from the Siskiyou Pattern include a predominance of triangular, concave-based projectile points and harpoon tips used primarily for hunting sea mammals (Connolly 1988).

The exchange of intellectual capital across cultural linguistic groups also related to developing agroforestry systems beyond what has typically been described as "complex hunter-gatherers," with cultural uses of fire and horticultural stewardship practices for landscape-scale management strategies. Potentially more complex knowledge of the bio-region, ecological processes, habitats, and species populations allowed for a tradition of resource utilization that enhanced sustainability, rather than depleting ecosystem diversity. Some Tribal creation or similar earlier times stories tell of greed, overharvesting, abuse of resources, and mistakes made by some individuals—for which they were punished or ostracized. Stories such as these are used to inform humans about ecological stewardship and are used to guide our conduct today. These Tribal teachings and exchanges of knowledge (i.e., best practices and lessons learned), localized refinements and adjustments in human behavior and conduct, and conduct allowed ecological systems to be more resilient to human and broader climatic perturbations (Berkes and Turner 2006).

Climate and Fire Shape the Klamath

For the Pacific Northwest and Klamath Mountain regions, climate change has affected the interactions between humans, fire, and vegetation for over 9,000 years (Hildebrandt 2007). Fire potential, vegetation conditions, hydrology, fisheries, wildlife habitat, and species populations are all affected by changes in climate and thus influenced the immigration and emigration of various populations of Native Peoples over millennia. Thus, climate change also influences the movement of Tribal cultures, their languages, customs, adaptations, and survival practices (Connolly 1988).

Over the past 9,000 years, the climate has oscillated from warm and dry to cool and moist several times (Briles et al. 2005, see Climate Chapter 100–101). These shifts influenced species abundance and biodiversity. In the greater Pacific Northwest, circa 6,000 ybp, landscape coverage of forests expanded over former grasslands during the cooler and moister periods (circa 3,000–1,500 ybp). Then the region experienced a warming and drying period from circa 1,500–1,000 ybp (Medieval Climatic Anomaly), which negatively affected Tribal groups (800 C.E. –1350 C.E. in Jones et al. 1999). The region then transitioned to cool and moist condition from circa 1,000–250 ybp (Little Ice Age). Since then, the climate has been steadily warming with more droughts and intensifying weather events.

The biodiversity of the Klamath Mountain Region, circa 1850 C.E., has largely been attributed to the interaction of geological, climatic, and ecological processes. The cultural diversity of Tribal inhabitants reacted to and interacted with similar biophysical processes. However, human management outcomes as culturally-informed stewards, such as Tribal agroforestry, has added to biological process—particularly with the diversified uses of fire (Skinner et al. 2005).

Biophysical Diversity Promotes Cultural Diversity

Socio-cultural data derived from archaeological, ethnographic, and linguistic research is also used to reconstruct the evolution of Tribal agroforestry systems and cultural fire regimes. Many cultural traditions, customs, and practices are shared between Tribes because of their common relationship to the landscape. While this is not fully understood, linguists have described some language patterns within ancestral Hokan (Karuk, Shasta, and Achumawi-Pit River), Athapaskan (Tututni/Chetco, Tolowa, Hupa, Chilula, Whilkut, Mattole), Algonquin (Yurok, Wiyot), and Penutian (Takelma and Wintu) groups. These are language families among many others across western North America.

In northwest California alone, there were over 20 distinct languages representing at least five linguistic families circa 1800 C.E. (Hildebrant 2007, King et al. 2016). Likewise, Tribal codes, social customs, and class systems varied widely, resulting in complex, multi-faceted, socio-economic relationships and a range of effects on the landscape's ecosystems, species, and character.

The Klamath Mountains offers great potential for cultural niche specialization based on its complexity from the coast to the interior. The region has biophysical similarities to the Oregon coast (northern coastal vegetation on sandstone), Cascades (northern interior vegetation on volcanics), Sierra Nevada (southern interior vegetation on granite and serpentine), and California coast (southern coastal vegetation on sandstone and metamorphics). The high level of biophysical diversity from coastal beaches to river valleys to mountain ridges to montane meadows allows for the separation of groups by linguistic origin (King et al. 2016).

These fire-influenced environments promote conditions favored by Tribes or Tribal families—thus retaining ancestral familiarity to that type of vegetation and associated biodiversity. For example, the Penutian groups are closely associated with Sierra conifer forest vegetation across California and the Pacific Northwest. These groups tracked biophysical comfort zones of latitude or altitude, offering higher success in maintaining core territories based on ancestral familiarity of fire to vegetation, hydrology, fisheries, and wildlife populations.

Archaeological and ethnographic data for different areas of the Klamath Mountains suggests the development and evolution of cultural dependency on certain types of fire regimes. One example is how low-intensity/high-frequency fires within mixed-severity regimes directed certain cultures into select areas. Archaeological evidence points to an increase in the exchange of imported goods

The Karuk Creation Story and Tribal-environmental relationships:

"The rocks and the trees and the water and the air, the responsibility that I have, those are real relations. We have not forgotten that we are related and that we have responsibility. And at the same time, we give thanks to those other spirit people for helping to subsist us, and reminding them that we haven't forgot that we owe them something too"

— **Leaf Hillman, Karuk Elder**

between the coast and interior regions correlated with the Siskiyou Pattern and Tuluwat Pattern (Connolly 1988). The Tuluwat Pattern is associated with continuously occupied prehistoric and historic sites linked with people of Athapascan (e.g., Tolowa, Hupa, Chilula) and Algic (e.g., Wiyot and Yurok) origin. In other words, these resources could be fostered and exchanged because indigenous fire-stewardship practices promoted desired habitat conditions for Tribal agroforestry systems, travel, and trade.

The Tuluwat Pattern is represented by continued use of hopper bases (acorn, seed, and nut processing) and fish-net weights (complex fishing strategies), as well as advanced uses of stone, ceramic, and oils (King et al. 2016). It is debated whether the Athapascan- and Algic-speaking people, as immigrants, were responsible for importing the Tuluwat Pattern, which reflects cultural associations with coastal groups of the north Pacific Coast. Cultural settlements may also reflect the amount of fire use by Tribes of different cultural origins. The southernmost Athapascans (e.g., Hupa and Chilula), who are theorized to have imported the bow and arrow, were likely adapted to terrestrial prey like ungulates, small game, and heavily relied on fire and fire-induced effects on vegetation (Lake 2007). The northern most Athapascan groups (Tolowa and Tututni-Chetco) established earlier and adapted to marine and coastal subsistence strategies (see Tushingham and Bettinger 2013).

The ancestral Achumawi-Pit River, Karuk, Shasta, and Chimariko Tribal people, belonging to the Hokan linguistic group, have likely occupied regional territories before 8,000 ybp. The Penutian groups that established in the region's interior about 1,100 ybp probably became the Takelma and Wintu according to archaeologists, cultural historians, and linguists. The Algic and Algonquian linguistic group, spoken by the Wiyot and Yurok Tribes, likely migrated around 1,100 ybp into the lower Klamath River region before the Athapascan intrusion. The Athapascan cultural groups that were the predecessors of the Chasta Costa, Dakubetede, Gusladada, Taltuctunteda, Tututni-Chetco, Tolowa, Hupa, Chilula, Whilkut, Nongatl, Lassik, and Wailaki Tribes are believed to have initially inhabited southwestern Oregon circa 3,000 ybp. Some Tribelets likely broke off and migrated to northern California in the last 1,000 years (Connolly 1988). Differences in linguistic dialects, material culture, and subsistence practices are recognized between the more northern Tolowa and other Oregon Athapascans compared to the more southern Hupa-Chilula and other related California Athapascans.

Variations in Fire Potential

Prior to American colonization and settlement, lightning ignitions and Tribal burning were the main sources of fires (Lake 2013). Evidence supports ubiquitous fire use by American Indians for a multitude of reasons (Lake 2013). Natural fire regimes were influenced by Tribal fire uses since the first humans inhabited the region, although the spatial and temporal extent of cultural burning is not yet fully understood outside the indigenous community, largely due to the dismissal of Native oral tradition as folklore by western science. The landscape still reflects historic cultural fire regimes. Human (Tribal) fire use may be more responsible for legacy vegetation patterns than that of lightning in some areas of the Klamath Mountains (see Lake 2013). Broadscale historic fire frequencies are informed from

Table 2: Time periods of Tribal group establishment into the Klamath Mountain region circa A.D. 1850*.

Tribe circa 1850	Language family	Estimated Establishment	Location and River system inhabited
Coos	Penutian	Before 6,000 ybp	SW Or., Coos Bay coastal area
Takelma	Penutian	Circa 1,100 ybp	SW Or., upper-Rogue River
Tolowa	Athapascan	Circa 3,000 ybp	NW Ca., coastal Smith River
Tututni-Chetco	Athapascan	Circa 3,000 ybp	SW Or., coastal lower Chetco, Winchuck, and Rogue rivers
Gusladada	Athapascan	Circa 3,000 ybp	SW Or., Illinois River, lower-Rogue River
Taltuctunteda	Athapascan	Circa 3,000 ybp	SW Or., Galice Creek
Dakubetede	Athapascan	Circa 3,000 ybp	SW Or., Applegate River
Chasta Costa	Athapascan	Circa 3,000 ybp	SW Or., Mid lower-Rogue River and lower-Illinois River
Karuk	Hokan	Before 8,000 ybp	NW Ca., Mid-lower Klamath and Salmon rivers
Shasta	Hokan	Before 8,000 ybp	Northern Ca., Mid-Klamath River, Shasta, Scott and upper Salmon River systems.
Chimariko	Hokan	Before 8,000 ybp	NW Ca., middle Trinity River
Wintu	Penutian	Circa 1,100 ybp	Northern Ca., upper Sacramento River and Trinity River systems.
Wiyot	Algic	Circa 1,100 ybp	NW Ca., coastal, Humboldt Bay, Lower Mad and Eel rivers.
Yurok	Algic	Circa 900 ybp	NW Ca., coastal, lower-Klamath River.
Hupa	Athapascan	Circa 700 ybp	NW Ca., lower-Trinity River
Chilula	Athapascan	Circa 700 ybp	NW Ca., Redwood Creek
Whilkut	Athapascan	Circa 700 ybp	NW Ca., upper-Mad River and upper Redwood Creek
Nongatl	Athapascan	Circa 700 ybp	NW Ca., upper-Mad, Eel and Van Duzen rivers
Lassik	Athapascan	Circa 700 ybp	NW Ca., mid upper-Eel and North Fork Eel rivers
Wailaki	Athapascan	Circa 700 ybp	NW Ca., upper-Eel and North Fork Eel River
Yuki	Yukian	Before 8,000 ybp	NW Ca., South Fork Eel/Upper Russian River

*The information presented in Table 2 was derived from theories of estimates of time based on physical and linguistic evidence, and does not necessarily reflect the traditions and belief systems of Native American descendants of these groups. These Tribal people have their own creation accounts and perceptions of how and why they came to inhabit their ancestral territories (Lake 2007).

the charcoal and pollen in lake cores, and fire scars (years of fire events cross-dated among samples) from older trees (Figure 2), yet the source of ignition is not revealed. Thus, the spatial and temporal effects of American Indian fire use cannot be easily distinguished from those resulting from lightning-caused fires. This has contributed to the misunderstanding and minimization of Indigenous fire stewardship and cultural burning (Lake 2013).

Before the arrival of Westerners, Tribal cultures within the Klamath Mountains adopted numerous technologies that were both brought from other regions and developed locally. These included modifications to the local ecology through the use of fire at different times throughout the year. There were also changes in hunting technologies like projectile points (e.g., spears vs. bows and arrows) exemplified in the archaeological record by moving from atlatl spears (Glade Tradition) to bow and arrows (e.g., Tuluwat Pattern). Historical accounts strongly suggest cultural burning influenced game abundance, habitat quality, and range/forage conditions. While the archeological record should be interpreted with some caution due to incomplete understandings, some inferences can be made. One is that changes in projectile points could reflect specialization of hunting techniques, but not necessarily the amount of hunting or fire use. Fish-net sinkers could reflect increased specializations of fish harvesting and capture-strategy (Kroeber and Barrett 1960).

Climatic precipitation cycles affected fire regimes as well, which in turn affected watershed processes, hydrology, aquatic habitat quality, fish populations, and species distribution and abundancc. Whcn climatc was optimal for salmonids or

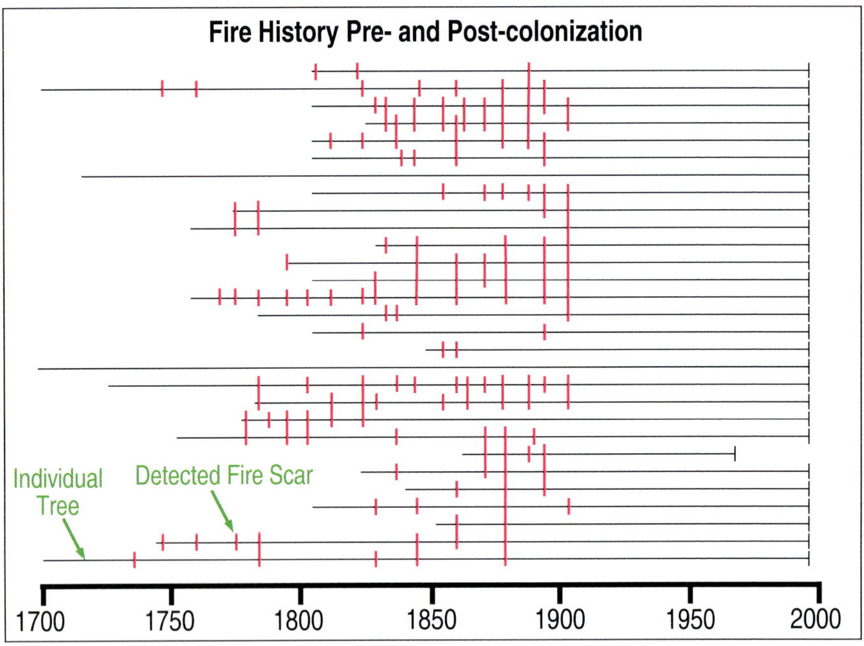

Figure 2: Data from the Clearview Study Site near Happy Camp, CA (2,000-3,000 ft, ~800 m). In 2004, numerous trees were sampled in a mixed-conifer forest to collect tree cores to determine regional fire history. They found that pre-contact, the fire return interval was circa 7-8 years with 75 percent of fires recorded in latewood or ring boundary (mid-summer to early fall). Fire was all but eliminated on the landscape post-contact. Adapted from Carl Skinner, unpublished reports.

other anadromous fish (e.g., Pacific lamprey, sturgeon), conditions were generally cooler and moister. Tribes adapted to the availability of food resources including anadromous and resident fishes, terrestrial game, plants, and fungi (Hildebrant 2007, Tushingham and Bettinger 2013). Ancestral Indigenous knowledge systems reflect subsistence strategies over warm/dry to cool/moist conditions that were likely applied to the Medieval Climatic Anomaly and the subsequent Little Ice Age (Jones et al. 1999, Crawford et al. 2015).

The extent and effects of historic Native American fire use on vegetation and wildlife habitat is not fully understood by most scholars. When describing the influence of fire on vegetation in creating historical fire regimes, some authors mention the contribution of Native American fire use, but often do not describe its extent (Skinner et al. 2006). Other authors demonstrate the extent or influence attributed to Native people for particular areas or habitats (e.g., Lake 2013). The influence of Native American fire use has been reported but the methods to define the actual extent of influence has not been fully attempted for the region or larger areas within the region (see Long et al. 2021). This is due to a cultural bias motivated by a number of factors including the resistance of non-native land managers to work with Tribes and some western scientists' lack of acknowledgment of indigenous knowledge and practices (Vale 2002).

The establishment of year-round villages increased opportunities for anadromous fish harvesting and long-term storage capacity. This fostered cultural niche specializations like the development of complex Tribal agroforestry systems, basketry/material technology, horticultural stewardship, and harvesting of plants or game (hunting or trapping). These adaptations required burning to influence the quality and abundance of favored resources of local habitats.

Cultural groups are described as either searchers or pursuers. Searchers seek a wide range of dispersed prey and pursuers seek specialized resources, predictable in their occurrences at particular times and places. Specialized cultural burning, with a sophisticated understanding of fire effects on vegetation, species population responses, and habitat quality could temporarily shift a proportion of a family or village's searchers to pursuers as a strategy to increase adaptive capacity and

> "Each Tolowa village had a headman, a prominent and prosperous man, who held rights to village hunting, gathering, and fishing territories as well as to valued verbal lore [i.e., Indigenous knowledge], and who possessed much dance regalia and other prestige wealth....headmen were able to organize more extensive food gathering than other families...long-distance trade in relatively-transportable valuables was, as markets usually are, a way of making wealth...While production and basic distribution were communal and egalitarian, some activities required ritual activation, on an annual basis, and this required cultural intervention, initiated by individuals."
>
> **— Understanding Tolowa Histories: Western Hegemonies and Native American Responses (Collins 1998)**

security. Examples of these shifts could occur between the pursuit of terrestrial resources like bear, elk, deer, small game, acorns, nuts, seeds, berries, greens, fungi, and medicines to the harvesting of anadromous fish such as salmon, lamprey eels, and sturgeon based on phenological timing of seasonal indicators (Kroeber and Barrett 1960, Anderson 2005).

Travel routes such as those radiating from villages to camps, water sources, resource use and subsistence areas, hunting/trapping and gathering places, sacred/ritual or ceremonial sites—connecting across territories—expanded Tribal agroforestry practices for utilitarian and religious purposes. Cultural burning developed to foster and diversify the natural potential of soils, plant communities, animal populations, and water availability.

Indigenous/Cultural Fire Regimes Sculpt Vegetation

Indigenous fire stewardship evolved from cultural "fire-regimes specific to certain ecosystems and plant communities created and maintained primarily by the specific and intended application of fire by indigenous people which may or may not have been in conjunction with natural fires ignited by lightning" (Lewis and Anderson 2002).

Cultural fire regimes historically affected the composition and characteristics of particular habitats, and especially the culturally defined resources therein. The distinguishing features and properties of cultural fire regimes include:

1. Alternate seasons of burning for different kinds of habitats (seasonality).
2. Frequencies with which fires are applied and reapplied over varying periods of time (frequency).
3. Corresponding intensities with which fuels can be burned affecting severity (intensity/severity).
4. The specific selection of sites to burn and, alternately, those that are not (specificity).
5. A range of natural and artificial controls that humans employ in limiting the spread of human-set fire, such as times of day, winds, fuels, slope, relative humidity, and natural fire breaks that affect the distribution and connectivity of fuels (topography/fuels).
6. Ignition patterns to promote desired fire behavior and associated intensities (ignition) (Lake 2007, 2013).
7. Burning for a range of resource objectives, including but not limited to: Reducing fuels and moderating subsequent fire intensity and severity, improving mobility/travel and foraging efficiency; creating more open, heterogenous habitats with close proximity of older growth and early seral age classes of plants; eliminating pests and diseases of food and basketry plants, rejuvenating plant growth a to promote the production of food, basketry materials, and tools/implements; increasing forage quality, basketry and tools/implements; cycling nutrients; creating biochar; and increasing water availability (Anderson 2005, Lake 2013).

Shifting climatic conditions likely prompted the use of fire by Tribal people to maintain predictable vegetation conditions across the Klamath Mountain landscape. Indigenous People modified lower-elevation vegetation and selectively modified mid- to high-elevation sites (i.e., ridge systems/travel routes, favored gathering areas, or fire dependent species' habitats). The skillful and intentional use of fire by Tribes to manipulate ecosystems directly and indirectly affected localized vegetation as well as wildlife, hydrology, and fisheries. The specialized and extensive use of fire developed into cultural fire regimes (Lake 2007).

Anadromous fish-centric cultural practices required Tribal People to observe and predict fire-hydrological effects on ecosystem processes. Key processes that affected fish habitat were fire, hydrology, woody material and sediment recruitment, and nutrient cycling. Over millennia, Tribal people developed an understanding of the relationships between the extent and severity of fire as it affected vegetational composition, structure, and function; transpiration; hydrology; and aquatic habitat quality (Lake 2013, Karuk Tribe 2016).

Some upslope fire-management practices can be linked to fishing techniques. For example, the Karuk World renewal ceremonies (Kroeber and Gifford 1952), the Yurok Kepel fish dam ceremony, and the Tututni salmon ceremony are all connected to fire, vegetation, hydrology, and atmospheric (smoke) effects to fisheries. The use of fire also enhanced plant growth in species used in fishing technology. For example, Iris grass (*Iris douglasiana* and *I. tenax* ssp. *klamathensis*) is found predominately in open habitats like grasslands and oak woodlands or openings in mixed hardwood-conifer forests. Iris fibers extracted from leaves are used to make cordage to construct nets, bags, snares, regalia/string, and other cultural items (Kroeber and Barrett 1960).

Tribal Agroforestry Across Space and Time

The use of fire in Tribal agroforestry systems resulted in the diversification of food, medicinal, and material resources as well as associated horticultural practices (Anderson 2005, Lake 2013). Over 80 percent of the Tribal material cultures (circa 1850 C.E.) are estimated to have depended on fire-induced habitat change (Tripp pers. comm. 2007, Lake 2013). Advanced systems of burning, horticultural treatments, and vegetation management occurred across all landscapes including coastal headlands and temperate rainforests, prairies and oak woodlands, interior mixed hardwood-conifer, and into higher elevations.

The greatest concentration of Tribal stewardship and burning occurred near villages, camps, favored resource areas, and long trail systems connecting these places across the landscape (Lake 2013, see Figure 3). Additionally, there are correlations between many Tribal sacred or spiritual areas with Tribal creation accounts, the establishment of villages, ceremonial sites, and unique geological or ecological convergences.

In the North Coast Ranges, many of the historical grasslands along coastal bluffs were burned by Tribes to maintain a greater proportion of complex early seral habitats. These modified sites produced and facilitated more food, medicine, and material resources of Tribal interest (Figure 4). The expansion of Tribal burning in grassland bluffs was similarly practiced within interior prairies, as well as oak woodlands, and even redwood-dominated forests. In the absence of fire, prairie systems transition into more fire-intolerant shrub, hardwood, and conifer-

Figure 3: Pre-1850 C.E. villages and trails of the First Peoples were widespread across the Klamath Mountains. This tree, on a now-defunct trail, was created as a marker along a well travelled route into the crossroads of ancestral Karuk, Yurok, and Tolowa territories. The montane chaparral in the foreground was once kept clear by using fire but is now an overgrown sign of the past.

dominated forests (Lake 2007). For example, the burning of the Bald Hills (now in Redwood National and State Parks) by the Chilula and Yurok facilitated access to many species of geophytes like Indian potatoes (*Camassia* spp., *Triteleia* spp., *Dichelostemma* spp., *Lilium* spp. (Anderson and Lake 2013). Shrubs like serviceberry (*Amelanchier* spp.) and California beaked hazel (*Corylus cornuta* ssp. *californica*) were managed for basketry and nuts. Redwood (*Sequoia sempervirens*) and other

Figure 4: Important food sources (from left to right) include huckleberries, hazelnuts, and strawberries.

conifers like Douglas-fir (*Pseudotsuga menziesii*), western hemlock (*Tsuga heterophylla*), and Sitka spruce (*Picea sitchensis*), which occur in riparian areas and on northerly/north-facing aspects, were burned less often.

In the coastal lowland forests of the western Klamath Mountains species like Douglas-fir, tanoak (*Notholithocarpus densiflorus*), madrone (*Arbutus menziesii*), chinquapin (*Chrysolepis chrysophylla*), and evergreen huckleberry (*Vaccinium ovatum*) were significantly influenced by Tribal agroforestry and burning practices. Tanoak acorns were, and are, the preferred food source. California black oak (*Quercus kelloggii*) was a secondarily preferred acorn. Madrone berries and chinquapin nuts were, and are, important foods while evergreen huckleberries were cultivated with burning to maintain open understories with light gaps and selectively pruned to foster flower/berry production. A variety of other understory plants are also used: ceanothus (*Ceanothus* spp.) for medicine and basketry; wild rose (*Rosa* spp.) for food; and Oregon grape (*Mahonia* spp.), pipsissewa (*Chimaphila umbellata*), and yerba buena (*Clinopodium douglasii*) for medicine. These areas also nurture a range of fungi, which are a nutritious food source (Anderson and Lake 2013).

Diverse forest types, soils, and topographic position (mid-slope benches or near ridges) facilitated a variety of agroforestry vegetation types. Tribally cultivated oak-dominated orchards favoring full-crown, wide-limbed, well-spaced trees were often at the edge of Tribal villages or family-owned resource areas pre-1850 C.E. These oak-dominated orchards included tanoak, black oak, canyon live oak (*Quercus chrysolepis*), Oregon white oak (*Quercus garryana*), chinquapin, madrone, and bigleaf maple (*Acer macrophyllum*). Associated with these orchards were other conifers like Douglas-fir, sugar pine (*Pinus lambertiana*), ponderosa pine (*Pinus ponderosa*), and incense-cedar (*Calocedrus decurrens*). At middle to higher elevations along ridgelines and on southerly aspects (often along travel routes), cultural burning and fuel wood practices favored sugar pine/manzanita (*Arctostaphylos*

Oak and Hazel Orchards

California beaked hazelnut

Prior to fire exclusion and suppression, many former California black oak woodlands—with California beaked hazel as a major understory associate—were culturally managed with fire and other Tribal horticultural practices. This vegetation type is a favored gathering area for acorns, hazelnuts, and hazel shoots. Open, early seral grassland-forest ecotones are also important areas for the hunting and trapping of high-value game. The intentional use of fire and promotion of black oak/hazel-dominated habitats expanded the distribution of this vegetation type across more suitable areas of the landscape, typically into lower to mid-elevations on southerly aspects. This vegetation type is considered critically important to wildlife species such as birds, native ungulates (black-tailed deer/elk), and pollinator insects (Long et al. 2021).

Cultural indicators for the initiation of burning include acorn/hazelnut pest-infestation levels, surface and ladder fuel loading (young tree/brush encroachment, reduced gathering efficiency), and shifts or changes in wildlife, plant, and fungi species. The restoration of this habitat type serves both the needs of regional Tribes and offers an opportunity for reintroducing fire as a socio-ecological process and achieving important conservation and Tribal eco-cultural revitalization goals.

Tanoak acorn is the preferred acorn for food among the Tribes of the Klamath Mountains. Many areas dominated by large, old, full crown, wide branching tanoaks were intentionally managed as orchards and have been tended for hundreds of years. Evidence of archaeological acorn-processing artifacts associated with large, old tanoak groves—as well as ethnographic and Tribal oral histories—substantiate that these habitats were nurtured orchards. Orchards can be at lower to mid-elevations, predominantly on northeasterly to easterly aspects and/or occurring on flatter sites. These areas are threatened by high severity wildfire (coupled with drought, increased fuel loading and competition from fire exclusion) and sudden oak death (*Phytophthora ramorum*). Additionally, many of these orchards are valued and prioritized as a broader Tribal landscape restoration strategy to enhance the growth of other hardwoods and conifers such as black oak, Oregon white oak, chinquapin, madrone, and conifers of Douglas-fir, redwood, and ponderosa and sugar pines.

Tanoaks and other fire-modified trees with larger, open-trunk cavities and hollowed sub-trunks and limbs provide important habitat for cavity-dependent wildlife (e.g., Pacific fisher [*Pekania pennanti*]). Hollowed trunks near Tribal gathering sites were also sometimes used as storage areas (i.e., Indian forest closets) for gear, firewood, and other items. Treatments that reduce tree density and encroachment around legacy hardwoods also increase the occurrence of widely spaced crown trees. Frequent burning (5-7 years) may enhance resistance to climate stressors such as drought and severe wildfire, as well as sudden oak death.

spp.). In the topographically complex Klamath Mountains, fires generally intensify near the upper ridges on southerly aspects (Skinner et al. 2006). Frequent burning kept the ridge trail systems clear, reduced surface and ladder fuels, and increased drought-tolerant/fire-adapted plant species. Sugar pines were tended for food, medicines, and materials, and manzanitas were tended for food (berries/sugar), wood (fuel/heating/ceremonies), and medicines (bark/leaves). Wildlife ranging from sought-after game to regalia species of birds, reptiles, and mammals also benefited from these open ridge systems, which they used for travel, foraging, and seasonal occupancy.

Many hardwood- and shrub-dominated communities—from the coast to high-elevation ridges—are influenced by Tribal burning and stewardship. Today, these practices continue to facilitate the production of California beaked hazel shoots for basketry, the stems for tools and lashing, and the nuts for food. Several species of huckleberry, (*Vaccinium ovatum, V. parvifolium, V. membranaceum,* and *V. deliciosum*), are important for their berries and for their stems, which are used to make tools/implements. Serviceberry is used for berries as food and shoots/stems for tools/implements. Ceanothus stems are used for basketry and the foliage and roots are used medicinally. Manzanita wood is used for heating and ceremony, the bark and leaves are used medicinally, and berries from a few favored species are used for sugar/food. Gooseberries and currants (*Ribes* spp.) are important as food and stems for utilitarian uses. Mock-orange (*Philadelphus lewisii*) limbs/stems are used for ceremonial and utilitarian arrows, tools/music implements, and pipes. Mountain mahogany (*Cercocarpus ledifolius*) stems are used for tools/implements. Blue elderberry (*Sambucus caerulea*) berries are used as food and the stems are used as musical instruments. When these plants are burned or trimmed correctly, pests and pathogens are reduced and the regrowth stimulates flowering, benefits pollinators, and increases fruit-nut-seed production (Anderson 2005).

Riparian areas near springs and along the banks of ponds, lakes, creeks, and rivers are favored for Tribal subsistence (fishing, hunting, and trapping), ceremonial (Tribal, village, family/individual), and recreational or personal practices. Flooding and fire are both influential ecological processes in these areas, affecting species dynamics and assemblages along rivers and larger stream systems. Additionally, Tribal uses and management of patches or areas within different riparian systems favor some species while affecting composition, and structural elements. Riverine cottonwood (*Populus* spp.), ash (*Fraxinus* spp.), and willow (*Salix* spp.) habitats are favored around villages, towns or housing, fishing camps, and ceremonial areas. Sandbar willows (*Salix exigua*) are used for basketry (shoots and roots), structures (meat-drying racks and shelters), and for other fishing and ceremonial practices. Along creeks, alders (*Alnus* spp.), mock orange, Pacific ninebark (*Physocarpus capitatus*), and other shrubs are utilized. At many springs or creeks, chain ferns (*Woodwardia fimbriata*)and maidenhair ferns (*Adiantum aleuticum*) are collected and used in basketry. These areas also serve as locations where acorn leaching takes place. Springs and smaller ponds in the high country are used for bathing, camping, and drinking.

Traditional Uses of Wildlife Species

Within the ecosystems and habitats addressed above, many resident and migratory wildlife occupy the region both year-round and seasonally. Many of these wildlife species are used as food, in making regalia, or in the manufacturing of tools, and implements.

Birds are harvested from a variety of habitats by methods including trapping and hunting. Birds of prey are used to make regalia for Tribal ceremonies. These raptors include the California condor (*Gymnogyps californianus*), white-tailed kite (*Elanus leucurus)*, eagles (*Aquila chrysaetos, Haliaeetus leucocephalus*), hawks (*Buteo* spp. and *Accipiter* spp.), and falcons (*Falco* spp.). Also used to make regalia are waterfowl like the tundra swan (*Cynus columbianus)* and various duck species. The feathers of colorful forest birds such as woodpeckers, warblers, hummingbirds, and others play a role in various dances. Some birds serve as important indicators of phenology and habitat quality (e.g., the number of birds singing) (Long et al. 2018, Long et al. 2021), and birds are often teachers in subsistence and ceremonial stories.

Mammals are of cultural importance to regional Tribes. As with other species, these furbearers have populations or life histories that are affected by fire, and Tribal knowledge of these species influenced agroforestry stewardship and related management strategies. Pacific fisher, Humboldt martens (*Martes caurina humboldtensis*), North American river otters (*Lontra canadensis*), American mink (*Neovison vison*), and weasel (*Mustela* spp.) are used for men's quivers, women's dress/hair items, and ceremonial pouches or cases. Ringtail cats (*Bassariscus astutus*) and gray foxes (*Urocyon cinereoargenteus*) are used for men's aprons/skirts. Smaller mammals such as bats, woodrats, deer mice, shrews, gophers, and moles also have prominence in stories and Tribal belief systems. Black-tailed deer (*Odocoileus hemionus columbianus*) and Roosevelt elk (*Cervus canadensis roosevelti*) are hunted and the hides are used for many purposes including drums; the tanned leather for dresses/regalia clothing; the sinew for thread; the hooves for rattles; and the antlers for many implements, tools, and regalia items (Figure 5). Grizzly (*Ursus arctos*) and black bears (*Ursus americanus*) are considered among some Tribes to be closely related to humans. Mountain lions (*Felis concolor*) are eaten, used for their hides, and considered to be good hunters and instructors for how to stalk prey. Tribes along the coast use marine mammals as well: whale baleen, teeth, and bones are used for rituals and the meat and blubber for food. Sea lions (*Zalophus californianus, Eumetopias jubatus*) hides are used for making drums and clothing-leather) while the teeth (ivory) is used for regalia. Harbor seals (*Phoca vitulina*) are also utilized in similar ways.

Many reptiles and amphibians are culturally and spiritually important and are used in making regalia and other utilitarian implements. The California mountain kingsnake (*Lampropeltis zonata*) and California kingsnake (*Lampropeltis californiae*) are revered for their strength in warding off evil and are associated with high-country spiritual training. Some Tribes use western rattlesnake (*Crotalus oreganus*) venom as poison on arrow points. The ring-neck snake (*Diadophis punctatus*) has monetary value (wrapped around dentaila shells). Lizards are prominent in stories

Figure 5. Hupa White Deerskin Dance, circa 1911. From the HCC Collection, Cal Poly Humboldt, Yerton Young Allen.

of courage and bravery, and northwestern pond turtle (*Actinemys marmorata*) shells are used as cases for carrying fire or holding valuable items. Rough-skinned newts (*Taricha granulosa*) are important in ritual beliefs, and some Tribes consider them to be guardians of sacred water sites such as mountain ponds and lakes. The coastal giant salamander (*Dicamptodon tenebrosus*) is considered to be the purifier of water by the Karuk and is a sacred indicator of water quality at springs, creeks, rivers, and ponds.

Marine, riverine, and terrestrial invertebrates are also used for their shells and as food. Mollusk shells of abalone, clams, limpets, and colorful snails are used in regalia, as forms of wealth, and personal adornment. Populations of freshwater mussels and crayfish have declined sharply in recent decades due to habitat degradation and loss severely curtailing the harvest and use of this important food source. Terrestrial snails (*Monadenia* spp.) are used for personal adornment, and slugs are prominent in stories. A variety of insects such as butterflies and moths are used (e.g., ceanothus silk moth cocoons are used in ceremonial rattles) and acknowledged as special. Ground and hive bees and wasps (e.g., yellowjackets), larva "grubs" and "measuring/inch worms" are central characters in many Tribal stories, as well as spiders, beetles, and other insects such as flies. Tribes have also taken note of a variety of non-native flies and other insects that came to the region with Euro-American settlement and colonization.

Conclusion: Tribal Cultures Then to Now

Over millennia, the integration of different Tribal cultures has sculpted the ecosystems of the Klamath Mountains. Knowledge systems bring new technologies and adaptive practices—one of the most important of which is Indigenous Fire Stewardship that includes cultural burning and horticultural practices. These traditions have evolved into complex Tribal agroforestry systems that shaped the environment and people of the Klamath Mountains.

Many Tribes across the Klamath Mountains—from the coast to interior, up the rivers, and across the mountains—have influenced and affected landscape-scale vegetation with the use of fire, stewardship practices, and associated cultural uses of the environment. From springs, creeks, ponds, lakes and rivers, across forests, prairies or meadows, and shrub fields, Tribal communities utilize a range of plants, fungi, animals, minerals/rocks, and other items to perpetuate and sustain their traditions and to fulfill spiritual and other cultural obligations and responsibilities. The region is a storied landscape that connects the past to the present.

There are benefits to recognizing ancestral ties of indigenous peoples to the Klamath Mountains when working with Tribes today. They have stewarded the land and resources since time immemorial. Integration of the Western and Indigenous forms of inter-disciplinary science and knowledge can improve understanding of the effects and diversification of many aspects of the environment. Tribes have affected and diversified many aspects of the environment including fire regimes and hydrologic processes, which have, in turn, sculpted plants, fungi, and animal populations.

It is through this Traditional Ecological Knowledge that we can all, today, increase our understandings of the long-term geological, ecological, and socio-cultural dimensions of climate, environmental, and human relationships that have and will continue to influence the region's ecosystem diversity and habitat complexity. Tribal communities continue to maintain, and thoughtfully claim, their ancestral connections to places across their aboriginal territories. These connections are the result of adaptive, complex, socio-ecological systems. Indigenous cultural lifeways and practices complement personal, familial, and community well-being—which all depend on intimate, close relationships with the environment.

Acknowledgments

Thanks to Leaf Hillman, Lisa Morehead-Hillman, and James Roscoe for their thoughtful review from a cultural perspective. Thanks to Jim Agee and Dylan Neubauer for reviewing the chapter for edits and clarity.

Resources:

Web-based media links for Tribal agroforestry, cultural burning, and landscape restoration strategies that are restoring these formerly Tribal managed forest and vegetation communities

DOI-BLM: Cultural Resources Overview-Northwestern California: https://www.blm.gov/sites/default/files/documents/files/North_West_Cultural_Resources_Overview_508-small.pdf

State of California-Native American Heritage Commission, Resources: http://nahc.ca.gov/resources/

Library of Congress-The First Peoples of California: https://www.loc.gov/collections/california-first-person-narratives/articles-and-essays/early-california-history/first-peoples-of-california/

Native Women's Collective: https://www.nativewomenscollective.org/regaliastoriescaliforniahistory.html

The Nature Conservancy-Cultural Burning: https://www.nature.org/en-us/magazine/magazine-articles/indigenous-controlled-burns-california/

Humboldt County-Hazard Plan, Cultural Resources Report: https://humboldtgov.org/DocumentCenter/View/1371/Chapter-6-Cultural-Resources-PDF

Literature Cited and Suggested Reading

Aikens, Melvin. 1993. *Archaeology of Oregon*. U.S. Department of the Interior, Bureau of Land Management, Oregon State Office. Portland, OR.

Anderson, Kat. 2005. *Tending the wild: Native American knowledge and the management of California's natural resources*. Berkeley, CA. University of California Press.

Anderson, M.K. and F.K. Lake. 2013. California Indian ethnomycology and associated forest management. *Journal of Ethnobiology*, 33(1), pp.33-85.

Anderson, M.K. and F.K. Lake. 2016. Beauty, bounty, and biodiversity: the story of California Indian's relationship with edible native geophytes. *Fremontia.* Journal of the California Native Plant Society, Sacrament, CA.

Berkes, F. and N.J. Turner. 2006. Knowledge, Learning, and the Evolution of Conservation Practice for Social-Ecological System Resilience. *Human Ecology*, Vol. 34, No. 4. Pages: 479-494.

Briles, C.E., C. Whitlock, and P.J. Bartlein. 2005. Postglacial vegetation, fire, and climate history of the Siskiyou Mountains, Oregon, USA. *Quaternary Research*, 64(1), pp.44-56.

Collins, James. 1998. *Understanding Tolowa Histories: Western Hegemonies and Native American Responses.* Routledge.

Connolly, Thomas J. 1988. A Culture-Historical Model for the Klamath Mountain Region of Southwest Oregon and Northern California, *Journal of California and Great Basin Anthropology.* 10, no. 2, 246–60.

Crawford, J.N., S.A. Mensing, F.K. Lake, and S.R. Zimmerman. 2015. Late Holocene fire and vegetation reconstruction from the western Klamath Mountains, California, USA: A multi-disciplinary approach for examining potential human land-use impacts. *The Holocene*, 25(8), pp.1341-1357.

Eshelman, Jason A. et al. 2004. Mitochondrial DNA and Prehistoric Settlements: Native Migrations on the Western Edge of North America. *Human Biology,* v. 76, no. 1, pp. 55–75.

Halofsky, J.E., D.L. Peterson,and Harvey, B.J. Changing wildfire, changing forests: the effects of climate change on fire regimes and vegetation in the Pacific Northwest, USA. *Fire Ecology.* 16, 4 (2020). https://doi.org/10.1186/s42408-019-0062-8

Hendryx, M. and E. Davis. 1991. *Plants and The People: the ethnobotany of the Karuk Tribe.* Siskiyou County Museum. Yreka, CA. 182 pp.

Hildebrandt, W.R. 2007. Northwest California: ancient lifeways among forested mountains, flowing rivers, and rocky ocean shores. In: Jones, T.L., Klar, K.A. (Eds.), *California Prehistory.* Altamira Press, Lanham MD, pp. 83–97.

Jones, T.L., et al. 1999. Environmental Imperatives Reconsidered: Demographic Crises in Western North America during the Medieval Climatic Anomaly. *Current Anthropology*, Vol. 40, No. 2, Pages: 137-170.

Karuk Tribe Department of Natural Resources. 2016. Karuk Tribe Climate Change Vulnerability Assessment.

King, J.H., W. R. Hildebrandt, and S.A. Waechter. 2016. Cultural Resources Overview for Northwestern California. DOI-Bureau of Land Management, Arcata/Redding Field offices, Report produced from Far Western Anthropological Research Group, Inc. Davis, CA.

Kroeber, A.L., S.A. Barrett, E.W. Winslow, and G.W. Hewes. 1960. Fishing among the Indians of northwestern *California. University of California Anthropological Records* 21(1): 1-210. Berkeley: University of California Press.

Lake, F.K. 2007. Traditional ecological knowledge to develop and maintain fire regimes in northwestern California, Klamath-Siskiyou bioregion: management and restoration of culturally significant habitats. Dissertation, Oregon State University, Corvallis.

Lake, F.K. 2013. Trails, fires, and tribulations: Tribal resource management and research issues in northern California. Occasion: *Interdisciplinary Studies in the Humanities*, 5, pp.1-22.

LaLande, J. 1991 The Indians of Southwestern Oregon: An Ethnohistorical Review. Department of Anthropology, Anthropology #9 Oregon State University, Corvallis, OR.

Lewis, H.T. and M.K.Anderson. 2002 (editors), In Stewart, O.C. *Forgotten Fires: Native Americans and the Transient Wilderness.* University of Oklahoma Press, Norman, OK. 364.

Long, J., et al. 2018. Volume 3—Synthesis of science to inform land management within the Northwest Forest Plan area. Gen. Tech. Rep. PNW-GTR-966. Portland, OR: US Department of Agriculture, Forest Service, Pacific Northwest Research Station: 625-1020. Vol 3., 966, pp.625-1020.

Long, Linda L., F.K. Lake, J.L. Stephens, J.D. Alexander, J.D. Wolfe. 2021. Using culturally significant birds to guide the timing of prescribed burns in the Klamath-Siskiyou bioregion. *Ecosphere* ECS20-0737.

Meighan, Clement W. 1955. *Archaeology of the North Coast Ranges, California.* Berkeley: University of California Archaeological Survey Reports 30:1-39.

Skinner, C.N., A.H. Taylor, and J.K. Agee. 2006. Klamath mountains bioregion. In: N.G. Sugihara, J.W. van Wagtendonk, J. Fites-Kaufmann, K.E. Shaffer, and A.E. Thode, editors. 2006. *Fire in California's ecosystems.* University of California Press, Berkeley. pp. 170-194.

Skinner, D. 2005. Viva La Difference: How and why Tribal forestlands are managed differently from federal forestlands. *Evergreen*: Pages: 21-24 Stein, S.J., Price, P.W., Abrahamson, W.G., and Sacchi, C.F. 1992. The effect of fire on stimulating willow regrowth and subsequent attack by grasshoppers and elk. Oikos, Vol. 62, No. 2: 190-196.

Sundahl, E., and W. Henn. 1993. Borax Lake Pattern Assemblages on the Shasta-Trinity National Forests, North-Central California. *Journal of California and Great Basin Anthropology*, 15(1). Retrieved from https://escholarship.org/uc/item/9z9530pm

Tushingham, S. and R.L. Bettinger. 2013. Why foragers choose acorns before salmon: Storage, mobility, and risk in aboriginal California. *Journal of Anthropological Archaeology*, 32(4), pp.527-537.

Vale, Thomas R. (ed.). 2002. *Fire, Native Peoples, and the Natural Landscape.* Washington, DC: Island Press.

Disclaimer: This chapter-publication does not represent an official position of American Indian Tribes. The author accepts all responsibility for content and accuracy.

Chapter 2: Western Science Arrives

James K. Agee

Alice Eastwood collecting and pressing plants.
Image courtesy of the Alice Eastwood Papers,
California Academy of Sciences Archives.

Introduction

The earliest human explorers of the Klamath Mountains were the Indigenous Peoples who have lived here for millennia. They have always had an intimate relationship with the land, forever relying on their surroundings for survival, health, and wellbeing. With a rich oral history, they pass this Traditional Ecological Knowledge from generation to generation. Euro-Americans essentially rediscovered the region's natural history, and applied the labels of western science to what they found. This chapter focuses on the growth of western science, in more of a written tradition, by Euro-American explorers in the Klamath Mountains.

18th and Early 19th Century Natural History Expeditions

The earliest explorations of the Klamath Mountain Region were complicated due to its remoteness and rugged terrain and that paradigm still continues today. Most early expeditions to western North America were restricted to ocean-going vessels that either skipped the northwest California coast altogether or briefly touched ashore (1785, La Pérouse Expedition; 1790-95, Vancouver Expedition; 1816, Kotzebue Expedition). The first overland Euro-American expedition was led, in 1828, by trapper Jedediah Smith, whose party of 20 men and 30 horses and mules crossed into the Klamath Mountains from the Central Valley near present-day Platina, into the Hayfork Valley, and down the South Fork Trinity River to the coast (Sawyer 2006). Eight years later, in 1836, fur trapper Thomas McKay collected 1,800 beaver from swampy Scott Valley, later claiming that Scott Valley had the highest concentration of beaver he had seen in a career trapping beaver widely across the West. None of these earliest explorations produced much lasting evidence of natural history.

The Wilkes expedition of 1841 was the first overland scientific exploration focused on natural history. The expedition split at the Columbia River, with some sailing south to Yerba Buena (San Francisco) and the rest, led by Lieutenant George Emmons, traveling south through the Willamette Valley, over Siskiyou Summit, and down into Shasta Valley on their way to rejoining the rest of the expedition moving up the Sacramento Valley. The expedition had several natural historians including a botanist, William Brackenridge; a zoologist, Titian Peale; and a geologist, James Dana. They noted the flora in the eastern Klamath Mountains was more diverse than anywhere else on the trip. Brackenridge described more than 40 new plant species and Peale noted pronghorn antelope, "marmots" (likely fishers), and California condors. He was particularly mesmerized by an American dipper walking on the bottom of a stream, so much so he forgot to collect it. While on the run from local Indians, Brackenridge collected a specimen of cobra lily (*Darlingtonia californica*), although it was not formally described until 1854 by famed botanist John Torrey. Geologist Dana gamely tried to describe the complex landscape he encountered, but beyond descriptions of "Shasty Peak," Black Butte, and Castle Crags, interpretation of what he saw was "nigh impossible" (Beidleman 2006).

Geology

In 1879, the U.S. Geological Survey (USGS) was created by Congress and comprehensive geological investigations were undertaken. One of the first western-educated geologists to focus on the northern California region was USGS geologist Joseph Diller. In 1886, he published a paper recognizing the boundaries of the Sierra Nevada, Cascade Range, and those areas west that he called the "Coast Range" (Diller 1886). By the early 20th century, he recognized the distinct nature of the Klamath Mountains and speculated that the area had been reduced to peneplains

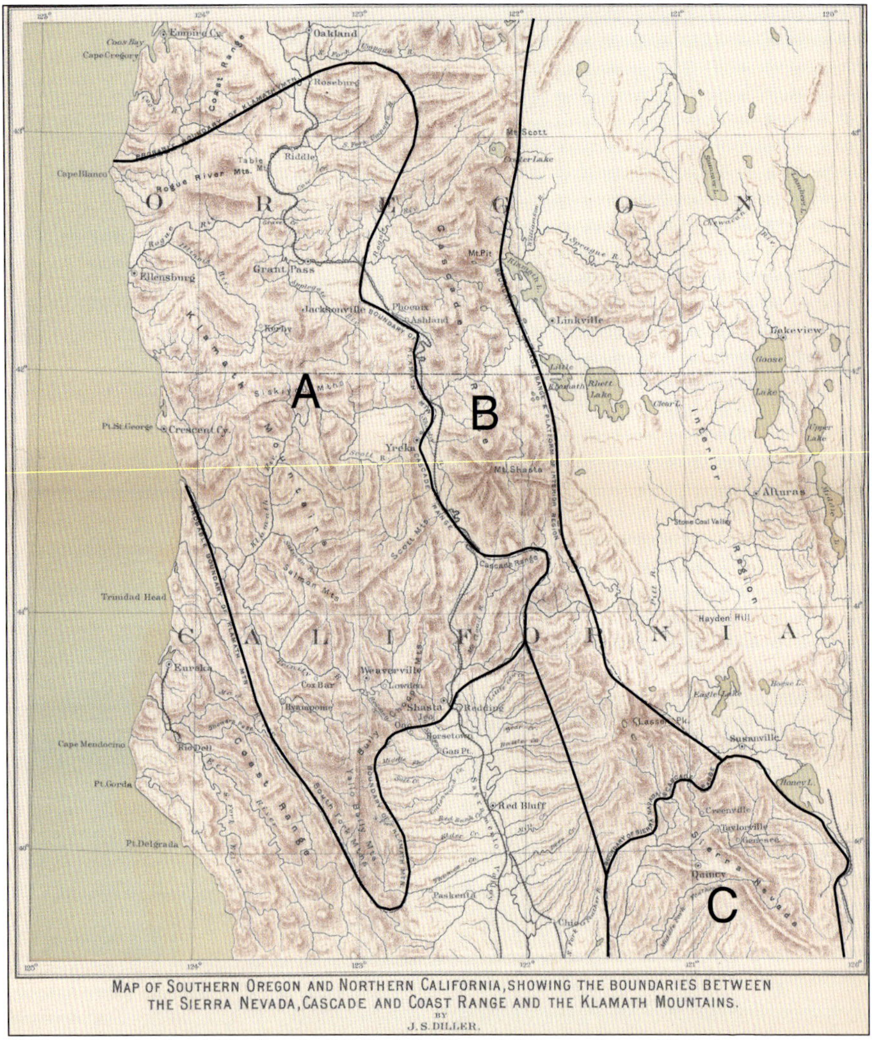

1894 map drawn by Joseph Diller from the U.S. Geological Survey's 14th annual report: "Preliminary Boundaries of the Sierra, Cascades, and Klamath Mountains." This is the first published map defining the Klamath Range along with detailed topography and major streams. Southern Pacific rail lines of California and Oregon are also mapped in an era long before highway and reservoir construction revealing the vast remoteness of the region. Note: many prominent localities have alternative names and spellings. Boundaries and letters have been added for clarity: A) Klamath Mountains B) Cascade Range C) Sierra Nevada.

(low-relief plains formed by protracted erosion) several times (Klamath peneplain inland, Bellspring peneplain nearer the coast), interrupted by faulting and uplift (Diller 1902). He noted that the upper portion of the principal divides had broad tops on which the relief was terraces or low, rolling hills, giving the impression of a dissected plateau. He also noted correlations between some geological formations in the Klamaths and some in the western Sierra Nevada. In 1906, he established a stratigraphic framework for the eastern Klamath Mountains (Diller 1906). These remain major contributions toward the geomorphic interpretation of the region (Aalto 2006).

Oscar Hershey was a contemporary of Diller, and although his published material was less prolific than Diller's, it had a profound impact on the understanding of Klamath geology—especially its metamorphic formations. He had completed some mineral surveys in the region but believed no systematic study had been done. In 1901, he noted that previous work in the Klamaths, including his own, was "of a rather desultory character." He thus set out to classify the region's metamorphic rocks (Hershey 1901). He classified the pre-Cretaceous rocks into "seven great formations," exclusive of the granitic intrusions, and provided the first detailed descriptions of the "Abrams Mica Schist" and "Salmon Hornblende Schist" (Hershey 1901). Hershey was also one of the first geologists to document the alpine glaciers of the Trinity Alps (Hershey 1900), including the glacially-caused pirating of Coffee Creek by the South Fork Salmon River at Big Flat.

William Porter Irwin. Image courtesy of the Geological Society of America.

The next seminal work in the Klamath Mountains was the regional geological synthesis of William Porter Irwin, also a USGS scientist (Irwin 1960). This bulletin represented "an enormous advancement in understanding the fundamental relationships between the Klamath Mountains and Coast Ranges," as well as among the diverse groups of rocks in the province. He also introduced the concept of four distinct lithic belts from east to west: (1) eastern Paleozoic belt, (2) central metamorphic belt, (3) western Paleozoic and Triassic belt, and (4) western Jurassic belt (see Geology Chapter 3). This, and subsequent work, established Irwin as the most significant contributor to the geology of the Klamath Mountains, evidenced by the special volume dedicated to him by the Geological Society of America for his fundamental contributions to the geology of western North America (Snoke and Barnes 2006).

Bob Sharp. Image courtesy of NASA JPL-Caltech (Michael Malin).

At the same time, Robert P. Sharp of the California Institute of Technology published a detailed analysis of glaciation in the Trinity Alps, with figures showing the extent of glacial episodes down many of the area's streams (Sharp 1960). While Hershey found three recent Holocene glaciers in the early 20th century, Sharp found only two by mid-century, and one of those disappeared by 2015 in the recent drought. Evidence of glaciation remains in Coffee Creek, Swift Creek, Canyon Creek, and the Stuart Fork—among others. Sharp expanded upon Hershey's identification of the pirating of Coffee Creek, noting "Any geologist working in the area who fails to report the diversion of the former headwaters of Coffee Creek into the South Fork of Salmon River at Big Flat will be characterized by his successors as totally blind."

Botany

Mount Shasta, part of the southern Cascade Range, attracted early naturalists, many of whom eventually ventured west into the Klamath Mountains. In 1852, Scottish botanist John Jeffrey, while focused on Mount Shasta, collected lodgepole pine (*Pinus contorta* ssp. *murrayana*) in the Siskiyous and in the Scott Mountains discovered foxtail pine (*Pinus balfouriana* ssp. *balfouriana*). In Shasta Valley he discovered the pine that was to bear his name, Jeffrey pine (*Pinus jeffreyi*) (Smith 2019). He also visited the Trinity Mountains in October of that year before returning to Oregon. In 1853, Dr. Albert Kellogg, a self-trained botanist, was one of seven who created the California Academy of Sciences in San Francisco. A specialist in oaks, the California black oak (*Quercus kelloggii*) is named for him. He described over 200 species in the coming decades, but perhaps his most lasting achievement was his hiring of two women co-curators, including Alice Eastwood, who made lasting contributions to the botany of the Klamath Mountains and California.

Willis Linn Jepson, famous University of California botanist of the early 20th century, noted that comparison of the high mountain flora of Mount Shasta to the flora of the Salmon Mountain region westward was of great interest to plant geographers (Jepson 1942). But not everyone found Mount Shasta a botanist's paradise. Alice Eastwood, new Curator of the Herbarium at the California Academy of Sciences in San Francisco, visited Mount Shasta in 1893 and found the flora "bleak, barren, and utterly disappointing," due to exposed rock, snow, scarce water, and shallow soils (Eastwood 1896).

The first large-scale Euro-American expedition in this area was the natural history survey of Mount Shasta in 1898 by C. Hart Merriam, Chief of the federal Biological Survey (Merriam 1899). Merriam described the flora and fauna of the mountain, categorizing them by the life zone concept he had pioneered in northern Arizona in 1890. He described the distribution of 215 trees, shrubs, and perennial herbs in the area. While he was able to compare the flora of Mount Shasta to that of the Cascades and Sierra Nevada, there was insufficient information to make comparisons to the remote, rugged, and densely forested Klamath Mountains.

Merriam sent many plant specimens for identification to Alice Eastwood. At the end of his Mount Shasta survey season, Merriam and two others trekked across the Klamath Mountains to the coast, collecting plant specimens along the way. Later, he extolled the virtues of the Klamath Mountains to Miss Eastwood, who

My Botanical Contest With Miss Alice Eastwood

Alice Eastwood had been Herbarium Curator at the California Academy of Sciences for seven years when in 1900, after a conversation with C. Hart Merriam, she was encouraged to visit the Klamath Mountains. She and her three companions walked from Redding with a string of horses, unsure if they could obtain stock closer to their destination: Canyon Creek, a tributary of the Trinity River upstream from Junction City. She documented the shrubs and trees they encountered along the way. Above Dedrick, they had to cut their own trail close to the bottom of the creek on the way up to Twin Lakes (now Canyon Creek Lakes). Two days later, they emerged at Twin Lakes and spent most of a week exploring the alpine country. Miss Eastwood published her account in the *Sierra Club Bulletin* (Eastwood 1902), and listed the shrubs and trees she collected in the upper Canyon Creek watershed.

In 2003, little more than a century later, and as a contribution to my book Steward's Fork (Agee 2007), I decided to challenge the indomitable Miss Eastwood by finding all her shrubs and trees and perhaps tacking on a few more. I arrived at the Canyon Creek trailhead soon after Labor Day. Pencil at the ready, I started up the trail and immediately spied elk clover (*Aralia californica*), a shrub Miss Eastwood missed. But this turned out to be a red herring, as elk clover is a tall herb, not a shrub. Undaunted, I continued up the trail and recorded about 20 species along the first mile. Before I reached Hound's-head Falls (now Lower Canyon Creek Falls), I discovered two species not on her list: ninebark (*Physocarpus capitatus*) and a buckwheat (*Eriogonum* sp.). I had an early lead.

I camped that night above the falls, and woke the next morning to a cold, hard rain. I still needed about 20 more species to complete my list. By the time I made it to the upper lakes, I needed just a few more, but I was wet and enveloped in a thick mist that obscured the cairn-marked trail around the west edge of the lakes. While I found mountain heather (*Phyllodoce empetriformis*), I realized that white heather (*Cassiope mertensiana*), which Miss Eastwood found on a windswept ridgetop, would not join my list. I was also not going to find alpine laurel (*Kalmia polifolia*), although I knew it would likely be at Kalmia Lake, farther up the canyon. While I was missing these two species, if I could find Labrador tea (*Rhododendron columbianum*) I would be assured of a tie with Miss Eastwood. Inclement weather forced me downstream where I found Labrador tea.

By the next day, I was back in civilization and began the final tally. Miss Eastwood had clearly listed ninebark, which I had simply overlooked, and she noted that she ignored low, shrubby plants such as buckwheat. Her list thus totaled 66 species and mine, without the missing two alpine species, totaled 64, so I posthumously declared Miss Eastwood the winner. But this was not ever intended to be a real contest. Its purpose was to show that at a landscape scale, species composition can stay remarkably stable, even over a century's time. While this was true for the 20th century, climate change in the 21st century is likely to disrupt that landscape stability in Canyon Creek and more widely across the Klamath Mountains.

exclaimed "It seemed as if life itself would lose its zest if these mountains could not be reached, their rugged peaks climbed, their botanical treasures collected, and their dangers and difficulties overcome" (Eastwood 1902). In 1900, she organized an expedition into the Trinity Alps and chose to explore Canyon Creek, a tributary of the Trinity River, in the summer. Her party of four obtained stock in Redding and walked through Lewiston, Junction City, and up to Dedrick where the expedition began in earnest. She described the vegetation zones they passed through and catalogued more than 100 shrubs and tree species identified along the route. Her passion for plant collecting and documentation is reflected in her actions during the great San Francisco Earthquake. She climbed into the destroyed Academy building up a damaged staircase and lowered by rope 1,500 type specimens that would otherwise have burned when the subsequent fire leveled the building—as it did much of the city.

In 1907, Willis Linn Jepson followed a 70-mile section of the Kelsey Trail from the Klamath River to the coast at Crescent City. He noted that it was "truly a frontier country" with "not only no railroads but no roads" (Wallace 2009). Several years earlier, Joseph Tracy, born and raised in Humboldt County, enrolled in botany at the University of California, Berkeley.

Jepson mentored him as a student assistant in the Herbarium, hoping he would help document the northwest California flora. Tracy graduated in 1903, but could not find work as a professional botanist in Eureka, instead working as a land surveyor for a bank. He continued his botanical work as an avocation, and was frequently spotted in the woods and hills of Humboldt and Trinity counties with his plant press. Tracy frequently sent specimens to Jepson. When Jepson published his seminal *Manual of the Flowering Plants of California* in 1923, he acknowledged Joseph Tracy for his contributions from Humboldt County. Over time, Tracy expanded his work to Del Norte County, Siskiyou County, and southwestern Oregon. He contributed more than 30,000 duplicate specimens to the University herbaria, many of which were distributed around the world. He had numerous species named for him including an aster (beaked tracyina, *Tracyina rostrata*, Tracy's beardtongue, *Penstemon tracyi*, and *Festuca tracyi* [now *Vulpia microstachys* var. *confusa*]). Before his death in 1953, he arranged for his large and significant personal herbarium to be donated to the University of California (Bacigalupi 1954).

Joseph Prince Tracy (right) and his nephew. Photo courtesy of the Tracy family.

By the mid-1930s, Jepson was preparing a multi-volume flora of California, a major expansion of his 1923 *Manual*, but he was always seeking more specimens

Ruby and Jepson Find A Milkweed

Ruby Van Deventer had impressed Willis Linn Jepson from the moment he met her in 1936 in his Berkeley campus office. She had compiled a list of 1,500 Del Norte County plants, and he saw in her an enthusiastic apprentice who could be very helpful in documenting the flora of the region. Later that year she sent him live specimens of Indian pipe (*Monotropa uniflora*) a rare saprophytic plant, which prompted Jepson to write that she would be "duly and appropriately immortalized" when *The Flora of California* was finally published (unfortunately he was never able to finish it).

Arthur and Ruby Van Deventer. Photo courtesy of Rick Bennett, co-editor of *A Rare Botanical Legacy*.

Jepson always gave credit where credit was due, and gave Ruby the "Dickens" if he thought she was mistaken about a plant identification. One such instance was the identification of narrow-leaved milkweed (*Asclepias mexicana*). Ruby had sent him a specimen and he had replied that she must have been mistaken about the place she had collected it, as it only grew in dry ground and foothills. Ruby did not argue with him, but when he visited her and her husband Arthur Van Deventer (known as Van) in 1937, Van drove them to a site in the northern Siskiyous called Poker Flat, which was known for a variety of unusual plants. Ruby led Jepson to a precipice in the area where the *Asclepias* grew, but she and Van said nothing. Jepson looked over the edge of the cliff and spotted it: "Well, well! It doesn't grow north of Tehachapi but there it is!"

Indian pipe (*Monotropa uniflora*). Photo by Michael Kauffmann.

It was the complexity of the botanical landscape that made Jepson exclaim to Ruby, "the flora of Del Norte has been mixed up with an eggbeater".

Note: This sidebar has been excerpted from Wallace (2009).

(Wallace 2009). He had received several from an amateur botanist, a young schoolteacher from Del Norte County named Ruby Van Deventer. When Ruby visited the Berkeley campus, she impressed him and he encouraged her to continue her work (Wallace 2009), and she eagerly returned to Del Norte County to continue collecting. Her husband, Arthur, who worked in natural resources, helped by painting watercolors of the plants she collected. Jepson and the Van Deventers continued to collaborate, with Ruby eventually contributing more than 400 specimens to his personal herbarium. He invited her to write the book for the Del Norte area. But Jepson never finished his opus and died in 1946, and the Van Deventers never finished their flora, either, before they died in the 1970s. Soon after, Ruby's grandchildren donated her field notes and 3,500 specimens to Cal Poly Humboldt.

While the biodiversity of the Klamath Region has been documented since the early 20th century, it took on added significance after Robert Whittaker published two articles summarizing the "central" significance of the Klamath Mountain flora (Whittaker 1960, 1961). He noted that the 3,540 taxa found in the region was comparable to only one other region in North America—the southern Appalachian Mountains. He explained this diversity as a result of the Klamath Mountains being a great meeting ground: its central location on the Pacific Coast; its continuity with other mountain systems; its diverse climate, geology, and topography; its complex disturbance regimes; and its long geological history (the region has supported flowering plants for the last 65 million years) (Sawyer 2006).

John O. Sawyer Jr. Photo by Michael Kauffmann.

One last botanical explorer is not really as "early" as many others in this chapter, but is deserving of mention. John Sawyer, Professor of Botany at Cal Poly Humboldt, was a nationally recognized authority on plant ecology, conifers, and the vegetation and flora of California, especially the Klamath Region. In 1970, he and Dale Thornburgh, a colleague at Cal Poly Humboldt, discovered a "miracle mile" in the Salmon Mountains in which 17 conifer species were found in a single square mile. That total has since increased to 18 with the recent addition of western juniper (*Juniperus occidentalis*) to the list. Sawyer died in 2012, but his natural history classic *Northwest California* (2006) will live on.

Zoology and Paleontology

Other than brief surveys, little was known about the fauna of the Klamath Mountains until Merriam conducted his survey of the flora and fauna of Mount Shasta (Merriam 1899), some of which was directly relevant to that grand mountain region to the west. He described the characteristics and distribution of 68 mammal and 136 bird species in the Mount Shasta vicinity and grouped them by life zones. There is little evidence that he described or collected faunal specimens when he traveled west through the Klamaths at the end of the season in 1898.

A young Annie Alexander. Photo from the University of California Museum of Paleontology

In 1902, two tusks were unEarthed during a hydraulic mining operation at the confluence of Union and Coffee creeks (Hershey 1903). Hershey speculated that they were from a woolly mammoth (*Mammuthus primigenius*) that somehow travelled to this remote high elevation (4,250 feet) site and became mired in the marshy soil surrounding a glacial lakelet, the remnant of a glacier that occupied Union Creek. Another important paleontological discovery was made in the eastern Klamaths by Annie Montague Alexander in the early 20th century. In 1900, and in her early 30s, she had been attending the paleontology lectures of Professor John C. Merriam (a cousin of C. Hart Merriam) at the University of California, Berkeley, and developed an interest in fossils (Stein 2001). In 1901, she accompanied Merriam to Fossil Lake in Oregon and then in 1902 and 1903 organized and funded trips to Bear Cove and Black Oak Canyon between the Pit River and Sulanharas Creek, in the area now partially covered by Lake Shasta. After thoroughly scouting the area, Annie uncovered the find of the trip, an ichthyosaur skeleton. Merriam proclaimed it a new species and named it in her honor, *Shastasaurus alexandrae*. Now *S. pacificus*, the Shasta lizard was a very large, ocean-dwelling reptile, up to 60 feet long and weighing up to 75 tons. A subsequent trip the next year yielded a six-foot, four-legged, semi-marine reptile of the Triassic period which was also named after Ms. Alexander (*Thalattosaurus alexandrae*).

Being independently wealthy as a Hawaii sugar cane heiress, Annie almost entirely funded the establishment of the Museum of Vertebrate Zoology (MVZ) in 1908 at the University of California, Berkeley (which was built in Faculty Glade, until 1930 when it was moved to the new Life Sciences Building). This enabled systematic collections to be amassed and studied. The first director was Joseph Grinnell, a prolific mammologist, who had been a confidant of Alexander's (together with C. Hart Merriam) during the planning of the MVZ. By 1911, Annie had also taken an interest in modern fauna, and with her amateur naturalist companion Louise Kellogg, proposed a collection trip sponsored by the MVZ to the Trinity Alps. The intent was to discover whether the fauna of the region differed from that of the "Humid Coast," Cascades, or Sierra Nevada, or was some blend of two or more of those regions. It was an expansion of the earlier work of Merriam at Mount Shasta. In the winter and spring of 1911, they made three trips, documenting 95 bird species, 142 mammal species, and had specimen totals of 449 birds and 976 mammals, all of which were deposited at the MVZ. They claimed to have found a new species of mountain beaver (*Aplodontia chryseola*) and subspecies of kangaroo rat (*Dipodomys californicus* ssp. *trinitatus*), but these were later lumped with the existing species. Kellogg published her findings in a 60-page monograph in the MVZ publication series (Kellogg 1916), and it was followed by an analysis of Kellogg's findings by Joseph Grinnell (Grinnell 1916). He noted that the species list was likely incomplete, but was a "known fauna," and concluded that the Trinity region showed "but weak faunal individuality," differing more from the "Humid Coast" than from the Sierras or Cascades, and the boreal fauna being somewhat closer to the Sierras than the Cascades.

Joseph Grinnell remains the most prolific, western-educated faunal scientist in California history, with over 550 publications (many of which were brief reviews or notes of sightings of birds or mammals in new locales) (Grinnell 1940). Most of his site-specific studies focused on the Sierra Nevada, but he did publish results of a bird survey in Siskiyou County (between Horse and Seiad creeks) in 1901-1902 (Anderson and Grinnell 1903). The study concluded that the area had a number of bird species usually not found together, the result of a "narrow line of mergence between the humid coast range and the arid Sierra fauna." Grinnell later published a number of separate comprehensive bulletins on the birds and mammals of California that were inclusive of the Klamath Mountains.

The unique assemblage of amphibians and reptiles of the Klamath Range have garnered interest from several generations of herpetologists, with most affiliated at some point with the University of California, Berkeley and its Museum of Vertebrate Zoology. Those who have pioneered our modern understanding of Klamath herpetofauna came in the 20th century and include Henry Fitch (UC Berkeley 1930s), Robert Stebbins (UC Berkeley; 1940s-2000s), Robert Storm (Oregon State University; 1940s-2000s), David Wake (UC Berkeley; 1960s-2021), and Bruce Bury (USGS; 1960s-present). Nobody has spent more time focused on the biogeography of amphibians and reptiles of the Klamath Mountains than Bruce Bury, emeritus zoologist with the U.S. Geological Survey. Bury has been studying northwestern pond turtles (*Actinemys marmorata*) in the Trinity River basin for over 50 years and shattered existing longevity records for the species by recapturing individuals that were marked 42-50 years earlier during his graduate work. Bury has also brought recognition to the Klamath region as a worldwide temperate amphibian diversity "hotspot" through various regional biogeographic studies and the discovery of many species range extensions (Bury 1970).

Conclusion

Early Euro-American explorers found the Klamath Mountains to be a tough place to explore. But they also found it to be a diverse region of inert and living things. Their discoveries laid the groundwork for discoveries still occurring today, such as the new salamander species recently found in the Scott River area. It will not be the last discovery made in this magical mountain region. The following chapters summarize what we know today, but there will surely be new discoveries that will encourage future editions of this book.

Literature Cited and Suggested Reading

Aalto, K.R. 2006. The Klamath peneplain: A review of J.S. Diller's classic erosion surface. Pp. 451-463 In: Geological Studies in the Klamath Mountains province, California and Oregon: A volume in honor of William P. Irwin. Geological Society of America Special Paper 410.

Agee, J.K. 2007. *Steward's Fork. A Sustainable Future for the Klamath Mountains.* University of California Press. Berkeley.

Anderson, M.P., and J. Grinnell. 1903. Birds of the Siskiyou Mountains: A problem in distribution. Proceedings of the Academy of Natural Science of Philadelphia 55 (Jan.): 4-15.

Bacigalupi, R. 1954. Joseph Prince Tracy: 1879-1953. *Madroño* 12: 190-192.

Beidleman, R.G. 2006. *California's Frontier Naturalists*. University of California Press. Berkeley.

Bury, R.B. 1970. A biogeographic analysis of the herpetofauna of Trinity County, California. *Journal of Herpetology* 4(3/4): 165-178.

Diller, J.S. 1886. Notes on the geology of northern California. *U.S. Geological Survey Bulletin* 33. Washington, D.C. 21 p.

Diller, J.S. 1902. Topographic development of the Klamath Mountains. *U.S. Geological Survey Bulletin* 196. Washington, D.C. 69 p.

Diller, J.S. 1906. Description of the Redding quadrangle, California. *U.S. Geological Survey Geological Atlas Folio* 138. 14 p.

Eastwood, A. 1896. The alpine flora of Mt. Shasta. *Erythea* 4(9): 136-142

Eastwood, A. 1902. From Redding to the snow-clad peaks of Trinity County. *Sierra Club Bull.* 4(1): 39-52.

Grinnell, H. 1940. Joseph Grinnell: 1877-1939. *Condor* XLII (1): 3-34.

Grinnell, J. 1916. An analysis of the vertebrate fauna of the Trinity region of northern California. *University of California Publications in Zoology* 12(14): 399-410.

Hershey, O.H, 1900. Ancient alpine glaciers of the Sierra Costa Mountains in California. *Journal of Geology* 8: 42-57.

Hershey, O.H. 1901. Metamorphic formations of northwestern California. *American Geologist* 27: 225-245.

Hershey, O.H. 1903. Some evidence of two glacial stages in the Klamath Mountains in California. *American Geologist* 31: 139-156.

Irwin, W.P. 1960. Geologic reconnaissance of the northern Coast Ranges and Klamath Mountains, California, with a summary of the mineral resources. *California Division of Mines Bulletin* 179.

Jepson, W.L. 1942. Early botanical ascents of Mount Shasta. Sierra Club Bull. 27(4): 23-30

Kellogg, L. 1916. Report upon mammals and birds found in portions of Trinity, Siskiyou, and Shasta counties, California, with description of a new *Dipodomys*. *University of California Publications in Zoology* 12(13): 335-398.

Merriam, C. H. 1899. Results of a biological survey of Mount Shasta California. USDA Division of Biological Survey 16. Washington, D.C.

Sawyer, J.O. 2006. *Northwest California: A natural history*. University of California Press. Berkeley.

Sharp, R.P. 1960. Pleistocene glaciation in the Trinity Alps of northern California. *American Journal of Science* 258: 305-340.

Smith, J.P. 2019. *A Catalog of the Vascular Plants of Northwest California*. California NativePlant Society. Sacramento, CA.

Snoke, A.W., and C.G. Barnes (eds.). 2006. Geological Studies in the Klamath Mountains Province, California and Oregon: A Volume in Honor of William P. Irwin. *Geological Society of America Special Paper* 410.

Stein, B. 2001. *On Her Own Terms. Annie Montague Alexander and the Rise of Science in the American West*. University of California Press. Berkeley.

Wallace, D.R. 2009. Plant explorers of Del Norte County: A botanical friendship. Pp. 9-26 In: Bennett, R., and S. Calla (eds.). *A Rare Botanical Legacy*. The Contributions of Ruby and Arthur Van Deventer. Heyday Books. Berkeley, California.

Whittaker, R.H. 1960. Vegetation of the Siskiyou Mountains, Oregon and California. *Ecological Monographs* 30(3): 279-338.

Whittaker, R.H. 1961. Vegetation history of the Pacific Coast states and the "central" significance of the Klamath region. *Madrono* 16(1): 5-23.

Chapter 3: Geologic History

Mark Bailey

Serpentine meets gabbro in the Red Trinities.
Photo by Michael Kauffmann.

Rocks: The Ambassadors of Time

The origin of the Klamath Mountains is an epic tale that begins in its rocks. Although it has been a long, difficult, and ongoing challenge for scientists to decipher the geologic history of the Klamath Mountains, the processes involved are often simple to understand. The difficulty has come from the fact that many of the rocks have been distorted, stretched, and crumbled (Figure 1) during the mountain building processes—forming a complex and varied assemblage of rock bodies and structures. Additionally, most of these rock outcrops are covered by deep soils and dense forests and only "peek out" locally.

Rocks are like messages in bottles. Cast into a vast sea of time, they drift unnoticed for eons. With luck they "wash up" at our feet to speak to us across the great gulf of time. They are the Ambassadors of Time. They record the essence of events long past—even of times before life existed. Rocks contain the keys to understanding the beginnings of the ground on which we stand. Their grains tell of molten masses of magma at great depths in the Earth, of sunny tropical beaches awash in the waves that existed in the distant past, and of the emergence of our ancestors from the rocky world around us. Rocks are inherently interesting and with a basic foundation of understanding, true wonders are revealed. The Klamath Mountains offer the perfect opportunity to explore these wonders.

For countless generations, people have traversed the Klamath Mountains. Yet the geologic understanding of the Klamath is new and still emerging. Hidden within their rocky crags lie the clues to how these mountains came to be. We now know that hundreds of millions of years have passed as the magnificent Klamath Mountain sub-ranges (see maps on page *ii*) developed into what they are today. One glance at the scope and ruggedness of these mountains helps explain the difficulty in untying the Gordian Knot of Klamath Mountains geology.

Historically, the area has been poorly understood by scientists because of its rugged terrain and isolation. Road building was difficult and maintenance is an ongoing challenge (Figure 2). Towns were few and small. It was, essentially, terra incognita to outsiders. Beginning in 1849, the California Gold Rush opened the landscape to adventurers and scoundrels who perpetrated a long series of disastrous actions on both the Native People and the landscape of these mountains. They soon discovered that the auriferous (gold-bearing) geology of the Klamath Mountains was nearly identical to that found in the northern Sierra Nevada, and a new gold rush descended upon the nearly impenetrable landscape and its Native People. With this, the dawn of modern geological thought had arrived in the Klamath Mountains.

Figure 1. This image, taken in the Western Klamath terrane along Clear Creek in the Siskiyou Wilderness, hints at the convoluted and difficult-to-interpret history of these mountains. Area shown is 6 feet across. Photo by Mark Bailey.

The Rocky Underworld

The Klamath Mountains are composed of nearly all the different kinds of rocks commonly found on Earth—and others that are not so common. They are a hotbed of information about ancient seas, volcanoes, and continents. They also speak of the crystal-studded depths of the Earth where rocks can bend and move without breaking—where rock will flow over millennia under the great pressure of its neighbors above and the great heat from those below.

Figure 2. Big French Creek landslide. 40 miles west of Weaverville on Highway 299. This landslide, one of the largest in Northern California, was triggered by heavy rains in early 2016 and closed Highway 299 on and off for months. It attests to the instability of some of the rocks in the Klamath Mountains. In this case, it is the jumbled mélange of ocean floor sediments in the Western Hayfork terrane. Photo by Mark Bailey.

Downcutting by streams, faulting, and landslides are some of the natural means by which these rocks have become exposed. Road building has also opened up the "underworld" of rocks in the many roadcuts carved through this steep country. It is in these locations that we can stand and peer at the roots of a range of mountains. The roadcuts along CA Highway 299 offer a wonderful sideways glimpse into the history of these peaks (Figure 3). Traversing this highway brings you in direct contact with a broad swath of the rocks containing the story of the origin of these mountains.

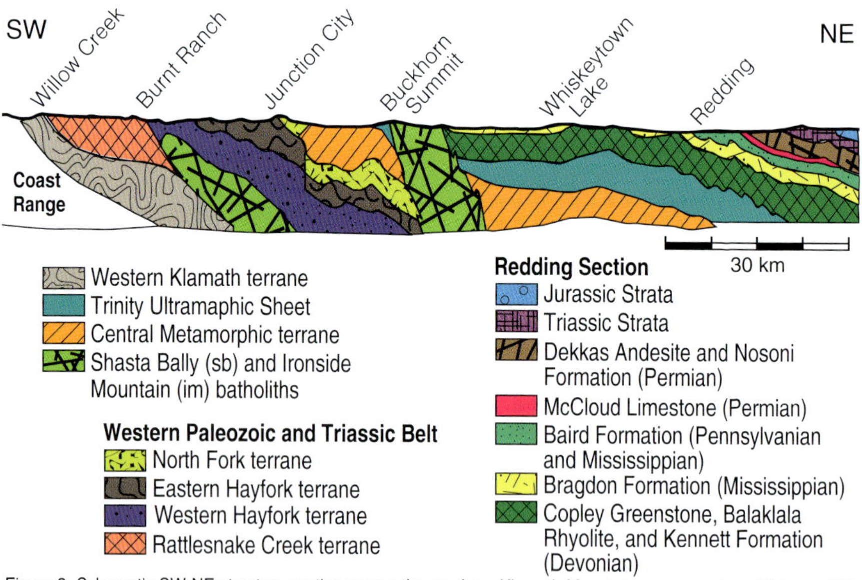

Figure 3. Schematic SW-NE structure section across the southern Klamath Mountains as seen along Highway 299. Adapted from Irwin (1997).

In the Beginning: The Formation of the Earth

Let's pause for a moment and look at the bigger picture. The Earth formed from the accretion, or accumulation, of many planetesimals and smaller bodies that collided in the spinning disk of our infant solar system. Imagine the heat generated by collisions at tens of thousands of miles per hour! This heat contributed to the melting of the entire mass of the Earth. Planetesimals also contained radioactive elements, such as uranium, that themselves give off large amounts of heat as they decay. This molten, rotating ball started to differentiate or separate out into layers of differing density. Most of the heaviest elements sank to the Earth's core. Ultimately, three basic layers—core, mantle, and crust developed—though there are many subdivisions within these.

Earth's crust, which averages a few tens of miles thick, is not quite as rigid as it seems. The Earth itself is nearly 8,000 miles across, so the crust is thin in comparison. Even though it has been more than 4.5 billion years since its formation, the Earth still contains vast amounts of heat. This heat is constantly leaking through the thin crust and, in the process, produces some interesting and dangerous effects—volcanoes, melting of rocks below the surface, and the flexing, cracking, and buckling of the crust. This heat, combined with gravity, is the engine behind the formation of the Klamath Mountains.

It's About Time

During a human lifetime, we may live long enough to witness 90 or so orbits around the sun. So, what does a billion—or even a million years—really mean? The entirety of our day-to-day experience is but the blink of an eye when compared to the span of Deep Time (geologists usually call it *geologic time*). On a geological scale, almost nothing noticeable will occur in a human lifespan (Earthquakes, volcanic eruptions, and a few other things excepted). But what of the lifespan of a mountain? The uplift and grain-by-grain disintegration of a mountain range takes millions of human lifespans. Getting a grasp on the immensity of Deep Time is fundamental to understanding geology and the rocky story it tells.

The rocks of the Klamath Mountains have experienced a vast span of time. The oldest radiometrically dated rock units in the Klamath Mountains are 1.033 billion years old (Lindsley-Griffin et al. 2008). That's about 40 million human generations! These oldest rocks are large, preexisting blocks of granite-like rock that were incorporated into more recently formed rocks found near Yreka. Some zircon crystals, later integrated as sediment grains into younger sedimentary rocks in the Klamath Mountains, are far older still—about 3.2 billion years old (Lindsley-Griffin et al. 2008). For comparison, Earth's oldest known rock, the Acasta Gneiss of northern Canada, is 3.96 billion years old. The oldest known Earthly mineral grains are also zircons found in Western Australia and are 4.4 billion years old.

Rocks are aggregates of individual mineral grains. Some of these individual grains can survive the weathering and erosion of their bedrock hosts because they are tougher and more durable. It is these durable grains that become the sediment of sedimentary rocks in the Klamath. In the case of zircon, these nearly indestructible grains can be far older than the sedimentary rocks of which they are a part. Zircons can survive being eroded out of their host rock and deposited as a sand grain through several such cycles. They would have originated in preexisting ancient continental rocks that have long since been decomposed and their surviving grains washed into the sea as sediment by long vanished rivers. This makes the mineral zircon an invaluable tool in dating geologic events involving those rocks that contain them. Additionally, fossils found in the Yreka area have been dated to well over 500 million years old (Lindsley-Griffin et al., 2008).

Nothing Happens in Isolation

The Earth as a whole is a startlingly active planet. From the beginning, worldwide, mountain ranges have developed and eroded away completely on a continual basis. All of the world's mountain chains have been generated by shifting lithospheric plates and their incessant bumping into one another. These lithospheric plates, which cover the entire outer part of the Earth, are made largely from the cooler upper mantle attached to the crust at their tops. When they shift relative to one another, the resulting stress can be transmitted across vast areas.

Long before the present Klamath Mountains existed, processes were occurring elsewhere along the western shore of what we would one day call North America. Large rocky masses (terranes) were smashed onto the growing continent, later to become the basement rocks of the Klamath and other areas. Sedimentary rocks were then deposited onto these basement rocks. Mountain ranges formed that, in some cases, have since been completely eroded away. Over time, sediment from these mountains spilled out into the surrounding seas, carried by river systems long since vanished, to form extensive oceanic sedimentary layers. These can be seen today in places like the Grand Canyon and Death Valley. What is even more astonishing is that these very sediments can be "sourced" to their places of origin. In the case of the Klamath Mountains, some zircon crystals in these sediments have originated from startlingly distant places, and will be discussed later.

The formation of the Klamath Mountains was part of an ongoing, dynamic reshuffling of the entire Earth's surface. We feel this reshuffling every time there is an Earthquake or a volcanic eruption. This process is still actively underway, and many more mountain ranges can be expected in the future.

Plate Tectonics and the Making of Mountains

It is yet unclear why the lithosphere fragmented into pieces of various sizes we now call tectonic plates, but asteroid impact has been implicated (O'Neill et al. 2020). Because of the constant transfer of internal heat, these rocky crustal plates move, raft-like, on top of the mantle. Some grow via searing magma bubbling to the surface at their boundaries. Other plates buckle and slide under neighboring plates in the process of subduction. Taken together, these processes of plate building by volcanism, destruction by subduction, and shearing along great faults like the San Andreas, are called plate tectonics. Look closely at Figure 4. This is at the heart of our Klamath Mountains story.

About 540 million years ago the western continental margin of ancient North America, future home of the Klamath Mountains, was geologically placid. In the west, this quiet geologic span lasted a *very* long time—from 1 billion to 400 million years ago. This western continental edge was covered by shallow, warm seas in which vast limestone and sandstone deposits formed (Figure 4a). The spilling of these vast sedimentary deposits into the sea was hastened by the lack of sediment-retaining land plants, which had yet to evolve. The location of the present-day Klamath Mountains had a similar geography to our modern Atlantic coast, where shallow waters cover a broad, flat continental shelf (Blakey and Ranney 2018).

This was the Cambrian Period—the dawn of the Paleozoic Era. It was a time when life was about to witness a dramatic evolutionary transformation. The warm, shallow sea that covered this area was ideal for supporting the now famous Cambrian Explosion, which resulted in a stunning variety of multicellular animals with hard parts arising for the first time on Earth. The massive sediment deposits forming there were also ideal for burying and preserving remains of animals that lived in that ancient sea (Coppold and Powell 2006). While the Klamath Mountains were still in the distant future, nothing lasts forever, and the western continental margin was about to experience some drastic changes.

The End of the Placid West and the Rise of the Cordillera

Many of the oldest rock units in the Klamath Mountains were deposited in the ocean as sediments. It seems that some of these ancient rocks were originally formed far to the south and west of where they are today and have similarities to rocks found in Australia, Siberia, and China, implying a direct relationship (Lindsley-Griffin et al. 2008). This connection is cloudy and distorted by time and tectonics, but it does underscore the fact that not all of the Klamath rocks originated at the location of the present mountains themselves. The relentless driver of crustal change—plate tectonics—has been actively moving large pieces of Earth's lithosphere great distances. The history of the Klamath Mountains presents an impressive, though incomplete, record of these travels.

It was during Devonian Period (circa 400 million years ago) that rock started colliding and fusing with the western edge of what would become North America. Prior to this, the western margin of the continent was open sea (Figure 4). These large masses of colliding rock were pieces of island arc volcanoes (Figure 5). Sediments eroded from these volcanoes as well the basement rocks upon which they sat. These were rocks that were not subducted below North America but were, in-

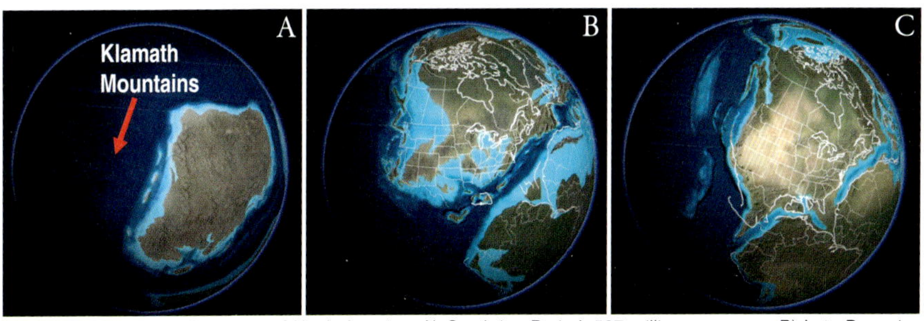

Figure 4. Paleogeographic maps of North America. A) Cambrian Period, 537 million years ago. B) Late Devonian period, 365 million years ago. C) Jurassic period, 173 million years ago. Numerous future terranes (in this case, arcuate strings of volcanic islands) approached the western edge of the continent in B) and C) near where the Klamath Mountains would one day rise. Notice also the very unfamiliar appearance of the continent at each time. Images adapted from Earth Viewer app, Howard Hughes Medical Institute.

Table 1: Geologic timeline of the Klamath Mountains

Era	Period	Age	Events
Cenozoic		66-0	Geomorphic events shaping modern Klamaths occur in last few million years
Mesozoic	Cretaceous	145-66	Dinosaurs flourish; Bulk of pluton emplacement; Extensive regional metamorphism; Condrey Mountain window; Chicxulub impact—80% species become extinct
	Jurassic	201-145	Western Hayfork terrane; Rattlesnake Creek terrane; Western Paleozoic and Triassic Belt; Josephine ophiolite; Smith River subterrane; Dry Butte subterrane; Nevadan Orogeny; Western Klamath terrane; Breakup Pangea; Terrane accretion begins
	Triassic	252-201	Fort Jones terrane; Eastern Hayfork terrane; Pangea forms; "The Great Dying" Permian/Triassic mass extinction—90-96% of *all species* die-off
Paleozoic	Permian	299-252	North Fork terrane; Tertrapods flourish; Redding subterrane forms
	Carboniferous	359-299	1st fully adapted land animals; Low atmos O_2; McCloud limestone corals
	Devonian	419-359	Central Metamorphic terrane; Rises of the Fishes; 1st seed; Late Devonian Mass Extinction-->70% *all species* die off; Eastern Klamath terrane
	Silurian	445-419	1st wood; Arthropods move onto land
	Ordovician	485-445	"Cambrian Explosion"-Rapid diversification of multicellular lifeforms→; Yreka subterrane forms; Trinity subterrane forms; Trilobites emerge
	Cambrian	541-485	First SEX!; 1st Plants; General regional trend of terrane ages is East →West
Precambrian	Proterozoic	2500-541 Ma	Oldest Klamath rock/breakup of Rodinia
	Archean	4.0-2.5 Ga	Oldest Klamath zircons formed; First unicellular life ~3.8 Ga; Earliest atmospheric O2
	Hadean	4.6-4.0 Ga	Formation of the Earth Differentiation into core-mantle-crust

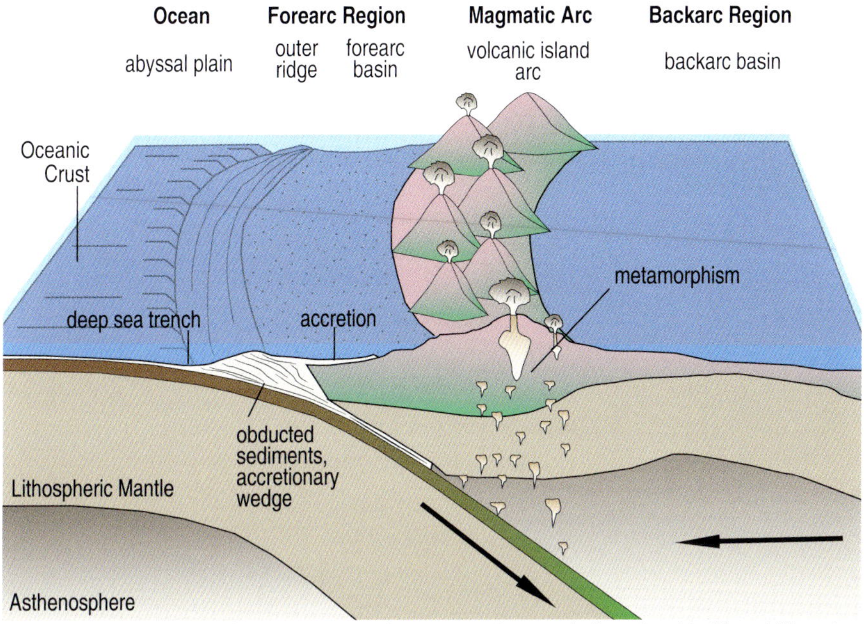

Figure 5. A summary of most of the important tectonic processes relevant to the formation of the Klamath Mountains. Adapted from Blakey and Ranney (2018).

stead, scraped off the top of the plate as the rest dived beneath the North American plate. We call these plastered-on chunks of thickened lithosphere *accreted terranes*.

Numerous locations in the Klamath Mountains have fossil occurrences. Many are obscure and not the type that "jump out" at you when walking across a rock outcrop. There are some particularly interesting coral reef fossils of Permian and Devonian age (Harden 2004). It was about this time when things started to radically change on the western continental margin. During the Devonian Period, many of the plants and animals familiar to us today were still millions of years in the future. Amphibians had yet evolved to walk on land. The shape and location of the continents would have been utterly unfamiliar to us.

Belts, Terranes, and Their Accretion Formed the Klamath Mountains

It is largely in the Klamath Mountains that the idea of accreted terranes, (often simplified to "terranes") was first developed circa 1960 (Snoke and Barnes 2006). This groundbreaking work is an example of what has become known as the accretionary model of mountain building and continental growth. These terranes were recognized and subdivided based on a variety of characteristics such as: rock type, stratigraphy, age, mineral deposits, and tectonic history—and, in some cases, all of these.

Research in the Klamath Mountains led to the development of more general theories about how continental margins and some mountain ranges grow by the assembling of accreted terranes (or the glomming-on of successive masses of rocky materials from elsewhere) (Figure 6). It was this process of terrane accretion that built the Klamath Mountains (Snoke and Barnes 2006).

Geology

Most of these rocks are of oceanic origin, and many are types associated with island arc-type volcanism. We know this because volcanic rocks of the same sort are seen in modern island arcs, such as the Aleutian and Japanese archipelagos.

This accretionary process resulted in huge chunks of rocks being wedged against the North American plate over a period of tens of millions of years as the tectonic plate on which they were traveling dove beneath the North American plate during subduction (Figure 6). Their accretion was vaguely similar to scraping the mud off your shoes on a sharp edge. In the Klamath Mountains, however, at least ten accreted terranes were smashed against and sutured to North America, giving rise to the geological complexity we see today (Irwin 1989).

Each of these ten terranes is separated from the neighboring terranes by low angle thrust faults and stacked somewhat like shingles on a roof (Alt and Hyndman, 2000). These thrust faults are the result of stresses from the once active sliding surfaces of the subduction zones that wedged these rocks against the North American continent in the first place.

The northern Sierra Nevada shares both similar rock types and ages with the Klamath Mountains and was also assembled, in part, by terrane accretion. The Blue Mountains of central/eastern Oregon may also have a similar origin based on both the compositions and ages of their rocks. If we look at the entire western edge of the North American plate, from Mexico to Alaska, we see a similar rocky character—a commonality in the building of the western North American continent as a whole and the North American Cordillera in particular (Blakey and Ranney 2018) (Figure 7).

Sometimes these complex terranes are simplified and grouped into units called belts, referring to their belt-like or north-south elongated character (Irwin 1972, Snoke and Barnes 2006). There has been some disagreement as to what constitutes a specific terrane in the Klamath Mountains, and their rocks have thus often been merged into different terranes or split into subterranes. Nonetheless, the basic ter-

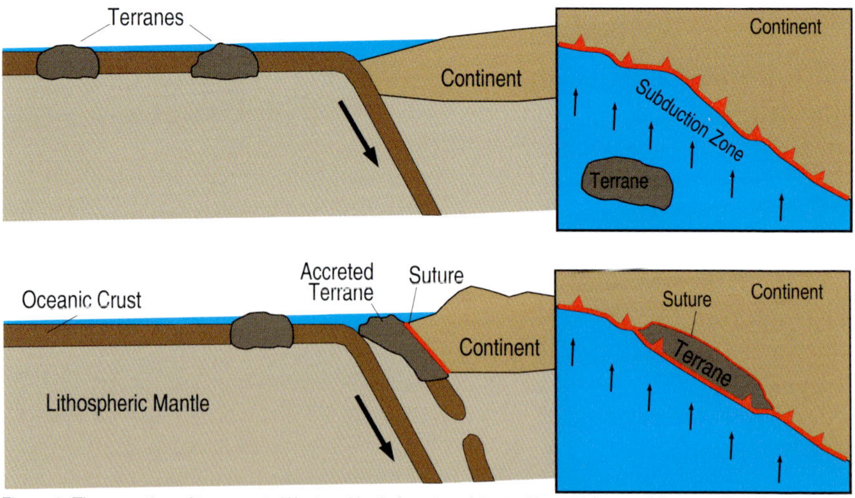

Figure 6. The accretion of terranes to Western North America. Adapted from Blakey and Ranney (2018).

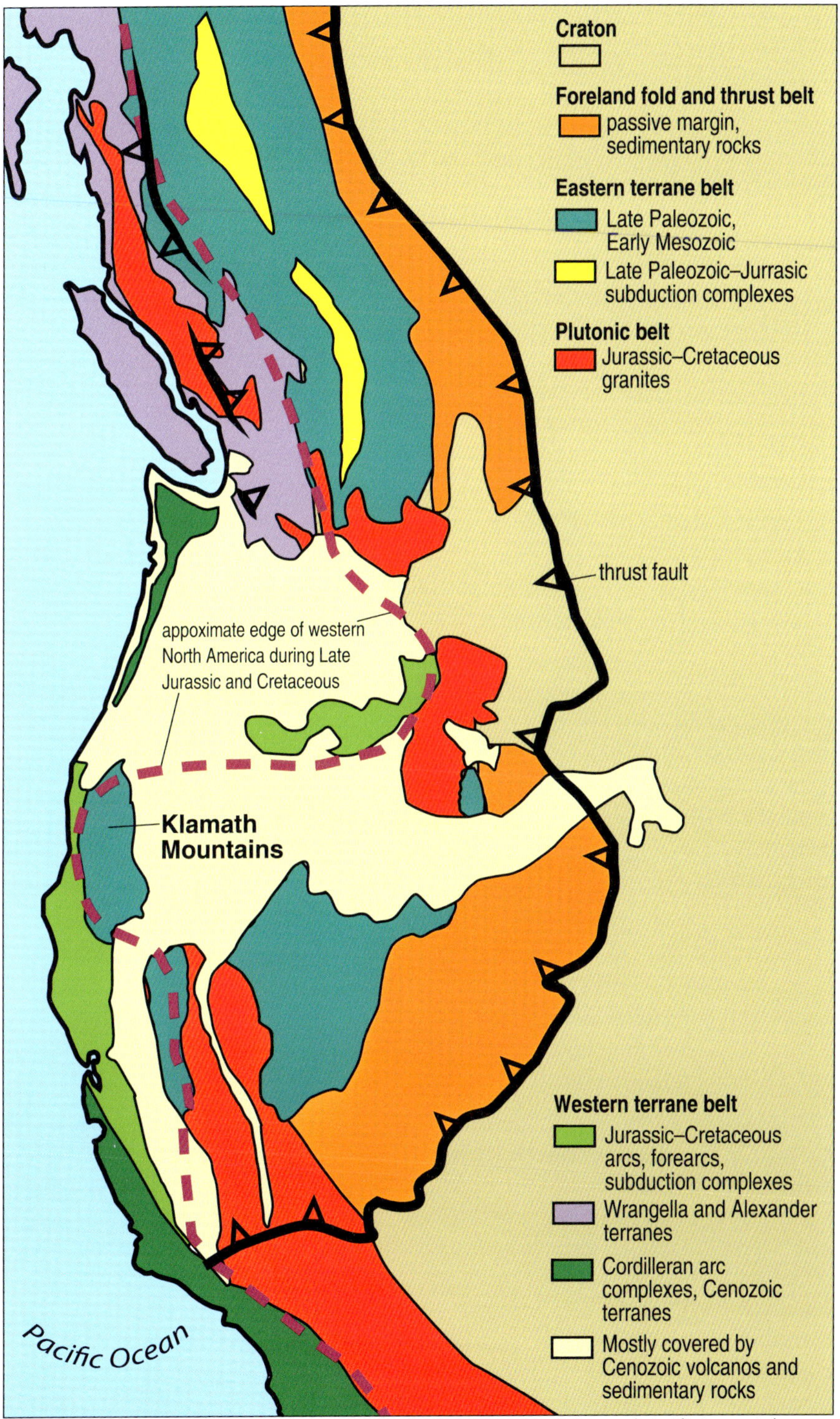

Figure 7. The Cordillera of Western North America showing the relationships between what at first seem to be unrelated localities. Belts shown are composed of numerous accreted terranes. Adapted from Blakey and Ranney (2018).

rane concept, their names, and relevance remain consistent in describing how these rock units were added onto the growing North American plate over time. These terranes will be described in detail later.

By applying basic geologic principles and techniques, such as superposition, field mapping, and radiometric dating—a picture of the Klamath Mountains has now emerged of the growth of a series of mountain ranges over a period of nearly 300 million years (Irwin 1989).

Mapping the Mountains

Geologic field mapping is one of the primary means by which geologists make sense of an area and field mapping is at the core of our understanding of the Klamath Mountains. Many geology graduates aim to do this sort of work upon entering the profession. With birds singing, flowers blooming, and wind tousling one's hair—it's an idyllic, naturalist's dream come true (sort of). The problem is rocks are only locally exposed in the Klamath Mountains. Most are buried under forests and soil, so geologists must be content with obtaining fragmentary information about the various rock types.

Rocks are exposed at outcrops, and it is to these outcrops, magnet-like, that all geologists are drawn. This is where the pieces of the puzzle are collected. The pieces are then assembled on a geologic map.

Geologic maps show where on the landscape each identified rock type is found. These visible outcrops are usually traced on topographic maps so each rock locality is accurately depicted in relation to the others. Each rock type or group of rocks is designated by a distinctive color, which is where much of the power of the map lies. When a regional map is complete, the map colors reveal patterns in the rocks. The geologist begins to see the connections between the rocks at different localities, sometimes inferring how separate rocks are actually parts of the same rock mass, only partially exposed here and there. With each new mapping project, a larger picture of geologic history begins to emerge from the fragmentary puzzle.

Many geologists have worked in the Klamath Mountains to decipher its complex geology. Each gathered parts of the puzzle. Most recently, it was William P. Irwin who assembled all the pieces of the puzzle that revealed the story of the Klamath Mountains in his "Geologic Map of the Klamath Mountains, California and Oregon," published in 1994 (Figure 9).

Klamath Mountains rock patterns are distinctive and informative, as shown in the highly simplified version of Irwin's original map (Figure 8). First, there is the arcuate, or curved, nature of the rock units. Second, each of these curved areas is made of similar or related rocks. These curved regions are mostly rocks of similar age. Third, dark black lines (which represent faults) bound most of these grand, rocky arcs. The separate rock units are in "fault contact" with their neighbors. Finally, the ages of the rock units get younger as you travel from Redding, in the east, to Eureka in the west. Irwin's original map, actually much simplified itself, is considerably more complex (Figure 9). Its legend has been omitted for simplicity's sake.

Types of Rocks in the Klamath Mountains

Despite the fact that there are a huge variety of rocks composing the Earth's crust, all of this rocky menagerie can be simplified into just three classes of rocks: igneous, sedimentary, and metamorphic. Identification of many rocks requires mineralogical analysis, so a rock may not be immediately identifiable without close examination.

Igneous Rocks

Igneous rocks are those that have formed from molten or liquid rock (magma) as it cooled and solidified. There are many examples of this rock type throughout the Klamath Mountains, and they are central to our understanding of the geologic province and the sequence of events that formed it. On the basis of their origins, these rocks can be subdivided into two groups: intrusive igneous and extrusive igneous (Table 2).

Intrusive Igneous Rocks

Intrusive igneous rocks are those that solidify from magma underground, or at depth, and are referred to as plutons when in large masses. Because these rocks cool slowly, the magma can stay liquid longer and slowly crystallize into comparatively large mineral crystals. Sometimes these crystals become quite large but, more commonly, their crystals are much smaller but still easily distinguishable in a hand sample. Crystals of many sizes are found within the intrusive rocks of the Klamath Mountains (Figure 10).

A large number of plutons are distributed across the Klamath Mountains. Some of these masses are huge—many miles across—and have rather uniform compositions throughout. Batholiths are very large intrusive bodies (>36 mi^2) that are amalgams of chemically distinct plutons (Figure 8). Other intrusive rock bodies are much smaller and are called stocks, dikes, and sills.

Table 2. Igneous rock identification.

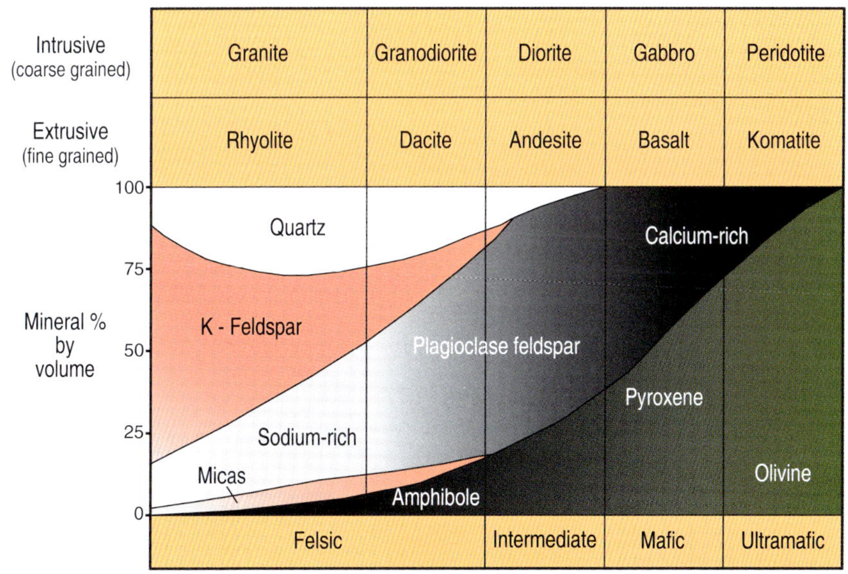

Geology

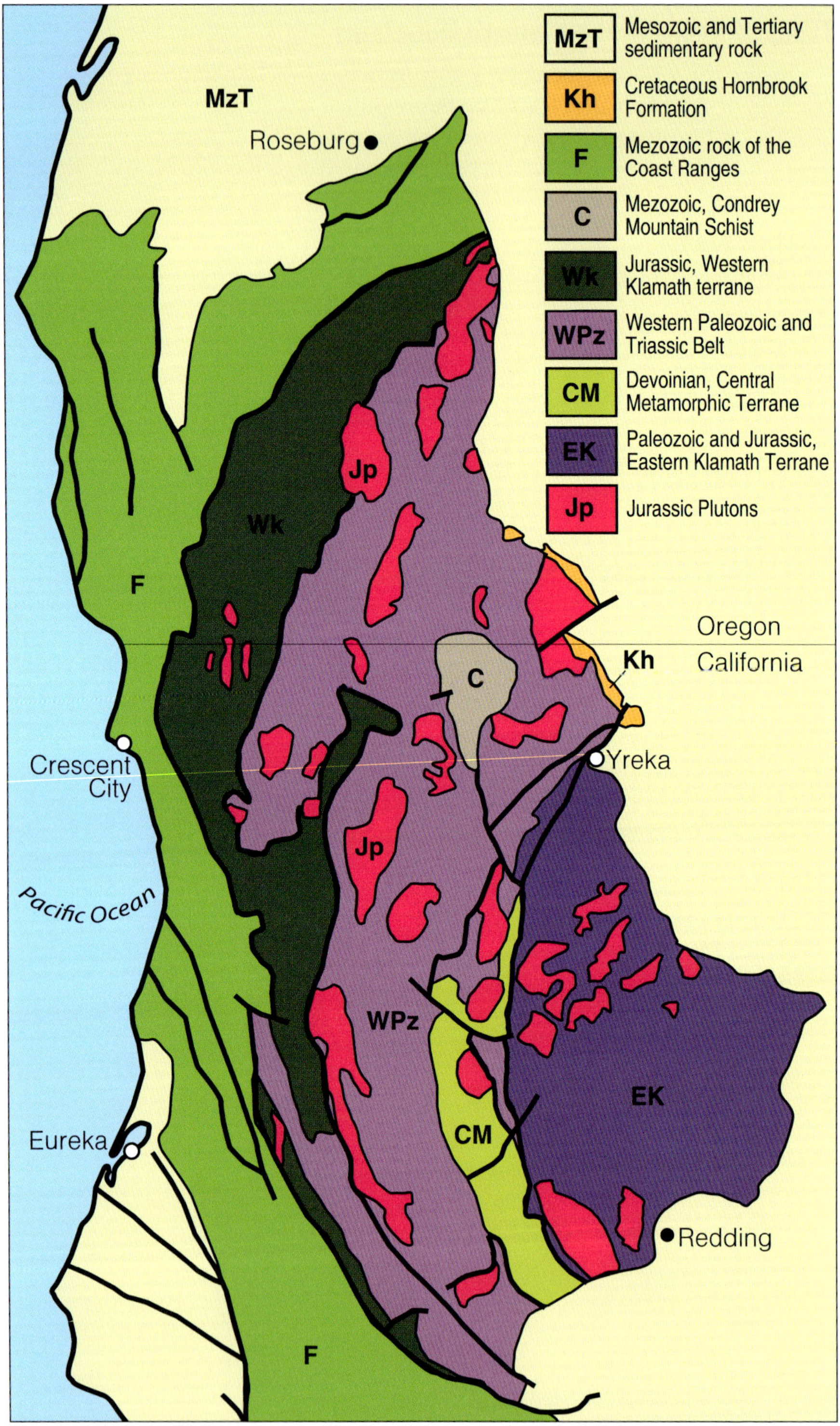

Figure 8. A highly simplified geologic map of the Klamath Mountains after Irwin (1994).

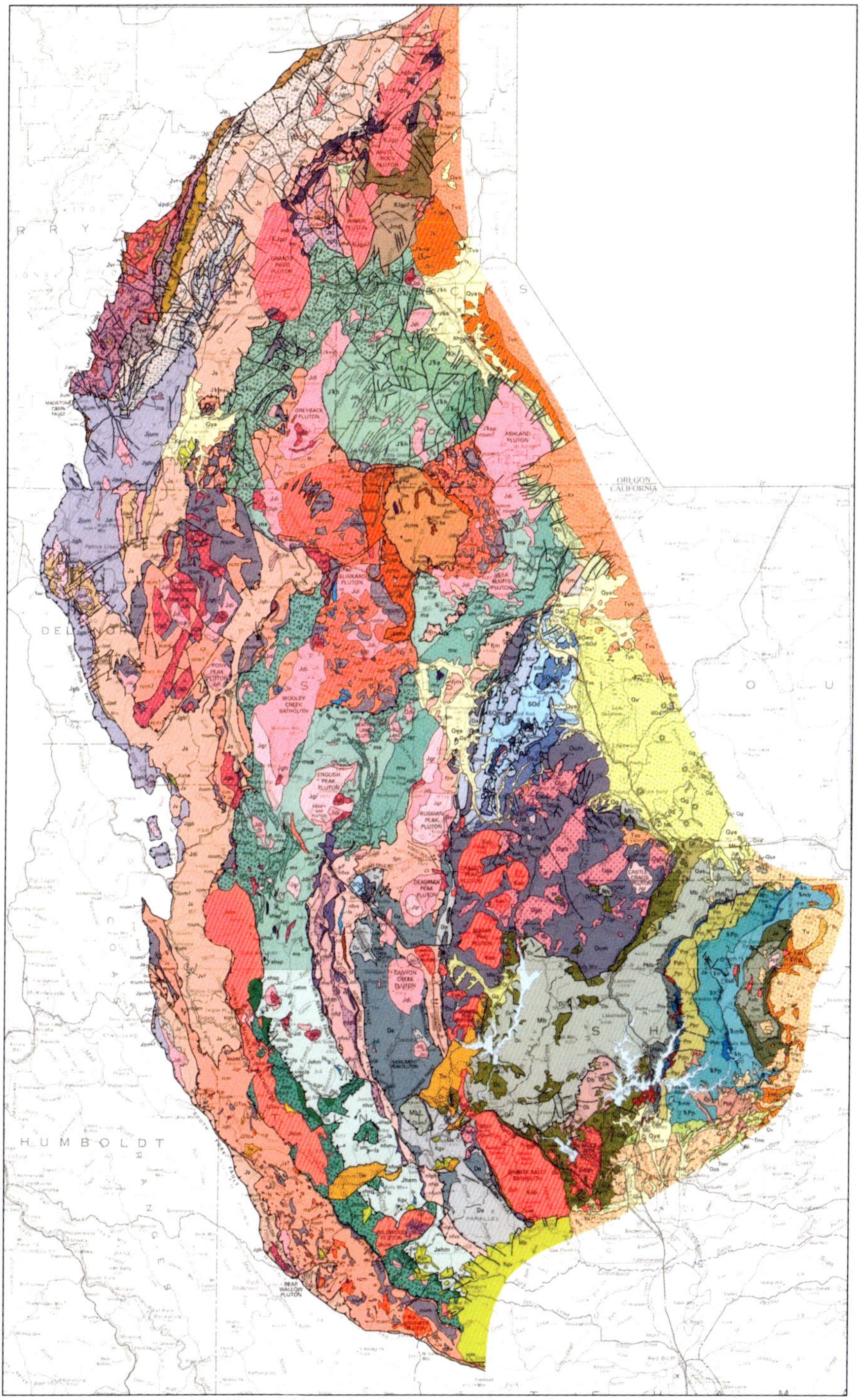

Figure 9. Geologic map of the Klamath Mountains, California and Oregon (Irwin 1994). This more detailed map provides a hint of the level of complexity to which the Klamath Mountains have been mapped. Numerous maps covering smaller areas have been assembled throughout the region with far greater detail. Map courtesy of the United States Geological Survey.

Figure 10. Some examples of Klamath Mountains intrusive igneous rocks. Top: A Diorite with uncharacteristically large crystal size from Stuart Fork drainage. Left: Tonalite, from the Canyon Creek pluton. Right: Diorite from lower Wooley Creek. Notice 1cm cube for scale. Photo by Mark Bailey.

Intrusive rocks include the iconic granite, a term referring to a specific mineralogical composition (see Table 2) that is often used loosely for "granite-like" rocks. Intrusive rocks come in a number of types, based upon their compositions. In the Klamath Mountains, these intrusive rocks are most commonly granodiorite, which has a lower percentage of silica than does granite. Granodiorite is also common in the western Sierra Nevada and the similar ages of these and Klamath granitic rocks provide a possible link between the Sierra and the Klamath Mountains (Figure 11).

Many people think of granitic rocks when they conjure an image of tall mountain peaks. There is a reason for this in that mountain ranges, by their very nature, are actively eroding. This erosion scours off miles of overlying rocks, exposing the deep core rocks of the system. Sometimes these core rocks are, indeed, granitic.

There are many granitic mountain ranges but also many exceptions to this. The Himalaya, Canadian Rockies, Cascade Range, and Appalachians are examples of ranges in which granite does not dominate the surface exposures. So, too are the Klamath Mountains (Figure 8).

Figure 11. Granodiorite, from the Russian Peak pluton, one of the more common types of granitic rock in the Klamath Mountains (left). Notice 1cm cube for scale. Outcropping of Russian Peak pluton from which the sample at left was extracted along the Sawyers Bar Road just west of Etna. Photos by Mark Bailey.

Castle Crags

The Castle Crags. Photo by Michael Kauffmann.

With accretion events occurring along the continent's edge at plate boundaries to the west, water was carried deep into the Earth with the diving tectonic plate. This water became heated to a high temperature and contributed to the melting of the rocks above the sinking plate. The resulting magma then rose towards the Earth's surface, collecting into pockets that later solidified into the plutons seen today. In other, similar events farther to the east, the magma reached the surface and lava erupted, forming volcanoes like Mount Shasta and numerous others in the high Cascades.

About 160 million years ago, a batch of molten rock developed beneath Earth's surface and began to solidify. This formed what would one day become the granitic rocks of Castle Crags. Rising up through the much older Ordovician-aged (443–490 million year old) Trinity ultramafic sheet, a part of the Trinity subterrane, the intrusion "cooked" these surrounding rocks where it came into contact with them.

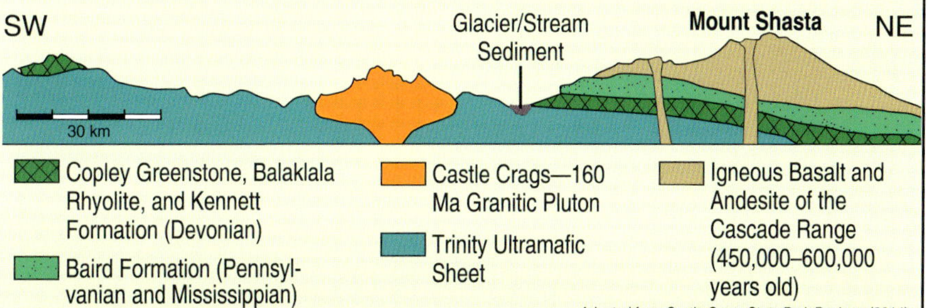

Adapted from Castle Crags State Park Bochure (2014).

The dramatic rock of the Castle Crags are the result of this intrusion of granitic magma. In the time since that intrusion, erosion has removed the overlying rock, exposed the now solidified intrusion below, and shaped the spires and domes into what we see today.

Castle Crags State Park is within the Klamath Mountains geomorphic province, though on the far eastern edge. The eastern Klamath Mountains are built from the oldest rocks in the range, with the newest rocks to the west.

Next time you are traveling the Interstate 5 corridor, take time to stop and marvel at the 160-million-year-old pluton we call the Castle Crags.

Granitic rocks, by their very presence, indicate a re-worked, preexisting, then melted rock. To get a granitic rock, a mix of preexisting rock types (such as sedimentary, metamorphic, and other igneous rock types), as well as water and other volatiles, play significant roles. As rock melts at depth, it may incorporate all or some of those surrounding rocks and the water they contain (which may be considerable) into the melt. Crystals form in the melt as it cools. Some of these concentrate near the bottom of the molten blob. All of this significantly changes the chemistry of the remaining melt. This may then solidify into granitic rock. Ultimately, the granitic rock type you stand on will depend upon the chemistry of its original melt and the rock types it may have melted its way through on the way upward in the crust. All of these "preconditions" are expected at tectonic plate boundaries.

Despite the magnificent peaks in the Klamath Mountains like Thompson Peak, Mount Ashland, English Peak, Russian Peak, and the Castle Crags, a look at the exposures of granitic rocks on the geologic map (Figure 8) will quickly reveal that the bulk of the rocks of the Klamath Mountains are non-granitic. This is not to say, by any means, that they are not important to our story. They are.

When the Klamath plutons were examined closely, a striking variation in compositions and ages became evident. The oldest granitic rock in the region is the Pit River Stock of the Eastern Klamath terrane at 246 Ma (Ma refers to millions of years ago, as Ga refers to giga, or billions of years ago). For the most part, however, the bulk of the Klamath granitics have a much younger range of ages.

One group, consisting of the Shasta Bally, Grants Pass, Canyon Creek, Caribou Mountain, and a few other plutons, are younger than virtually all of the other rocks of the Klamath Mountains at about 136–140 Ma. Another group, including the Chetco/Illinois River plutons, are mostly in the mid-140s Ma range.

The Wooley Creek suite dates to about 160 Ma. It includes some iconic plutons such as Ashland Peak, Wooley Creek, Russian Peak, English Peak, and Grayback Mountain. The Ironside Mountain batholith, the largest in the Klamath Mountains, is a long cluster of plutons, all about 170 Ma (Allen and Barnes 2004).

All these become important in deducing the ages of other Klamath rocks. Obviously, a pluton must intrude into an already existing (i.e., older) rock mass. Thus, if we can date a pluton we can say that it is younger than the surrounding rock it has intruded. This has come in handy when trying to put minimum ages on the jumbled, contorted, and confusing masses of rock they intrude.

Extrusive Igneous Rocks

Extrusive igneous rocks are those whose magma reaches to or very near the surface. Because these lavas cool quickly at Earth's surface, they crystallize at a much faster rate. This rapid cooling keeps their mineral crystal size small. Because of their fine grain size, it is often difficult to distinguish individual mineral crystals in hand samples. The extrusive rocks of the Klamath Mountains tell a distinct origin story. The most common types are andesite and basalt (Figure 12) which, in itself, is a profound clue to the origin of the Klamath Mountains. Andesite forms in volcanic island chains. Its presence speaks of volcanic island arcs and intertwines tightly with the story of the Klamath Mountains. Basalt can flow from massive crustal fissures and produce vast volumes of lava flows that cover thousands of square kilometers, as in the *Traps* of India and Siberia. Thankfully, this type of flood basalt eruption has not occurred during historic times. Basalt can play another fundamental role as it issues from mid-ocean ridges at seafloor spreading centers. It is during this process that it builds the oceanic crust that covers much of Earth's surface. As a result, it is always present in ophiolite sequences (see Ophiolite sidebar, page 78).

Figure 12. Diabase sample from near Salmon Mountain in Trinity Alps Wilderness. This type of rock forms in shallow intrusions of basaltic composition that solidify underground in dikes and sills. Notice 1cm cube for scale. Photo by Mark Bailey.

Other common extrusive rocks, although much less common in the Klamath Mountains, include rhyolite and dacite. All of these are common in the High Cascades and Great Basin to the north and east of the Klamath Mountains.

Sedimentary Rocks

Sedimentary rocks are those that form from the debris of preexisting rocks. Their particles are typically the weathered and eroded pieces of older igneous rocks, sedimentary rocks, metamorphic rocks, or bits of shell produced by marine organisms. The sediments from these precursor rocks were buried under neighboring sediments, deeper and deeper, and were then compressed and cemented together into sandstones, shale, mudstones, and limestone (Figure 13).

The fact that we see them now at locations well above sea level in the Klamath Mountains attests to the restlessness of Earth's crust. Sedimentary rocks are common original members of the suite of rocks that built the Klamath Mountains and, as a group, they crop out in distinct bands across the region—although mostly not in the horizontal layers that you see in the Grand Canyon. The processes that formed the Klamath Mountains have done a good job of taking these originally horizontal beds of sediment and twisting them up into slanted and distorted layers (Figure 14).

Figure 13. Limestone from North Fork terrane on Highway 299. Photo by Mark Bailey.

Although distorted and, in many cases, heated up and metamorphosed into new rock types, the layering of sedimentary rocks is often visually distinctive from great distances and can be seen in many roadcuts and stream channels across the Klamath Mountains.

Klamath Mountain Caves (Caverns)

Almost anywhere limestone exists in the Klamath Mountains, so, too, do caves. Actually, "cave" is the frequently misused term for most of these underground openings dissolved from surrounding limestone or marble. Caverns would be the correct term for such dissolved features but "cave" has stuck and seems permanently attached to the names of most of these caverns. Oregon Caves, Bigfoot Cave, Samwel Cave, and Potter Creek Cave (all misnamed) are among the 197 known caverns in the Klamath Mountains. Bigfoot Cave in the Marble Mountains is among the largest caverns with a total of 12 miles of passages that reach 1,200 feet in depth. Many contain rare and unusual cave faunal assemblages (Elliott et al. 2017).

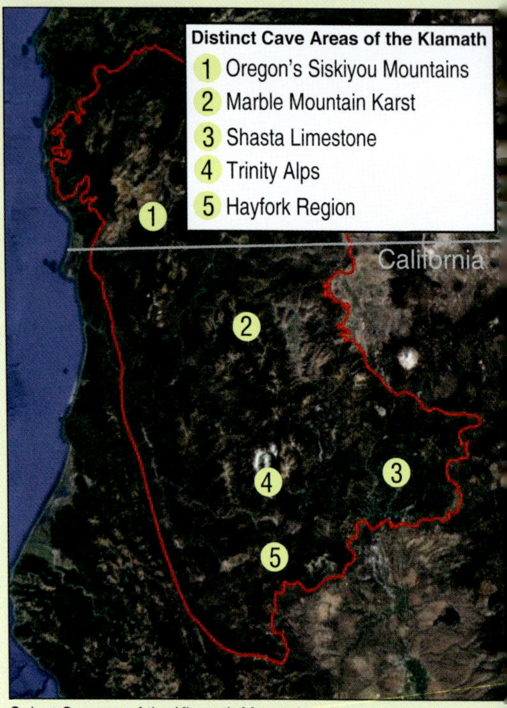

Select Caverns of the Klamath Mountains.

The McCloud limestone of the Eastern Klamath terrane, is perhaps best known today for its (correctly named) roadside attraction of Shasta Caverns. It also contains other important caverns, several of which have produced copious amounts of Pleistocene fossils that have contributed significantly to our

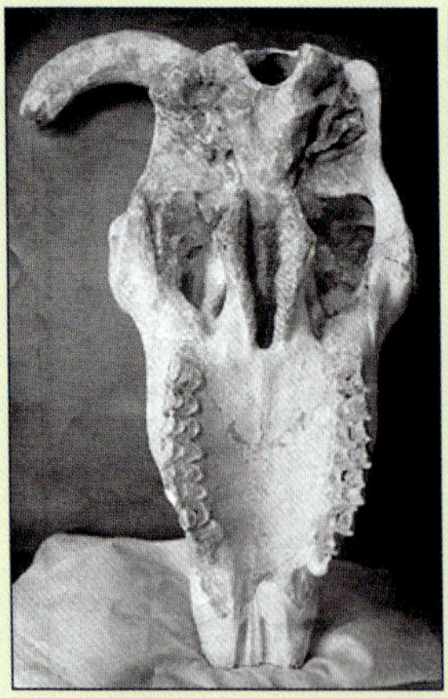

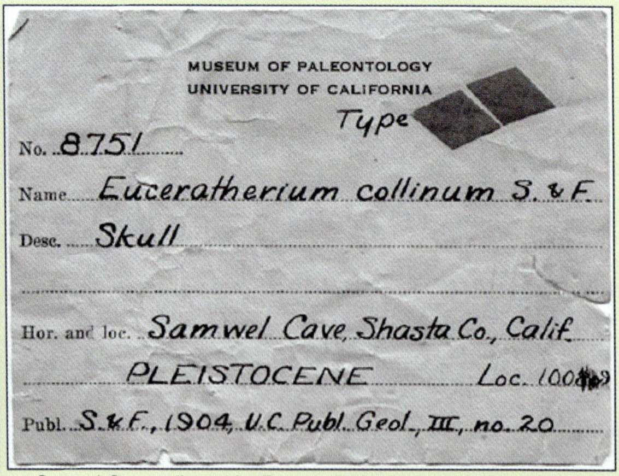

MUSEUM OF PALEONTOLOGY
UNIVERSITY OF CALIFORNIA
Type
No. 8751
Name Euceratherium collinum S. & F.
Desc. Skull
Hor. and loc. Samwel Cave, Shasta Co., Calif.
PLEISTOCENE Loc. 100
Publ. S. & F., 1904, U.C. Publ. Geol., III, no. 20

Left: A Pleistocene shrub ox skull from Samwel Cave collected by Eustace Furlong around 1904. **Right**: The original specimen label in the UCMP collections. Images courtesy of the University of California Museum of Paleontology, Berkeley, California.

Entrance to Samwel Cave. Photo by Mark Bailey.

understanding of Ice Age animals and climate. Caves are both the perfect trap for wayward animals and the perfect place to preserve their remains.

Archaeologists have long noticed this and there have been a large number of spectacular cave discoveries involving early humans. Chauvet, Lascaux, and many other caves in France and Spain have preserved magnificent Ice Age paintings on their walls. In the late 19th century, Samwel Cave, Potter Creek Cave, and others in the McCloud limestone, were discovered in the search for early humans in California.

Although little human evidence was ultimately discovered, a plethora of fossil Ice Age creatures was. These California fossil discoveries rank second only to the Tar Pits of Rancho La Brea in Los Angeles for their variety and level of preservation of Pleistocene megafauna, or large mammals (as well as many smaller creatures including a large variety of rodents). Mammoth, mastodon, short-faced bear, grizzly bear, dire wolf, shrub ox, two species of giant ground sloth, mountain goat, horse, camels, condor, and numerous other creatures are among those found in these caves. Most were dated to 17,000–24,000 year range—practically yesterday—during the heart of the Ice Age in the Klamath Mountains (Feranec et al 2007).

Most of these named species went extinct regionally within a few thousand years from when their remains were deposited, even though most were also widespread throughout North America. Climate change likely contributed to their demise, as most of the megafauna went extinct around the world at nearly the same time as the Ice Age came to an end. This makes these fossil discoveries an important window into climate change and related extinctions—a sobering thought for our time. Perhaps coincidentally, humans arrived on the scene at nearly the same time that these extinctions occurred. As it turned out, the changing climate was beneficial to humans, resulting in our worldwide expansion. Nonetheless, it has proven difficult to determine the exact cause of the animal extinctions. It was probably a number of influences working at the same time.

Figure 14. Sedimentary layering is evident in this photograph of Marble Mountain. Photo by Michael Kauffmann.

Metamorphic Rocks

Metamorphic rocks are nearly everywhere in the Klamath, attesting to the tortured geologic history of these mountains. Often formed of unusual mineral assemblages and composed of swirling, recrystallized rock types, metamorphic rocks are often among the most visually striking of all rocks (Figure 15). These shape-shifting rocks are the epitome of survivors. Created by pressure and heating of any previously existing rock type (even other metamorphic rocks), they respond by rearranging their crystal structures into new and characteristic minerals, sometimes without actually melting.

Some of these new metamorphic minerals can be dated to the time at which these changes occurred, thus recording an imprint of shifting global and local tectonic events. The ubiquitous metamorphic rocks of the Klamath Mountains (Figure 16) have been altered from their progenitor rocks (or "protoliths") offering a window into the origins of the mountain range. They become the sometimes flashy rocks with names like marble (from previous limestone), slate and phyllite (from shale), greenstone (from almost anything, but especially lava), schist (also from almost any previous rock), gneiss (from granitic or sedimentary rocks), and an all-time favorite—serpentinite (from deep-seated iron and magnesium-rich igneous rocks).

Figure 15. Evidence of metamorphism. Left: Schist/gneiss. Salmon Schist Formation, Central Metamorphic terrane exposed above East Weaver Lake. Photo 2 feet across. Right: A hodgepodge of granitic/metamorphic rocks interacting. Exposed along Clear Creek, Siskiyou Mountains. Western Klamath terrane. Photo 5 feet across. Photos by Mark Bailey.

Figure 16. A record of metamorphism displaying evidence of a stressful past. Top left: Phyllite from Galice Formation near Orleans, Top right: Skarn from near Scott Bar, in/near Rattlesnake Creek terrane. Center left: Greenschist from Galice Formation, Eastern Siskiyous. Center right: Greenstone/schist from Galice Formation, Eastern Siskiyous. Bottom left: Marble/dolomite from Salmon River, Galice Formation. Bottom right: Unweathered serpentinite from Josephine Ophiolite in the western Siskiyou Mountains. Notice 1cm cube for scale. Photos by Mark Bailey.

Geology

Ophiolite Sequence and Ultramafic Rocks

An ophiolite sequence is a characteristic suite of rock types that represent a cross section of the crustal and upper mantle rocks which comprise an oceanic plate. Ophiolite sequences are composed primarily of ultramafic rock with a covering of seafloor crust and oceanic sediment on the top. Several of the Klamath terranes have well-exposed ophiolites in various states of preservation.

Rocks are characterized by their chemical origins. As the Earth was forming and in a molten state, the heavier minerals sank inward and the lighter floated to the surface. These included, among others, minerals rich in magnesium (Mg) and iron (Fe). If a rock contains abundant magnesium and iron, it tends to be dark and heavy. These are referred to as mafic. Rocks even richer in these elements are referred to as ultramafic. Because ultramafic rocks are not at all like the types typically formed near Earth's surface, they stand out as "foreign" and can be the telltale signs of their oceanic and deep-seated origins. They also speak of dynamic pressures that pushed them up to be exposed at the surface.

When an oceanic basin is narrowed and closed off by converging tectonic plates, some of the deepest rocks tend to be sheared and pushed up during the subduction process. There is a huge amount of force behind a moving plate that is kilometers-thick and perhaps hundreds or thousands of kilometers-wide. This force distorts the deep basement of the oceanic crust along with its cover of oceanic sediments and may well force these rocks to the surface. As subduction is occurring, most of these underlying rocks, along with a large volume of seawater, are consumed and sink into the mantle, perhaps to be recycled as volcanic rocks some distance from the subduction zone or as plutonic rocks beneath the arc volcanoes. If we are lucky, in a stream bed or a mountain side, we can see a frozen record of this dynamic process in the form of an unsubducted ophiolite sequence.

Pictured is a cross-section of a classic ophiolite sequence, showing characteristic rock types with depth.

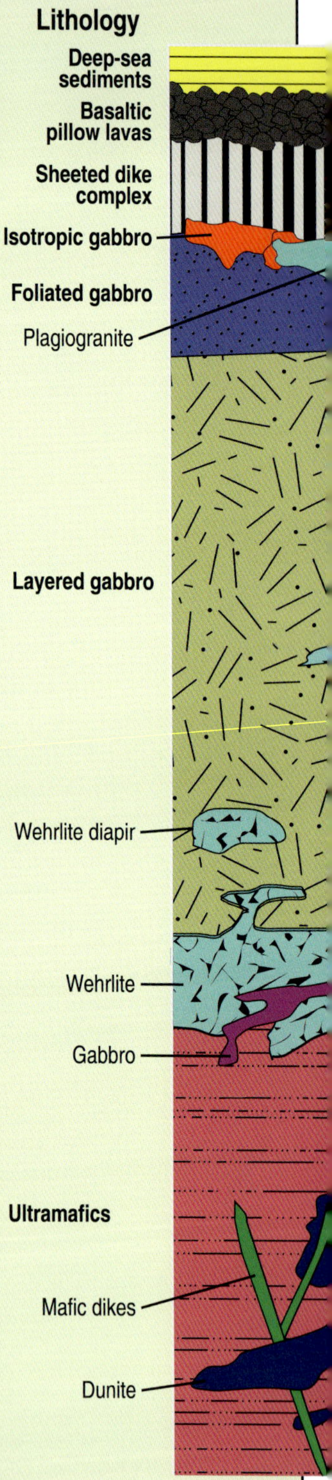

Terranes of the Klamath Mountains

The accreted terranes of the Klamath Mountains (Figure 8 and Table 3) are a variable and complex collage of rocks in nearly every imaginable state of discombobulation. They have been squashed, twisted, rotated, heated, and squeezed in ways that one could expect from rocks formed in and around a long-lasting series of subduction zones in the western part of North America where pretty much all of the region's rocks have had similar twisted histories.

As research techniques have improved, terrane ages have become increasingly well defined. Fossil assemblages help to establish the range of time over which many of the sedimentary rocks originated. The igneous rocks are readily dated using radiometric techniques. Additionally, increasingly fine resolution in radiometric dating has vastly improved our understanding of the sequence of events.

As previously mentioned, most of the terranes in the Klamath Mountains are separated by low-angle thrust faults. Recognizing these faults is one of the primary ways that terranes are differentiated from each other in the Klamath Mountains. Thrust faults, typical of subduction zones, are products of compressional forces produced by a terrane-carrying plate subducting under the North American continent or by a wider-scale regional compression. As each new terrane was accreted onto the continent, a series of thrust faults developed that separated the newer terranes from the older.

Eastern Klamath terrane

The eastern-most part of the Klamath Mountains, near Redding, is where the beginnings of the Klamath Mountains are revealed. The first of the Klamath Mountain terranes—the Eastern Klamath terrane (EKT)—was accreted to the North American continent here (Figure 6). These are the oldest rocks in the Klamath Mountains. Irwin (1989) and others subdivided the EKT into the Yreka subterrane, the Trinity subterrane, and the Redding subterrane. The western extent of this terrane is clear and locally crops out as the Bully Choop thrust-fault zone. The true eastern extent of the Eastern Klamath terrane is unknown because it is buried by the much younger volcanic rocks of the Cascade Range and sediments of the Great Valley.

Yreka subterrane

This is a complex and jumbled stack of late Proterozoic (Ediacaran) through Upper Silurian (570 Ma- ca. 400 Ma) sedimentary strata and numerous other rock types. A noteworthy discovery within these rocks (near Gazelle) was the unmistakable fossil imprint of Ediacaran cyclomedusoid fossils. Cyclomedusoids are cryptic organisms. Nothing quite like them exists on Earth today. It is not known exactly what type of organism they were: animal? plant? colonial bacteria? They remain enigmatic. The Ediacaran period (635–502 Ma) is when the first multicellular creatures evolved. Thus, these oldest rocks of the Klamath Mountains can be traced to the age of the first multicellular organisms.

These Klamath fossils are most similar to those of similar age found in Australia or northern China (which were also located in different places at the time), a testament to moving plates and a dynamic Earth. At the time, much of what would

Table 3. A comparison of lithic belts as defined by Irwin (1989) and subsequent terrane nomenclature for the Klamath Mountains. The chart shows an approximate age sequence from East to West (top to bottom) that roughly corresponds to decreasing age from East to West.

<table>
<tr><th colspan="2">Terrane</th><th colspan="2">Subterrane</th><th>Age (Ma)</th><th>Made of</th></tr>
<tr><td colspan="2" rowspan="6">Western Klamath</td><td rowspan="3">Smith River</td><td>Condrey Mountain</td><td>220-117</td><td>Schist from metasedimentary and metavolcanic rocks, probable Galice affinity, mostly greenschist</td></tr>
<tr><td>Galice Formation</td><td>150</td><td>Indicates offshore w/turbidites, sandstones, argillite mix w/ shale, some volcanics</td></tr>
<tr><td>Josephine Ophiolite</td><td>157</td><td>Sheeted dikes, pillow basalts, pyroxenite, ultramafic rocks, harzburgite</td></tr>
<tr><td colspan="2">Rogue Valley</td><td>157</td><td>Basaltic and andesitic volcanic/volcanoclastic interbedded w/ Galice Formation U-Pb dates 157 Ma. Cut by 150 Ma dikes</td></tr>
<tr><td colspan="2">Briggs Creek</td><td>?</td><td>amphibolite</td></tr>
<tr><td colspan="2">Dry Butte</td><td>Jur</td><td>Mostly Jurassic igneous rocks: plutons, cumulate gabbro, volcanics, ultramafic, peridotite</td></tr>
<tr><td colspan="2">May Creek</td><td colspan="2">-</td><td>146</td><td>Schists derived from what appear to be metamorphosed basalts and associated igneous rock, sandstone, limestone (now marble)</td></tr>
<tr><td rowspan="5">Western Paleozoic and Triassic Belt</td><td>Rattle-snake Creek</td><td colspan="2">-</td><td>252-145</td><td>Mostly mélange, metamorphosed mélange, serpentinized ultramafics, much older fragments included in mélange</td></tr>
<tr><td>Western Hayfork</td><td colspan="2">-</td><td>177-156</td><td>Sediments characteristic of seafloor washed off volcanic islands</td></tr>
<tr><td>Eastern Hayfork</td><td colspan="2">-</td><td>210</td><td>Disorganized mélange of clay and chert—fragments typical washed off active continental edge</td></tr>
<tr><td>North Fork</td><td colspan="2">-</td><td>310-265</td><td>Ophiolite complex interbedded with chert. Paleomagnetism indicates formation in low tropics</td></tr>
<tr><td>Fort Jones</td><td colspan="2">-</td><td>220</td><td>Mostly metamorphosed volcanic and sedimentary rock—blueschist facies</td></tr>
<tr><td colspan="2">Central Metamorphic</td><td colspan="2">-</td><td>399-380</td><td>Mostly Abrams and Salmon schist derived primarily from metamorphosed sedimentary and mafic volcanic rocks</td></tr>
<tr><td colspan="2" rowspan="3">Eastern Klamath</td><td colspan="2">Redding</td><td>419-170</td><td>Derived from sedimentary and volcanic rock, also included biologically significant McCloud limestone</td></tr>
<tr><td colspan="2">Yreka</td><td>570-400</td><td>Important Ediacaran cyclomedusoids fossil-bearing sediments, seafloor sediments</td></tr>
<tr><td colspan="2">Trinity</td><td>600-460</td><td>Ophiolite complex, serpentine</td></tr>
</table>

become the North American continent was not yet assembled and is referred to as Laurentia. We now know that in the late Proterozoic, the continents then present were locked up, Pangaea-like, in an older and long since fragmented supercontinent (of which Laurentia was a part) we call Rodinia. The configuration of those continents was also different compared to today's geography (Lindsley-Griffin et al. 2008). You should now be getting a good idea of the complexity geologists have faced deciphering the Klamath Mountains.

Geology

Redding subterrane

The Redding subterrane is built from 16 formations of both sedimentary and volcanic origin. Fragments of limestone-bearing fossil coral reef and the presence of andesitic volcanic rocks indicate an oceanic and volcanic origin. Dating of granitic plutons intruding the limestone and andesite, and identification of distinctive fossil assemblages, indicate that these rocks represent a nearly 200 million-year time interval from the Devonian to the Middle Jurassic (tables 1 and 3).

The McCloud limestone is of special interest, in part because of its numerous and important cave discoveries (see Caves sidebar, page 74). However, entombed within the McCloud limestone (Figure 17) there are also preserved (and now extinct) fossil horn corals and other early Permian sea creatures. Of similar age to limestone found in northern Nevada, the McCloud contains dissimilar coral species. With the two rock units being close geographically today, it seems obvious that coral reefs forming in close proximity should have similar species compositions through intermixing. They do not. This implies that the McCloud reefs formed at a great enough distance from those in Nevada to contain a different assemblage of organisms. Since corals reproduce by the budding of free-floating (or planktonic) polyps that spread great distances in sea currents, this Klamath coral conundrum implies the reefs of the McCloud limestone, like the rocks in the Yreka subterrane, originated thousands of miles from their current locations. Little evidence of long journeys exists for all of the other terranes in the Klamath as most of the Klamath terranes appear to have been largely constructed from materials that originated near the continent's western edge. There are some intriguing exceptions to this, however. Some sedimentary detrital zircons, for example, seem to have originated from distant localities but most Klamath terranes likely formed close offshore (Lindsley-Griffin et al. 2008).

Figure 17. McCloud limestone along Shasta Reservoir in the Eastern Klamath Mountains.

Earth's Magnetism and the Rotating Klamath Mountains

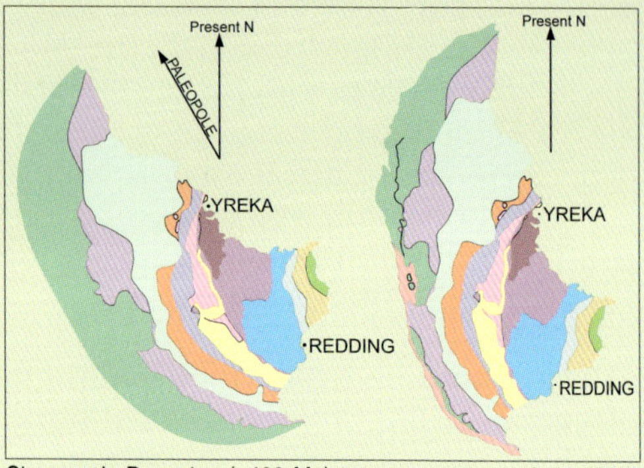

Since early Devonian (~400 Ma) to present, some of the Klamath terranes have rotated clockwise by as much as 110 degrees. Other, more recently accreted terranes have not rotated nearly as much. There has been a wide range of rotations for the various terranes through this time. Left is late Jurassic (~150 Ma), right is present. From Irwin and Mankinen (1998).

The Earth's magnetic field will cause ferromagnetic mineral grains (those containing iron and other elements) to align like compass needles with the current magnetic field as they crystallize from molten rock. These grains line up in a way similar to how a magnetic tape records information. If geologists can measure the magnetic alignment in minerals of a series of rocks, and if they can accurately date those rocks, a narrative can be developed for the orientation of the Earth's magnetic field at a particular time in the geologic past. This paleomagnetic narrative can then be used to determine how rocks were oriented in relation to their present orientations.

Using these data, Irwin and Mankinen (1998) deciphered the rotational history of the Klamath Mountains and have revealed that during the Jurassic period the entirety of the Klamath Mountains rotated by at least 65 degrees clockwise, though it remains unclear how consolidated the assemblage of terranes was at this time. Perhaps the rocks of the more westerly terranes were still offshore. Many of the sedimentary particles within the more westerly terranes clearly came from a landmass that lay mainly to the east. This is demonstrated by the occurrence of sedimentary grains of greater age on the continental sides of these terranes. It is such discoveries that, in part, enable us to construct maps of different geologic times as seen in Figure 8. Everything to the west of the Klamath Mountains was younger or yet to be formed. Since this discovery, amazing detective work has continued and enabled researchers to conclude that the Western Klamath terrane has rotated 99° ± 10° clockwise, the Western Hayfork terrane has rotated 73.5° ± 10.6° clockwise and the Eastern Klamath terrane has "only" rotated 38.6° ± 14.4° clockwise. This suggests these terranes were not connected during a portion of this rotation or they would all show similar rotations.

Half of the Western Hayfork terrane's rotation and two-thirds of the Western Klamath terrane's rotation must have occurred before accretion to the Eastern Klamath terrane. A sequence of accretionary events has emerged from these magnetic studies. The terranes of the Klamath Mountains were once separated and rotating independently from those that are their neighbors today (Mankinen et al. 2013). Additionally, using paleomagnetic signatures in the Redding subterrane rocks, there is another peculiarity. The Redding subterrane has rotated in the opposite sense to the nascent North American continent to the east, suggesting again that at the time of rotation they were not yet connected.

Trinity subterrane

The Trinity subterrane more or less defines the western edge of the Eastern Klamath terrane. Driving north from Trinity Center on Highway 3 over Scott Mountain Summit, you are largely traveling on the Ediacaran to Late Ordovician-aged Trinity subterrane. The characteristic orange-rich colors of this suite of rocks are typical of the way ultramafic rocks weather when exposed to surface conditions. Rich in iron and magnesium, these normally dark rocks weather into lovely orange hues that set them apart from most other rock types. Many of these deep-seated rocks have been extensively modified by metamorphism into serpentinite—a slick, greenish rock type whose color helps to identify it as ultramafic rock, even from a great distance. These serpentine rocks are metal-rich and nutrient-poor, setting the evolutionary stage for striking and wonderfully unique plant species, which are discussed in the Plant Communities chapter. The Trinity subterrane is composed of ultramafic rocks and gabbro that display some classic characteristics of an ophiolite assemblage. Ophiolites are important rock types (see sidebar) that originate at great depth and are abundant in the Klamath Mountains.

Central Metamorphic terrane

The Central Metamorphic terrane (CMT) is exposed on the western side of the Eastern Klamath terrane and, like the other terranes in the Klamath Mountains, is in fault contact with its neighbors. Structurally, the Eastern Klamath terrane has ridden up and over the top of the CMT, putting the older rocks in fault contact with the younger CMT rocks below.

Composed primarily of metamorphosed sedimentary and mafic volcanic rocks, the Central Metamorphic terrane is typical of the kind of rock and the type of metamorphism found in a subducting oceanic volcanic arc. It subducted below the Trinity ultramafic rocks during the Middle Devonian period (380 Ma). Radiometric dating of the Salmon Schist and the Abrams Schist, the two major rock formations of the CMT (Irwin, 1994) produced ages of 390–399 Ma for the Salmon Schist (Hotz 1979) and 380 Ma for the Abrams Schist (Lanphere and Irwin, 1987) (Figure 18).

Figure 18. Prominent outcrops of the Central Metamorphic terrane. The Abrams Schist (left) at the junction of highways 299 and 3 in Douglas City, and the Salmon schist (right) at Oregon Summit just west of Weaverville. Photos by Mark Bailey.

Western Paleozoic and Triassic Belt

The Western Paleozoic and Triassic Belt is composed of four or five recognized terranes and is the most extensive of the rocky belts curving through the Klamath Mountains (Irwin 1972). This belt is also an amalgamation of numerous rock formations that differ in both composition and age. These rocks are mostly sedimentary, mafic plutonic, and volcanic rocks with a total thickness of up to 30,000 feet (Davis et al. 1979).

Small limestone outcrops occur in numerous places throughout the Western Paleozoic and Triassic Belt and yield valuable age-determining fossil ammonites (a shelled, squid-like creature that is now extinct but is superficially like the modern nautilus) and corals, as well as other useful age-determining fossils. These outcrops are generally deformed, recrystallized, and of limited size. They also tend to be lens-shaped and show a strong signal of tropical ocean environments.

From east to west, the terranes of the Western Paleozoic and Triassic Belt are the Fort Jones terrane, North Fork terrane, Eastern Hayfork and Western Hayfork terranes, Rattlesnake Creek terrane, Condrey Mountain window, and May Creek terrane (see tables 1 and 3).

Fort Jones terrane

The rocks of this terrane are mostly metamorphosed volcanic and sedimentary rocks. The beautiful blueschist rocks found in the eastern part of the terrane have been dated to circa 220 Ma (Hotz 1979). Much of the rest of the Western Paleozoic and Triassic Belt was later "overprinted" with younger greenschist metamorphism that occurred about 158 Ma, resetting its radiometric clock to the age of the younger minerals. This implies that at least two major metamorphic events occurred during the formation of this entire belt. It is likely that this represents an initial subduction episode, followed by a dying down of subduction for perhaps 62 million years, and then a reactivation of the subduction zone.

While this metamorphism was taking place, a major reshuffling of plates was underway farther east. The Atlantic Ocean was just beginning to form by rifting open as Pangaea (the giant supercontinent that followed its much older predecessor, Rodinia), began to break up, creating the separate North American, African, and European continents that we know today (Blakey and Ranney 2018).

North Fork terrane

This smaller (60 miles long) terrane is built primarily from an assemblage of rock types typical of an ophiolite sequence or ophiolite complex (see Ophiolite sidebar, page 78). Ophiolite complexes are an important component of the Klamath Mountains, where they are unusually large and well exposed. Uranium-lead dating gives an age for the North Fork terrane ophiolite as late Paleozoic (265–310 Ma). Cherts are also present and have been dated to the same time. Chert is a very hard, silica-rich sedimentary rock forming in deep-water sediments settling to the ocean bottom. Some of this sediment contains the bodies of radiolarians, a tiny planktonic animal that has a "shell" or test made of silica. These have distinctive morphologies, evolving noticeably through time. They can be used to date a rock they are found in as each form was distinctive to a specific time. Amazingly, these cherts also have a paleomagnetic stamp on them, indicating they were formed at around 8.6°N or 8.6°S. The rocky terranes of the Klamath Mountains have been quite the travelers but, except in the case of the Eastern Klamath, it appears that most of the terranes were formed near the ancestral North American continent and, perhaps, traveled with the continent as a whole through an impressive range of latitudes.

Eastern Hayfork terrane

The Wilson Point thrust fault separates the Eastern Hayfork terrane from the Western Hayfork terrane. This fault occasionally reveals itself as a zone of serpentinite that has oozed up along the fault plane. The Eastern Hayfork terrane (dated to 210 Ma) is composed of a structurally disorganized and confusing mélange. Mélange is typified by a jumble of diverse rock fragments that forms from debris tumbling down a continental shelf near active continental edges, such as that which existed in the Klamath Mountains at the time they were being assembled (and like today off the present west coast). This mélange is made mostly of chert and clay-rich argillite members (Wright and Fahan 1988). Clay is typically a weathering product of continental rock.

Western Hayfork terrane

The Western Hayfork terrane has been dated to about 156–177 Ma (Irwin, 1997a). Layered volcano-sourced (volcanoclastic) sediments in its lower section have been metamorphosed by the titanic pressures and heat generated during mountain building. Sediments parental to these rocks were washed off the land areas onto the sea floor surrounding their volcanic source rock (Blakey and Ranney 2018), destined to become the sedimentary and metasedimentary rocks of the Klamath Mountains. The upper (younger) portions of the Western Hayfork terrane include limestone and mud typical of offshore sedimentation. These rocks formed somewhere off the west coast at a time when, inland, dinosaurs lumbered about in their prime. The age of metamorphic minerals in these rocks indicate at least two major metamorphic events at circa 168 Ma and 150 Ma—well after the time of deposition of the Eastern Hayfork terrane sediments (Wright and Fahan, 1988).

Rattlesnake Creek terrane

This widespread terrane is composed of a hodge-podge of various rock types with similar ages but highly variable compositions. Within this mélange, fossil bearing sedimentary rocks are found. The fossils include radiolarians, embedded in their cherty graves and identified as being of Triassic and Jurassic age (252–145 Ma). Limestone bodies in the Rattlesnake Creek terrane also contain sparse Devonian (419–359 Ma) fossils (Irwin 1997a), revealing a rather wide age range.

May Creek terrane

The May Creek terrane is another schist-dominated terrane largely made of what appear to be metamorphosed basalts and associated igneous rocks, as well as metamorphosed sandstone and limestone (now converted into marble). Its age, as of this writing, is unknown but its rocks are distinctive enough (compared to the surrounding terranes) to view it as a separate entity.

Figure 19. Outcrops of the Western Klamath terrane along Highway 199 west of the Collier Tunnel (left) and Highway 96 north of Orleans. Both of these outcrops expose sedimentary rocks of the widespread Jurassic-aged Galice Formation, which overlies the Josephine Ophiolite. Photos by Mark Bailey.

Western Klamath terrane

The Western Klamath terrane (WKT) is a long, arc-shaped assemblage—spanning about 180 miles—from the Yolla Bolly Mountains north to the Umpqua River near Canyonville. Being the last and most westerly of the Klamath terranes, it is composed of the Dry Butte subterrane (an igneous-rich unit composed of ultramafic, granitic, and volcanic rocks), Briggs Creek subterrane (composed largely of amphibolite, but also including some schist and chert), Rogue Valley subterrane (a volcanic and chert-rich assemblage), and the Smith River subterrane.

Smith River subterrane

Perhaps the most intriguing of the four subterranes in the WKT is the Smith River subterrane. It is composed of the Josephine Ophiolite and overlain by the 164–162 Ma Galice Formation (Figure 19). The metagraywacke sandstones and turbidite-rich sediments (sediments indicative of undersea landsliding) of the Galice Formation interfinger with the Rogue Valley volcanic rocks to the north. They are also believed to have been metamorphosed repeatedly and exposed as the Condrey Mountain window to the east (Bill Hirt, pers. comm.).

Condrey Mountain window

Straddling the California-Oregon border, the Condrey Mountain window (figures 8 and 9), is mostly schist. Derived primarily from metasedimentary rocks, it also includes metavolcanic rocks. Radiometric age dates vary uncharacteristically widely, ranging from 220 to 117 Ma). The Condrey Mountain rocks are made mostly of greenschist, a comparatively low-pressure, low-temperature metamorphic rock type. Its place in the landscape seems somewhat enigmatic.

The schist crops out as a round area in the middle of the otherwise arcuate trend of the neighboring terranes. Exposed by doming up and erosion of the overlying rock, it appears likely that the source rock is the Galice Formation of the Smith

Figure 20. The Smith River with a well-weathered exposure of the Josephine Ophiolite (left). The mouth of Bluff Creek exposing the same Josephine Ophiolite rocks over 40 miles away. This attests to the extent of the Josephine Ophiolite exposure. Notice the characteristic orange color of weathered ultramafic rocks typical of ophiolite sequences when exposed. Photos by Mark Bailey.

River subterrane that has been overridden by thrust faulting and, in the process, metamorphosed several times into schist by intense compressional stresses.

To simplify, the WKT is largely an ultramafic body of rock covered in many places by sedimentary rocks of the Galice Formation and volcanic rocks of the Rogue River Formation. One of the "kinder" aspects of the Western Klamath terrane, unlike many of the rocks in the more easterly terranes, is that it is largely composed of readily identifiable rocks.

Josephine Ophiolite

The Josephine Ophiolite (Figure 20) deserves special mention for its completeness and size. The basal peridotite unit of the Josephine Ophiolite is chemically distinctive and has the almost whimsical name of harzburgite, an ultramafic rock composed mostly of olivine and calcium pyroxene. Being exposed across >300 mi^2, the Josephine Ophiolite is one of the largest, most complete of the ophiolites exposed at the surface of the Earth (Harper 1984). The Smith River has excellent exposures of this distinctive rock assemblage in its bed.

Assembled circa 157 Ma, the Josephine Ophiolite has been chemically interpreted as the seafloor basement that was generated adjacent to a volcanic arc. Most ophiolites on Earth have been formed at mid-ocean ridge spreading centers, but the chemical fingerprint of the Josephine Ophiolite indicates it was formed in a backarc basin (see Figure 4) of an island arc system (Harper,1984). Ophiolites (as well as other rock types) and their age-relationships with those of the northern Sierra Nevada strongly suggest a common history (Figure 21).

The Western Klamath terrane is separated from the Coast Ranges by the low angle South Fork Mountain thrust fault. This large thrust fault defines the boundary between the Coast Ranges of California and Oregon, and the Klamath Mountains.

Geology

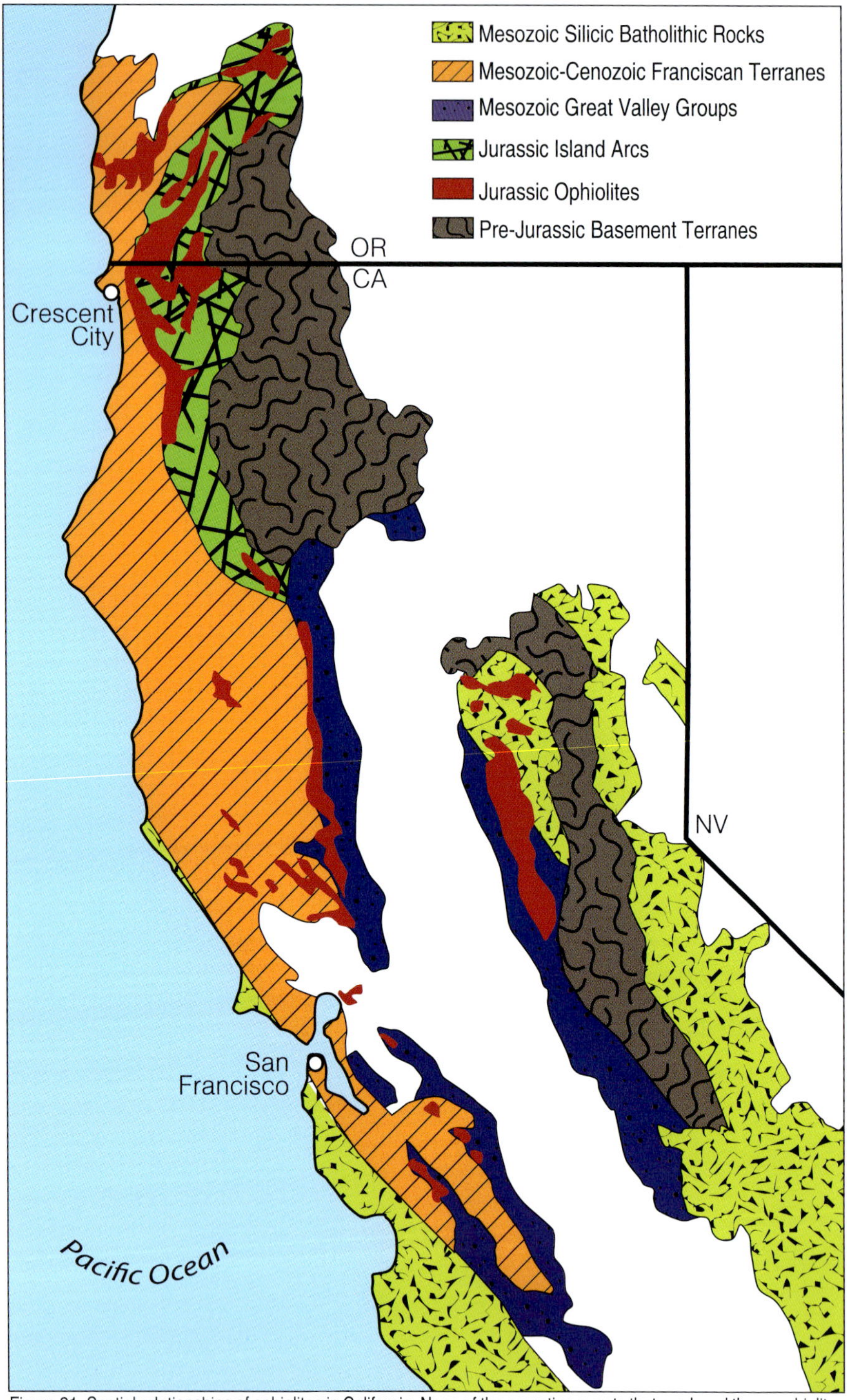

Figure 21. Spatial relationships of ophiolites in California. None of the accretion events that produced these ophiolites occurred in isolation, and a connection between the Klamath Mountains and the northwestern Sierra Nevada seems evident, as shown here. Figure created and generalized from numerous sources.

The Uplift of the Klamath Mountains

One of the long-standing questions about the Klamath Mountains is: when did they actually become mountains? It turns out to be a difficult question with a somewhat confusing answer. Was it during or after the terranes became assembled as a whole?

An impressive amount of volcanism is apparent in the rocks of the Klamath Mountains. In virtually all of these cases, the volcanic cones have been long erased by relentless erosion and subduction events. Many of these volcanic peaks may very well have surpassed Mounts Shasta and Rainier in height. The roots of these offshore island arc volcanoes are still evident to geologists, and it is partly from these rocks that the convoluted history of the Klamath Mountains has been reconstructed. Additionally, a compelling narrative of Klamath uplift has emerged from the study of stream profiles, the measurement of erosion rates, the mapping of eroded former surface remnants, the reconstruction of plate tectonic stresses, and several other lines of evidence.

The Nevadan Orogeny

Between 155 and 145 Ma, a comparatively rapid and spectacular series of mountain-building events took place in western North America. This was the Nevadan Orogeny. It seems no coincidence that during this time the supercontinent Pangaea was breaking apart. Many of the rock units that comprise the Sierra Nevada and the Klamath Mountains were accreted to the continent (see Figure 4). Both by age and composition, the granitic rocks of the Klamath Mountains appear somewhat similar to those in the western Sierra (Figure 21). This suggests that the forces exerted on the continental edge by the Nevadan Orogeny likely played a significant role in both the production of the Klamath and Sierra Nevadan plutons.

Even so, the concept of the Nevadan Orogeny and its sibling, the Siskiyou orogeny, may not have been the only main mountain-building events of the Klamath Mountains. Increasingly finer resolution of the sequence of events that formed the Klamath Mountains has emerged from increasing numbers of radiometric dates of rocks throughout the Klamath Mountains. We now know that many of the iconic granitic rocks were intruded both before, during, and after the time of the Nevadan Orogeny (Hacker and Ernst 1993).

Some of the emplacement of these plutonic rocks, and the associated metamorphism of preexisting rocks, occurred well before the Nevadan Orogeny. It was near the end of the Nevadan Orogeny (140 Ma), that nearly all the island-arc magmatism ceased and the terranes had finished forming. The rocks of the Klamath Mountains had now stopped being assembled. This may have been caused by the subduction of colder material underneath the Klamath Mountains (Hacker and Ernst 1993). From this point on, compression, extension, and associated faulting worked together with erosion, to ultimately produce the mountains we see today.

Not a One-time Event but a Series

During the Paleogene Period (66–23 Ma), there was apparently enough relief in the Klamath Mountains to enable streams to erode and wash out enough sediment to form extensive deposits offshore. These sediments were destined to become part of the Coast Ranges of our day. Geologists infer that during the Eocene Epoch

(56–23 Ma), there was some sort of mountain system where the present-day Klamath Mountains are but those mountains have been eroded completely away since that time (Aalto 2006).

Although not part of the Klamath Mountains, nearshore marine sediments of the Wildcat Group exposed along the Eel River at Scotia contain detrital mica crystals that strongly suggest an origin not in the Klamath Mountains but much farther to the northeast in the Idaho batholith. The ancestral Snake River was the likely source of these sediments, as this is the river that currently drains that source area (Aalto 2006). Apparently, the Snake River entered the ocean near Eureka after crossing the eroded, low-elevation surface of the Klamath Mountains! This sediment from the ancestral Snake River was then choked off during the Oligocene Epoch (~30 Ma). A reasonable cause for this cessation would be the beginning of renewed uplift of the Klamath Mountains, but this was only of limited duration (Aalto 2006).

The Klamath Peneplain

Diller (1902) was likely the first to observe an ancient erosional surface called a peneplain in the western Klamath Mountains. A peneplain is a gently undulating, almost featureless plain produced by fluvial (stream) erosion. Over geologic time, the land is reduced to base level (sea level), leaving so little gradient that essentially no more erosion could occur.

There is clear evidence to support the idea of this Miocene (23–5.3 Ma) peneplain (Aalto 2006). The remnants of the Klamath peneplain can actually be seen in the relatively uniform height of the eastern Klamath Mountain peaks and the steady decline in altitude of their sibling peaks as one moves west and approaches the western Klamath and Coast Range mountains (Figure 22). Additional sedimentary evidence is revealed in the relatively young (~5 Ma, late Miocene–early Pleistocene) marine sediment of the Wimer Formation that is preserved in remnant outcroppings east of Crescent City that sit *atop* the eroded remnants of this peneplain (Irwin 1997b).

Additional evidence supports a more mountainous topography for the southwestern Klamath Mountains at the time the peneplain existed to the north. The Weaverville Formation is composed of much younger (middle Miocene) sedimentary rocks deposited on top of the older Eastern Klamath terrane. These sedimentary rocks are from lake deposits. Ironically, an ancient lake (or lakes) existed in Trinity County 20 million years before today's reservoirs. Within these lake sediments pollen is found that originated from trees which surrounded this ancient lake. Analysis of the pollen grains from various levels in these lacustrine sediments shows that mixed-species montane conifer forests existed at the presumed time of the peneplain to the north and west. If montane forests of that time conformed to the elevational range of such forests today, then this implies higher elevations at that time in the southern Klamath Mountains (Piotraschke et al. 2015). Vegetational ranges do, however, change with changing climate. Perhaps montane forests grew at lower elevations 20 million years ago. How tall were these mountains then? We may never know for sure but their eroded remains were washed out into the ancestral Pacific to become, in part, today's Coast Range mountains.

The Ice Age in the Klamath Mountains

There is abundant evidence of glaciation in the Klamath Mountains. In light of the time span and detail involved with all the other geology so far, glaciation seems somewhat like the flea on the elephant's back. To a large extent however, glaciers have left one of the most visible, and recent, imprints on the landscape. One need not look far to see the telltale signs of glaciation. These include knife-edged ridges or arêtes; U-shaped valleys; sharp peaks or horns; bowl-shaped lakes or tarns; moraines; and rounded valley heads or cirques. The Marble Mountains contain numerous small lakes of glacial origin but nowhere is the signature of recent glaciation as distinctive as in the Trinity Alps.

Beginning about 2.5 Ma, conditions in what would be California became suitable for glaciation to begin (Guyton 1998). The further back you go to look for glacial evidence, the more muddied the view is because successive waves of glacial activity largely obliterated the evidence of earlier glacial episodes. Because the Klamath Mountains are shorter in stature than parts of the Sierra Nevada, they did not experience the same intensity of glaciation as did the Range of Light. However, their proximity to the ocean allowed for a considerable amount of glacial ice to accumulate. You would be suitably impressed if you went back in time 20,000 years (during the last glacial maximum) to the valleys of Stuart Fork, Canyon Creek, Coffee Creek, or Swift Creek. Some glaciers in the Klamath Mountains advanced downstream from Thompson Peak (8,936 feet) to as low as 2,450 feet as is evidenced by the moraines these glaciers left behind after melting away (Guyton 1998). Moraines are the piles of rocky debris carried and pushed along by advancing ice. As one walks up the lovely trails along the above-mentioned streams, moraines indicating several advances and retreats become obvious by their curved, rock-pile mounds.

During the late phase of glaciation, there were at least 30 valley glaciers in the Trinity Alps and several cirque glaciers. Swift Creek hosted the largest of these at roughly 18 mi^2 in size. Stuart Fork held a 16.7 mi^2 glacier and Canyon Creek had a 9.2 mi^2 glacier. Some exceeded 13 miles in length. More recent glaciers were all less than 10 miles long (Sharp 1960).

Characteristic glacial features on the landscape at Emerald and Sapphire lakes in the Trinity Alps. Note the distinctive bowl-shaped cirque lakes, or tarns, carved by glacial ice at the heads of mountain glaciers. Photo by Melinda Bailey.

One modern glacier remains in the Trinity Alps (See Glacier sidebar, page 130) on the cool, shaded north slopes of Thompson Peak. The Grizzly Glacier has experienced rapid melting in recent years (Garwood, et al 2020).

Geology

Figure 22. Remnants of the Klamath Peneplain as seen from the Siskiyou Mountains looking east into the Marble Mountains (on horizon). Notice the comparably even elevations of most peaks all the way across the scene. Russian Peak visible at far right in distance. Photo by Mark Bailey.

The Start of Uplift of Today's Klamath Mountains

The Klamath Mountains we see today may be as young as 5 million years old (Anderson 2008). A story of repeated exposure of the diverse rocks of the ancestral Klamath Mountains, their erosion to a peneplain, their submergence below sea level, and their most recent uplift and re-emergence has sound footing and makes a fascinating story. Those parts of the Klamath where the marine Wimer Formation remnants have been found must have been at or below sea level at that time. Anderson (2008) and others measured the tilt of this remnant surface (up to 3°), suggesting that at least parts of the Klamath have had an uplift rate of between 0.1 mm/yr and 0.4 mm/yr. This corresponds to over 6,000 feet of uplift during the 5 million years since the deposition of the Wimer Formation—a believable number when looking at today's elevations in the Klamath Mountains.

As mentioned earlier, stream profiles and the remnants of erosional surfaces caused by the incision of numerous rivers can also shed some light on the uplift history. Streams respond to uplift by increased incision into the bedrock. High river channel steepness points to continuous ongoing uplift. Knickpoints are areas along a stream course with locally steeper gradients. The central Klamath Mountains have evidently experienced the highest incision rates in the range as inferred from numerous knickpoints along the stream courses.

Faulting and differences in the rock type through which a stream incises complicate the picture. An additional complication arises from isostatic uplift of the region. Isostatic uplift occurs when mass is removed by erosion or other means and the area no longer experiences the same amount of weight pressing down on the rocks. The rocks respond by slowly rebounding upward. It has been estimated that the western Klamath Mountains alone have lost nearly *3,300* mi^3 of rock through erosion (Anderson 2008). Rebound from this loss has also contributed to uplift.

A River Runs Through It

The fact that mountains wear away and reemerge helps explain how current rivers—the Klamath River, Rogue River, and most of the other river systems of the Klamath Mountains—cross through the range from the east (see the Water chapter for more on this Water Gap). It is difficult to explain how a river could cut its way through and cross an existing mountain range unless that mountain range was not there during the river's youth when it established its course. Subsequently, as the mountains rose again from worn down vestiges, these rivers were able to keep up with this uplift by eroding the bedrock at a similar rate. These events have also produced a geologic treasure chest that has generated considerable interest.

Mining in the Klamath Mountains

Western mining history often implies that these activities are noble, romantic, and inevitable. They were far from it. "The Law" was often far away and miners were disinclined to follow laws anyway, no matter how weak those laws were. Sadly, this is largely still the mindset of many modern miners. Miners were often an unruly bunch with no regard for the health of original occupants or the land. *All* of these mines were on the homelands of a wide variety of indigenous peoples and cultures who had tended the land with reverence for countless generations. These people were a part of the land and as these lands were destroyed, so too were many of the original inhabitants. As a result of the westerner's greed for gold (and other minerals), many of these cultures were fragmented. Displacement of the original occupants, murder, rape, enslavement, genocide, and ecocide were the rule—and the often untold story of mining in the Klamath Mountains and elsewhere. This legacy continues to this day in the minds and lives of the indigenous populations as well as on the landscape. From the Native perspective, there was absolutely nothing romantic or noble about what was perpetrated on them by an inflow of outsiders interested only in taking whatever they could. Despite these injustices, Native Peoples and cultures still persist here—a tribute to their endurance and strength.

But despite this sordid history, mining was important to the deciphering of the geology of the Klamath Mountains. Much of the initial geological study of the Klamath Mountains centered on the extraction of valuable minerals with gold being of particular and sustained interest. With the 1848 discovery of gold at Sutter's Mill in the Sierra Nevada, prospectors soon noticed the geologic similarity between the Sierra Nevada and the Klamath Mountains while continuing their relentless search for gold. More than half of all Klamath gold came from placer deposits (those where gold has been concentrated by stream erosion) (Silberman and Danielson 1991). What they couldn't have realized at the time was that these two geologic provinces, at least in part, have a common origin driven by the engine of shifting tectonic plates and the associated mountain-building processes that are evident throughout the 4,000-mile length of North America's Cordillera.

Gold Dredging

There are hundreds of mines in the Klamath Mountains. Obvious remnants of placer gold mining can be readily found to this day and are a constant reminder of the disrespect towards the land and its inhabitants that occurred. Notable examples are evident along Canyon Creek and on Highway 299 from near Junction City to Del Loma and beyond. The distinctive piles of boulders heaped high in the river beds are the waste produced as miners used dredges of various types to sift through stream gravels, seeking the gold that had been eroded out of the surrounding mountains and deposited as sediment in the river channel (Figure 23). Other woeful examples are evident in the piles of boulders that become partially flooded by Trinity Reservoir along Highway 3 north of Trinity Center and along the Scott River near Callahan.

Hydraulic Mining

A good example of another destructive mining strategy is found at the La Grange Mine at Oregon Mountain Summit on Highway 299 west of Weaverville. Water for this process was generated by damming the outlet of Sapphire and Emerald lakes in the Trinity Alps at the headwaters of Stuart Fork, 29 miles away. A decaying dam remains today. A giant water cannon, called a monitor, was sprayed at the ancient gold-bearing gravels under high pressure during this hydraulic mining (Figure 24). The resulting slurry was channeled through large sluice boxes to extract the gold. The remaining slurry of spent gravel was routed out into the stream channel with ecologically smothering consequences (Figure 25). The La Grange Mine was among the largest of placer mines in the United States. These auriferous gravels were deposited and sat unnoticed for millions of years—as they also did in the Sierra Nevada. They were actively mined from 1852 through the 1940s, producing about 390,000 ounces of gold. At today's values, that is worth nearly $500 million. As has often been the case, the economic value of the minerals extracted is FAR below the cost of fixing the damage done. This damage has been lasting and, typically, it has been left to those who were there before and those that

Figure 23. Gold dredge at Junction City, along Highway 299, sometime around 1950. Notice large piles of boulder tailings and self-made lake from the operation. Trinity River channel is beyond. This same dredge was later refurbished and moved to the upper Trinity River where it made the distinctive rock piles at the north end of Trinity Reservoir just before the lake was created. Photograph courtesy of the Trinity County Historical Society.

Figure 24. Hydraulic mining using a monitor at the La Grange Mine during the early 1900s. Photograph courtesy of the California Geological Survey.

Figure 25. **Left:** Tailings from the La Grange Mine (scarred hillside in background) filling the canyon of Oregon Gulch. Notice comment on photo: "over 200 feet deep." Photograph courtesy of the Trinity County Historical Society. **Right:** The same location in 2018. Notice notch, center distance, where the current highway 299 cuts through. Much of the sediment has since been eroded away. Photo on right by Mark Bailey.

follow these catastrophic practices to try to restore normal ecological function to the battered environment. Little, if any, of these lands have been repatriated to the original native occupants. Unfortunately, many of the mining practices and laws that allowed for such destruction are still on the books.

Along most of the rivers in the Klamath Mountains, the sedimentation from these activities is so extensive that, today, this waste is being mined for gravel at many locations. The old stream channels are emerging with the extraction of the previous waste gravel, now used for concrete and road building. It seemed logical that if gold were found in streams, then it must have come out of the surrounding mountains. These invaders hauled equipment, animals, and pretty much anything you can think of into the steepest, most remote of mountain slopes looking for the mother lode. Few got rich and many more suffered greatly.

The emplacement of the many granitic plutons scattered throughout the Klamath Mountains not only helped to "stitch" the terranes together, but also produced mineral-rich fluids. Groundwater, made hotter than normal by circulating through fractures near these hot molten masses, dissolved gold, platinum, silver, and a wide variety of other metals. Circulating upward through fractures and cooling, the fluids finally solidified as precious metal deposits in these mineral-filled cracks, or veins. These lode deposits were the ultimate sources of the placer gold. A large portion of these fluids would solidify into quartz—that beautiful, white, hard mineral that is a signpost for the presence of other minerals. Although much of this mineral wealth has already been depleted in the Klamath Mountains, there are still minable deposits remaining. Mining companies have been trying for years to extract nickel- and chrome-rich areas, located on public lands, in the drainages of the Smith River, Illinois River, and more. These ultramafic deposits, associated with seafloor spreading centers (those features that build ophiolite sequences) are rich in chrome, nickel, copper, and other metals and have been mined in the past. The so-called massive sulfide deposits, formed by undersea hot springs, are much sought after. Indigenous as well as non-indigenous activists have correspondingly fought such renewed activity, with mixed success.

A notorious case is that of the Iron Mountain Mine near Redding. The tailings from this mine are visible to the north from Highway 299 driving west from Redding. Mining of massive sulfide deposits often opens nightmarish environmental consequences. In this case, pollutants include heavy metals such as copper, zinc, and cadmium, along with sulfuric acid waste. The runoff water from this mine is the most acidic surface water ever measured (USGS 2016). The EPA has dedicated $1 billion to this Superfund site to prevent these devastating pollutants from entering surrounding waterways, such as Lake Shasta and the Sacramento River. Examples such as these pose many questions as to what sort of mining activities, if any, should or should not be allowed in the Klamath Mountains and elsewhere going forward. After considerable and successful pressure was placed on the state government by among others, the Karuk Tribe, the state of California has banned even small-scale gold dredging in Klamath streams because the sediments produced smother the eggs of spawning salmon.

Geology

Epilogue

The Klamath Mountains have had a convoluted geologic history. Little of the original geological structures and rocks remains in pristine condition. The assemblage of these magnificent mountains has obscured or destroyed many of the geological clues that scientists use to interpret the past. But the rocks, our Ambassadors of Time, don't lie—they just sometimes confuse us. Enough of an imprint of these tumultuous geologic processes has remained in the rocks and valleys and peaks for geologists to interpret the large-scale events that transpired over these millions of years. Much work remains to be done.

Chapter Reviewers:

- Bill Hirt, Steve Sillett, Jim Agee, Dylan Neubauer, Lisa Hillman, and Leaf Hillman.

Photo Contributors:

- Mark Bailey, Melinda Bailey, Michael Kauffmann, and Trinity County Historical Society

Literature Cited and Suggested Reading

Aalto, K.R., 2006. The Klamath Peneplain: A review of J.S. Diller's classic erosion surface. *Geological Society of America Special Paper 410.*

Allen, C.M., and C.G. Barnes. 2004. Ages and cryptic sources of Mesozoic plutonic rocks in the Klamath Mountains, California and Oregon. *Geological Society of America Special Paper 410.*

Alt, D., and D.W. Hyndman. 2000. *Roadside Geology of Northern and Central California.* Missoula, MT: Mountain Press Publishing Company.

Anderson, T.K. 2008. *Inferring bedrock uplift in the Klamath Mountains Province from river profile analysis and digital topography.* Texas Tech University: Master of Science thesis.

Blakey, R.C., and W.D. Ranney. 2018. *Ancient Landscapes of Western North America A Geologic History with Paleogeographic Maps.* Cham, Switzerland: Springer.

Coppold, M., and Powell, W. 2006. A Geoscience Guide to the Burgess Shale, Geology and Paleontology in Yoho National Park. *The Burgess Shale Geoscience Foundation*, Field, British Columbia, Canada.

Davis, G.A., C.J. Ando, P.H. Cashman, and L. Goullaud. 1979. Geologic summary of the central Klamath Mountains, California. *Geological Society of America Map and Chart Series* MC-281.

Diller, J.S. 1902. T*opographic Development of the Klamath Mountains.* U.S. Geologic Survey, Bulletin 196.

Elliott, W.R., J.R. Reddell, D. C. Rudolph, G.O. Graening, T.S. Briggs, D. Ubick, R.L. Aalbu, J. Krejca, S.J.Taylor. 2017. The Cave Fauna of California. *Proceedings of the California Academy of Sciences, July, 28, 2017, Vol. 64, Supplement 1.* Institute for Biodiversity Science and Sustainability.

Feranec, R.S., J.L. Blois, and E.A. Hadly. 2007. Preliminary Radiocarbon Chronology and Paleoecological Analysis of the Small Mammals in Samwel and Potter Creek Caves, Shasta County, California, and its Relation to the End-Pleistocene Extinction. *Radiocarbon.* Vol. 49, No. 1: 117-121.

Garwood, J.M., A.G. Fountain, K.T. Lindke, M. van Hattem, and H.J. Basagic. 2019. 20th century retreat and recent drought accelerated extinction of mountain glaciers and perennial snowfields in the Trinity Alps, Klamath Mountains, California, USA.

Graham, J.T., 2013. Late Holocene Glacial Advances in the Klamath Mountains, Northern California, Determined From 10Be Cosmogenic Exposure Dating and Dendrochronology. California State University, Northridge. Master of Science thesis.

Guyton, B. 1998. *Glaciers of California. California Natural History Guides, 59.* Berkeley, CA: University of California Press.

Hacker, B.R., and W.G. Ernst. 1993. Jurassic orogeny in the Klamath Mountains: A geochronological analysis. *Mesozoic Paleogeography of the Western United States.* Book 71: 37-60.

Harper, G.D. 1984. The Josephine ophiolite, northwestern California. *Geological Society of America Bulletin,* vol. 95: 1009-1026.

Hershey, O.H. 1903. Some evidence of two glacial stages in the Klamath Mountains in California. *The American Geologist.* Vol. XXXI: 139-156.

Hotz, P.E. 1979. Regional Metamorphism in the Condrey Mountain Quadrangle, North-Central Klamath Mountains, California. *U.S. Geological Survey Professional Paper 1086.*

Irwin, W.P. 1972. Terranes of the western Paleozoic and Triassic belt in the southern Klamath Mountains, California. *U.S. Geological Survey Professional Paper 800-C:* C103-C111.

Irwin, W.P. 1989. Terranes of the Klamath Mountains, California and Oregon. *Tectonic evolution of northern California: Sausalito to Yosemite National Park, northern California, June 28-July9, 1989.* Field Trip Guide T108.

Irwin, W.P. 1994. Geologic Map of the Klamath Mountains, California and Oregon. *U. S. Geological Survey Miscellaneous Investigation Series.*

Irwin, W.P. 1997a. Geologic traverse across the Klamath Mountains and northern Coast Ranges—westward along highway 299 from Redding to highway 101 near Arcata, California. *Prepared for GEOMAR and University of Kiel field trip, June 14, 1997.*

Irwin, W.P. 1997b. Preliminary map of selected post-Nevadan geologic features of the Klamath Mountains and adjacent areas, California and Oregon. *U.S. Department of Interior U.S. Geological Survey Open-File Report 97-465.*

Irwin, W.P., and E.A. Mankinen. 1996. Rotational and accretionary evolution of the Klamath Mountains, California and Oregon, from Devonian to present time. *U.S. Geological Survey Open File Report 98-114.*

Irwin, W. P. and E.A. Mankinen. 1998, Rotational and Accretionary Evolution of the Klamath Mountains, California and Oregon, from Devonian to Present time: U. S. Geological Survey Open-File Report 98-114, https://pubs.usgs.gov/of/1998/0114/.

Lanphere, M.A. and W.P. Irwin. 1987. In Search of the Abrams Post Office. *California Geology.* May, 1987: 99-103.

Lindsley-Griffin, N., J.R. Griffin, J.D. Farmer. 2008. Paleogeographic significance of Ediacaran cyclomedusoids within the Antelope Mountain Quartzite, Yreka subterrane, eastern Klamath Mountains, California. *The Geological Society of America Special Paper 442.*

Mankinen, E.A., C.S. Gromme, and W.P. Irwin. 2013. Paleomagnetic contributions to the Klamath Mountains terrane puzzle—A new piece from the Ironside Mountain batholith, northern California. *Tectonophysics* 608: 401-407.

O'Neill, C., S. Marchi, S. Zhang, and W. Bottke. 2017. Impact-driven subduction on the Hadean Earth. *Nature Geoscience* volume 10, pages 793–797.

Piotraschke, R., Cashman, S.M. Furlong, K.P. Kamp, P.J., M. Danisik, and G. Xu. 2015. Unroofing the Klamath Mountains—Blame it on Siletzia? *Lithosphere* vol. 7, no. 4: 427-440.

Sharp, R.A. 1960. Pleistocene Glaciation in the Trinity Alps of Northern California. *American Journal of Science* vol. 258: 305-340.

Silberman, M.L. and J. Danielson. 1991. Geologic setting, characteristics, and geomorphology of gold-bearing quartz veins in the Klamath Mountains in the Redding 1 x 2 degree quadrangle, northern California. *U.S. Department of Interior U.S. Geological Survey Open-File Report 91-595.*

Snoke, A.W. and C.G. Barnes. 2006. The development of tectonic concepts for the Klamath Mountains province, California and Oregon. *Geological Society of America Special Paper 410.*

U. S. Geological Survey, 2016. Iron Mountain. https://ca.water.usgs.gov/projects/iron_mountain/environment.html

Wright, J.E. and M.R. Fahan. 1988. An expanded view of Jurassic orogenesis in the western United States Cordillera: Middle Jurassic (pre-Nevadan) regional metamorphism and thrust faulting within an active arc environment, Klamath Mountains, California. *The Geological Society of America Bulletin,* vol. 100: 859-876.

Chapter 4: Climate

Rosemary Sherriff, Justin Garwood
and Michael Kauffmann

The orographic effect seen in
the Kalmiopsis Wilderness.

Introduction

The Klamath Mountains offer a distinctive, Mediterranean-type seasonality characterized by wet, cool winters and dry, warm summers. High inter-annual variation in precipitation (rain and snow) is the norm, with most precipitation occurring between November and April, followed by a prolonged dry season from May through October. Across the vast range, however, there is a high level of variability. From west to east, moisture and temperature gradients are created by proximity to the Pacific Ocean. In the western Klamath, rainforests are fed by wet and cold winter storms that move in from the north Pacific. These storms drop over 100 inches of rain and snow in Del Norte County, California, and Josephine County, Oregon, but when the storms reach the lower flanks of the eastern Klamath much less rain and snow fall. There is also a steep elevational gradient that influences the spatial pattern of both temperature and precipitation. Low-elevation coastal forests rarely receive snow, while the coastal Siskiyou Mountains hold snow into spring and feed the fast-flowing Smith, Chetco, and Rogue rivers. Inland mountains are much higher and receive much more frequent snow from storms—creating a deep snowpack. In the Pleistocene, glacier-carved canyons were created in these areas, which now contain hanging subalpine lakes that stay frozen into June--feeding cold mountain streams late into summer.

Paleoclimate

Today, Earth's climate is influenced by a combination of natural variability and human-induced factors. Ocean currents, which drive all major climate systems, have been in place for millions of years. Carbon and other greenhouse gases in the atmosphere are increasing, which is causing the planet to warm at accelerated rates over the last hundred years. To comprehend the effects human-induced climate change is having on Earth, it is important to model for future climate and interpret those of the past. Past climates are conveniently viewed on three time scales (years before present [ypb]): (1) the last 11,700 years; (2) the Pleistocene Epoch back to 2.6 million years ago (Ma); and (3) the pre-Pleistocene period prior to 2.6 Ma. Because of the relevance to the Klamath Mountains of today, this section focuses on (2) the Pleistocene and (1) the last 11,700 years. We focus on future climate models in the book's last chapter. Also see the Paleoclimate Reconstruction sidebar.

The Pleistocene (2,580,000–11,700 ybp)

The Pleistocene is the geological epoch that lasted from circa 2.6 million to 11,700 ybp, which included Earth's most recent period of repeated glaciations. The end of the Pleistocene corresponds to the end of the last glacial period. Climate regimes from this time frame have been reconstructed using, among other methods, interglacial sediments and plant fossils. Additionally, by coring ice in Antarctica, estimates of atmospheric CO_2 over the last 800,000 years inform us about past global temperature fluctuations (Figure 1). These changes, in turn, affected past biotic communities. Fossil plant locations are used for mapping concentration centers of flora—which aids in the recreation of temperature and precipitation estimates across landscapes. Based on molecular concentrations in the atmosphere coupled with fossil records, paleoclimatologists approximate that circa 120,000 years ago, the Klamath Mountains were covered by a species-rich forest composed

Paleoclimate Reconstruction in the Klamath Mountains

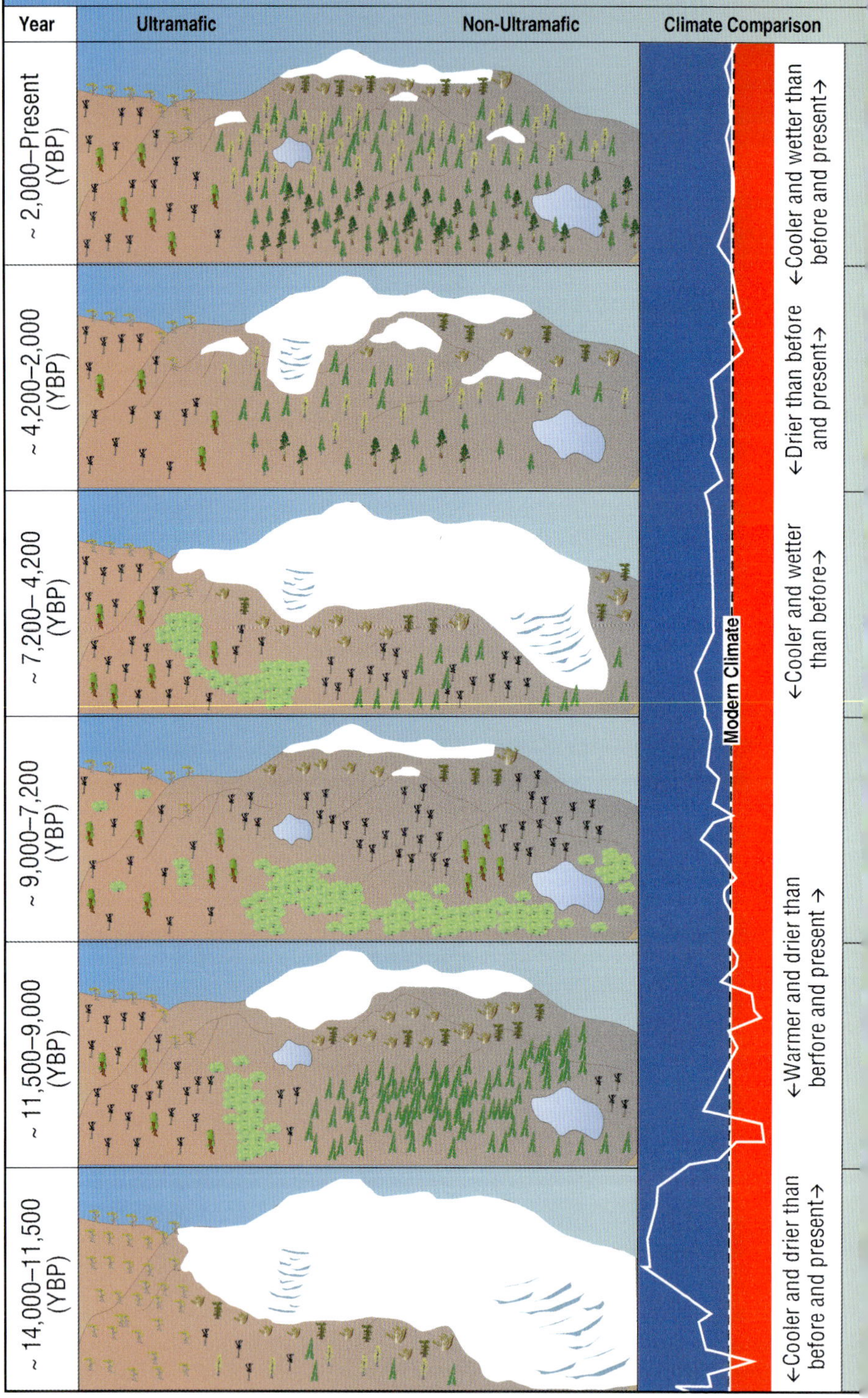

Notes: See Paleoclimate subsection text to understand how this information was collected by researchers. Pictures and text herein summarize those data (Briles et al. 2011).

Fire Return Interval	Vegetation Pattern Summary
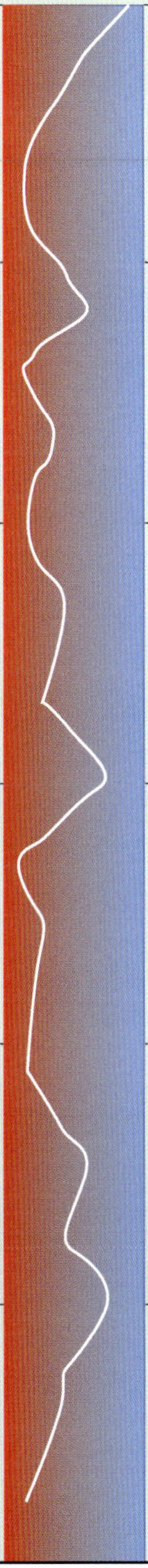	Climate became cooler and wetter than before present. Douglas-fir (*Pseudotsuga menziesii*) and true firs (*Abies* spp.) reached peak abundance for the Holocene. At the same time, forest canopies gradually became more closed.
	Climate became drier than before and present with more conifer species entering the pollen record. The abundance of chaparral species diminished to present levels. Pollen abundance of true firs peaked, and Douglas-firs became more abundant across California. Mountain hemlock (*Tsuga mertensiana*) became well established in most high-elevation forests.
	Climate became cooler and wetter again allowing for many conifer species to appear in the upper elevations for the first time in the Holocene—although shrubland species maintained a strong presence. Ultramafic soils remained relatively consistent in their vegetation patterns. The fire record shows an increase in charcoal accumulation rates, which suggests an increase in fire frequencies.
	Climate was warmer and considerably drier, allowing for the expansion of oak and other shrubland species. Fire frequencies increased but low rates of charcoal accumulation suggest a frequent, low-intensity fire regime. Ultramafic vegetation remained similar to those patterns established before. Overall, the pollen record closely resembles those from montane forests in the Klamath and Cascade ranges today.
	Climate was warmer and drier than before and present. Fire frequencies remained low, but individual fires may have been more intense due to increased forest fuel loads. Common juniper (*Juniperus communis*) and Great Basin sagebrush (*Artemisia tridentata*) pollen suggests increases in shrub communities. Ultramafic substrates increased in diversity from before with oaks and other Cupressaceae becoming more common in the pollen record.
	Climate was cooler and drier than present. Vegetation on non-ultramafic soils included open, subalpine parkland with low fire frequencies and fuel biomass. Ultramafic vegetation was dominated by Jeffrey pine (*Pinus jeffreyi*). Paleoclimate models suggest a northward shift in the jet stream brought wetter conditions to the Pacific Northwest.

of broad-leaved evergreens and a selection of conifers that thrived in a climate which was slightly warmer in the winter and cooler in the summer than the climate today.

Based on these and other data we also know that, on a global scale, there were numerous major glacial events as well as many minor glacial events during the Pleistocene. The severe climatic fluctuations had major impacts on the fauna and flora of the Klamath Mountains. With each massive ice sheet advance, large areas of the continents were depopulated of life. Tremendous stress was placed on the plants and animals being pushed in front of the advancing glaciers.

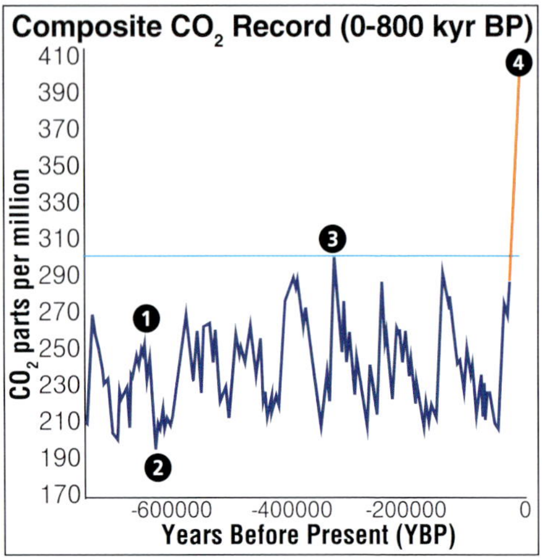

Figure 1. Carbon dioxide record for the past 800,000 years before present. 1) Warm period (interglacial). 2) Ice age (glacial). 3) Highest previous carbon record. 4) 409 ppm hit in 2019. Data from NOAA Climate.gov; NCEI.

Large ice sheets never completely covered the Klamath Mountains during the Pleistocene. There were at least four glacial substages during the Pleistocene in the Trinity Alps (Sharp 1960). During these four substages, more than 30 individual glaciers—with at least one measuring over 13.7 miles long—carved peaks of the Klamath Mountains into the landscapes we see today (Sharp 1960). Glaciers were relatively small in this region compared to other mountain regions of California and the Pacific Northwest. Nevertheless, glaciers were important drivers of landform change that sculpted landscapes and promoted biodiversity. During major glacial events, the lower elevations remained habitable, but vast areas in the higher elevations were covered in snow and ice for long periods of time (see Ice Age in the Klamath Mountains, page 91). These ice-covered landscapes had little, if any, biotic diversity. Because the unglaciated areas offered habitat for creatures being pushed south and refugia for those pushed out of the higher elevations, these pockets were like an Ark of diversity. When the upper-elevation Klamath glaciers receded, the fittest flora and fauna moved from these pockets to repopulate novel landscapes. Natural selective pressures allowed plants and animals to migrate in, find haven, and remain for millennia.

The Last 11,700 Years

To interpret the more recent climate regimes, scientists rely on records found in ice cores, lake-sediment cores, tree rings, coral signatures, cosmogenic rock dating, and packrat middens. Using these sources, past climates are reconstructed from physical, chemical, or biological proxies that provide information on environmental changes such as precipitation, atmospheric carbon levels, and air and ocean temperatures. These data are then used to reconstruct records and offer insights into rates of climate, vegetation, and fire-frequency change over time. In the Klamath Mountains, most paleoclimatic data have been generated through the analysis of pollen and charcoal concentrations in lake-sediment cores (e.g., Mohr et al. 2000, Wanket 2002, Daniels et al. 2005, and Briles et al. 2008 and 2011).

Once removed, layers of the sediment cores are analyzed under a microscope, and organic matter can be dated using the properties of radiocarbon dating (carbon-14). Using the unique signatures of pollen in each layer, researchers reconstruct the compositional and structural changes in forests around each lake over time. Charcoal concentrations in each layer are used to interpret fire frequencies over time. The Paleoclimate Reconstruction in the Klamath Mountains sidebar summarizes a sample lake-core study from the Klamath to summarize shifts in climate, vegetation, and fire frequency since the end of the Pleistocene.

Paleoclimatic studies have enhanced our understanding of the complex climatic history of the Klamath Mountains. In recent decades, the remnant glaciers of the Trinity Alps (within the Klamath Mountains) occur entirely below the tree line at an elevation of 8,070 feet, which is far lower than other glaciated areas in California (Sharp 1960, Garwood et al. 2020). During the time period known as the Little Ice Age (LIA), which occurred from the early 14th century through the mid-19th century, mountain glaciers expanded in the Northern Hemisphere. During the LIA, there is evidence of three glacial advances across 600–700 years in the Klamath Mountains (Sharp 1960). Over the last century, substantial retreat occurred in all of the six known LIA glaciers, with only two remaining as of 2009 (Grizzly and Salmon glaciers). Unfortunately, due to the severe 2012–2015 drought, the Salmon Glacier disappeared and the Grizzly Glacier retreated to 1.67 ha, which is a 97 percent loss of total glacial area since the 1880s (Garwood et al. 2020). Interpreting the paleoclimate of the Klamath Mountains is integral to understanding today's biodiversity and predicting future changes.

Regional Microclimates Today

The Klamath Mountains abut the North Coast region of California, which is the wettest part of California and receives 55 in. (140 cm) of precipitation on average (1981–2010; Grantham 2018). Precipitation primarily comes into the Klamath Mountains via large storms, often associated with "atmospheric rivers" that track eastward from the Pacific Ocean. These storms contribute an average of 40 percent of the annual snowpack in California (Guan et al. 2013, Grantham 2018). In the Klamath, the average snow line is circa 5,000 ft. (1,524 m) as seen in Figure 2a. The Pacific Ocean also moderates summer and winter temperatures in the western portion of the Klamath Mountains, which ranges on average around 43–44 °F (6–7°C) in the winter to low- to mid-70's°F (low 20s°C) in the summer (e.g., Willow Creek; Figure 2b), whereas interior temperatures can often fall below 32°F (0°C) in the winter and maximum temperatures often reach 100°F (38°C) in the summer (e.g., Dunsmuir; Figure 2b). During the warm season, temperature inversions are not uncommon during widespread or large fire events due to a combination of factors including the adjacency to cool, moist air from the Pacific, the rugged mountain topography, fire weather, and trapped smoke. Temperature inversions can lead to considerable differential effects on fire behavior both below

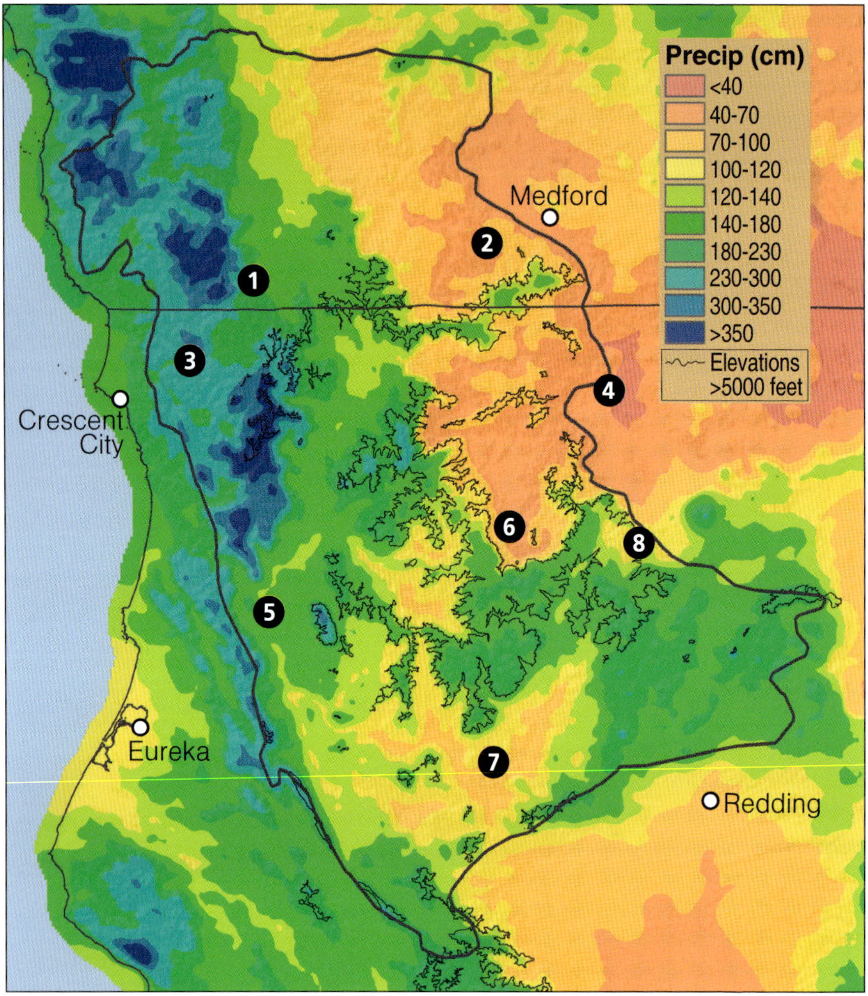

Figure 2a. The 5,000-foot contour roughly represents the elevation where winter precipitation accumulates as snow and remains until the spring melt. Precipitation data is from the 30-year normals (1981 to 2010) PRISM re-analysis product, PRISM Climate Group, Oregon State University; https://prism.oregonstate.edu/normals. Compare to graphs at right.

and above the inversion. Observations by fire-management personnel suggest that inversions in the Klamath Mountains often occur at an elevation of approximately 4,265 ft. (1,300 m) (Estes et al. 2017), which is characteristic and relatively unique to this region.

The complex mountain topography, coupled with a strong west–east climatic gradient, leads to substantial local variability in climatic conditions throughout the Klamath Mountains. Notably, the west–east atmospheric circulation from the Pacific Ocean and the topographic variability produce the wettest conditions in the northwest (e.g., Gasquet and Cave Junction), seconded by the southeast (e.g., Weaverville and Dunsmuir), while the driest conditions occur in the northeast margins of the region (e.g., Yreka, Callahan; Figure 2b). The driest conditions

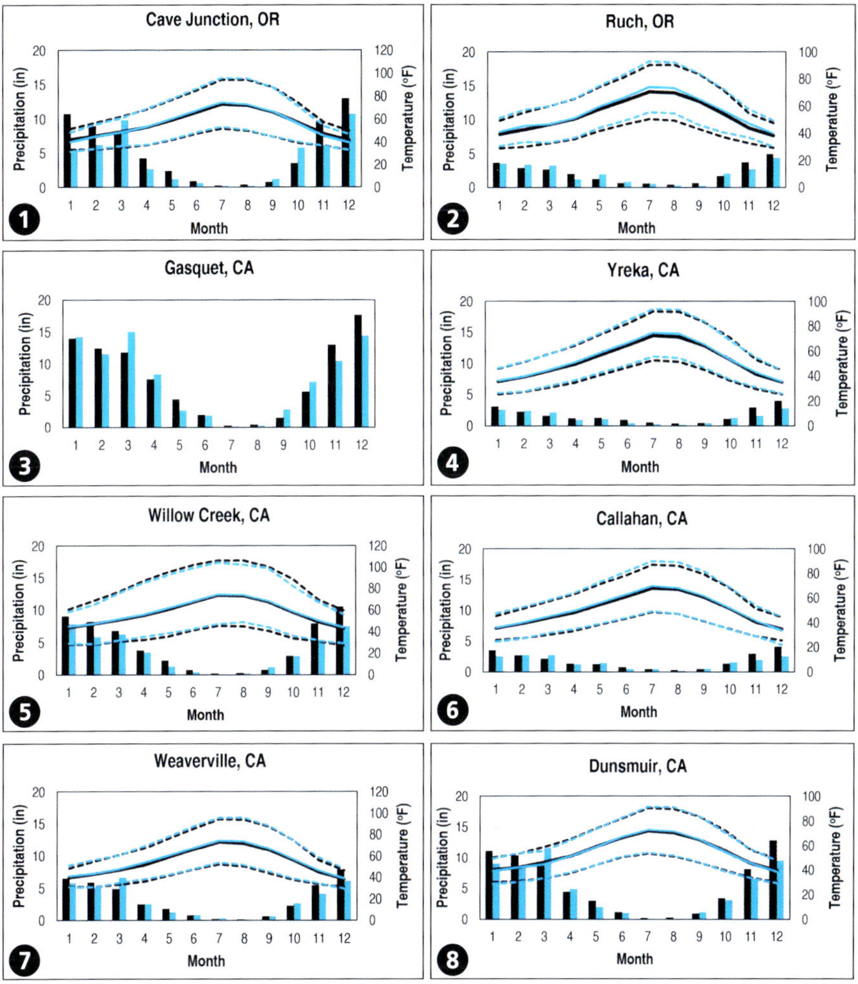

Figure 2b. Regional weather station normals for 1981 to 2010, compared to 2011 to 2020. Black = normals (1981 to 2010), blue = normals for 2011–2020 (or up to whatever year it ends in that period), bars = precip, solid line = aveT, dashed lines min/maxT. Data from NCDC (NOAA).

occur along the northeast transition zone of the Klamath Mountains into the southeastern Cascade Mountains, which transitions north into the Pacific Northwest. The lack of instrument stations along elevation gradients limits our ability to evaluate actual long-term climate trends by elevation. Nevertheless, there are strong orographic effects throughout the region that consistently lead to less precipitation in the valleys and canyons than at higher elevations and on the leeward side of the mountain range (also see the Camp Six sidebar this chapter). Likewise, the orographic uplift of moist air masses leads to high precipitation amounts in the westernmost portion of the region as the first major mountain range encountered by southwesterly winds moving northeast from the Pacific Ocean.

Climate Trends Over the Past 100 Years

Temperature

California's annual temperatures have increased on average by about 1.5°F in the last century, however temperature changes across the Klamath Mountains that extend into southern Oregon have been highly variable. In the western portion, increases in annually averaged mean daily temperatures have been under 1°F over the last century. In interior locations, such as Orleans, average temperatures have increased by 2°F in the period from 1931-2014 (Butz et al. 2015).

Comparing weather stations with available data for recent decades illustrates that stations show at least slight increases in minimum, average, and maximum monthly temperatures since 2011, compared with the prior two decades (Table 1). Throughout the Klamath Mountains, monthly maximum temperatures show the most consistent increases across all stations throughout the year. The magnitude of temperature increases has been greatest in the warm summer months, increasing the frequency of extreme heat events that impact terrestrial and freshwater ecosystems (Grantham 2018). Thus, there are important trends illustrating warmer (and drier) conditions throughout the region.

Precipitation

El Niño is the most significant driver of climate variability and the only reliable source of seasonal to interannual prediction of western North American climate (Jong et al. 2016). However, El Niño has inconsistent effects in the Klamath Mountains as they are in a regional transition zone between positive (wet) to negative (dry) effects in the southwestern U.S. and the Pacific Northwest, respectively.

Table 1. Difference in total monthly precipitation (Precip), and monthly minimum (MinT), average (AveT), and maximum (MaxT) temperature comparing 2011–2019 to 1980–2010 for stations with available data. Positive (negative) values indicate an increase (decrease). The table below highlights the average difference for each variable across all 12 months. * denotes significant differences comparing monthly values for 2011–2019 to 1980–2010 (paired t-test; $p < 0.01$). Instrumental station data acquired from the National Climate Data Center (NCDC), with the exception for Ruch, OR, and Willow Creek, CA, where data used for comparison were only available through the Western Regional Climate Center (WRCC).

Location	Precip (in)	MinT (°F)	AveT (°F)	MaxT (°F)
Cave Junction, OR	-1.0	0.5	-0.1	-0.8*
Ruch, OR	0.0	3.1*	2.4*	1.7*
Klamath Falls, OR	-0.4*	0.2	1.1*	2.0*
Yreka, CA	-0.3	1.8*	1.3*	0.9*
Orleans, CA	-0.7	0.6*	0.8*	1.1*
Callahan, CA	-0.2	0.0	1.1*	2.1*
Dunsmuir, CA	-0.6	0.0	-0.1	-0.3*
Willow Creek, CA	-0.9*	2.2*	0.9*	-2.8*
Weaverville, CA	-0.2	0.9*	1.3*	1.6*
Whiskeytown, CA	-0.7	-0.1	0.5	1.1*
Average	**-0.5**	**0.9**	**0.9**	**0.7**

The Mother of All Storms

Camp Six Lookout circa 1948 from the Gasquet Ranger District Archives.

In 1923, the United State Forest Service built a lookout located at the northeast end of Lower Coon Mountain, about 16 air miles east of Crescent City and five air miles east–southeast of Gasquet—just inside the western boundary of the Klamath Mountain Geomorphic Province in California. The lookout at Camp Six was used in the summer months to watch for wildfires, but throughout the year it was a weather-monitoring station managed by the California Department of Water Resources (CDWR).

A settlement agreement between the U.S. Forest Service and the Sierra Club led to the establishment of monitoring system for logging impacts, including a system of 12 rain gauges in the watershed. Fresh out of UC Berkeley, with a masters in soil science and forest watershed management, Michael Furniss was tasked with managing the monitoring system, including the weather data at Camp Six. The station had high-tech rain gauges that continuously recorded rainfall and included alter shields to lessen the winds that can blow rain and snow past the gauges. Michael's data from December 1981 reported 81.90 inches of rain—a record for California. In a three-day stretch in January 1982, in another nearby gauge, 45 inches were recorded, including a whopping 23 inches in 24 hours! For the entire weather year an astounding 254.9 inches of rain was recorded at Camp Six, a continental U.S. record that still stands. All of these records are, more than likely, safe for our lifetime.

Just down the hill, 3,200 feet lower and five miles to the west, Gasquet reported a *mere* 145 inches for the year. These data tell a variety of stories but speak to the importance of the orographic effect in the Klamath Mountains, where mountainous terrain lifts and cools the air and causes it to drop its load of moisture. Sometimes storms stall out on upper watershed bowls and drop extreme amounts of rain or snow. The January 1982 record of 23 inches in 24 hours supports the idea that a storm cell got "stuck" against the Ship Mountain ridge. During these three days, there was no significant downstream flooding in the nearby trunk streams of Jones and Hurdygurdy.

The National Weather Service (NWS) got wind of this record and contacted Michael to check its veracity. When he shared his original records from the Camp Six, Bear Basin, and Ship Mountain gauges (other nearby locations) the NWS deemed the January 1982 storm the "Mother of all storms" for the continental U.S.

The impact of El Niño rainfall is strongest in southern California. This contrasts with the Klamath Mountains (and throughout northern California and southern Oregon), where there is no clear pattern of moderate-to-strong El Niños contributing consistently to wet or dry conditions (Jong et al. 2016). However, very strong El Niños tend to favor a greater chance of heavy precipitation.

In the Klamath Mountains, like in other areas of northern California and southern Oregon, most of the precipitation and the wettest winter storms derive from atmospheric rivers. Typically, only a few large atmospheric rivers bring most of the precipitation in any given year—making them critical for water resources (Dettinger et al. 2011, Guirguis et al. 2019). However, predictability is low and water storage fluctuates widely from year to year depending on the frequency and intensity of atmospheric river events. Likewise, these storms are also capable of transporting large amounts of moisture and heavy precipitation, which can lead to widespread flooding (e.g., Ralph et al. 2006, Neiman et al. 2008, Leung and Qian 2009, Konrad and Dettinger 2017, Guirguis et al. 2019).

Recent research shows that different atmospheric states moderated by interannual (El Niño Southern Oscillation [ENSO] and Pacific Decadal Oscillation) and subseasonal (Arctic Oscillation, Pacific North American Pattern, Western Pacific Oscillation, and the Eastern Pacific Oscillation) modes of large-scale climate variability lead to the behavior of atmospheric rivers that are critical for predictability

A summer thunderstorm above Coffee Creek in the Trinity Alps Wilderness. Photo by Michael Kauffmann.

for precipitation and flood control (Guirguis et al. 2019). Wet conditions in 2016 were greater than those observed over the last 70 years of record in northern California, and were at least in part due to the persistence of atmospheric rivers characterized by a southward storm track and southerly orientation, and are associated with the negative phase of subseasonal regional teleconnection patterns (Guirguis et al. 2019). This pattern was also the case for prior heavy rainfall events. These findings are important because they emphasize that regional teleconnection modes are more influential than ENSO in modulating atmospheric river precipitation. They also demonstrate how, subseasonally, the regional modes can act to reinforce or counter the effects of ENSO (i.e., El Niño wet conditions).

Evaluation of recent trends in precipitation—and the amount of precipitation that falls as snow within the Klamath basin—show increasing variability, greater precipitation at higher elevations falling as rain instead of snow, and accelerated snowpack melt leading to earlier peak runoff and lower base flows (e.g., Butz et al. 2015). Comparing weather stations throughout the Klamath Mountains shows on average that March precipitation has increased slightly at most stations since 2011 compared to the prior 20 years (1980–2010), while in all other months the average monthly precipitation has decreased slightly since 2011 for most stations.

Using a coupled climate and land-surface model, researchers analyzed snow cover and "snow water equivalent" (SWE)—a proxy for the amount of water (snow depth x density)—for historical and projected future periods in California and southern Oregon (Flint et al. 2013). They found that between 1951 and 1980 and between 1981 and 2010, the average spatial extent of snow (snow cover) on April 1st has declined from 60–50 percent in areas exceeding 3,000 ft in elevation--with the greatest loss (SWE; to 500 mm/yr) occurring in the Klamath Mountains, compared to other snowpack regions in the Sierra and southern Cascade Mountains (Flint et al. 2013). Further analyses also support these findings and show major reductions in SWE in the Klamath Mountains from 1955–2016 (Mote et al. 2018).

Drought

Drought is a common, recurring feature in California. The 2012–2015 California drought led to the most severe moisture deficits in the last 1,200 years (Griffin and Anchukaitis 2014). The development of a persistent, high pressure system in the north Pacific, coined the "ridiculously resilient ridge" by climate scientist Daniel Swain, is considered to be the primary cause of precipitation deficits during the drought (Seager et al. 2015, Swain 2015). The ridge steered storms northward, away from California, which suggests that the persistence of the ridge was possibly reinforced by melting of arctic sea ice and sea-surface temperature anomalies (Wehner et al. 2017, Grantham 2018). Despite the record-breaking drought, paleoclimatic records have shown that even longer periods of drought—i.e., megadroughts that span decades to centuries—have occurred in California in the past (Malamud-Roam et al. 2007, Mann and Gleick 2015).

Lightning Fire Storm

Lightning is common in the Klamath Mountains but not as predictable as it is in other western mountain ranges. Summers in the Klamath Mountains are dry and experience long periods without precipitation. If lightning does come to the region it is often after weeks or months of rain-free periods. These thunderstorms often involve dry lightning—which heightens fire's chance of ignition.

On July 29, 2014, a dry lightning storm ignited 26 fires in Siskiyou County, including the Whites Fire in the Salmon River watershed. While these events were preceded by record dry conditions, they burned relatively slowly for the first ten days. On August 10th, the fires unexpectedly jumped to the North Fork of the Salmon and exploded. The next day, a large pyrocumulonimbus formed above a blow-up and sent lightning to the northwest—starting approximately 30 new wildfires in the southern Marble Mountain Wilderness (Scott Harding, pers. comm.). Ultimately, these fires merged into the July Complex and burned an estimated 134,056 acres. Another notable lightning-caused fire was the 2002 Biscuit Fire, which involved 500,000 acres in southwest Oregon. It began in July and continued to burn into November.

Not all lightning strikes result in fire. When strikes do cause fire they are often on a small scale—affecting single trees or a small patch of land. The collection of strikes, or even single-tree fires, burn a mosaic across the landscape. These complex patterns create mixed-severity disturbance across the vegetation and, through this variable disturbance, nurture biotic diversity (see Fire Chapter).

A blow-up during the 2014 Whites Fire on Tanners Peak formed a pyrocumulonimbus cloud, created its own lightning, and started 30 new wildfires. Photo by Scott Harding.

Thunderstorms and Lightning

The NOAA Weather Forecasting Office recently put together a quantitative analysis of thunderstorm activity and ground lightning flashes from 1995–2019 (Keene 2019). The analysis included the frequency of observed cloud-to-ground lightning flashes and the total observed cloud-to-ground lightning flashes per hour and week of the year for regional areas in the U.S. From 1995–2019, there were 100.4K total lightning flashes detected throughout the Klamath Mountains. Late May through early August have substantially more lightning than any other time of the year, with the majority occurring from July 9–15. Interestingly, during this period the majority of lightning strikes (>40 percent of all years) occurred between the hours of 2–8pm (>50 percent of all years between 4–6pm) (Figure 3).

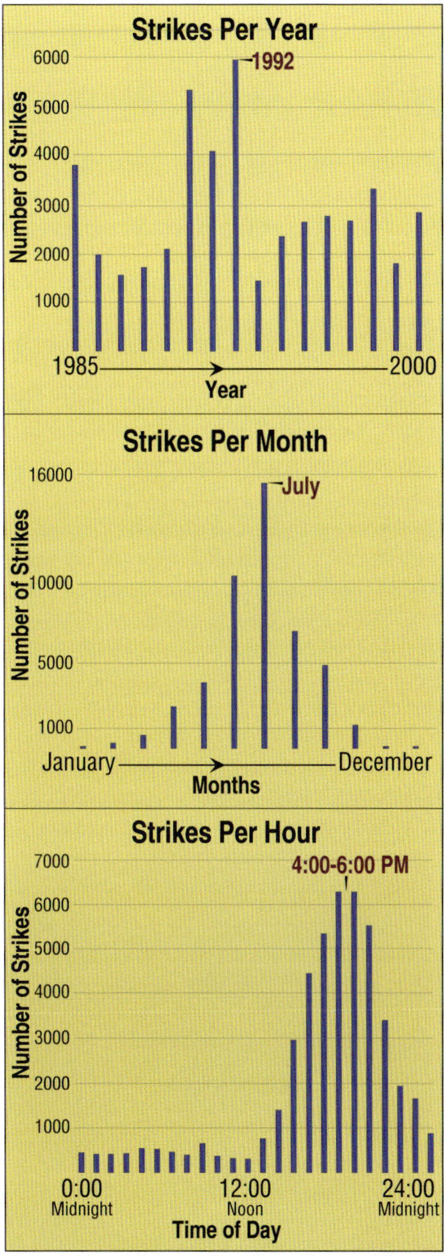

Figure 3. Klamath lightning data adapted from van Wagtendonk and Cayan (2008).

Geographic analysis within the Klamath Mountains illustrates that most of the lightning activity occurs throughout the eastern portion of the range (Figure 4). The western portion has the least amount of thunderstorm and lightning activity, which is related to the extent of the marine layer. Nighttime/morning thunderstorms tend to be drier (with less rainfall) than afternoon storms, and occur more frequently in the northwest (e.g., Kalmiopsis Wilderness) and northcentral portions of the region compared to other areas. This is also the area with strong offshore/northeast winds that are often accompanied by low relative humidity, contributing to significant fire spread during these "strong thermal trough" patterns. Perhaps not surprising, this is also the area where some of the largest lightning-ignited wildfires have occurred over the last two decades (e.g., the Chetco Bar, Klondike, Biscuit fires).

Climate

Climate

The Snow Forest and Long-term Snowpack Trends

Snow line is the lower elevational boundary on a landscape, above which winter snow accumulates and persists throughout the spring. In the Klamath Mountains, the snow line ranges from 4,500–5,000 feet depending on location. The amount of accumulated snow depends on a variety of factors including soil moisture content, air temperature, precipitation patterns, storm frequency, and even regional plant abundance. In addition, geography plays a key role in how these factors act together. Before winter storms arrive, soil moisture determines the amount of initial runoff and accumulation. Another consideration is the texture and density of the snow. In the Klamath, the western part of the range is lower in average elevation, and therefore receives less snow—but the snow here is usually denser and thus has a higher water content. As storms move to the eastern Klamath, the higher average elevation means more snow that is, on average, less dense with a lower water content. This regulates the speed at which snowpack melts and how much water is released during winter warming events.

We estimate the landmass area above 5,000 feet where deep winter snowpack normally accumulates—is ~1,490 mi^2 or 12.4% of the total province. This subalpine snow forest is uniquely shaped by snow accumulation and avalanches. It is garnished with *krummholz* conifers and hundreds of snow-adapted plant species. Snow here also provides a protective layer during extreme winter conditions for hibernating and active animals, such as amphibians, reptiles, and rodents. Snow-fed lakes, ponds, and meadow systems in the snow forest are essential habitats for a variety of rare flora and fauna including showy raillardella (*Raillardella pringlei*), Cascades frogs (*Rana cascadae*), and spinytail fairy shrimp (*Streptocephalus coloradensis*). See also Glacier Sidebar, page 130. A blanket of snowpack also keeps soil moist by covering it—ultimately influencing the onset of fire season, fire prevalence, and fire severity.

Although the snow forest is a small fraction of the Klamath Range, its massive snow storage supplies half of the annual runoff for many parts of the region (Mount 1995). As air temperature incrementally rises across spring and summer, vast amounts of meltwater are released. This annual pulse of cold water profound-

February 2006 snow line in the Marble Mountains in upper Wooley Creek as seen from Medicine Mountain. Photo by Michael Kauffmann.

ly affects physical, chemical, and biological processes governing the overall balance and health of Klamath ecosystems from the subalpine to the lowest river reaches. For example, snowmelt runoff directly contributes to the timing and magnitude of stream discharge, and to the thermal diversity of mountain lakes, wetlands, and streams throughout the range. Unique runs of Chinook salmon and steelhead have evolved with the snowmelt runoff pulse by migrating in the spring and over-summering in cooler tributaries normally fed by snowmelt deep into the summer months before spawning in the fall.

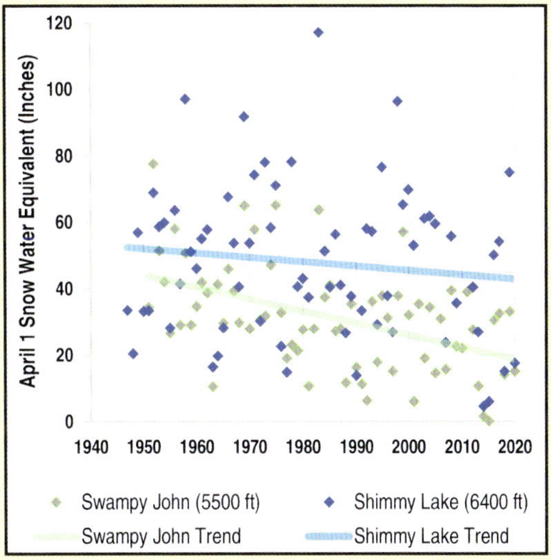

The snow-water equivalent for two data stations in the Klamath Mountains. Notice the difference in the trend lines as compared to the station's elevation over the past 70 years. Data for the graphs was extracted from the California Data Exchange Center (2021).

Snowpack has variable density, so the amount of water in the snow, or snow water equivalent (SWE), is much more important than the depth alone. Between 1950 and 2020, the average SWE on April 1st—the annual date used to define a water year—was 27.8 inches across the Trinity, Scott, and Shasta river basins (CDEC 2021). The deepest recorded April 1st snow across these stations was 228% of normal in 1983. During this record year, Shimmy Lake snow depth measured 261 inches, or 21.7 feet on April 1st. By May 1st, 1983, the depth measured 263 inches, or 21.9 feet, which stands as the snow depth record for the Klamath Mountains since the earliest measurements began in 1936. The lowest recorded April 1st snowpack for the region was 3% of normal in 2015, followed by 9% of normal in 2014. These two years are exceptional outliers, with the nearest April 1st low being 34% in 1963.

Digging into long-term trends, snowpack is declining in the Klamath Mountains, as seen in other mountain regions across the western U.S. (Mote et al. 2018). Both fall and spring are experiencing warmer temperatures, so snow is arriving later and is melting off sooner while snow lines are increasing in elevation. For example, the Swampy John (5,500 feet) station has ~50% less April 1st snow on average than when record keeping began around 70 years ago. In contrast, it appears that the nearby Shimmy Lake station, occurring 900 feet higher in elevation (6,400 feet), now has ~25% less April 1st snow on average over the same period. The loss of snowpack, and the general aridification of the west induced by rapid climate change, will continue to impact subalpine and riverine ecosystems—especially rare subalpine conifer populations, which are more prone to drought and wildfire, as well as cold-water—dependent species such as salmonids.

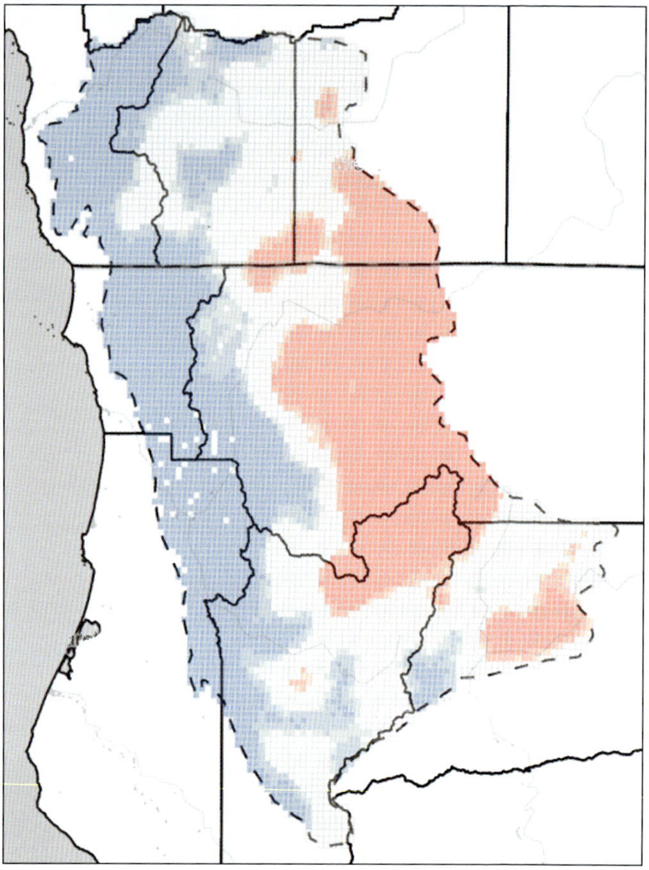

Figure 4. Hotspots of lightning activity from 1995–2019. Red indicates areas with the highest frequencies, blue indicates areas with the lowest frequencies, and white indicates a transition from low to high.

Data source: Vaisala National Lightning Detection Network (NLDN) data. Analysis and map by Noel Keene, NOAA Weather Forecasting Office, Medford, OR.

Climate in the Anthropocene

The Klamath Mountains experience a strong Mediterranean-type seasonality with a high level of regional variability related to their proximity to the Pacific Ocean and the steep elevational gradients. Climate change, the dominant force affecting the planet, is shifting seasonal and annual weather patterns, although the Mediterranean seasonality will continue into the future. For the North Coast region of California, which includes most of the Klamath Mountains, average annual maximum temperatures are likely to increase by 5–9°F through the end of the 21st century (Grantham 2018). Interior regions will experience the greatest degree of warming. Projections specifically for the Klamath Mountains indicate temperature increases of up to 3.5–4.5°F in the next 50 years (as compared to the average from 1971-1999; Garfin et al. 2014; KDNR 2016). Projections of precipitation are more variable, but annual precipitation for the Klamath and the North Coast regions are generally projected to be within or near the historical range of variability for the 21st century. However, precipitation is also expected to have greater variability, and more intense storms are expected to occur within a shorter wet season (Swain et al. 2018). A shorter wet season (i.e., earlier spring drying and later onset of fall rains) will result in prolonged dry seasons and reduced soil moisture conditions, even if annual precipitation totals remain relatively consistent. This also means greater variability in streamflow and lower/warmer low flows, which will impact biota and water resources.

Future predictions include a higher likelihood of extreme wet years and extreme dry years (drought), and the "average" rainfall year will be less common. The potential for a dramatic increase in the frequency of wet and dry year extremes has been termed "precipitation whiplash," and would create a new climate regime for the region (Swain et al. 2018). Likewise, it is expected that less precipitation will fall as snow and total snowpack will be a small fraction of its historical average. Collectively, the current and projected increase in temperature along with shorter wet seasons and loss of snowpack, will lead to larger climatic water deficits and will enhance the duration and magnitude of droughts (Diffenbaugh et al. 2015, Cayan et al. 2018). Climatic water deficits that may increase by 10-19% mid-century and 16–32% by the end of the century translate to lower soil water availability and increasing stress experienced by vegetation and ecosystems (Micheli et al. 2018).

Streamflow patterns are also expected to shift with expectations of lower water storage and summer stream base flows. Already trends have shown summer streamflow declines in northern California and southern Oregon over the last half-century (Asarian and Walker 2016). The most consistent declines in monthly streamflow are for August through November, with 73% of sites in this study showing a declining trend in September streamflow. Streamflows in the dry season are expected to decline, and peak winter flows in the winter are likely to increase (Grantham et al. 2018). Specifically, January is predicted to experience the greatest increase in flows, while May flows are predicted to have the greatest declines. From June–September and November, the range of predicted flow change bounds the historical average, with some climate models corresponding to increasing and others to declining flow. Likewise, wildfires will continue to be a major disturbance in the region. Future wildfire projections due to the climate changes described above suggest a longer fire season, an increase in wildfire frequency, and an expansion of the area susceptible to fire.

Acknowledgments

Thanks to Michael Furniss for his historical account of Camp Six. We also thank Noel Keene for his analysis of lightning data, and Matthew Kidwell, Charles Smith, and Troy Nicolini from the NOAA Weather Forecasting Offices in Eureka, CA and Medford, OR for their additional expertise. Funding support for this chapter was provided to Rosemary Sherriff by the National Science Foundation (BCS-1853903). Thanks to Jim Agee and Dylan Neubauer for reviewing the chapter.

Literature Cited and Suggested Reading

Asarian J.E., Walker J.D. 2016. Long-Term Trends in Streamflow and Precipitation in Northwest California and Southwest Oregon, 1953-2012. *JAWRA* 52:241-261.

Briles, C., C. Whitlock, and P.J. Bartlein. 2008. Regional and local controls on postglacial vegetation and fire in the Siskiyou Mountains of northern California and southern Oregon. *Palaeogeography, Palaeoclimatology, Palaeoecology* 265: 159-169.

Briles, C., C. Whitlock, and P.J. Bartlein. 2011. Holocene forest development and maintenance on different substrates in the Klamath Mountains, USA. *Ecology* 92: 590-601.

Butz, R.J., S. Sawyer and H. Safford. 2015. A summary of current trends and probable future trends in climate and climate-driven processes for the Six Rivers National Forest and surrounding lands. Internal USFS document.

Cayan, Daniel R., David W. Pierce, Julie F. Kalansky. 2018. Climate, Drought, and Sea-level Rise Scenarios for the Fourth California Climate Assessment. California's Fourth Climate Change Assessment, California Energy Commission. Report number: CCCA4-CEC-2018-006.

CDEC. 2021. California Data Exchange Center. Regional Station Group Snow Course Data. https://cdec.water.ca.gov/cgi-progs/prevsnow/COURSES. Accessed 4/3/2021.

Daniels, M.L., R.S. Anderson, and C. Whitlock. 2005. Vegetation and fire history since the Late Pleistocene from the Trinity Mountains, USA. *Holocene* 15: 1062-1071.

Dettinger, M.D., F.M. Ralph, T. Das, P.J. Neiman, D.R. Cayan. 2011. Atmospheric rivers, floods, and the water resources of California. *Water* 3:445–478.

Diffenbaugh N.S., D.L. Swain, D. Touma. 2015. Anthropogenic warming has increased drought risk in California. Proceedings of the National Academy of Sciences 112:3931-3936.

Estes, B.L., E.E. Knapp, C.N. Skinner, J.D. Miller, and H.K. Preisler. 2017. Factors influencing fire severity under moderate burning conditions in the Klamath Mountains, northern California, USA. *Ecosphere* 8(5):e01794.10.1002/ecs2.1794

Flint L.E., A.L. Flint, J.H. Thorne, and R. Boynton. 2013. Fine-scale hydrologic modeling for regional landscape applications: the California Basin Characterization Model development and performance. *Ecological Processes* 2:25.

Garfin, G., et al., 2014: Ch. 20: Southwest. Climate Change Impacts in the United States: The Third National Climate Assessment, J. M. Melillo, Terese (T.C.) Richmond, and G. W. Yohe, Eds., U.S. Global Change Research Program, 462-486. doi:10.7930/J08G8HMN.

Garwood, J.M., A.G. Fountain, K.T. Lindke, M.G. van Hattem, and H.J. Basagic. 2020. 20th century retreat and recent drought accelerated extinction of mountain glaciers and perennial snowfields in the Trinity Alps, California. *Northwest Science* 94: 44-61.

Grantham, T. 2018. "North Coast Region Report" in California's Fourth Climate Change Assessment. Publication number: SUM-CCC4A-2018-001.

Grantham T.E.W., D.M. Carlisle, G.J. McCabe, J.K. Howard. 2018. Sensitivity of streamflow to climate change in California. *Climatic Change*. https://doi.org/10.1007/s10584-018-2244-9

Griffin D., and K.J. Anchukaitis. 2014. How unusual is the 2012 – 2014 California drought? *Geophysical Research Letters* 41:9017-9023.

Guan B, et al. 2013. The 2010/2011 snow season in California's Sierra Nevada: Role of atmospheric rivers and modes of large-scale variability. *Water Resources Research* 49: 6731-6743.

Guirguis, K., et al. 2019. Atmospheric rivers impacting Northern California and their modulation by a variable climate. *Climate Dynamics* 52: 6569–6583.

Guyton, B. 1998. *Glaciers of California*. University of California Press, Berkeley.
Jong, B., M. Ting, and R. Seager. 2016. El Niño's impact on California precipitation: seasonality, regionality, and El Niño intensity. *Environmental Research Letters* 11: 054021.
Karuk Tribe Department of Natural Resources. 2016. Karuk Tribe Climate Change Vulnerability Assessment.
Keene, N. 2019. Lightning Climatology. NOAA Weather Forecasting Office, Medford, OR. Weblink. Accessed 07/29/2020.
Konrad C.P., M.D. Dettinger. 2017. Flood runoff in relation to water vapor transport by atmospheric rivers over the western United States, 1949–2015. *Geophysical Research Letters*. https://doi.org/10.1002/2017GL075399
Leung L.R., Y. Qian. 2009. Atmospheric rivers induced heavy precipitation and flooding in the western U.S. simulated by the WRF regional climate model. *Geophysical Research Letters* 36:L03820.
Malamud-Roam F., Dettinger M., Ingram L.B., Hughes M.K., Florsheim J.L. 2007. Holocene Climates and Connections between the San Francisco Bay Estuary and its Watershed: A Review. *Journal of San Francisco Estuary and Watershed Science* 5:1-28.
Mann M.E., and P.H. Gleick. 2015. Climate change and California drought in the 21st century. Proceedings of the National Academy of Sciences 112:3858-3859.
Micheli E., C. Dodge, L. Flint, T. Comendant. 2018. Climate and Natural Resource Analyses and Planning for the North Coast Resource Partnership: a technical memorandum summarizing data products.
Mohr, J.A., C. Whitlock, C.N. Skinner. 2000. Postglacial vegetation and fire history, eastern Klamath Mountains, California, USA. Holocene 10: 587-601.
Mote, P.W., S. Li, D.P. Lettenmaier, M. Xiao, and R. Engel. 2018. Dramatic declines in snowpack in the western US. Climate and Atmospheric Science 1, 2.
Mount, J. F. 1995. *California rivers and streams*. University of California Press, Berkeley and Los Angeles, CA.
Neiman P.J., F.M. Ralph, G.A. Wick, J.D. Lundquist, and M.D. Dettinger. 2008. Meteorological characteristics and overland precipitation impacts of atmospheric rivers affecting the West Coast of North America based on eight years of SSM/I satellite observations. Journal of Hydrometeorology 9:22–47.
Ralph F.M., P.J. Neiman, G.A. Wick, S.I. Gutman, M.D. Dettinger, D.R. Cayan, and A.B. White. 2006. Flooding on California's Russian River: Role of atmospheric rivers. Geophysical Research Letters 33:L13801.
Sawyer, J.O. 2006. *Northwest California: a natural history*. University of California Press.
Seager R, Hoerling M, Schubert S, Wang H, Lyon B, Kumar A, Nakamura J, Henderson N. 2015. Causes of the 2011-14 California drought. *Journal of Climate* 28:6997-7024.
Sharp, R. 1960. Pleistocene glaciation in the Trinity Alps of northern California. *American Journal of Science* 258: 305-340.
Swain D.L., B. Langenbrunner, J.D. Neelin, and A. Hall. 2018. Increasing precipitation volatility in twenty-first century California. Nature Climate Change Early View.
Swain D.L. 2015. A tale of two California droughts: Lessons amidst record warmth and dryness in a region of complex physical and human geography, pp. 9999-10003.
van Wagtendonk, J.W., D.R. Cayan. 2008. Temporal and Spatial Distribution of Lightning Strikes in California in Relation to Large-Scale Weather Patterns. *Fire Ecology* 4, 34–56.
Wanket, J. 2002. Late quaternary vegetation and climate of the Klamath Mountains. PhD thesis. University of California, Berkeley.
Wehner M.F., Arnold J.R., Knutson T., Kunkel K.E., LeGrande A.N. 2017. Droughts, floods, and wildfires, Climate Science Special Report: Fourth National Climate Assessment, Volume I. Washington, DC.

Chapter 5: Mountains of Water

Tim Palmer and Justin Garwood

Rafting the Rogue River, Oregon.
Photo by Tim Palmer.

Introduction

Water is a key abiotic factor contributing to the grandeur and biodiversity of the Klamath Mountains. Along with geology and tectonic uplift, water has had a profound influence in shaping the region over eons—with powerful ancient rivers and massive subalpine glaciers of the Pleistocene Epoch that ended over 10,000 years ago. Today, the western portion of the range receives rainfall topping 100 inches in wet years with total yearly precipitation here among the highest in the U. S. Other areas of the eastern Klamath Mountains receive so little rainfall they resemble the high-desert landscapes of the Great Basin province to the east. The complex mountainous topography absorbs and sheds immense amounts of precipitation from rain and snow—metering it out through dynamic surface and groundwater pathways. Water flows from ancient, glacially shaped ridges where deep winter snowpack melts into hanging cirque basins with sponging fen meadows, lakes, ponds, and springs. Thousands of tiny streams also emerge from mountain slopes, maintaining massive, dendritic drainage networks that accumulate in the Klamath, Sacramento, Rogue, and other major rivers. From deep, polished bedrock canyons to broad alluvial valleys, Klamath riverscapes are the scenic byproduct of the massive erosive power of water.

Water also has connected the Klamath Mountains to other regions over the eons resulting in unique faunal assemblages in each Klamath watershed. The Rogue, Klamath, and Pit rivers represent three of the six rivers in western North America that flow through water gaps where ancient rivers cut through mountain ridges—in these cases through the Cascade Range—before flowing into the Klamath Mountains en route to the Pacific. This chapter highlights the diversity of lotic (flowing) and lentic (still) waters along with the intricate ecosystems created and maintained by water throughout the Klamath Mountains.

Streams and Rivers

Water in the Klamath Mountains creates a network of streams and rivers of extraordinary importance and natural wealth unlike any other in America. From the Rogue River in the north to the Trinity River in the south, this suite of waterways is one of the premier assemblages of relatively undeveloped rivers nationwide. Most of these have been designated in the National Wild and Scenic Rivers System for a continuous north–south expanse of 260 miles (418 km)—the greatest such length anywhere. Several of the West Coast's few remaining natural river mouths are found here and lack the harbors, levees, and riprap that armor other major river mouths between Mexico and Canada. Streams flow from some of the largest wilderness areas on the West Coast, forming a region-wide matrix of aquatic systems revered for their intricate biological qualities, their appealing recreational values, and their important contributions to local life and economies.

The Annual Pulse of Klamath Drainages

Rivers in the Klamath Mountains receive precipitation from a varied mix of rain and snow. Runoff can occur in quick response to storms or more slowly through snowmelt and intricate groundwater systems. Individual sources of runoff are separated over space and time, with each type and proportion governed largely by the physiographic components of each watershed. Geology, tectonics, soil

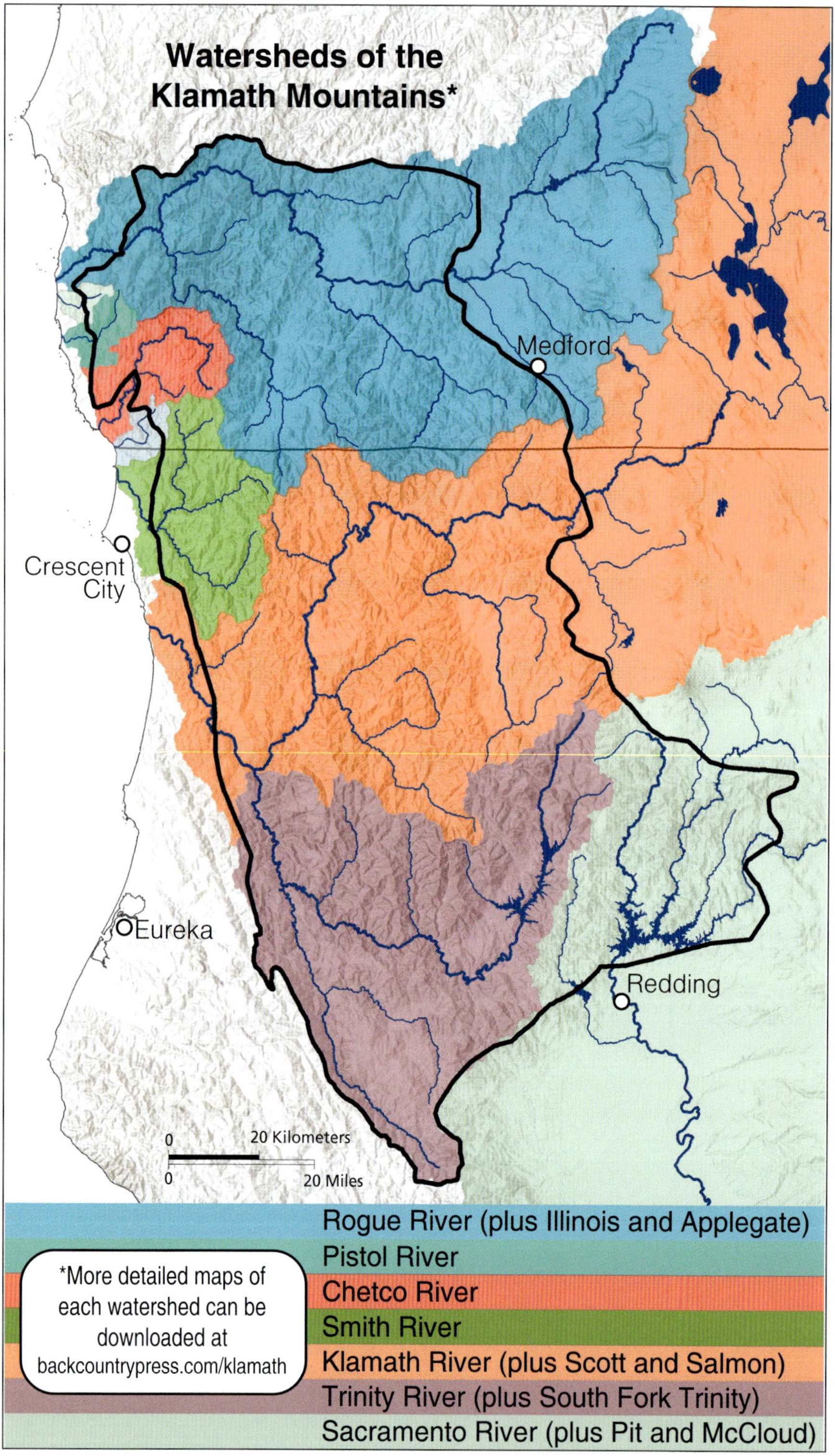
Watersheds of the Klamath Mountains*
Medford
Crescent City
Eureka
Redding
0
20 Kilometers
0
20 Miles
Rogue River (plus Illinois and Applegate)
Pistol River
Chetco River
Smith River
Klamath River (plus Scott and Salmon)
Trinity River (plus South Fork Trinity)
Sacramento River (plus Pit and McCloud)
*More detailed maps of each watershed can be downloaded at backcountrypress.com/klamath

development, ridge elevations, forest cover, and climate all influence the source, amount, and timing of runoff in a watershed throughout a water year—the 12-month period spanning annual wet and dry seasons between October 1st through September 30th.

The Smith River (719 mi²; 1,862 km²) and the Salmon River (751 mi²; 1,945 km²) have similar watershed areas and are separated by only 37 miles (60 km) but each behaves dramatically different throughout a typical water year. The contrasts in precipitation and runoff regimes in these two systems exemplify distinctions that occur in watersheds throughout the Klamath Mountains.

The Smith drains the high-elevation flanks of the western Siskiyou Mountains, which are uniquely positioned in the maritime climate just east of the Pacific Ocean. This orographic effect generates an unrivaled rain machine, with the Siskiyou Mountains receiving the brunt of strong winter Pacific storms—some hauntingly classified as "atmospheric rivers." In fact, the Smith-Chetco region is one of the wettest spots in the conterminous U.S., averaging over 100 inches (254 cm) annually (see page 96–97, Climate chapter). In contrast, the Salmon River is farther inland, receiving around 49 inches (124 cm) annually, after winter Pacific storms wring out a substantial amount of water before they arrive. As a result, the rain-soaked Smith produces over twice the average annual total discharge as the Salmon. Due to the Smith's shallow soils on serpentine geology, runoff is rapid, and the river responds with sharp rise-and-fall sequences after each storm event throughout the rainy season as seen in the 2018 hydrograph in Figure 1. The Salmon, however, drains some

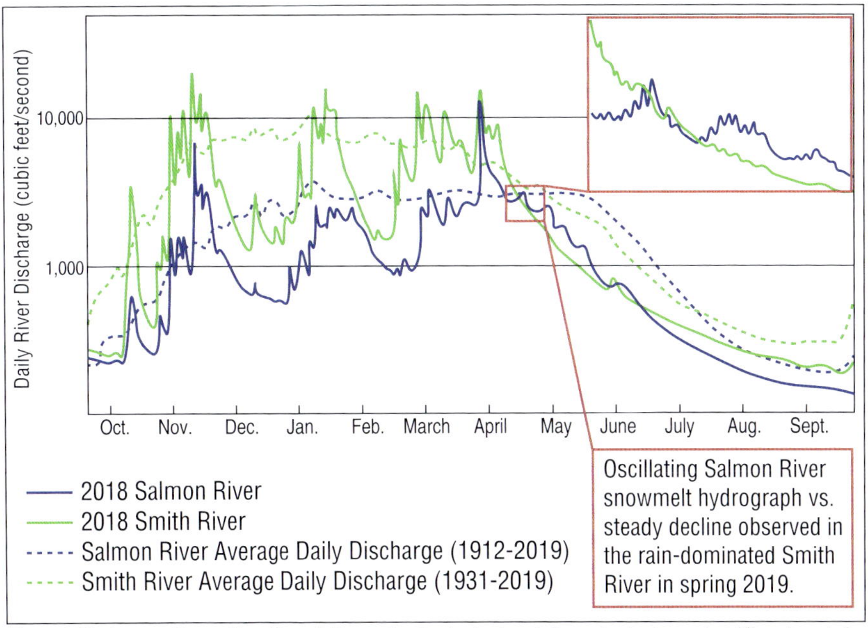

Figure 1. Discharge patterns of two Klamath Range river basins having similar watershed areas but different precipitation- and runoff-associated regimes. The Salmon River watershed (1,945 km²) receives rain and substantial snowmelt from the highest subalpine portions of the Klamath Mountains. In contrast, the Smith River watershed (1,862 km²) is located along the coast and drains the rain-soaked Siskiyou sub-range. The Smith River basin is among the rainiest locations in the conterminous United States. Note logarithmic scale. USGS Data: Jed Smith (Smith); Somes Bar (Salmon).

of the highest summits and most dramatic subalpine topography in the Salmon Range and northern Trinity Alps. Unlike the rain-dominated Smith, most winter precipitation in the high Salmon River drainage above ~4,500 ft (1,524 m) falls as snow, which can stockpile and reduces the magnitude of winter storm runoff as seen in the 2018 hydrograph in Figure 1. Many subalpine portions of Klamath watersheds can store massive amounts of snow and act as giant frozen reservoirs.

Basins that store substantial winter snowpack release a massive amount of cold meltwater from March–June with an April or May peak depending on the amount of annual winter snowpack and spring weather. After peak spring runoff occurs, river flows decline steadily with diurnal oscillations in discharge caused by variations in daily air temperature that influence snowmelt—as seen in the inset of Figure 1. Many widespread species, such as the foothill yellow-legged frog (*Rana boylii*) and salmonid fishes, have adapted their reproductive cycle to snow-fed rivers. The Klamath, Salmon, and Trinity rivers each have distinct runs of summer steelhead (*Oncorhynchus mykiss*) and spring Chinook salmon (*O. tshawytscha*) species that have evolved specifically to take advantage of snow-fed streams. These enigmatic fish migrate into the upper reaches and tributaries of these streams before low-water migration barriers develop. Here they discreetly over-summer in areas of exceptional water quality until they spawn in the early fall.

Watersheds Like No Others

Rogue River

Average flow: 6,622 cfs (163 cms)
Watershed: 5,159 miles² (13,361 km²)

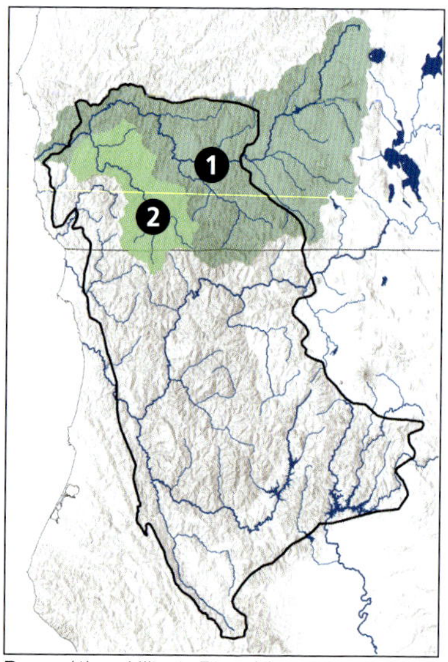

Rogue (1) and Illinois River (2) watersheds.

The Rogue is sourced at upper elevations of the Cascade Range with springs fed by America's deepest freshwater body, Crater Lake. A legendary wild river and deep in western lore, the Rogue arrives in the Klamath Mountains already laden with snowmelt from the higher and snow-capped Cascade Range. Then it courses through the 59-mile (95 km) long Rogue Valley and Interstate 5 corridor of towns and rural settlements, before reaching a magnificent roadless canyon. It flows through these wildlands for 34 miles (55 km) before emerging into its final run of steep, wooded terrain of the Klamath Mountains for another 32 miles (51 km) and meeting the ocean at Gold Beach.

The river remains one of the most productive salmon and steelhead fisheries on the West Coast, and the fall Chinook salmon season lures thousands of anglers to Gold Beach and upriver sites every year. In addition to the Klamath Basin, the lower Rogue is also one of few rivers in the Klamath Mountains with a spawning population of green sturgeon (*Acipenser medirostris*), an ancient species that migrates upriver from the ocean. The entire length of the Rogue through its Pacific-bound canyon is a key wildlife corridor.

Stream and River Ecosystems

From miniature pebblesnails (*Fluminicola* spp.), found where springs emerge from the Earth, to giant green sturgeon circling in deep, bedrock-lined river pools, the streams and rivers of the Klamath Mountains represent a complex network of aquatic and riparian plant and animal communities. Ecological communities vary widely across Klamath streams, both along their length and laterally along their banks. Some of the most productive forests in the world thrive on the floodplains of these streams and rivers, where their water supply depends on stream-related groundwater.

Stream ecosystems are structured by abiotic factors including stream size, local geomorphic processes dictating the shape of the stream (e.g., steep, bedrock gorges vs. broad, alluvial floodplains), and sediment supply with its resulting habitat-forming characteristics. Local biological communities have many species with specialized niches—especially macroinvertebrates—resulting in stream ecosystems that vary from headwater springs down to broad, alluvial rivers. However, some species can be found across the whole network—such as the American dipper (*Cinclus mexicanus*)—a small, amphibious bird that forages on fishes, amphibians, and invertebrates in habitats ranging from broad rivers up to headwater glacial lakes.

Streams of the Klamath provide essential migration and dispersal corridors for aquatic and terrestrial fauna along narrow, moisture-rich riparian zones that bisect a much drier landscape. Although anadromous salmon and steelhead runs are now a remnant of their historical size, Klamath Mountain rivers hold some of the finest runs occurring this far south on the continent. After spawning and dying, these fish transfer abundant, ocean-derived nutrients into inland stream and riparian ecosystems benefiting countless plant and animal species. Large mammals, including black bears (*Ursus americanus*) and river otters (*Lontra canadensis*), also focus seasonally on adult salmon and steelhead prey.

These same streams can also act as major dispersal barriers for other fauna, such as terrestrial snails, fishes, and plethodontid salamanders. Isolation leads to restricted gene flow and local diversification. For example, anadromous salmonids are widespread in Klamath streams but are eventually restricted by increased stream gradients or natural landform barriers. Many genetically isolated populations of rainbow trout and coastal cutthroat trout (*Oncorhynchus clarkii*) have been discovered above substantial barriers—revealing an intricate relationship between local fish distributions and past geomorphic events.

An adult American dipper feeds a salmon fry to a fledgling. Photo by Roberta Olenck, www.neverspook.com.

Water

Lost Creek Dam, at the river's interface between the Cascade foothills and the Klamath Mountains, has buried spawning habitat of spring Chinook salmon and altered flows since the 1970s by reversing the natural hydrograph with lower releases in winter and spring and higher flows in summer. Other impediments to the Rogue downstream come from urban growth with municipal discharges, clearcut logging, the legacy of gold mining, and the cumulative effects of settlement leading to higher water temperatures.

Dramatic improvements were made between 2008 and 2010 with the removal of three antiquated dams; Gold Ray, Gold Hill, and Savage Rapids. One of the path-breaking dam removal efforts nationwide, this allowed salmon to pass more freely and to reach upstream spawning beds as far as Bear Creek in Ashland, which had not seen a robust Chinook salmon run in many decades. Other programs are underway to plant trees on the river and tributary floodplains with the intent to lower the troublesomely high water temperatures that have resulted from withdrawals, warm return wastewater, urbanization, farming, and logging.

A recreational hotspot, the wild reach of the Rogue below Grave Creek is the most popular multi-day whitewater river trip on the West Coast. The Bureau of Land Management issues permits and limits use to 140 boaters daily in the wild and roadless reach between Grave Creek and Foster Bar. This was one of the pioneering permitting systems established on recreational rivers of the West. The entire free-flowing reach of 154 miles (248 km) from Lost Creek Dam to the Pacific at Gold Beach makes an excellent extended river journey—one of the longest possible on the West Coast.

The 85 miles (137 km) of the Rogue from the Applegate River confluence below Grants Pass to Lobster Creek above Gold Beach was among the original eight rivers designated in the National Wild and Scenic Rivers System in 1968. Headwater reaches in the Cascade Mountains were added in 1988, and in 2019 a bill was under consideration for designation of sixteen additional tributaries to the lower canyon. These are important as cold-water refuges that would be diminished if plans to log in those basins are approved.

Illinois River

Average flow: 1,257 cfs (36 cms)
Watershed: 380 miles^2 (985 km^2)

The Illinois is the Rogue's largest tributary and flows entirely from the Siskiyou Mountains, a sub-range of the Klamath Mountains. Like the Rogue, the Illinois begins in remote terrain and forests, traverses rural settlements and towns in its mid-reaches, then finishes with a tumultuous descent featuring intense whitewater in one of the wildest canyons on the West Coast.

The Illinois harbors important salmon and steelhead runs, and tributaries—including Lawson, Silver, Indigo, Briggs, and Klondike creeks are among the Rogue basin's most important for imperiled coho salmon (*Oncorhynchus kisutch*). Water quality is excellent, though further development in the Illinois Valley of southwest Oregon could imperil this in the future. Highly prone to wildfires, much of the Illinois basin has burned during mega-fires in 2002 and 2017.

The most immediate threat to the Illinois basin comes from proposals to stripmine for nickel at the headwaters of tributary Rough and Ready Creek. A congressionally legislated permanent withdrawal of areas from new mining claims is crucial to halting this imminent threat to the Illinois' pristine water quality and its important salmon and steelhead runs.

The Illinois' lower canyon is legendary among whitewater paddlers. The 31-mile (50 km) run takes two or three days and includes the Class V Green Wall Rapid as the big challenge. The river is run in winter to late spring, with caution to approaching storms; the Illinois can rise insanely from 1,000–20,000 ft^3 per second in a day or so. A hiking trail traverses the lower canyon country, but most of its mileage probes tributary basins and mounts high ridgelines.

Chetco River

Average flow: 2,244 cfs (64 cms)
Watershed: 353 miles^2 (914 km^2)

Little known outside southwest Oregon, the Chetco River is sourced, like the Illinois, in the heights of the Siskiyou sub-range of the Klamath Mountains. Instead of flowing directly westward from high peaks, the Chetco takes a circuitous course north, west, south, and finally conclusively west to the Pacific at Brookings. This is the wildest West Coast river south of the Olympic Mountains in Washington, with the upper three-quarters almost entirely within Rogue River–Siskiyou National Forest and mostly in the Kalmiopsis Wilderness. There are no dams on the river or its primary tributaries, and no bridges or land development occurs until the lower 10 miles (16 km).

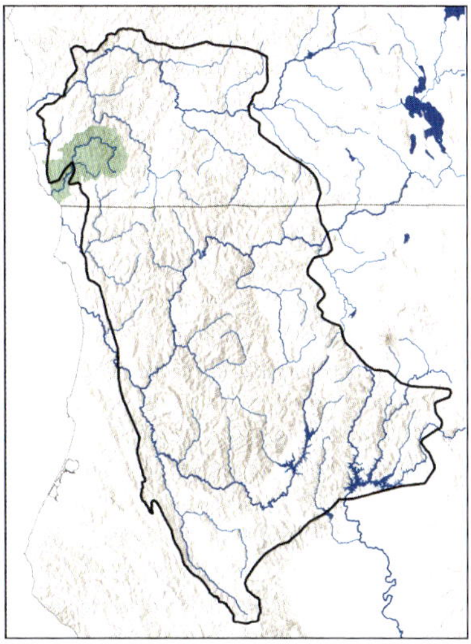

Chetco River Watershed.

Crystal clear waters are atributed to little watershed disturbance, other than the 2002 Biscuit Fire (pictured at right), and to a bedrock- and a boulder-strewn riverbed, followed by vast cobble bars along the lower third of the river's mileage. In Chetco Gorge, 3 miles (5 km) above the South Fork, showy chert boulders were formed undersea from skeletons of microscopic radiolaria that accumulated over geologic time as sedimentary strata—much like limestone does—now striped in Earthy tones of red, white, and gray on impressive boulders. A hydrologic phenomenon, the small watershed can briefly produce as much water as a strong-flowing

Looking into the Chetco River watershed from high in the Kalmiopsis Wilderness. Photo by Michael Kauffmann.

Klamath; winter discharge is typically 50 times summer flows and can spike from 1,000–60,000 ft^3 per second in a few days.

The upper basin is a center of plant endemism, with rare wildflowers and forbs resulting from ophiolite formations originally formed undersea as a former oceanic crust overlying peridotite, or upper mantle rock. The Kalmiopsis Wilderness is named after *Kalmiopsis leachiana*, a rare and showy flower. The nation's northernmost naturally occurring redwoods are found along trails at Loeb State Park, 7 miles (11 km) east of Highway 101.

The Chetco has strong salmon and steelhead populations relative to other coastal streams and remains a popular sport fishery with storied Chinook salmon still reaching 60 pounds (27 kg). Coho salmon also spawn in tributaries. Below the South Fork, the lower river offers four recreation access points that are popular for gravel bar camping, fishing, and swimming during a hot summer when the river's clear waters make for delightful refreshment.

Smith River

Average flow: 3,746 cfs (96 cms)
Watershed: 719 miles^2 (1,862 km^2)

The Smith is the most undeveloped and protected river remaining in California, and the largest completely undammed river from source to sea in the state. Much of the watershed features serpentine formations of the Josephine Ophiolite (see Geology chapter), and large areas are suited to rare plant life that thrives in harsh settings and in soils containing metals and minerals that are not conducive to typical forest growth. These conditions also allow carnivorous plants to survive on infertile soils by capturing and digesting insects. Similar to the Chetco River, the sparsely vegetated and thin, rocky soils of the Smith shed water from winter storms faster than any other watersheds in the Klamath. However, the Smith is twice the size of the Chetco and commonly rises insanely fast—from 5,000–80,000 cfs within 24 hours. In less than three days after peaking the Smith will be flowing at under 10,000 cfs. These extreme discharge fluctuations are unmatched throughout the Klamath Mountains and highlight the natural pulse of undammed rivers.

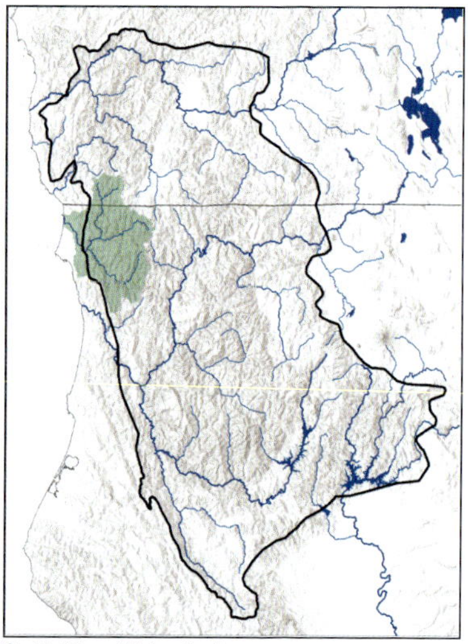

Smith River Watershed.

Waterfalls in winter along the North Fork Smith River. Photo by Tim Palmer.

Highway 199 parallels much of the Middle Fork and its rugged, forested canyon. The larger South Fork comes from higher elevations of the Siskiyou Wilderness where snowmelt lingers and contributes to the basin's highest flows through spring. The 28-mile (45 km) North Fork is wilder, with no road paralleling its length and just one remote crossing, which is downstream from the Oregon border. This free-flowing gem begins in Oregon's Kalmiopsis Wilderness.

The Smith River is clearly one of the West Coast's finest winter steelhead and salmon strongholds. It also maintains California's largest coastal cutthroat trout population—dwarfing many other populations occurring to the north where the species is more widespread. Green sturgeon sightings have become more common in deep pools of the lower river in recent years and a small juvenile was detected in 2022 by Jolyon Walkley of CDFW 2022—they are reproducing in the basin.

Much of the Smith's quality is attributed to nearly all the watershed being in public ownership as the Smith River National Recreation Area, with legislated priorities on land and water protection. In addition, all of the 18-mile (29 km) mainstem and 320 miles (515 km) of tributaries were designated as National Wild and Scenic Rivers in 1981. This marked the first time that the streams of virtually an entire river basin were together designated in the National Wild and Scenic Rivers System as part of a watershed-wide approach.

Klamath River

Average flow: 17,010 cfs (482 cms)
Watershed: 15,689 miles^2 (40,630 km^2)

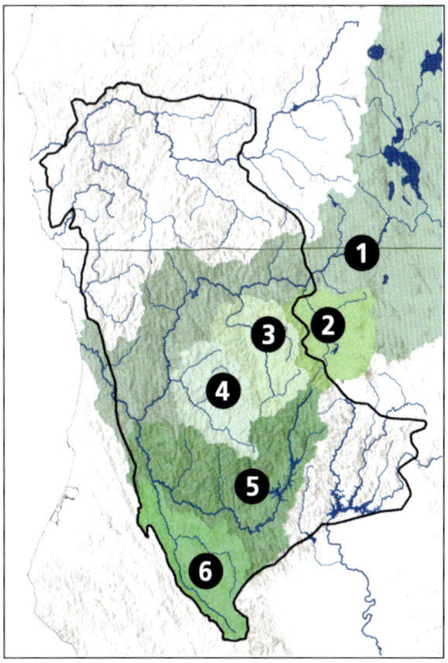

The Klamath River watershed includes the Klamath (1), Shasta (2), Scott (3), Salmon (4), Trinity (5), and South Fork Trinity (6) rivers.

The Klamath River headwaters reach even farther inland than those of the Rogue—to the much drier east side of the Cascades. In distant geologic eras, the proto-Klamath drew water from much farther eastward than it does today, including the Snake River basin of southern Idaho. With today's sources north of the Modoc Plateau, the upper river gathers spring-fed tributaries across the volcanic landscape, transects the Cascade Range in a great gorge entrenched between Crater Lake and Mount Shasta, then zigzags through the Klamath Mountains. Epic in the riverscape of America, the Klamath is the third largest river on the West Coast south of Canada, behind only the Columbia and Sacramento. Its free-flowing 193 miles from Iron Gate Dam to the Pacific constitute one of the longest unimpounded mileages on the West Coast, and the river has historically been one of the highest-producing salmon fisheries in the continental U.S.

While the Klamath lacks long reaches of wild passage devoid of roads or development, both are typically back from the banks or on high slopes above the water, leaving the river with the feeling and the look of wildness. Ironically, because it lacks the true wilderness found along the neighboring Rogue, the Klamath sees far less boating and recreational use. Nonetheless, its free-flowing length makes an epic extended trip of two weeks, usually split between upper and lower portions separated by the unrunnable Ishi Pishi Falls near Somes Bar, which requires road-portage by vehicle for through-boaters. This is California's premier long-distance river trip and one of the great extended journeys of America for rafters, kayakers, and capable canoeists. Sections of the river downstream from Happy Camp are closed annually on certain dates for the Karuk Tribe's World Renewal Ceremony and other spiritual ceremonies.

For all its free-flowing length, the Klamath also has six dams in its upper reaches. Two of these dams provide support irrigation for the Bureau of Reclamation's Klamath Project, while the other four dams are exclusively for hydropower generation. These dams have blocked anadromous fish from hundreds of miles of habitat and imperiled native suckers that are important to local Tribes. The dams and their reservoirs also significantly warm the river's water and contribute to massive outbreaks of blue-green algae (including *Microcystis aeruginosa* and *Aphanizomenon flos-aquae*), which are toxic to swimmers, dogs, and other life. Although the Klamath's clear-water tributaries dilute the algae downstream, toxic algae health warnings have sometimes been posted all the way to the Pacific Ocean.

Recognizing these and other problems, including the economic viability of modernizing its Klamath Hydroelectric Project, the owner, PacifiCorp, has agreed to dismantle its four troublesome dams on the Klamath. Dam removal is anticipated to begin in 2023, although the regulatory process and the complexity of the removal could cause short-term delays. Once completed, this will be the world's largest dam-removal project. Further work addressing irrigation withdrawals, return flows, and watershed restoration will also be necessary to more fully restore the Klamath.

An alluvial point bar along the Salmon River near Somes Bar. Photo by Michael Kauffmann.

Scott River

Average flow: 627 cfs (18 cms)
Watershed: 806 miles^2 (2,088 km^2)

A major tributary to the middle Klamath River in California, the Scott's headwaters lie in high terrain of the Marble Mountain, Scott, and Salmon sub-ranges of the Klamath Mountains. Runoff then courses through abandoned gold-mine dredging in the upper main stem, and the irrigated farmlands of the Scott Valley, before re-entering the mountains and plummeting through a steep whitewater gorge and into the Klamath River about 30 miles (48 km) west of Interstate 5. The Scott River holds the Klamath's key coho salmon spawning grounds, though difficult conditions have only grown worse over time with increased diversion for irrigation in the Scott Valley (see Beaver sidebar in this chapter). Along with the Shasta River, which joins the Klamath upstream, the Scott offers one of the basin's most important restoration opportunities.

Salmon River

Average flow: 1,792 cfs (51 cms)
Watershed: 751 miles^2 (1,945 km^2)

The second largest tributary to the Klamath River after the Trinity, the Salmon is fed by high-elevation headwaters of its North Fork and South Fork. Flowing from the heart of the Marble Mountain Wilderness Area, the 37-mile long North Fork is slightly smaller than the South Fork in watershed size and flow (484 cfs average flow). The South Fork drains the highest peaks of the northern Trinity Alps and Salmon Mountains, flowing 39 miles to join the North Fork in the appropriately named Forks of Salmon. With an average flow of 677 cfs, the South Fork typically holds its runoff later into summer due to its greater area of snow-covered mountains. The vivid green water and deep transparent pools of the Salmon support one of the Klamath watershed's strongest remaining runs of wild spring Chinook salmon. However, these rare fish are in peril and populations have declined significantly in recent years.

A premier whitewater rafting and kayaking run, the Salmon's entire main stem is included in the National Wild and Scenic Rivers System, as is the North Fork, the South Fork below Cecilville, and Wooley Creek. The continuous run of the combined North Fork and main stem, along with the Klamath from the Salmon confluence to the ocean, totals 100 miles and is the second longest waterway designated as Wild and Scenic from source to sea.

Trinity River

Average flow: 4,866 cfs (138 cms)
Watershed: 2,969 miles^2 (7,690 km^2)

This largest Klamath tributary is a major river by itself, draining 20 percent of the Klamath's massive watershed. The Trinity pulls much of its water from the Klamath sub-range called the Trinity Alps. A tributary, Grizzly Creek, flows from the north face of Mount Thompson and is fed by the only remnant glacier remaining in the greater coastal mountain complex south of Washington's Olympic Peninsula.

The upper river is blocked by the massive 400-ft-tall (122 m) Trinity Dam, which forms Trinity Reservoir and the Lewiston Dam and lake directly below. The

The Last Glacier in the Klamath Mountains

Glaciers are frozen reservoirs, accumulating snowpack and ice during the winter, and releasing meltwater throughout the summer melt season. This annual process not only maintains glaciers but also unique riparian and aquatic ecosystems at elevations far above the headwaters of nearby subalpine streams. The two highest-elevation perennial surface waters in the Klamath Range begin as glacier and snow meltwater above 8,000 ft (2,450 m), cascading down north-facing slopes of the Trinity Alps crest. While these streams originate within one kilometer of each other, their respective watersheds are separated by 285 stream kilometers. The Salmon Glacier stream drains to the South Fork of the Salmon River while the Grizzly Glacier stream feeds the North Fork of the Trinity River. Both meet where the Trinity River enters the Klamath River at Weitchpec.

Until recently, Grizzly and Salmon glaciers of the Trinity Alps have represented the lowest and westernmost glacier ice in California, persisting 1,700 feet below all other California glaciers—only found on Mount Shasta and across the southern Sierra Nevada. Here, in the coldest and most rugged niche of the Klamath Mountains, snow accumulation far exceeds that in the surrounding landscape. Unique topographic properties of Thompson and Caesar Cap peaks produce exceptional snowpack necessary to maintain glaciers at these low elevations. Conducive features include northeast aspects, tall headwalls, snow avalanching, and windblown snow recruitment from the summits. Furthermore, the proximity of this high cluster of peaks to the Pacific Ocean produces unrivaled regional precipitation levels.

The massive glaciers of the Pleistocene Epoch (spanning approximately ten thousand to two million ybp, see Geology Chapter) shaped the rugged pyramid summits, excavated hundreds of bedrock-lined cirque lakes, and carved dozens of long-shadowed, U-shaped canyons found across the Klamath Range today. During the Little Ice Age (LIA), a ~600-year period lasting from around 1300 to

Aerial photo of Grizzly and Salmon glaciers taken around 10 a.m. on September 19, 2013. Notice the glaciers tucked below steep northeast faces of Thompson Peak and Caesar Cap already partially shaded in the morning light. This topographic shading has played a vital role in maintaining these two glaciers in the Klamath Range over 130 years after the end of the Little Ice Age in the 1880s—a cooler period which began over 700 years ago. Salmon Glacier melted away by the fall of 2015 during an extreme multiyear drought leaving the retreating Grizzly Glacier as the only surviving feature from the Little Ice Age. Photo by Justin Garwood.

the 1880s, glaciers were much smaller, clinging to the north slopes of the highest summits of the Trinity Alps. At least six LIA glacier locations and three LIA glacial substages have been identified in the Trinity Alps based on remaining rocky terminal moraines (Sharp 1960, Graham 2013, Garwood et al. 2020). Grizzly and Salmon glaciers represent the largest of the LIA glaciers and the only two that persisted into the 21st century. Unfortunately, beginning during the catastrophic five-year California drought of 2012–2016, Salmon Glacier melted away and Grizzly Glacier partially broke apart (stagnated) by fall 2015 (Garwood et al. 2020). Grizzly Glacier still barely persists in 2022 and is expected to melt away within the next decade due to prolonged drought and climate change.

Melissa Reneski admiring a cavernous wall of blue ice at the toe of Grizzly Glacier after a summer of rapid retreat, October 2013. Photo by Ken Lindke.

While the demise of these small Klamath glaciers is seemingly insignificant from a global perspective, they offer a unique local climate signature by measuring Earth's responses to changes in temperature and precipitation. These changes will resonate in the diversity of the regional flora and fauna that depend on cold, relictual habitats. Species in these habitats are still being discovered, like ice-loving beetles (see *Nebria* sidebar page 263) and ice crawler (*Grylloblatta* spp.) insects found to be endemic to Grizzly Glacier. Coldwater stream niches fed by glacier and perennial snow meltwater also support genetically distinct stenothermic stream fauna, including the world's highest populations of coastal tailed frogs (*Ascaphus truei*) and relictual populations of spring Chinook salmon and summer steelhead.

Trinity Reservoir stores water for diversion by tunnel to the Sacramento River basin and southward to corporate agribusiness in the southern Central Valley and to southern California. These diversions depleted up to 90 percent of the river's flow after the dam was finished in 1962 , but reforms of the 1992 Central Valley Project Improvement Act retained 43 percent of the river's natural runoff in the Trinity system. Since then, corporate farming interests have repeatedly tried to take back a larger share of the Trinity's flow. An intensive restoration effort supported by the 1992 law seeks to minimize negative effects of the withdrawal and has pioneered creative approaches to reinstate river health.

A major tributary in its own right, the 87-mile South Fork is the longest completely undammed river in California. Immediately to its west, South Fork Mountain parallels the river in a remarkably straight, 35-mile-long ridge marking the fault separating uplifts of the Klamath Mountains and Coast Range. Before heavy logging, the South Fork was a legendary fishery, but clearcutting and road construction have decimated what was once a champion steelhead and spring-run Chinook stream—with a run of 11,000 fish plummeting to near zero. Logging was brought to a halt with a court case recognizing the Forest Service's obligation to protect salmon-bearing waters, and the stream has since been on the road to recovery.

Sacramento, McCloud, and Pit rivers

Average flow: 22,490 at mouth (637 cms)
Watershed: 26,050 miles^2 (55,296 km^2)

The Sacramento and its tributaries, the McCloud and Pit rivers, flow briefly through a far eastern arm of the Klamath Mountains lying adjacent to the Cascade Mountains immediately north toward Mount Shasta, eastward, and also south toward Lassen Peak. The Klamath Mountains segment of the watershed includes the Sacramento's South Fork of 10 miles followed by 36 miles to the backwaters of Shasta Reservoir—California's largest, holding 4.5 million acre-feet behind the 602-foot-tall dam. Interstate 5 follows the upper river corridor, though the highway is mostly on mountain slopes far above, leaving the river in a whitewater tumult through deep forest. Salmon and steelhead of the upper river were eliminated by the dam, leaving the upper Sacramento as a noted trout stream today.

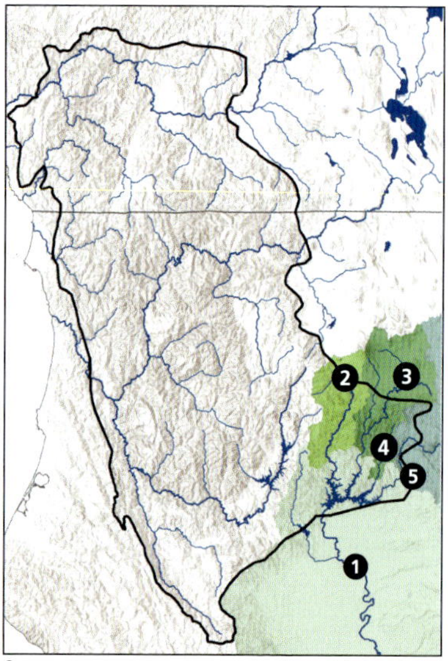

Sacramento River Watershed includes the Sacramento River (1), upper Sacramento River (2), McCloud River (3), Sulanharas Creek (4), and (5) Pit River.

Like the Sacramento, the tributary McCloud River takes the vast majority of its runoff from the formidable hydrologic feature of Mount Shasta. This 14,180-foot landmark and its volcanic sub-summit, Shastina, are two of the three highest peaks in the entire Cascade Range. The 77-mile-long McCloud surges from springflows south of the great mountain and flows into the backwaters of Shasta Reservoir. Legendary trout waters here include spectacular waterfalls and crystalline whitewater. The lower McCloud is currently threatened by plans of the federal Bureau of Reclamation to raise the height of Shasta Dam, which would impound farther up the McCloud, Pit, and Sacramento rivers. The state of California opposes this plan, and its outcome is uncertain as of this writing.

The Pit River begins far to Mount Shasta's northeast, taking a circuitous course through the northern extent of the Basin and Range province of the interior West, then swelling with the runoff of the southern Cascade Range, finally flowing for 35 miles through the eastern arm of the Klamath Mountains and into the backwaters of Shasta Reservoir. The Pit's total length (including its South Fork) is larger in volume than the upper Sacramento where they join in the reservoir, and the mileage of the Pit is far greater than that of the upper Sacramento above the confluence.

Lakes, Ponds, and Peatlands

One of the exceptional experiences while visiting the upper elevations of the Klamath Mountains is arriving at a glacially formed lake. Cool, glistening waters complement the polished rocks at the shore—composing the perfect platform for a dip followed by a bask in the high summer sun. The Klamath is decorated with many fine lakes, but not all lentic waters in the Klamath are of glacial origin. Of the 345 natural lakes in the Klamath Mountains greater that one acre in size, 89 percent (306) were created by glacial forces. The other 11 percent were formed by processes including landslide slumps and depression basins with origins in the mountain building tectonic processes themselves. Ponds also form when low-elevation meandering river channels change course—leaving behind severed backwaters or abandoned arc-shaped channels called oxbow ponds. These dynamic, river-shaped, still waters can be further characterized and maintained by American beavers (*Castor canadensis*).

In the upper elevations, thousands of ponds—from a few square feet to over an acre—are clustered in distinct geographic zones across the range. The origins of small ponds follow similar patterns to the lakes. Most are a product of glacial action followed by a combination of landslide slumps, mountain building, and river geomorphic processes. Peatlands, also known as fens, are primarily found in formerly glaciated regions and serpentine geology throughout the Klamath Mountains. Often generically referred to as wet meadows or springs, these pond- and stream-laden wetlands are created by the slow, anaerobic decay of wetland plants fed by a stable water source, such as a spring or a lake edge. Here, thick layers of peat can build up over the eons establishing an intricate and stable biological community starkly different from the landscape around it. Hidden waterbodies, such as limestone cave ponds, and even the small, liquid-filled stems of the California pitcher plant (*Darlingtonia californica*), support their own distinct aquatic communities separate from all others in the Klamath Mountains.

Despite being locally abundant in many areas, natural lakes, ponds, and peatlands make up only a small fraction of the landscape. These isolated wetlands are, more importantly, vessels of biodiversity—with each aquatic community uniquely shaped by geology, soil development, climate, hydrology, water chemistry, wetland

The Hidden Lives of Beavers in Klamath Streams

Big Slough, a tangled chain of oxbow wetlands threaded by slow-moving slough channels, is a relic example of how the greater Scott River Valley riverine-wetland matrix looked and functioned before the upper valley was bucket dredged and the lower valley was converted to well-drained, fertile agricultural lands. This complex wetland is fed by the terminal reaches of Patterson and Kidder creeks where they abruptly fan out from confined canyon channels into the broad Scott Valley just west of the mainstem Scott River. Slough channels flow north for 4.5 mi (8 km) before entering the Scott River. Though relict oxbows survive here today, one key element is largely missing from the historical structure and function of Big Slough, the greater Scott Valley, and other broad, alluvial river floodplains of the Klamath Range—an abundance of beavers.

A relic portion of Big Slough surrounded by agriculture fields where Kidder Creek enters the Scott River Valley, CA. Note faint historic channel evidence in agricultural lands. Image from Google Earth, 2017.

In 1836, early Klamath explorers wandered into a verdant valley on the eastern edge of the range and found the finest beaver grounds they had ever seen. Today we know this region as the Scott Valley, but these early explorers called it Beaver Valley (Goldfarb 2018) and the Beaver River (Gudde and Bright 2004)—now the Scott River. In a single year around 1850, a large party of fur trappers claimed to have caught 1,800 beavers in Scott Valley and shipped their pelts to Europe to make hats (Tappe 1942). Others followed, and by the end of the 1800s beavers were noted as rare, and in 1929 beavers were locally extirpated from 80 years of persistent trapping. Prior to trapping, beavers ranged across much of California and Oregon. Valleys in the Klamath Mountains had abundant populations, including the Applegate River valley which the Takelma native people called "the beaver place" because the animals were so prevalent.

Beaver in the Scott Valley. Photo by Charnna Gilmore.

American beavers (*Castor canadensis*) are nocturnal, semi-aquatic rodents best known for building dams and lodges across North America. Along with the Eurasian beaver (*C. fiber*), they are the second-largest rodent in the world after the tropical capybara. Beavers are unique geomorphic and hydrologic disturbance agents in rivers where they can have a profound influence on creating and vigorously maintaining side-channel and oxbow ponds in broad, braided river valleys. When families establish a territory, they can build intricate systems of dam and migration channels. These persistent rodents ensure that water surface elevations are maintained year-round for den access, woody riparian vegetation is maintained for foraging, while a network of ponds and migration channels offers protection from predators. Beavers are a keystone species that provide extensive ecosystem services for aquatic communities. They create water-storage systems that increase summer–fall stream flows and water quality, reduce bank erosion and channel degradation, and create aquatic and riparian habitats for a broad suite of aquatic and terrestrial organisms including caddisflies, fishes, turtles, waterfowl, and bats, to name a few.

Example of a complex river beaver bank lodge with multiple burrow entries in the Smith River. These wood and root laden reef-like features attract aquatic organisms, especially juvenile salmonids and lamprey. Photo by Marisa Parish Hanson.

Most imagine beaver dams that span channels of slow-moving water. However, beavers in the Klamath region today mainly live in rivers, where, instead of dams, they excavate cavernous bank-lodges into loamy, root-bound riverbanks (Parish Hanson 2016). When necessary, they will also build small seasonal dams in side-channels and smaller streams to maintain summer water levels near dens. These dams are often later destroyed with the onset of winter flows. Beaver habitats in rivers are much more cryptic than those of pond-building beavers and are best seen with a mask and snorkel swim along riverbanks. These bank lodges offer reef-like habitats—especially for juvenile salmonids, amphibians, reptiles, river otters, and freshwater mussels, to name a few. The most common evidence of beaver is chewed willow stumps.

In its historical state, the Scott Valley's countless beaver dams would have impounded a tremendous amount of snowmelt runoff from the surrounding mountains that was slowly metered out to the Klamath River basin throughout the summer and fall. This was surely a massive and fertile rearing ground for juvenile salmonids and most likely offered the most extensive waterfowl habitat in the Klamath Range. When the beavers were removed from the valley, dams collapsed and so did the wetlands, leading to a cascading decline of a legacy wetland ecosystem.

Stewards are trying to increase numbers and ecological function through restoration efforts in the Scott Valley and other Klamath basins. In Sugar Creek, a tributary to the Scott River that drains from the Russian Wilderness, the landscape is being transformed by placing logs and slowing the flow to improve beaver habitat—so that Beaver Valley may return to some of its former glory.

size and shape, and landscape position. Terrestrial organic matter and inorganic nutrients from the surrounding drainage basin greatly influence an individual wetland's productivity—especially in basins with large drainage areas relative to the receiving waterbody. Elevation and aspect are also important factors in wetland productivity.

High-elevation lakes in the Klamath tend to be oligotrophic—with exceptionally clear, oxygenated waters due to a lack of plant and animal productivity. In contrast, lower-elevation lakes surrounded by thick mixed-conifer and hardwood forests are often mesotrophic—characterized by intermediate nutrient levels, increased plant and animal productivity, and greater fluctuations in oxygen levels. Dystrophic lakes and ponds are yellowish-brown from the leaching of humic acids (tannins) from the breakdown of organic matter, such as conifer needles or leaves, blown in or transported to the lake by inlet streams, snow avalanches, and melting snowpack. These transparent, tea-colored waters are generally small and common throughout the montane and subalpine conifer belts of the Klamath Mountains. Shorelines of dystrophic lakes and ponds are dominated by sedge (*Carex* spp.) peatlands, often with catchments surrounded by dense forest. Rich organic matter provides a constant supply of humic acid that lowers the pH and effectively reduces biodiversity compared to clear lakes. As with fen peatlands, dystrophic waterbodies are carbon sinks. Unlike peatlands, carbon is converted and stored in dissolved and particulate forms. Algae-rich eutrophic lakes, resulting from excessive nutrients, are rare in the Klamath and are more associated with artificial ponds and lakes. This general pattern has its exceptions—some higher-elevation waters are surprisingly productive due to local factors like soil and forest development.

The wetland-to-land ecotone further enriches local wetland-associated foodwebs in the Klamath by the mutual exchange of energy and resources between wetland and terrestrial habitats such as forests and upland meadows. Upward of 15 species of bats feed on wetland insects, while a myriad of terrestrial insects pol-

Upper Twin Lake in the Siskiyou Mountains is a non-glacially formed mesotrophic lake. Photo by Justin Garwood.

linate wetland plants. From a pack of metallic whirligig beetles (Gyrinidae spp.) frantically gyrating on the water surface to a raft of weary northwestern pond turtles (*Actinemys marmorata*) stealthily basking just under the water surface, Klamath wetlands offer an endless supply of discovery. Timestamps can be found preserved in the lake-ooze. Treasures including pollen grains, charcoal, diatoms, and fossilized invertebrates reveal lake age, past climates, fire regimes, and major floristic shifts in the regional landscape (see Paleoclimate sidebar pages 100-101).

Trinity, Whiskeytown, Shasta, Pit, McCloud, Siskiyou, and Applegate reservoirs store massive volumes of water, but because they are reservoirs, we omitted them to focus on naturally occurring waters.

Origins and Geography of Lakes, Ponds, and Peatlands

Glacial Lakes and Ponds

The Pleistocene lasted ~2.5 million years with four major glacial substages, ending nearly 10,000 years ago (Sharp 1960). When the Holocene began, glaciers receded and left behind the jeweled fruits of their labor—hundreds of turquoise lakes and thousands of rock garden ponds (Figure 2). From the perspective of geologic time 10,000 years is just a blink, so these waterbodies are relatively new features of the Klamath landscape—a surprising revelation for anyone who has stared into the bottomless blue depths of a mountain lake.

Most glacially formed waterbodies are concentrated in the highest-elevation ridges and summits of the eastern Klamath Range across the Shasta-Trinity Mountains, Trinity Alps, and the Salmon, Scott, and Marble mountain sub-ranges. But there are outliers. Black Rock Lake (6,220 ft, 1,896 m) is one of the lone glacially formed lakes in the southern Klamath Mountains—tucked in the north shadows of the nearly 8,000-foot Yolla Bolly crest. This is 42 air miles (68 km) south of the nearest glacial lake in the Trinity Alps. Not surprisingly, just above the lake is a

This glacially carved lake in the western Klamath Mountains is only 4,800 feet in elevation. Photo by Michael Kauffmann.

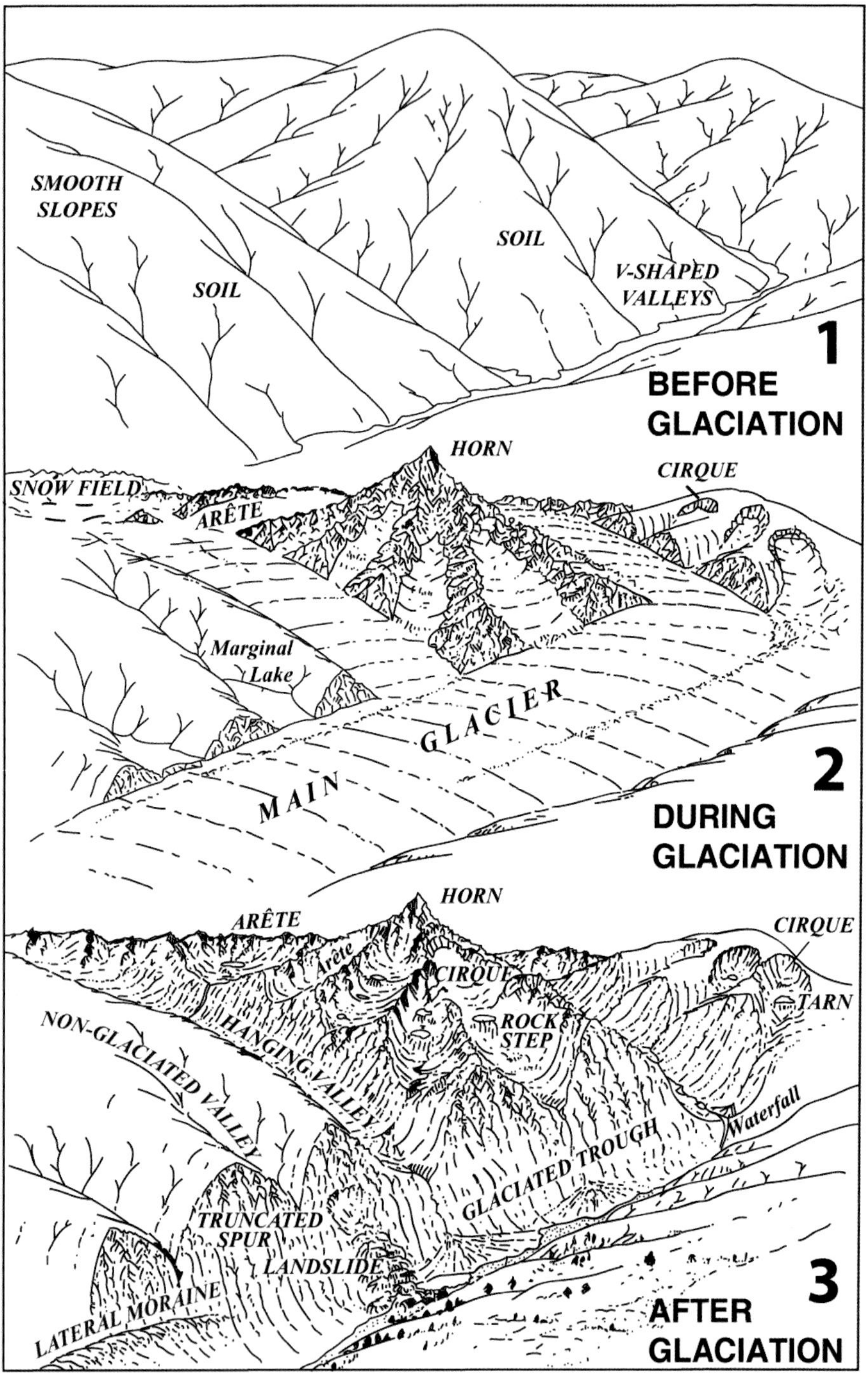

Figure 2. Generalized illustration depicting the major changes that occur to a mountain landscape before, during, and after glaciation. From Hopson and Clynne (2019).

relict stand of mountain hemlock (*Tsuga mertensiana*), some 42 milies (68 km) south of next-nearest stand of the same species!

The Siskiyou Mountains are on average lower-elevation than the eastern Klamath's cluster of high peaks and hold a remarkable, albeit much less extensive, glacier lake district. Extending farther westward compared to other interior lake regions, the Smith River, Illinois River, and Chetco River basins hide glacier-formed lakes near the Pacific Ocean in the northern reaches of the Klamath Mountains. Vulcan Lake (3,897 ft, 1,187 m) and its small companion Lish Lake (3,767 ft, 1,148 m) are the lowest glacial formed lakes in the Klamath Mountains and only 17 miles (28 km) northeast of the Pacific Ocean. Babyfoot Lake (3,950 ft, 1,203 m) is the northern-most glacial lake and 9 miles (15 km) northeast of Vulcan Lake. Both Vulcan and Babyfoot lakes drain through the Kalmiopsis Wilderness into the remote and rugged Chetco River. Just north, in the mainstem Rogue River basin, crest elevations dramatically decrease. This topographic shift results in the northernmost Klamath Range having had fewer Pleistocene glaciers and thus few lakes and ponds today (Figure 3).

Most glacial lakes (also known as tarns) are tucked into small, rock-lined amphitheaters called cirques. These cirque basins occur on high, north-facing slopes shadowed by jagged ridges and horns. The adjacent south slopes are often gently

Summary of the largest, deepest, and highest-elevation from a total 306 glacial lakes greater than one acre in size from the Klamath Mountains. Average glacial lake size is 6.9 acres (2.8 ha), while the average glacial lake elevation is 6,150 ft (1,875 m).

Lake	Subrange	Imperial; Metric
Largest		
Caribou	Trinity Alps	72 acres; 29 hectares
Ukonom	Salmon Mountains	65 acres; 26 hectares
Cliff	Salmon Mountains	55 acres; 22 hectares
Castle	Shasta-Trinity	49 acres; 20 hectares
Sapphire	Trinity Alps	48 acres; 19 hectares
Deepest		
Sapphire	Trinity Alps	200 feet; 61 meters
Grizzly	Trinity Alps	180 feet; 55 meters
Cliff	Salmon Mountains	175 feet; 53 meters
Smith	Trinity Alps	167 feet; 51 meters
Man Eaten	Salmon Mountains	112 feet; 34 meters
Highest-Elevation		
Upper Deadfall 1	Scott Mountains	7,800 feet; 2,378 meters
Upper Deadfall 2	Scott Mountains	7,757 feet; 2,364 meters
Lois	Trinity Alps	7,612 feet; 2,321 meters
Billy-Be-Damned	Trinity Alps	7,603 feet; 2,318 meters
Little Crater	Scott Mountains	7,586 feet; 2,312 meters

sloped due to the lack of glacial action. During the Pleistocene ice ages, high-elevation north-facing slopes created the perfect cold niche for hundreds of small—but impressive—cirque glaciers.

The only south-facing cirque lakes in the Klamath are found in the Canyon Creek basin of the Trinity Alps—where a steep-walled valley glacier grew south, fed by accumulating snow below the highest cluster of peaks in the range. This region, centered on Thompson, Sawtooth, and Caribou peaks, was a glacier factory during the Pleistocene. Dozens of massive ice-stream valley glaciers sculpted the region with U-shaped canyons, pinnacle arêtes, and waterfalls. This slow, violent process left cirque lakes perched above hanging valleys overlooking densely forested canyons. Examples of this are seen today in the Stuart Fork, Trinity River, Swift Creek, and North Fork Salmon River. Collectively, cirque and valley glacial forces left 306 glacial lakes over an acre in size and thousands of pond and fen wetlands (see Ice Age in the Klamath Mountains sidebar, page 91).

The deepest part of a cirque lake is at the base of the cliff headwall with the shallowest part at the outlet. This happens because the weight of a glacier is concentrated at the headwall cliff. With glacier ice consistently moving downslope, solid rock was fractured for thousands of years across multiple glacial sub-stages. Rock fragments, known as till, became frozen in the glacier ice matrix where they functioned as embedded chisels in a giant-gravity-powered ice-conveyor belt that plucked away and polished the surrounding bedrock as it moved. Lake Anna and Grizzly Lake (Trinity Alps) and Man Eaten Lake (Salmon Mountains) have smoothed bedrock lips at their outfalls with conspicuous grooves and chatter marks in the rock. These spectacular grinding patterns indicate the direction ancient glaciers moved. Till debris eventually worked through the terminuses of glaciers and was deposited into massive refuse piles of jagged boulders and cobbles known as glacial moraines. Just about every cirque basin in the Klamath Mountains has conspicuous moraine piles left behind as evidence for us to discover and piece together local glacial extents.

Glacial polish on rocks adjacent to Lake Anna, Trinity Alps Wilderness. Photo by Justin Garwood.

Most cirque lake outfalls in the Klamath Mountains are composed of a combination of polished bedrock and moraine debris. There are also some examples of proglacial lakes, where the lake is entirely impounded by the damming action of a massive, semi-circular, terminal moraine left behind after the glacier retreated. These lakes often do

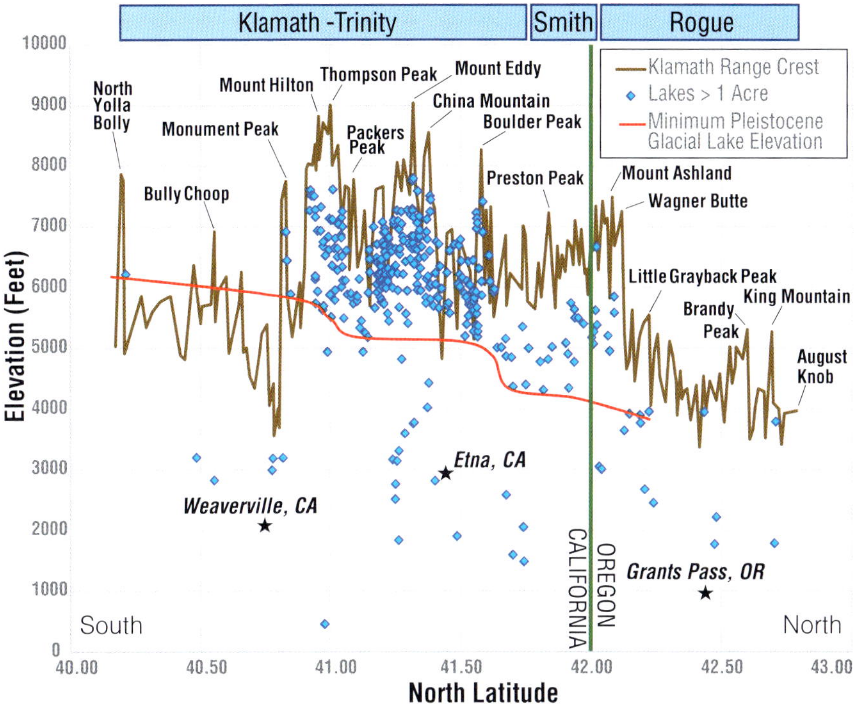

Figure 3. Profile of the Klamath Mountains from south to north showing the elevational crest relative to 345 natural lakes > 1 acre in size. Most lakes are found among the highest elevations and formed during the Pleistocene ice age with the last major glaciation. All lakes below or north of the red line have non-glacial origins.

not have outlet streams—rather, their waters emerge as hillslope fen springs far below the lake. Classic examples of this type are Little Crater Lake below Mount Eddy, Washbasin Lake and Summit Lake in the Trinity Alps, and South Sugar Lake in the Salmon Mountains. There are also examples of glacial lakes that occur together like a string of beads connected by streams—known as paternoster lakes. These lake groups can be found in small basins, like the Rush Creek Lake series in the Trinity Alps, or in large basins carved by massive valley glaciers such as the striking Mirror-Sapphire-Emerald lake chain in the Trinity Alps–Stuart Fork basin. The Marble Mountains also host an impressive five-lake paternoster chain from Dogwood Lake down to Deep Lake.

Ponds created by glacial forces do not receive the same attention as lakes but likely surpass lakes in total water surface area. Their sheer numbers—easily in the thousands—illustrate the consequential role these diverse wetlands have in shaping unique aquatic plant and animal communities throughout the Klamath Mountains, especially where lakes are rare. Although cirque glaciers created hundreds of lake basins we see today, many more glacier-carved cirque basins do not have lakes—but rather contain high hanging valleys skirted by dense conifer stands that are drained by pond-laden and botanically rich fen meadows. Regardless of the outcome, glaciers were a profound force in shaping the unique aquatic resources and resulting biodiversity we see in the subalpine Klamath today.

Non-glacial Montane Lakes and Ponds

Non-glacial montane lakes include those formed by landslide slumps and tectonic processes and are rare in the Klamath Mountains, representing only 39 (11 percent) of the 345 natural lakes greater than one acre. Non-glacial lakes also occur at much lower elevations than their glacial cousins, with average elevations (3,014 ft, 919 m) half that of average glacial waters. These lakes are also small and shallow, averaging 4 acres—nearly 40 percent smaller than the average glacial lake. Given these physical characteristics, the morphometry and resulting ecological communities of these lakes are more aligned with large, shallow ponds than with deeper lakes. Immediate shorelines are commonly lined with sedges and bushy shrubs such as willows (*Salix* spp.) and western azalea (*Rhododendron occidentale*). The hydrology is diverse, with some lakes such as Bummer Lake in the Smith River watershed drying completely each year. Others maintain stable surface levels benefiting from ample spring sources, while many lakes fall in between by partially drying down each summer or fully drying in droughts. Most do not have perennial outlet streams, and others occur as ridgetop depressions—so like glacial lakes, these waters have not supported native fish populations in the Holocene.

Similar to glacial lakes, most montane lakes and smaller ponds occur in clusters, which indicates formation by similar local geological processes. All but a few of these lakes are found along the moisture-rich western flank of the range where geology is a jumbled mélange of Coast Range and Klamath rocks folded by faulting—and is thus prone to landslides, hillslope slumping, and local subsidence at mountain tops from Earthquake activity. Small western Klamath lake districts occur in the South Fork Trinity River, the Klamath River near Bluff Creek, the South Fork Smith River, and in the western and north Rogue River basin.

Unlike glacial waters, where a nearly synchronous birth of lakes began after glaciers receded ~10,000 years ago, montane lakes formed on separate timelines based on local tectonic and landslide processes. Some lakes are young, while others are far older than glacial lakes. For example, a lake-sediment aging study at Twin Lakes in the Klamath River basin yielded a 19.5 ft (5.9 m) vertical sediment core from the lakebed. Despite its impressive core sample length, this was likely not the complete sediment record. The oldest organic material in the sample was radiocarbon dated to 22,578 ybp (Wanket 2002). There are some lakes of unknown age that appear ancient because their open waters are slowly being overtaken by sediment, vegetation mats, and peat accumulation from dead plant material along their shallow shorelines. Walking on these mats is risky yet fun—like standing on a sinking raft. These slow, reliable, soil-building processes will eventually convert these waters to depressional peatland meadows. Great examples of these dying lakes include Maddox Lake in the South Fork Trinity River, and the northern Lake of the Woods in the Rogue River basin northwest of Agness.

Despite their rarity and small size, montane lakes and their accompanying ponds occurring below 5,000 ft (1,500 m) host the greatest aquatic biodiversity and arguably the greatest productivity per unit area over any other aquatic ecosystem in the Klamath Mountains. Most mid-elevation lakes contain patches of western yellow pond-lily (*Nuphar polysepala*) which sometimes blankets the water surface by mid-summer. *Nuphar* provides vertical substrate and cover for aquatic species throughout the water column akin to the giant subtidal kelp forests of the Pacific Ocean. Many silt-lined lakes and ponds also contain unusual grass-like aquatic plants called quillworts (*Isoetes* spp.) that reproduce like ferns by producing spores. Large, downed conifer logs are also abundant and are well-preserved features in these waters—often serving as a raft for basking northwestern pond turtles or as botanically laden barges for mosses, plants, and shrubs. Many rare aquatic species occur in western Klamath lakes and ponds, such as the northwestern pond turtle and fairy shrimps (Anostraca). Some widespread coastal species also reach their easternmost limits here, including the northern red-legged frog (*Rana aurora*), the northwestern salamander (*Ambystoma gracile*), and the northwestern garter snake (*Thamnophis ordinoides*).

Most of these lakes and ponds are difficult to find and are not celebrated destinations like the picturesque glacial lakes. However, a muddy walk around one of these productive waters will turn anyone into a curious biophiliac, channeling adventures akin to an Amazonian swamp exploration—except these temperate lakes are surrounded with mixed-conifer and hardwood forests. If one is brave enough to don a mask and swim in one of these waters, a thick soup of aquatic life—including large zooplankton predators like scarlet-colored *Diaptomus* copepods, mayflies, caddisfly larvae, glass-like phantom midges, fairy shrimps, dragonflies and damselflies, small leeches, ramshorn snails, predacious diving beetles, toe biter bugs the size of your big toe, newts, anuran tadpoles, garter snakes, and turtles will bombard you. What will be clear to you in these murky waters is the fact that these montane aquatic environments of the western Klamath Mountains are compact vessels holding immense biodiversity worth celebrating.

Fluvial Ponds

Lentic waters created by abandoned alluvial river channels are rare and localized in a few broad river valleys of the Klamath Mountains today. Prior to the 20th century, these dynamic wetlands were likely common in areas such as the Rogue River Valley near Grants Pass, Applegate Valley, Illinois River Valley near Cave Junction, Hayfork Valley, and especially the greater Scott Valley where extensive remnant clues of oxbow and braided side-channel ponds remain imprinted throughout agricultural fields. Most appear to have been lost to a combination of factors including flood control measures resulting in confined stream channels, agriculture and other floodplain development, runoff capturing reservoirs on the Klamath and Trinity rivers, catastrophic changes in alluvial valleys, local bucket-dredge gold mining (Figure 23 in Geology), and dramatic declines in beaver populations. Today, oxbow and side-channel ponds are still being created and destroyed by natural processes across the Klamath Mountains, albeit at a much lower frequency than in the past due to loss of floodplain connectivity coupled with historic beaver population declines. Today, stream restoration projects are recreating these lost habitats to benefit juvenile salmonids, but overall natural processes need to be restored as well in order to provide any lasting benefits to fish and other wildlife.

Peatlands

Peatlands are unique biotic communities primarily created by biological processes rather than abiotic means. The Klamath Mountains have two types of peatlands, including fens and basin peatlands. Fens are perennial wetland seeps and meadows on slopes that accumulate rich organic peat soils through the annual cycle of plant decay over long periods. Most Klamath wetland meadows are fens because at least a portion has a perennial water source, with sedges (*Carex* spp.) dominating the vegetation. Basin peatlands are like fens, but develop along the edges of lakes and ponds, where soils are saturated year-round.

The formation of peatlands requires a combination of processes that most commonly occur in flat areas in both tropical and boreal regions. Suitable peatland-forming conditions are generally limited in mountain landscapes (Cooper et al. 2012) within temperate regions because of variable topography, geology, and even water chemistry. An estimated 0.2 percent of the land area in Sequoia and Kings Canyon national parks are covered by peatlands (Hopkinson et al. 2013). In temperate mountains, rare peatlands form over millennia if perennial soil saturation, low mineral-soil deposition, erosion rates, and net storage of soil carbon resulting from plant productivity complement each other perfectly (Moore and Bellamy 1974) (Figure 4). In arctic and alpine environments, the formation of peat is often associated with peat moss (*Sphagnum* spp.). In temperate regions such as the Klamath Mountains, peatlands are dominated and formed by sedges.

TOP: A hillslope peatland is interspersed with spring mounds. TOP INSET: These fen types, decorated with California pitcher plants (*Darlingtonia californica*), are common on ultramafic rock, seen here in the Siskiyou Wilderness. BOTTOM: Isinglass Lake in the Marble Mountains is an excellent example of a basin fen formation. BOTTOM INSET: This basin holds the only known population of great sundew (*Drosera anglica*) in the Klamath. Photos by Michael Kauffmann.

Fens are broadly defined as a type of peatland that forms in concave depressions or on slopes with streams flowing through them. Bogs, another peatland type, are not influenced by ground or surface water but are instead hydrated by precipitation. By this definition, only fens occur in the Klamath Mountains—there are no true bogs. *The Jepson Manual* uses overlapping terms to describe these habitat types. These include bog, fen, sphagnum swamp, sphagnum bog, boggy site, and peatland. Taken together, these terms characterize oligotrophic (nutrient-poor) peatlands, where the saturated vegetation is either quaking (it shakes when you traverse it) or floating (it undulates when you traverse it). Because, for example, sphagnum produces acidic conditions or bedrock geology influences pH, Klamath fens vary from acidic to basic with others being neutral to alkaline (Leppig 2004). The carnivorous California pitcher plant (*Darlingtonia californica*) thrives in nutrient-poor fens on serpentine geology by relying on terrestrial invertebrate prey.

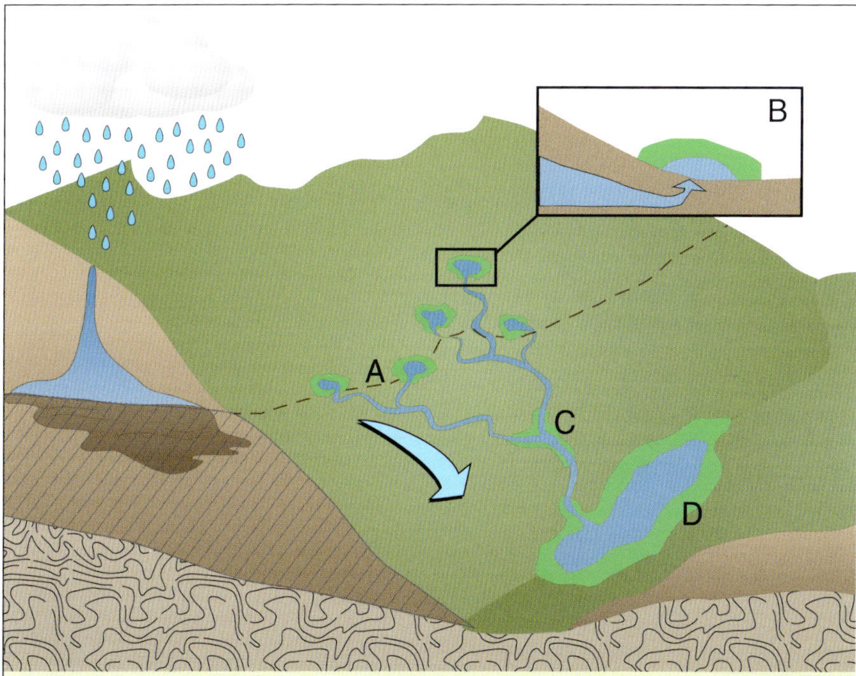

Four major geomorphic settings identified for California fens

A) Bedrock contact fens are fed by upwelling where more permeable rock overlies less permeable rock, forcing water out of the ground.

B) Spring mounds form at localized points of groundwater discharge and are often small.

C) Hillslope fens form where groundwater discharges to the surface from glacial moraines, talus, alluvium, and hillslope-exposed bedrock fractures

D) Basin peatlands originate as lakes or ponds that fill with plant material.

Adapted from Wolf and Cooper (2015).

Figure 4.

The fact that fens occur in the Klamath Mountains is exceptional when you consider they are here despite a Mediterranean-type climate that keeps rain away in the summer months. The botanical legacy of Klamath fens is built from plant species from across western North America. Annual plants from the temperate and subtropical regions build the fen's foundations and are decorated by a dominant layer of perennials and shrubs with an affinity for northern boreal or arctic regions.

Peatlands intercept runoff from rain, snow, and groundwater storing it throughout thick peat layers like a giant sponge. This water capture dramatically slows the runoff by effectively creating a shallow groundwater aquifer that regulates streamflow, temperature, and water quality throughout the dry summer months. Worldwide, peatlands are the largest natural terrestrial carbon sink—sequestering more CO_2 than all other types of vegetation combined making these wetland types among the most valuable ecosystems on Earth (IUCN 2017). Although Klamath peatlands cover a small fraction of the land surface overall, their carbon storage capacity is substantial given that peat in this region can accumulate to several meters deep.

Although Klamath peatlands are relatively unproductive for aquatic invertebrates, amphibians such as the Cascades frog (*Rana cascadae*) and the rough-skinned newt (*Taricha granulosa*) use perennial fen ponds for breeding, foraging, and overwintering. Under deep snowpack during the winter, peatland spring-fed ponds provide stable temperatures around 41°F (5°C) for hibernating amphibians (Garwood 2009). Reptiles like the common garter snake (*Thamnophis sirtalis*) and northern alligator lizard (*Elgaria coerulea*), are common predators in peatlands. Ground-nesting sparrows such as the ubiquitous dark-eyed junco (*Junco hyemalis*) and the rarer vesper sparrow (*Pooecetes gramineus*) build cryptic nests perched just above the peat in matted clumps of dried sedges. Small mammals including shrews (*Sorex* spp.), voles (*Microtus* spp.), and pocket gophers (*Thomomys* spp.) thrive here as well during the summer, with extensive tunnel systems on the surface hidden in the thick carpet of sedges. These small mammals also find insulation and food in the upper peat layers during long winters. Pocket gophers often build elaborate subnivean tunnel systems used for winter foraging. Many snow tunnels become densely packed with subsurface tunnel tillage—later revealed as bizarre soil cylinders that are perfect tunnel molds on the peatland surface after snowmelt.

Acknowledgments

Many thanks to Darren Ward, Scott Harding, Alison O'Dowd, and Jim Agee for reviewing this chapter. Gordon Leppig shared his extensive knowledge on peatland formation and ecology. Discussions with Tim Bailey on Pleistocene lake formation informed our glacial lake mapping across the Klamath Mountains. Charnna Gilmore, Marisa Parish Hanson, and Eli Asarian provided feedback on beavers.

Photographic contributions

Thanks to Roberta Olenck, Charnna Gilmore, Ken Lindke, and Marisa Parish Hanson.

Literature Cited and Suggested Reading

Cooper, D.J., R.A. Chimner, and D.M. Merritt, 2012. Mountain wetlands of North America. In: Batzer, D. and Baldwin, A. (eds.) *Wetland Habitats of North America: Ecology and Conservation Concerns*, University of California Press, Berkeley, USA, 313–328.

Crum, H.A. 1992. A focus on peatlands and peat mosses. University of Michigan Press. Ann Arbor, MI.

Garwood, J. 2009. Spatial ecology of the Cascades frog: Identifying dispersal, migration, and resource uses at multiple spatial scales. [Thesis]. Cal Poly Humboldt, Arcata, CA.

Garwood, J., A. Fountain, K. Lindke, M. van Hattem, and H. Basagic. 2020. 20th century retreat and recent drought accelerated extinction of mountain glaciers and perennial snowfields in the Trinity Alps, California, USA. Northwest Science 94:44-61.

Goldfarb, B. 2018. Beavers, rebooted. Science 360 (3693):1058-1061.

Graham, J. T. 2013. Late Holocene glacial advances in the Klamath Mountains, northern California, determined from 10Be cosmogenic exposure dating and dendrochronology. M.S. Thesis, California State University, Northridge.

Gudde, E. and W. Bright. 2004. California place names: the origin and etymology of current geographical names. Fourth edition. University of California Press, Berkeley, USA.

Hopson, F., and M. Clynne. 2019. *Geology of the Lassen Country*. Backcountry Press. Kneeland, CA.

Hopkinson, P. et al. 2013. A natural resource condition assessment for Sequoia and Kings Canyon National parks Appendix 13 - Meadows. Natural Resource Report NPS/SEKI/NRR-2013/665.13, National Park Service, Fort Collins, Colorado, 103 pp.

IUCN 2017. Peatlands and Climate Change. International Union for Conservation of Nature. https://www.iucn.org/resources/issues-briefs/peatlands-and-climate-change

Lanman, C.W., K. Lundquist, H. Perryman, J.E. Asarian, B. Dolman, R.B. Lanman, and M.M. Pollock. 2013. The historical range of beaver (*Castor canadensis*) in coastal California: An updated review of the evidence. California Fish and Game 99:193–221. https://nrm.dfg.ca.gov/FileHandler.ashx?DocumentID=78258

Leppig, G. 2004. Rare plants of northern California coastal peatlands: patterns of endemism and phytogeography. pp. 43-50, in: Brooks, M.B., S.K. Carothers, and T. LaBanca, editors. The Ecology and Management of Rare Plants of Northwestern California: Proceedings from a 2002 Symposium of the North Coast Chapter of the California Native Plant Society. California Native Plant Society, Sacramento, CA.

Moore, P. and D. Bellamy. 1974. *Peatlands*. Springer-Verlag. New York, NY.

Mount, J. F. 1995. *California rivers and streams*. University of California Press, Berkeley and Los Angeles, CA.

Palmer, Tim. *Field Guide to California Rivers*. 2012. University of California Press, Berkeley.

Palmer, Tim. *Field Guide to Oregon Rivers*. 2014. Oregon State University Press, Corvallis.

Parish Hanson, M. 2016. Beaver bank lodge use, distribution and influence on salmonid rearing habitats in the Smith River, California. [Thesis]. Cal Poly Humboldt, Arcata, CA.

Rydin, H. and J. Jeglum. 2013. *The Biology of Peatlands*. Second edition, Oxford University Press, Oxford, UK, 432 pp.

Sharp, R. 1960. Pleistocene glaciation in the Trinity Alps of northern California. *American Journal of Science* 258:305-340.

Tappe, D. 1942. The status of beavers in California. State of California, Department of Natural Resources, Division of Fish and Game, Sacramento, CA.

U.S. Bureau of Land Management (BLM). 1998. Applegate-Star/Boaz Watershed Analysis. Medford District, Ashland Resource Area. Medford, Oregon. 204 p. https://www.blm.gov/or/districts/medford/plans/files/starboazwa_acc.pd

Wanket, J. 2002. Late Quaternary vegetation and climate of the Klamath Mountains. PhD Thesis, University of California.

Wolf, E.C. and D.J. Cooper. 2015. Fens of the Sierra Nevada, California, USA: patterns of distribution and vegetation. *Mires and Peat, Volume 15* (2014/15), Article 08, 1–22.

Chapter 6: Fire Ecology

Jeffrey Kane

A prescribed surface fire moves through oak leaves in the interior lowland forests. Photo by Jeff Kane.

Introduction

The phenomenon of fire is an important ecological and cultural process that shapes the natural history and biodiversity of the Klamath Mountains. Fire is an essential driver of plant community dynamics. High severity fire can serve as a disturbance that "resets" plant communities to early successional stages. However, recurring lower severity fire serves as a means of stabilizing plant communities. Fire also provides many ecological services such as nutrient cycling and wildlife habitat (e.g., snags and coarse woody debris). Humans in the region have long benefited from the practice of burning for a variety of cultural values. Preserving the integrity of Klamath Mountain ecosystems is contingent on promoting and maintaining the appropriate fire regimes.

Fire is a chemical reaction that combusts live and dead plant material, referred to as fuel, in the presence of sufficient heat and oxygen. The interaction of weather, topography, and fuels influence fire behavior—or the pattern of movement and intensity of fire across a given area over time. Generally, there are three types of fire that can occur alone or in combination: surface fires, crown fires, and ground fires. Surface fires burn through the surface fuels (e.g., litter, grasses) at relatively low intensities (amount of energy released during a fire) and rates of spread (speed of the fire). Crown fires carry through the canopies of plants (shrubs and trees) and are typically much more intense and rapidly moving than surface fire. Ground fires predominately burn at a slow rate in the organic soil layers of duff or peat. Recurring patterns of fire types and behavior in a given area contribute to the development of fire regimes and resultant vegetation structures over time (Figure 1). While patterns of fire have shifted over time, abrupt disruption of fire regimes have numerous negative impacts that affect ecological and human communities.

Paleofire

The accumulation of pollen, macrofossils, and charcoal in sediments of Klamath Mountain lake basins provides a longer-term perspective of the fluctuating vegetation and fire activity (e.g., Briles et al. 2011). Erosion of upslope soils serves as a continuous source of sediment that accumulates in the bottom of montane lakes creating distinct layers—with deeper sediment representing conditions further back in time. Identification of pollen grains and macrofossils (plant parts embedded in the sediment) provide information about the dominant plant species present in the area while the relative abundance of charcoal deposited reflects changing levels of fire activity (frequency of fire) in the region (see Paleoclimate sidebar page 100–101). The paleofire record of the Klamath Mountains extends back through the Holocene to the end of the last ice age (Late Pleistocene), about 20,000 to 15,000 years before present (ybp).

Time periods of higher fire activity typically coincided with warmer and drier periods while lower fire activity coincided with cooler and sometimes wetter periods. These fluctuations were broadly driven by climatic changes partly associated with shifting levels of solar insolation due to orbital and axial shifts of the Earth. Based on sediment records from Bolan Lake, the climate 17,000 to 15,000 ybp was much colder and drier with low fire activity. Forests in this region were open and mostly composed of mountain hemlock (*Tsuga mertensiana*) and sagebrush

(*Artemisia* sp.). The earliest records for ultramafic sites indicate that they were largely composed of open stands of yellow pines, most likely Jeffrey pine (*Pinus jeffreyi*), with an herbaceous understory that also had low fire activity. Fire activity declined further during the Younger Dryas chronozone (12,900 to 11,750 ybp), a time period associated with a relatively abrupt 5°F (~3°C) cooling event, with increased abundance of fir (likely *Abies concolor* and *Abies* x *shastensis*) and white pine (likely *Pinus monticola* or *Pinus lambertiana*) on non-ultramafic substrates.

During the early Holocene (11,700 ybp), a substantial warming event coincided with an expansion and intensification of the northeastern Pacific subtropical high-pressure system that contributed to more regional aridity. Over the next two thousand years forests on non-ultramafic substrates shifted from less fire tolerant species like mountain hemlock and fir to more fire tolerant species, such as Cupressaceae, possibly incense-cedar (*Calocedrus decurrens*), Port Orford-cedar (*Chamaecyparis lawsoniana*), or Baker cypress (*Hesperocyparis bakeri*) as well as understory shrubs, such as huckleberry oak (*Quercus vaccinifolia*) and serviceberry (*Amelanchier* sp.). Forests on ultramafic substrates showed similar increases in Cupressaceae and

Figure 1: **TOP**: During a fire heat is transferred to living tree tissues. This happens through (1) convection of smoke (2) conduction via smoldering combustion (3) bole heating and charring (4) radiation with flames and heat consuming live needles and killing buds. **BOTTOM**: Due to heat transfer during the fire a variety of injuries occur to tree parts including roots, stems, vascular cambium, and foliage. Results include (5) crown scorch with bud survival via convective and radiative heat (6) radiation to thin bark that provides little insulation (7) soil heating causes root injuries (8) thicker bark insulates from heat and tree may survive (9) crown scorch and bud kill occurs via convective and radiative heat. Illustration courtesy of Robert van Pelt.

huckleberry oak, but also had a decrease in Jeffrey pine. Shifts to more pyrophytic (fire-adaptive) vegetation were reinforced by increases in fire activity in the region until about 7,000 ybp. From 7,000 to about 5,000 ybp, fire activity fluctuated but declined overall and fire tolerant species became less abundant.

Fire activity fluctuated during the early portion of the late Holocene (4,500 to 2,000 ybp) but increased afterwards with true fir and Douglas-fir (*Pseudotsuga menziesii* var. *menziesii*) becoming a larger component of forests on non-ultramafic sites. The current assemblage of vegetation types in the Klamath Mountains established about 4,000 to 3,000 ybp in the late Holocene. Ultramafic sites concurrently experienced lower and more stable fire activity with a reduction in incense-cedar and huckleberry oak—likely due to a higher density of Jeffrey pine. Fire activity declined from this time period forward due to decreased solar insolation and a weakening of the northeastern Pacific subtropical high-pressure system. Conditions were slightly drier and warmer during the Medieval Climatic Anomaly (~1,000 ybp) and slightly cooler and wetter conditions during the Little Ice Age (~500 ybp), but these fluctuations did not result in meaningful shifts in fire activity.

Historic fire

Historic fire regime patterns (prior to Euro-American settlement) for the Klamath Mountains have been largely determined from fire scar data obtained from trees. Trees that experience fire, hot enough to kill the cambium (part of a tree that produces phloem and xylem just under the bark) at the base of a tree but not kill the tree, creates a fire scar that can more easily record subsequent fires. Through the use of dendrochronology (study of tree rings), the year and season of the fire can be retrospectively determined. The tree wound responses—and sometimes the presence of char—along particular tree rings, can indicate the specific year that a fire event occurred. Fire history studies using fire scars are limited to low and mixed severity fire regimes and can extend back hundreds of years (about 300 to 400 years in the Klamath Mountains).

In the Klamath Mountains, fire history studies indicate that fire during the pre-Euro-American settlement period was relatively frequent, with median fire return intervals (years between fire) ranging from 2 to 42 years (e.g., Taylor and Skinner 1998, 2003). Most of the existing fire scar studies have occurred in mixed-conifer forests with a majority indicating that fires for a given area occurred every 12 to 19 years. In some instances, the fire return interval is slightly longer for riparian forests and forests closer to the coast (e.g., Wills and Stuart 1994, Stuart and Salazar 2000, Skinner 2003). It is likely that ecosystems below 2,000 ft (~600 m) in elevation had shorter fire return intervals and ecosystems above 5,000 ft (~1,500 m) in elevation had much longer fire return intervals—although fire history of these ecosystems in the Klamath Mountains have not been well studied.

The size of a fire is highly variable and dependent upon the conditions present within a given year—estimates of historic fire size are poorly understood. A fire history study in mixed conifer forests near Hayfork, California estimated that fire size ranged between 60 and 3,700 acres (25 and 1,500 ha) with an average of about 100 ha (250 ac; Taylor and Skinner 2003). Majority of the fire scars occurred within the latewood or on the tree ring boundary, between growing seasons, indicating most fires occurred in mid-summer or early fall (>80 percent). Fewer fires were recorded on the earlywood that is formed in the spring or early summer. Late summer and early fall are also time periods when dry lightning storms occur with fuel conditions sufficiently dry to ignite and sustain fires.

Historical fire severity estimates are also difficult to precisely define. The fact that many forests contain old, fire-scarred trees indicates that the fire severity must have been low to moderate for most forest ecosystems (Figure 2). Estimations of fire severity from a fire history study at Thompson Ridge (Taylor and Skinner 1998), near Happy Camp, indicates that lower elevation (2,180–3,180 ft [664–970 m]) mixed-conifer ecosystems had primarily low severity fires (~50–90 percent) with patches of moderate severity (<40 percent) and even fewer high severity patches (<25 percent). Ecosystems at higher elevations tend to coincide with a greater proportion of moderate and high severity patches. Severity patterns also vary with aspect and other topographic features. At Thompson Ridge, the upper portion of slopes and ridgetops on the southern and western side had the highest proportion of high severity. Lower slopes and north- or east-facing aspects predominantly have low severity fire. These severity patterns are consistent with other regions in the Klamath Mountains (Agee 1991). Middle elevation forests typically have greater patchiness in fire severity with lower and higher elevations having greater homogeneity.

It is important to state that the historic and paleo fire records of the Klamath Mountains include active management by Native American Tribes—and these

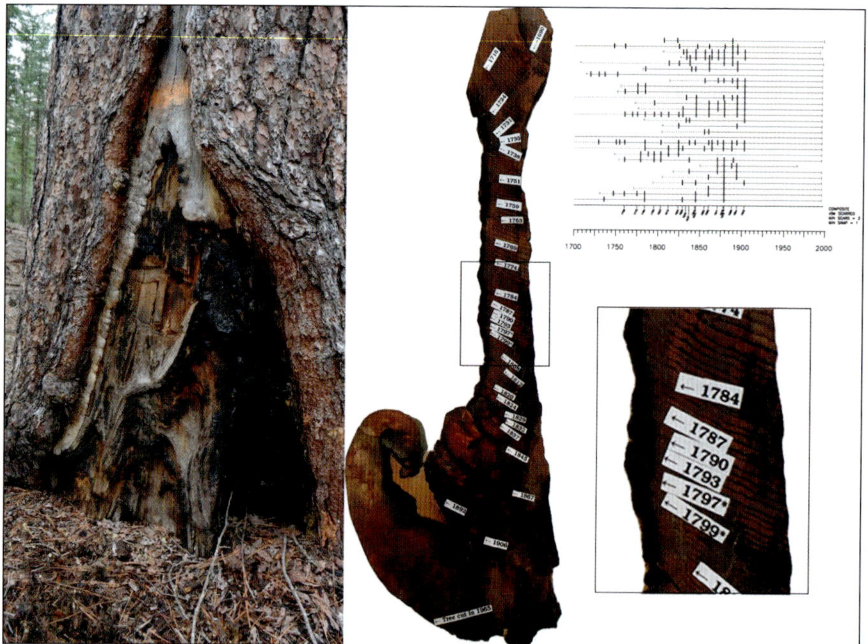

Figure 2: Fire scar studies provide information on the historical role of fire in the Klamath Mountains. Left- Picture of a fire-scarred sugar pine with many visible fire scars near Hayfork, CA (J.M. Kane). Center-crossdated fire scar slab of a sugar pine from Mule Creek (west of Trinity Reservoir) with inset (lower right) highlighting frequent fire from the late 1700s (J.K. Agee). Upper right- fire chronology from the Clear Creek watershed near Happy Camp, CA (C.N. Skinner).

practices continue today (see First People chapter). Much of the Klamath Mountain natural history is, in part, a legacy of past and current Native American activities. It is difficult to ascertain the degree of influence Native American Tribes historically had on the region, but their "fingerprint" on the landscape is likely quite extensive (Knight et al. 2022). The traditional knowledge and practice of fire management of Native Americans lives on and many Tribes are revitalizing the traditional use of fire. Native Americans use fire in many capacities and vary that use by the season, frequency, and intensity depending on particular objectives—such as basketry materials, food production, pest management, ceremonies, and much more.

Modern fire

Euro-American settlement of the Klamath Mountains began in earnest following the discovery of gold in 1849. From this time forward, the region experienced substantial changes to the landscape. Some of the early settlers adapted the use of fire from Native Americans to clear land and manage livestock. Eventually, Native American burning was almost entirely halted for a variety of reasons including death from disease or being forcibly displaced. Historic records and fire scar data reinforce that the early settlement period resulted in the dramatic reduction of fire throughout the region (Figure 2, page 32). In some areas however, intentional and unintentional ignitions generated by settlers resulted in less frequent but larger, more contiguous (less patchy) fires (Fry and Stephens 2006).

In the early 20th century, many foresters perceived wildfire as damaging to forests in the western U.S. These perceptions soon affected federal forest policy decisions. While some advocated for the beneficial role of "light burning" fires, these voices ultimately lost out to fire suppression practices that aimed to "protect" forests from fire. These practices were formalized as policies that aimed to extinguish and prevent all fires. The effectiveness of these policies was slower to take hold in the Klamath Mountains compared to other regions due to remoteness and steep terrain, but ultimately became the overarching regional dogma. Fire suppression policies continued for many decades and, in some form, still persist today. It is important to state that fire exclusion co-occurred with extensive logging that iteratively cut large sections of the low- to mid-elevation old-growth forests. These practices continued into the late 1980's and early 1990's. Many of the harvested areas were either planted or naturally regenerated, creating dense, young forests that were much more susceptible to high severity fire (Odion et al. 2010).

Despite the increasingly aggressive efforts to suppress fires in the Klamath Mountains throughout much of the 20th century, the frequency of large fires increased in the mid 1900s. Mean fire size, that escaped initial attack, in the Klamath Mountains increased from 1,500 to 4,500 acres (610 to 1,820 ha) during this time period (Miller et al. 2012). This increase in fire activity over the past few decades is largely attributable to increasing temperatures and greater aridity associated with climate change. Unlike other parts of California and the western U.S., increased fire activity has not included an observed increase in fire severity in the Klamath Mountains, but the size of high severity patches has increased (Odion et al. 2004, Skinner et al. 2018) and individual fires have had substantial levels of high severity (e.g., 2008 Panther Fire, south of Happy Camp). Areas that have burned twice over the past 27 years have also not shifted in their fire severity over time, suggesting that topographic factors are still the primary control of fire severity in the region (Grabinski et al. 2017). While the fire activity in the Klamath Mountain region has increased, it remains below pre-Euro-American settlement levels and a substantial fire deficit persists.

Topography and Fire weather

Topography is a major factor that influences fire regimes of the Klamath Mountains. Steep slopes encourage more intense fires. Northern aspects receive less solar radiation that results in cooler and wetter conditions than southern aspects, thus, northern aspects typically have lower fire behavior (Agee 1991, Taylor and Skinner 1998).

Lightning ignitions have caused the majority of fires during the modern period. Generally, lightning strike density increases from the western Klamath to the eastern part of the range as well as from lower to higher elevations (van Wagtendonk and Cayan 2008). In some years 100s to 1000s of lightning strikes can occur in one event. On June 21st, 2008, thousands of lightning strikes resulted in 11 individual fires (e.g., Panther Fire, Siskiyou Complex, Ukonom Complex) that eventually coalesced to burn over 190,000 acres (76,890 ha) in the Klamath Mountains. Yet, years with large areas burned from lightning ignitions are not always years with numerous lightning strikes.

Active fire seasons and extreme fires in the Klamath Mountains are often associated with three weather systems (Skinner et al. 2018). Two of these systems align with the Pacific high-pressure center and cold fronts that typically pass from the northwest to the southeast during the fire season. Strong southwesterly to westerly prefrontal winds are generated on the southern edge of a passing cold front, while strong northerly to northeasterly postfrontal winds are generated following the passage of a cold front. The other weather scenario that commonly facilitates

Figure 3: Smoke inversion above the Klamath River in 2017. Photo by Michael Kauffmann.

extreme fire conditions is when a north Pacific subtropical high-pressure system is aloft with dry air masses that subside and create warm and stable conditions that readily dry fuels. These conditions, in combination with the steep topographic gradients of the Klamath Mountains, can create a strong temperature inversion where mid-slope positions are warmer and drier than the valleys (Figure 3). In this situation, smoke will accumulate along the thermal belt of the inversion. This moderates fire behavior in the valley bottoms because the smoke layer inhibits solar radiation and further drying below the inversion (Estes et al. 2017).

Fire activity in many regions are also related to synoptic (large-scale) oceanic/atmospheric circulation patterns. The most familiar circulation pattern that affect fire activity in the western U.S. is called the El Niño Southern Oscillation—which is related to sea surface temperature changes in the southern Pacific Ocean that tend to occur over a three to five year cycle. During the positive, or warm, phase (El Niño), warmer conditions contribute to increased fire activity in the Pacific Northwest. During the negative, or cool, phase (La Niña), drier conditions are associated with increased fire activity in the southwestern U.S. Fire activity in the Klamath Mountains does not have a strong linkage with either of these phases—likely due to the region being situated within the pivot zone between the north and south dipole of the El Niño Southern Oscillation (Trouet et al. 2009). So in some years, fire activity may resemble conditions to the Pacific Northwest and, in other years, is more similar to conditions in the southwestern U.S. Conditions conducive to large fire years in the Klamath Mountains are associated with drier and less stable atmospheric conditions that coincide with the positive phases of the Pacific/North American and the Pacific Decadal Oscillation circulation patterns (Trouet et al. 2009). The alignment of the positive phases of these two climatic phenomena is associated with the development of a persistent high-pressure ridge along the Pacific coast during the winter that limits the amount and timing of precipitation to the region.

Plants as Fuel

In addition to topography and weather, fuels strongly influence the behavior of fires in the Klamath Mountains. Fuels include live and dead plant material that provide energy to ignite and sustain combustion. The regional plant diversity contributes to the diversity of fuel composition and structure present. The fire effects that result then feedback to influence vegetative structures and biological legacies (e.g., gaps, dense patches, large woody debris) that supply fuels for future fires and habitats of other organisms. In other words, fuels are more than fodder for fire, but they are an important ecological component to the Klamath Mountains natural history.

The flammability of fuels is based on particle and fuel bed properties. At the particle level, moisture content, chemical composition (e.g., terpene content), heat content (stored energy within the fuel), and density (mass per unit volume) are major factors of fire behavior. Both live and dead fuels vary considerably in their moisture content (Kane and Prat-Guitart 2018). Fuel moisture content of new foliage typically starts around 300 percent (proportion of water mass to dry mass) which slowly declines over the growing season to around 80–120 percent, as soil moisture decreases and chemical constituents (fats, carbohydrates, etc.) of the foliage increases. These lower live fuel moisture levels are fairly sustained over subsequent years for older foliage on conifers and other evergreen species. Fine dead fuel moisture is often between 5–15 percent during the fire season and are most influenced by changes in relative humidity and temperature. Warmer and drier air yields lower dead fuel moisture that increases potential ignitability and combustion of fuels. Dead fuel particles are classified based on the time it takes to gain or lose moisture relative to the equilibrium moisture content (stable moisture content when exposed to a constant temperature and humidity) or what is referred to as the time-lag concept. Smaller diameter particles (0.25 in [< 0.64 cm]) are considered in the 1-hr time-lag class because it takes about one hour to gain or lose moisture to reach two-thirds equilibrium moisture content. Fuel particles below 3 in (7.62 cm) in diameter are considered fine fuels (1-100 hr), which have a strong influence on fire behavior. Large, or coarse, fuels greater than the 3 in (7.62 cm) or 1000 hr fuels have a greater influence on fire effects, such as soil heating and smoke production. The amount, composition, and arrangement at the fuel bed level also exert an important influence on fire behavior and effects in the Klamath Mountains. Areas that have abundant fine dead fuels that are moderately dense contribute to faster rates of spread and longer flame lengths.

Fire Regimes

Fire regimes represent an organizational principle that provides an ecological and evolutionary context for Klamath Mountain ecosystems. In the way that climate represents the culminating patterns of day-to-day weather fluctuations in an area, fire regimes reflect the cumulative pattern of individual fires over time for a given area (Agee 1993). Relatively consistent fire regimes develop over time based on the characteristic seasonality, intensity, severity, size, patchiness, and frequency of fire. Fire regimes are multidimensional and can be classified in many ways, but are often simplified into three coarse categories: 1) low-severity, 2) mixed-severity,

Fire Ecology of the Baker Cypress

(LEFT) unopened and (RIGHT) opened serotinous cones of the Baker cypress. Photo by Jeff Kane.

Baker cypress (*Hesperocyparis bakeri*) is a rare conifer species with 5 of 11 metapopulations located in the Klamath Mountains. The species is the northernmost naturally occurring cypress in the world, reaching this point in southern Oregon, just outside the Klamath Mountains, near Flounce Rock north of Medford, Oregon. Within the Klamath Mountains, other stable and often prolific populations grow north of Seiad Valley, the upper reaches of the East Fork of Elk Creek southeast of Happy Camp, and in a few other isolated areas of the Marble Mountain Wilderness. It is a regionally distinct conifer by virtue of the airy-splayed bluish-green foliage held on wispy branches that—near the tops of the trees—are often upturned. The spherical, serotinous cones develop first with pointed umbos that become rounded with age. The bark of older trees has a distinctive reddish, flaky appearance and the trunk has a well-defined taper.

This species has a suite of traits that facilitate and respond well to stand replacing fire including shade intolerance, short stature, dense crown, lower branch retention, and serotinous cones. It typically co-occurs with highly flammable shrubs including greenleaf manzanita (*Arctostaphylos patula*) and tobacco brush (*Ceanothus velutinus*). The species' characteristics promote higher severity fires that result in greater seedling regeneration and reduced competition from non-serotinous conifer species. In the instance where stands have experienced a long absence of fire, other conifers can establish and eventually overtop and kill Baker cypress due to competition for light, moisture, and nutrients. Seeds require bare mineral soil to germinate and do not readily establish without fire. Large fires have burned many of the Baker cypress metapopulations in the Klamath in recent years. This has raised concerns that short-interval fires may contribute to population losses or declines if fire recurs in stands that have not yet produced cones (< 25 years).

A stand of Baker cypress in the Seiad Baker Cypress Botanical Area on the Klamath National Forest. In 2017, the Seiad and Horse Valley fires burned as a stand-replacing fire through much of this population. Photo by Michael Kauffmann.

and 3) high-severity (Figure 4) fire regimes (Agee 2007). All three occur in the Klamath Mountains. In this context, fire severity refers to the proportion of living biomass (trees, shrubs, grasses, etc.) killed or consumed in a fire. Fire severity is often inversely related to fire frequency so that high frequency fire regimes typically have low severity fires and low frequency fire regimes have high severity fires. Yet, there are some ecosystems where low intensity fire can result in high severity.

Fire regimes reflect the interaction of climatic, topographic, edaphic, biotic, and anthropic features in an area. The modern Mediterranean-type climate of the Klamath Mountains is characterized by a prolonged summer dry season. Most of the precipitation occurs between the late fall and early spring. This region also has a maritime influence due to the proximity to the Pacific Ocean—creating a strong west to east moisture and temperature gradient. The topographic complexity interacts with climate to create local variability in temperature and precipitation that can affect fire regimes. Steeper slopes will generally support high severity fire regimes, while less steep areas tend to have low- to mixed-severity fire regimes. However, fire regimes are also dependent on the traits of the dominant vegetation. For example, species with thinner bark tend to be associated with higher severity fire regimes whether fires are of low or high intensity. The presence of ultramafic substrates also influences fire regimes. The lower productivity of ultramafic sites limits tree density and fuel accumulation that promotes and sustains a lower severity fire regime. Additionally, the historic fire regime of the area was influenced by the fire activity of Native American Tribes. Areas that were closer to Tribal villages, cultivation sites, or along travel corridors likely had more frequent, low severity fire regimes.

Low severity fire regimes are associated with lower intensity (individual fires generate less energy from combustion) and higher frequency (fires occur more often) fires. In the Klamath Mountains, low severity fire regimes are typically characteristic of ecosystems located in hotter and drier areas that occur within the interior lowlands into the low- and mid-montane ecological zones. Common examples of ecosystems with low severity fire regimes in the Klamath Mountains includes prairies (grasslands), oak woodlands, redwood (*Sequoia sempervirens*) forests, Douglas-fir/tanoak (*Notholithocarpus densiflorus* var. *densiflorus*) forests, and ponderosa pine (*Pinus ponderosa*) forests. Prairies with low severity fire regimes are also located within the upper montane and subalpine ecological zones.

The lower montane ecological zone also contains chaparral and other shrublands that are characterized by a high severity fire regime, where fires tend to kill or consume much of the aboveground vegetation during fire. Some examples of shrub species include chamise (*Adenostema fasciculatum*), whiteleaf manzanita (*Arctostaphylos viscida*), Brewer oak (*Quercus garryana* var. *breweri*), toyon (*Heteromeles arbutifolia*), Lemmon's ceanothus (*Ceanothus lemmonii),* and deer brush (*Ceanothus integerrimus).* Adjacent and intermixed within these shrublands are scattered individuals or stands of knobcone pine (*Pinus attenuata*) or ghost pine (*Pinus sabiniana*), which are also characterized by high severity fire regimes.

Ecosystems in the mid to upper montane ecological zone have a mixed-severity fire regime due in part to high topographic variation and high diversity in woody

Figure 4: Fire regimes vary. TOP: Low severity fire in a stand of white fir. MIDDLE: Mixed severity in a stand of Jeffrey pine. BOTTOM Stand replacing, high severity fire in the mid-elevation montane conifer forest. Photos by Michael Kauffmann

plant composition and structure. Mixed-severity fire regimes typically contain a combination of low, moderate, and high severity patches. A diverse array of conifer and hardwood forests have a mixed-severity fire regime in this zone. Most conifer forests contain Douglas-fir in combination with other species such as sugar pine (*Pinus lambertiana*), knobcone pine, incense-cedar, white fir (*Abies concolor*), and Shasta fir (*Abies* x *shastensis*), western white pine (*Pinus monticola*), California black oak (*Quercus kelloggii*), tanoak, California bay (*Umbellularia californica*), and madrone (*Arbutus menziesii*). Wetter forests along rivers and streams in this zone similarly have a mixed-severity fire regime but are slightly denser and can contain other species such as big-leaf maple (*Acer macrophyllum*), canyon live oak, Pacific yew (*Taxus brevifolia*), giant chinquapin (*Chrysolepis chrysophylla*), and Port Orford-cedar. Forests in the mid to upper montane zone with ultramafic soils are composed of Jeffrey pine, incense-cedar, and Port Orford-cedar. The low productivity of these sites results in a lower severity fire regime. The exception is for Baker cypress, which has a high severity fire regime (see sidebar page 157). Montane chaparral and other shrublands also occur within the upper montane ecological zone. Greenleaf manzanita (*Arctostaphylos patula*), tobacco brush (*Ceanothus velutinus*), shrub tanoak (*Notholithocarpus densiflorus* var. *echinoides*) and bush chinquapin (*Chrysolepis sempervirens*) are three of the common shrub species associated with montane chaparral. Dense stands of montane chaparral generally have a high severity fire regime but are more often intermixed with mixed conifer forests with a mixed-severity fire regime.

Fire and Plants

It is difficult to prove that plants have evolved particular traits that are strictly adapted to fire regimes. However, the selective force of fire over millions of years has contributed to the development of adaptive strategies in plants to persist in fire prone regions. Many plants in the Klamath Mountains are dependent on fire and have suites of traits that correspond to the historic fire regime where they occur (Figure 5).

Tree species found in low and mixed severity fire regimes, such as redwood, Oregon white oak (*Quercus garryana* var. *garryana*), California black oak, ponderosa pine, and sugar pine, accrue sufficient bark early on to insulate and protect the cambium from fire. These species also have longer needles or leaves that form more optimally dense litter beds that enable fire to spread rapidly with shorter residence times (e.g., Engber and Varner 2012). Ponderosa pine, Jeffrey pine, and sugar pine also self-prune (lose their lower branches) which elevates their crowns and limits the transition of surface fires into the crown. Seeds of these species also benefit from fire because the consumption of litter exposes mineral soil required for germination and seedling survival. Younger individuals of incense-cedar, Douglas-fir, white fir, and Shasta fir are often susceptible to fire but with increased size they accumulate bark as they grow larger that can protect the cambium. However, these species do not sufficiently self-prune and have short needles or leaves, creating a compacted litter bed that promotes longer duration burning, traits more commonly associated with higher severity fire regimes.

Figure 5: Common fire-adaptive traits of plants in the Klamath Mountains. **Top-left**: greenleaf manzanita resprouts with skeletons from previous fire present. **Top-right:** serotinous cone of a knobcone pine. **Center-left**: Jeffrey pine with evidence of self-pruned branches. **Center-right**: post-fire beargrass flowering. **Bottom-left:** low intensity fire burning through a mixed-hardwood litter of California black oak and tanoak. **Bottom-right**: large fire scar on a large living tanoak.

Low- to mid-elevation woodlands and forests containing Pacific madrone are typically characterized by a low-to-moderate severity fire regime. Madrone are uniquely fire-adaptive with exceedingly thin bark but relatively large leaves that form a lower density litter bed. The resultant litter is relatively resistant to carrying fire (Varner et al. 2017), possibly due the chemical composition of the dead leaves. The poor flammability of the litter may aid in the protection of the thin-barked bole. However, Pacific madrone can also vigorously resprout from the base and crown following a fire. Seeds of Pacific madrone prolifically germinate from the soil seed bank following fire and may have smoke or heat-triggered germination.

The poor leaf flammability and mat-like form of prostrate ceanothus (*Ceanothus prostratus*) promotes persistence during fire and may indirectly facilitate the protection of other species growing within its mat. Many hardwood species in these forests and woodlands prolifically resprout or epicormically sprout following fire. Beargrass (*Xerophyllum tenax*) and some other herbaceous perennials respond by prolifically flowering after fire. Similarly, blueberry and huckleberry (*Vaccinium* spp.) will produce prolific flowers and fruits in the second year following fire.

Many tree species growing within high severity fire regimes have other traits and strategies including thin bark, low growing crowns, and poor litter flammability. Some tree species also have serotinous cones, or cones that remain closed and accumulate in their crown. When a fire occurs, the adult plant is typically killed, but the fire also volatizes the binding resins or disrupts the hydraulic conductivity that otherwise maintains cone closure. After the fire passes, the cone scales slowly reflex, release the seeds, and seedlings prolifically establish over the next two years. Knobcone pine and Baker cypress are regionally prominent examples of serotiny.

Other species, such as Brewer spruce (*Picea breweriana*), ghost pine, Engelmann spruce (*Picea engelmannii*), and yellow-cedar (*Callitropsis nootkatensis*) have thin bark and readily die in a fire. These species can persist in fire refugia, or areas that have low amounts of fuel, moist microclimates (e.g., lake basins, canyons), or other microsites (e.g., rocky outcrops) where fire cannot readily penetrate. Otherwise, these species mostly rely on recolonizing from unburned edges of adjacent forests or refugia over extended fire-free periods.

Lower elevation chaparral stands have numerous strategies to persist after high severity fire. Species such as chamise and toyon prolifically resprout following fire. Chamise has a specialized storage organ called a lignotuber that protects meristematic buds and allows resprouting after fire. The highly volatile leaves, low foliar moisture content, and retention of dead branches facilitate high flammability in stands of chamise. Other shrubs like whiteleaf manzanita, bush poppy (*Dendromecon rigida*), and deerbrush (*Ceanothus integerrimus*) do not resprout and the parent plant is readily killed by fire, but the germination of seeds, stored in the soil over many years (soil seed bank), are often stimulated by heat or smoke from the fire.

Fire in chaparral also results in the establishment of herbaceous post-fire ephemerals, or species that are largely only present for a short period following fire. Other perennial herbaceous geophytic species have a corm or bulb (storage organs) which are usually only present in a vegetative form, but are stimulated to prolifically flower in the year following fire. These species include soap plant (*Chlorogalum pomeridianum*) and whitestem frasera (*Frasera albicaulis*).

Fire and Animals

Animal species in the Klamath Mountains are both directly and indirectly reliant on fire regimes and the resultant plant communities and structures. Direct impacts include the influence of fire on animal survival and behavior. Often, concern arises surrounding the survival of animals during the passage of fire. Studies indicate that many animals in fire-prone ecosystems can survive fires. If fire regimes have sufficiently shifted away from historical conditions, or extreme fire weather conditions promote uncharacteristically intense fires, animals can be killed directly by fire. However, the more profound effect of fire on animals is through the maintenance and formation of habitats.

Fire is one of the main ecological processes that shapes biological legacies that serve as habitat for animals. Tree death and damage incurred by fire can promote or create hollows and cavities for birds and mammals. For instance, fishers (*Pekania pennanti*) commonly use mature California black oaks with cavities as den sites (Long et al. 2016). Large snags and dead woody debris generated from fires also provide habitat for a myriad of other animals. Fire-created snags provide a source of large woody debris that is deposited into streams and rivers—creating important aquatic habitats for fish and invertebrates. Stream and river nutrient loads are often enhanced by fires through the deposition of scorched leaves and killed trees. These deposits increase invertebrate populations which ultimately provide food for salmonids and other aquatic organisms.

Smoke trapped in a thermal inversion creates beneficial circumstances for fish bearing streams and rivers. This thick smoke reduces solar radiation that limits water heating and may increase water flow by reducing evaporative demand in the atmosphere and lessening transpiration via trees below the inversion (David et al. 2018). These conditions are particularly helpful to salmonids that require cooler water in freshwater streams and rivers during the summer to avoid lethal water temperatures and decrease susceptibility to pathogens.

Areas that experience frequent lower intensity fire promote an open forest structure that fosters high understory plant diversity. Many of these plants then provide food resources for invertebrate species. Additionally, hardwood trees that resprout following fire provide an excellent source of food for browsers such as Columbian black-tailed deer (*Odocoileus hemionus columbianus*) and Roosevelt elk (*Cervus canadensis roosevelti*). Fire also increases the nutritional value of grasses, which improves survival and fecundity of grazers.

The mixed-severity fire regime in the Klamath Mountains creates a diverse assemblage of vegetative structures including large openings, shrublands, patchy forests, and dense forest stands. Given that many animals have specific habitat needs, a landscape with a varied vegetative structure will contain higher animal diversity (e.g., Fontaine et al. 2009). Additionally, some species that are associated with late successional (old-growth) forest conditions also require varied forest structures. Recent evidence indicates that, in addition to requiring dense patches of older trees for nesting and roosting, the northern spotted owl (*Strix occidentalis caurina*) also utilizes edge habitats associated with open forests or shrublands for foraging that are created by fire. Given the status of this endangered species, promoting mixed severity fire regimes that retain old-growth forest components will likely aid in its recovery.

Future of Fire in the Klamath Mountains

Sustaining and supporting the natural history of the Klamath Mountains into the future must consider the role of fire to promote resilience during a period of rapid change. Climate projections include continued increases in average temperature, more variable annual precipitation, drier summers, and earlier springs—all of which will contribute to further increases in fire activity over the next century. This increase will likely result in both positive and negative outcomes depending on the magnitude of change and mitigation efforts employed. Generally, the increased size of fires will facilitate the reintroduction of fire into areas that have long been fire excluded—reducing fuel loads, promoting more early seral vegetation, and developing greater spatial heterogeneity.

The increase in fire activity will also likely result in undesired changes. For instance, the size of large, high severity patches in the region is projected to increase, potentially shifting many conifer stands to shrublands or hardwoods (Serra-Diaz et al. 2018). A projection that has been supported by recent findings within reburn areas, especially under more extreme fire weather conditions (Taylor et al. 2021).. If these larger, high-severity patches exceed the distance conifer seeds can disperse, the rate of recovery will be further inhibited—compounding hotter and drier conditions that further limit seedling survival and time to maturation (Tepley et al. 2017). This shift may also include the local extirpation of fire sensitive species, such as Brewer spruce—a Klamath Mountain endemic. Yet, these circumstances may also provide opportunities for greater use of managed wildfires and prescribed fire to promote desired conditions and effects in a way that sustains fire regimes and the beloved natural and cultural history of the Klamath Mountains.

Literature Cited and Suggested Reading

Agee, J.K. 1991. Fire history along an elevational gradient in the Siskyou Mountains, Oregon. *Northwest Science* 65:188–199

Agee, J.K., 1993. *Fire Ecology of Pacific Northwest Forests*. Island Press, Washington, D.C.

Agee, J.K. 2007. *Steward's Fork: A Sustainable Future for the Klamath Mountains*. University of California Press. Berkeley, California.

Briles, C.E., C. Whitlock, C.N. Skinner, and J. Mohr. 2011. Holocene forest development and maintenance on different substrates in the Klamath Mountains, northern California, USA. *Ecology* 92:590–601.

David, A.T., J.E. Asarian, and F.K. Lake. 2018. Wildfire smoke cools summer river and stream water temperatures. *Water Resources Research* 54:7273–7290.

Engber, E.A., and J.M. Varner. 2012. Patterns of flammability of the California oaks: the role of leaf traits. *Canadian Journal of Forest Research* 42:1965–1975.

Estes, B.L., E.E. Knapp, C.N. Skinner, J.D. Miller, and H.K. Preisler. 2017. Factors influencing fire severity under moderate burning conditions in the Klamath Mountains, northern California, USA. *Ecosphere* 8:e01794.

Fontaine, J. B., D.C. Donato, W.D. Robinson, B.E. Law, and J.B. Kauffman. 2009. Bird communities following high-severity fire: Response to single and repeat fires in a mixed-evergreen forest, Oregon, USA. *Forest Ecology and Management* 257:1496–1504.

Fry, D.L., and S.L. Stephens. 2006. Influence of humans and climate on the fire history of a ponderosa pine-mixed conifer forest in the southeastern Klamath Mountains, California. *Forest Ecology and Management* 223:428–438.

Grabinski, Z.S., R.L. Sherriff, and J.M. Kane. 2017. Controls of reburn severity vary with fire interval in the Klamath Mountains, California, USA. *Ecosphere* 8:e02012.

Kane, J. M. and N. Prat-Guitart. 2018. Fuel Moisture. Pages 1–13 in S. L. Manzello, editor. Encyclopedia of Wildfires and Wildland-Urban Interface (WUI) Fires. Springer International Publishing, Cham.

Knight, C.A., C.V. Cogbill, M.D. Potts, J.A. Wanket, and J.J. Battles. 2020. Settlement-era forest structure and composition in the Klamath Mountains: reconstructing a historical baseline. Ecosphere 11:e03250.

Knight, C.A., L. Anderson, M.J. Bunting, M. Champagne, R.M. Clayburn, J.N. Crawford, A. Klimaszewski-Patterson, E.E. Knapp, F.K. Lake, S.A. Mensing, D. Wahl, J. Wanket, A. Watts-Tobin, M.D. Potts, and J.J. Battles. 2022. Land management explains major trends in forest structure and composition over the last millennium in California's Klamath Mountains. Proceedings of the National Academy of Sciences 119:e2116264119.

Long, J. W., M.K. Anderson, L. Quinn-Davidson, R.W. Goode, F.K. Lake, C.N. Skinner, 2016. Restoring California black oak ecosystems to promote Tribal values and wildlife. Gen. Tech. Rep. PSW GTR-252. Albany, CA: U.S. Department of Agriculture, Forest Service, Pacific Southwest Research Station. 110 p.

McNamara, B.A., D.F. Greene, and J.M. Kane. 2019a. Strong dispersal limitation in postfire regeneration of Baker cypress, a rare serotinous conifer. *American Journal of Botany* 106:1566–1574.

McNamara, B. A., J.M. Kane, and D.F. Greene. 2019b. Post-fire fuel succession in a rare California, USA, closed-cone conifer. *Fire Ecology* 15:39.

Metlen, K. L., C.N. Skinner, D.R. Olson, C. Nichols, and D. Borgias. 2018. Regional and local controls on historical fire regimes of dry forests and woodlands in the Rogue River Basin, Oregon, USA. *Forest Ecology and Management* 430:43–58.

Miller, J. D., C.N. Skinner, H.D. Safford, E.E. Knapp, and C.M. Ramirez. 2012. Trends and causes of severity, size, and number of fires in northwestern California, USA. *Ecological Applications* 22:184–203.

Odion, D. C., E.J. Frost, J.R. Strittholdt, H. Jiang, D.A. Dellasala, and M.A. Moritz. 2004. Patterns of fire severity and forest conditions in the western Klamath Mountains, California. *Conservation Biology* 18:927–936.

Odion, D. C., M.A. Moritz, and D.A. DellaSala. 2010. Alternative community states maintained by fire in the Klamath Mountains, USA. *Journal of Ecology* 98:96–105.

Serra-Diaz, J.M., C. Maxwell, M.S. Lucash, R.M. Scheller, D.M. Laflower, A.D. Miller, A.J. Tepley, H.E. Epstein, K. J. Anderson-Teixeira, and J. R. Thompson. 2018. Disequilibrium of fire-prone forests sets the stage for a rapid decline in conifer dominance during the 21st century. *Scientific Reports* 8: 6749.

Skinner, C.N. 2003. A tree-ring based fire history of riparian reserves in the Klamath Mountains. Pages 116–119 in P. M. Faber, editor. Riparian habitat and floodplains conference proceedings. Riparian Habitat Joint Venture. Sacramento, CA, USA.

Skinner, C.N., A.H. Taylor, J.K. Agee, C.E. Briles, and C.L. Whitlock. 2018. Klamath Mountains Bioregion. Pages 171–193 in J. W. van Wagtendonk, N. G. Sugihara, S. L. Stephens, A. E. Thode, K. E. Schaffer, and J. A. Fites-Kaufman, editors. *Fire in California's Ecosystems*. University of California Press, Berkeley, California, USA.

Stuart, J.D., and L.A. Salazar. 2000. Fire history of white fir forests in the coastal mountains of northwestern California. *Northwest Science* 74:280–285.

Taylor, A.H., and C.N. Skinner. 1998. Fire history and landscape dynamics in a late-successional reserve, Klamath Mountains, California, USA. *Forest Ecology and Management* 111:285–301.

Taylor, A.H., and C.N. Skinner. 2003. Spatial patterns and controls on historical fire regimes and forest structure in the Klamath Mountains. *Ecological Applications* 13:704–719.

Taylor, A.H., L.B. Harris, and S.A. Drury. 2021. Drivers of fire severity shift as landscapes transition to an active fire regime, Klamath Mountains, USA. *Ecosphere* 12: e03734

Tepley, A.J., Thompson, J.R., Epstein, H.E., Anderson-Teixeira, K.J., 2017. Vulnerability to forest loss through altered postfire recovery dynamics in a warming climate in the Klamath Mountains. *Global Change Biology* 23:4117-4132.

Trouet, V., A.H. Taylor, A.M. Carleton, and C.N. Skinner. 2009. Interannual variations in fire weather, fire extent, and synoptic-scale circulation patterns in northern California and Oregon. *Theoretical and Applied Climatology* 95:349–360.

van Wagtendonk, J.W., and D.R. Cayan. 2008. Temporal and spatial distribution of lightning strikes in California in relation to large-scale weather patterns. *Fire Ecology* 4:34–56.

Varner, J.M., H.G. Kuljian, and J.K. Kreye. 2017. Fires without tanoak: the effects of a non-native disease on future community flammability. *Biological Invasions* 19:2307–2317.

Wills, R.D., and J.D. Stuart. 1994. Fire history and stand development of a Douglas-fir hardwood forest in northern California. *Northwest Science* 68:205–212.

Chapter 7: Cryptogams

Michael Kauffmann, Jordan Mayor, Dana York, Tom Carlberg, Eric Peterson, and Lucy Kerhoulas

Devil's matchstick lichen (*Pilophorus acicularis*) decorates a rock outcrop along the Smith River. Photo by Michael Kauffamnn.

Introduction

In 1883, August W. Eichler, a prominent plant taxonomist of the day, divided the plant kingdom into two groups: Phanerogamae and Cryptogamae. Phanerogams produce seeds and have obvious reproductive structures (like flowers and cones) while cryptogams are seedless, have inconspicuous reproductive structures, and reproduce via spores (see Spore sidebar). Eichler's original classification only included plants as cryptogams but the term has since expanded to include, among others, mushrooms and cyanobacteria (formerly known as blue-green algae). This collective group, while taxonomically incoherent with species from more than one kingdom, represent the most ancient lineages of land-dwelling species on Earth.

An astounding diversity of terrestrial cryptogams exists in the Klamath Mountains—represented by mosses, liverworts, lichens, ferns, forest mushrooms, and algae. From this region northward, cryptogams become increasingly significant in the forests of the Pacific Northwest due, in part, to the high moisture levels in the region. Moisture plays a key role in cryptogam reproduction.

The spore is the reproductive structure for all separate groups of cryptogams and is dependent on the presence of water for dispersal. Because of this, cryptogams are generally restricted to regions with more frequent precipitation and along perennial waterways in drier regions. It has been suggested that in damp coastal lowlands forests, like those found in the Coquille or Smith rivers, as much as half of the above-ground biomass is composed of cryptogams (Vitt et al. 1988). Further inland and upland into the Klamath Mountains, lichens decorate exposed rocks, ferns hide in north-facing rock crevices, and mushrooms flourish in the fall.

Cryptogams provide a wide range of ecosystem services including water retention in habitats, nutrient cycling, habitat creation, and food for animals. Bryophytes (mosses and liverworts) have developed amazing adaptations to capture and store water which benefits not only the long-term success of the species but also the ecosystems in which they live. Imagine the water storage potential of a forest with a bryophyte biomass of 50 percent! Cryptogams are also proficient at nutrient cycling and sequestration. For example, peat bogs hold evidence from atomic reactions from the 1900s in their many-foot-deep accumulations of peat.

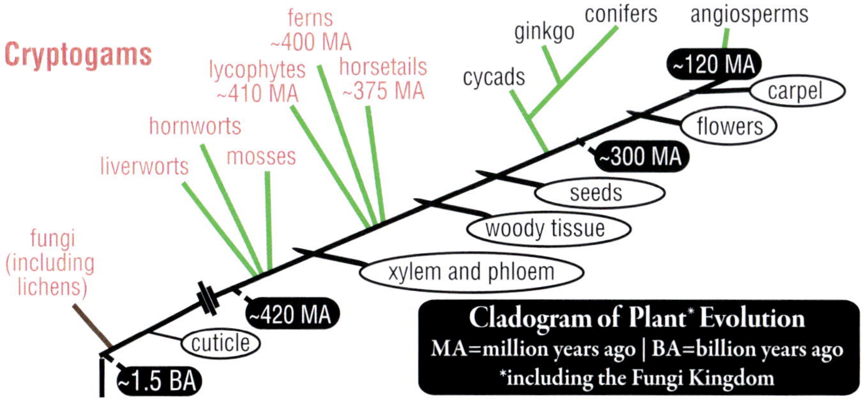

Cladogram of Plant* Evolution
MA=million years ago | BA=billion years ago
*including the Fungi Kingdom

Forest Fungi

Background

Globally, fungi represent a diverse group of over 100,000 described and ~1.5 million undescribed microorganisms that are both ubiquitous and critical for ecological functions of forests. These functions range from blooms of the saprotrophic *Cryptoporus volvatus* (see sidebar, page 268) which churn through burned forests accessing energy and macro-nutrients stored in dead trees after fire, to the biotrophic and symbiotic fungi that obtain energy from either killing or cooperating with living organisms (mostly plants).

Fungi are non-photosynthetic, primarily filamentous eukaryotes with absorptive nutrition, cell walls made of chitin, and reproduction through the emission of spores. They include not only filamentous fleshy fungi (mushrooms) or wood decomposers (shelf fungi), but also the many unicellular yeast-like fungi which ferment sugars in a wide range of habitats, including human intestines, floral nectaries, water filled traps of pitcher plants, and kitchens (e.g., baking, fermenting). Fungi also include lichens with one-fifth of all fungi estimated to be lichenized. The members of the Fungi kingdom include a great variety of growth forms all designed to disperse their spores across a wide variety of habitats in virtually every ecological niche[1]. Most large fleshy fungi encountered in the Klamath Mountains will fall into one of two main groups: the fleshy or woody mushroom and shelf forming members of the Basidiomycota and the cup forming Ascomycota[2] discussed further below. These broad groups are evolutionarily segregated at the broadest level by the design of their microscopic, spore-bearing structures—with either a club-shaped basidium typically bearing four reproductive spores or a tube-shaped ascus typically bearing eight spores.

Despite the categorization of fungi as micro-organisms, many fungi build large and elaborate spore-bearing fruiting bodies comprised of thousands of intertwined hyphal strands that are regularly visible—yet often highly transient. Paleo-historic peoples undoubtedly recognized and ate or used many mushrooms. Today we have documentation of such applications like the recent discovery of indigenous peoples applying *Rhizopus* spp. for the partial fermentation of foods in the New World (Henkel 2005), a practice known previously only in Asian fermentation processes (tempeh anyone?).

Beginning in 1729 with the book *Nova Plantarum Genera* by the Italian botanist Pier'Antonio Micheli (Alexopoulos et al. 1996) the scientific field of mycology (Gr. mykes = mushroom + logos = discourse), literally meaning a discourse of mushrooms, developed. While fungal-plant associations most likely predated roots, because early land plants did not have them, today symbiotic associations with roots may have been critical in enabling early Devonian plants to began

1. Did you know that estimates for the total diversity of fungi are often based on better known plant diversity? The ratios used 7:1 species of fungi for every plant on Earth (Hawksworth 2001, Brundrett and Tedersoo 2018) meaning there may be up to 25,000 species of fungi in the Klamath Mountain ecoregion!

2. Oddly enough, these two phyla seem to have settled upon separate seasons in which to fruit, with most basidiomycetes fruiting in fall and ascomycetes fruiting in the spring (especially after fires!).

growing in intertidal zones and gradually colonize the shores 450 million years ago (Remy et al. 1994). While liverworts do not have roots, such fungal roles are eerily similar to the arbuscular mycorrhizal fungi living on extant liverworts suggesting they haven't changed morphologically for millennia (Humphreys et al. 2010). Recent fossil discoveries from the Canadian Arctic have doubled the origin date of filamentous fungi to an astounding 1 billion years! This newly described fossil was named *Ourasphaira giraldae* by the paleo-mycologist team (Loron et al. 2019).

The story of how the estimated 1.5 million species of fungi (in the broad taxonomic sense) evolved and radiated across the Earth is not just fascinating, but also necessary to grasp the functional diversity of fungi in the Klamath Mountains. This is because most of the region's biodiversity, endemic species, and species vulnerable to climatic change are invertebrates, non-vascular plants, and fungi—not charismatic mega-flora. The many, persistently cool and moist, late-successional forests are responsible for cultivating the region's less-apparent biodiversity (Olson et al. 2012). Extensive forest elevation gradients, chaparral communities, and meadow/grassland ecosystems are the result of repeated glacial refugial contractions and expansions over the past 150,000–30,000 years (Whittaker 1960, 1961; Sawyer 2007)—conditions which allowed cool and moist microhabitats favored by dozens of conifer species and many hundred species of macrofungi.

The vast majority (Table 1) of the conifer species in the Klamath ecoregion form ectomycorrhizal (ECM) associations with hundreds of different fungal species that have moved and intermingled with the region's coniferous forests. Ectomycorrhizal forests generally contain much higher numbers of fungal taxa than their host plant communities (Tedersoo et al. 2014; van der Heijden et al. 2015). Studies from coastal pine forests along California's coast found that 15 ECM-forming species of fungi could colonize single trees and hundreds of species could be found within uniform stands of the same tree species (Saari et al. 2005).

Physiological ecology of forest fungi

So what are the fungi doing in our forests? In the Lichen section of this chapter we will learn more about the basidomycete and ascomycete fungi that make up the body of lichen host tissues. In Chapter 10 we learned that many fungi are pathogenic, attacking through wound entry points, burrowing beetle tunnels, or through root systems throughout the Klamath Mountain region. Because many fungi are adapted to the decomposition of wood or cellulose from a wide variety of plant materials, they possess a wide variety of enzymatic tools used to decompose

The Spore

Spores are small, one-celled (typically), reproductive units that form part of the life cycles of many plants, algae, fungi, and protozoa. Because spores evolved so long ago, they could be considered a less highly derived means for reproduction. But, the fact that they still persist after 425 million years speaks to the long-term success of the process. This mechanism arrived in early cryptogams to increase the progeny's dispersal success while also allowing for short periods of dormancy when conditions were suboptimal.

lignin-rich wood that would otherwise accumulate in our fire-prone forests. Such enzymes mine organic matter for mineral nutrients and attack the roots, leaves, and stems of living plants. Occasionally, pathogenic fungi have the ability to decimate trees and alter forest structures and carbon residency times. They make themselves known in the Klamath Mountains by fruiting in the fall during warm rains or in the spring along the melting snow lines and after fires have scorched trees and released nutrients.

ECM fungi absorb and deliver soil macronutrients in exchange for tree photosynthate and, owing to repeated evolution from saprotrophic ancestors, some decompose soil organic matter using oxidative mechanisms (Shah et al. 2016). Other commensal fungi may exist as endophytes inside pine and spruce needles, broadleaved tree leaves, or xylem cells for reasons that have largely escaped scientific understanding. ECM symbionts may enable coexistence among conifer trees through shared networks (Kennedy et al. 2003), or alter competitive advantages through specialization among smaller sets of species. Globally, ECM trees represent two percent of all plant species but constitute 80 percent of all tree stems outside of the tropics (Steidinger et al. 2019); those proportions are likely similar to the flora of the Klamath Mountain ecoregion.

In addition to ECM formers, other specialized morphotypes of mycorrhizal fungi are found within select plant lineages in the Klamath Mountains. For instance, within the Ericaceae family there are both an arbutoid-type of mycorrhizal associations with madrone and manzanitas (*Arbutus* and *Arctostaphylos* spp.) whereas the more general ericoid-type of mycorrhizal fungi are found in association with species of *Rhododendron*, *Gaultheria*, *Vaccinium*, and *Pyrola* (Smith and Read 2008). Therefore, the story of trees and shrubs in the Klamath Mountain ecoregion is best understood with a myco-centric perspective because the dominant plant communities in the Klamath Mountain ecoregion co-evolved with pathogenic and symbiotic fungi. It boggles the mind to think how their specificity for partnering with (or attacking!) one tree over another across the landscape could shape the plant community we see today.

Regional Mycology

Much of the mycological background in this section has dealt with understanding forest fungi from a global perspective. Here we now focus on the regional diversity of growth forms and select fungal species in the Klamath Mountains. Much of the taxonomic information is generalizable to the greater Oregon and California forests of the Klamath Mountains. If you would like to learn more, we recommend *Mushrooms Demystified* by Arora, *California Mushrooms* by Desjardin, and *Mushrooms of the Redwood Coast* by Siegel and Schwarz. Also, a handy California-based website was created by Wood and Steven (mykoweb.com). The mycoflora of most regions of the world, particularly in the tropics, is grossly understudied and this certainly applies to the diversity of cryptogams in the Klamath Mountains.

Fortunately, the Klamath Mountains are a unique, well-studied area. Late-successional forests in the Klamath Province were included in coordinated survey efforts under the Northwest Forest Plan. Strategic sampling across National Forest Lands took place over thirteen years (2000-01, 2003-06, 2009-10 and 2011-13) in a program known as the Survey and Manage program within the CA Northwest Forest Plan. Klamath region sampling was largely overseen by the U.S. Forest Service and associates, particularly Lisa Hoover of Six Rivers National Forest (retired), and Tom Carlberg, Scott Davison, Noah Siegel, and members of the Humboldt Bay Mycological Society. Their ongoing efforts have greatly expanded our understanding of mushroom diversity in the Klamath region—particularly in identifying rare species.

The author of this section (Jordan Mayor) had the privilege of participating in these survey efforts on our public lands throughout the coast range of Oregon and in the Klamath Mountains of CA in its early years (2001, 2003-04). Overall the Survey and Manage mushrooms surveys added at least 23 new fungal species to the CA Northwest Floristic Province, including the Klamath Mountains (Mycoportal 2018; Six Rivers National Forest, unpublished data) and added much needed understanding of natural history of the fungal species (i.e., tree host associations, forest stand characteristics, elevational preferences, etc.). Some taxa were found to be much more commonly encountered than previously thought (e.g., *Cantharellus subalbidus, Clavariadelphus truncatus, Craterellus tubaeformis, Gomphus floccosus, Otidea leporina, O. onotica, Phaeocollybia kauffmanii, P. olivaceae* and *Ramaria rubrievanescens*), whereas relative rarity[3] was confirmed in many other species (e.g., *Arcangeliella crassa, Bondarzewia mesenterica, Clavariadelphus pistillaris, Cortinarius cyanites, Phellodon atratus, Polyozellus multiplex, Rubroboletus haematinus,* and *R. pulcherrimus*). We hope this section will entice you to add to our distributional knowledge of Klamath Mountain mushrooms by participating in community science through documentation of our mycoflora on websites such as iNaturalist (https://www.inaturalist.org) and our project: The Biota of the Klamath Mountain Geomorphic Province.

To get started, you'll need rain gear, sturdy boots, a collecting kit, and a penchant for heading into the Klamath Mountains during potentially bad weather. A collecting kit should include sturdy containers and, if targeting gastronomic specialties—such as the delicious trumpet mushroom (*Craterellus cornucopioides*), Pacific golden chanterelle (*Cantharellus formosus*), the king or queen boletes (*Boletus edulis* var. *grandedulis* and *B. regineus*), or the illustrious black morels (*Morchella* spp.[4])—you will only need water resistant mesh bags and a collecting knife. If you're collecting to document biodiversity consider bringing plastic tackle boxes where you can nestle your specimens amongst moss for transporting. Back at home keep the boxes in the refrigerator (or outside) to make them last longer until you can identify your collections. Local mushroom club activities, regionally relevant books, and even university courses or workshops will catapult your mushroom ID skills as well. Try to team up with others for your fungal forays into the Klamath Mountains and don't be turned off by a bit of rain—instead enjoy the petrichor!

3. Rarity is harder to demonstrate than you might think! Mushrooms are ephemeral, sometimes persisting for only a few weeks in one location, so rarely encountering a fungus does not necessarily mean it is rare in the environment.

4. According to California Mushrooms, seven species of black morels are known from California – four are found following forest fires, two in non-burned forests, and one from urban landscapes.

no common name
Armillaria sinapina

Pinecone Amanita
Amanita magniverrucata

blewit
Clitocybe nuda

Silvery-Violet Cort
Cortinarius alboviolaceus

bird's nest fungus
Crucibulum laeve

witch's butter
Dacrymyces chrysospermus

Western Varnished Conk
Ganoderma oregonense

Smith's Slimespike
Gomphidius smithii

White Elfin Saddle
Helvella crispa

bear's head
Hericium erinaceus

sulfur shelf
Laetiporus gilbertsonii

no common name
Hygrophorus goetzii

Yellow Icicle Fungus
Mucronella flava

bleeding fairy helmet
Mycena haematopus

Western Yellow Mycena
Mycena oregonensis

bird's nest fungus
Nidula niveotomentosa

Brown Cup
Peziza arvernensis

Orange mock oyster
Phyllotopsis nidulans

fading pink coral
Ramaria rubrievanescens

no common name
Sclerotinia sulcata

eyelash cup
Scutellinia scutellata

no common name
Stropharia scabella

slippery Jack
Suillus tomentosus

turkey-tail
Trametes versicolor

Table 1. Mycorrhizal status of Klamath Mountain conifer species. 82% form ectomycorrhizal (ECM) associations.

Klamath Mountain Conifer Species	Family	ECM forming
White fir (*Abies concolor*)	Pinaceae	Yes
Grand fir (*Abies grandis*)	Pinaceae	Yes
Shasta fir (*Abies x shastensis*)	Pinaceae	Yes
Noble fir (*Abies procera*)	Pinaceae	Yes
Subalpine fir (*Abies lasiocarpa*)	Pinaceae	Yes
Pacific silver fir (*Abies amabilis*)	Pinaceae	Yes
Brewer spruce (*Picea breweriana*)	Pinaceae	Yes
Sitka spruce (*Picea sitchensis*)	Pinaceae	Yes
Engelmann spruce (*Picea engelmannii*)	Pinaceae	Yes
Douglas-fir (*Pseudotsuga menziesii*)	Pinaceae	Yes
Sugar pine (*Pinus lambertiana*)	Pinaceae	Yes
Western white pine (*Pinus monticola*)	Pinaceae	Yes
Foxtail pine (*Pinus balfouriana*)	Pinaceae	Yes
Ponderosa pine (*Pinus ponderosa*)	Pinaceae	Yes
Jeffrey pine (*Pinus jeffreyi*)	Pinaceae	Yes
Ghost pine (*Pinus sabiniana*)	Pinaceae	Yes
Lodgepole pine *(Pinus contorta* ssp. *murrayana)*	Pinaceae	Yes
Beach pine (*Pinus contorta* ssp. *contorta*)	Pinaceae	Yes
Knobcone pine (*Pinus attenuata*)	Pinaceae	Yes
Bishop pine (*Pinus muricata)*	Pinaceae	Yes
Western hemlock (*Tsuga heterophylla*)	Pinaceae	Yes
Mountain hemlock (*Tsuga mertensiana*)	Pinaceae	Yes
Incense-cedar (*Calocedrus decurrens*)	Cupressaceae	No
Port Orford-cedar (*Chamaecyparis lawsoniana*)	Cupressaceae	No
McNab cypress (*Hesperocyparis macnabiana*)*	Cupressaceae	No
Baker cypress (*Hesperocyparis bakeri*)	Cupressaceae	?
Common juniper (*Juniperus communis*)	Cupressaceae	Yes
Western juniper (*Juniperus occidentalis*)	Cupressaceae	?
Redwood (*Sequoia sempervirens*)	Cupressaceae	No
Western redcedar (*Thuja plicata*)	Cupressaceae	No

**Extirpated from Trinity Mountains in 1963. Data from Brundrett and Tedersoo (2018).*

Lichens

From brilliant chartreuse wolf lichens on the highest mountain conifers, valley oaks draped in dull-green beard lichens, and chaotic patterns and colors of a multitude of species on rock—lichens add diversity and broaden food webs of the Klamath Mountains. To appreciate lichens, one must first have some basic understanding of what they are. They are often mistaken for a variety of life forms that are more familiar to humans, particularly mosses. But, in fact, a lichen is unique in having stable mutualistic symbiosis between two or more partners that are wildly different. Stable symbiosis means that both partners benefit, but neither at the expense of the other.

Physiology

One of the most phenomenal aspects of the lichen symbiosis is how fundamentally different the partners are. In the process of categorizing Earth's living organisms, the first distinctions are the most profound—intended to separate those that are the *least* closely related evolutionarily. In a lichen symbiosis there is an entity from the fungal kingdom (a heterotrophic eukaryotic fungus without chlorophyll) paired with either an organism from the plant kingdom (an autotrophic eukaryotic green alga) or an autotrophic prokaryotic cyanobacterium—both of which produce food via photosynthesis. Sometimes, a lichen is a symbiosis involving all three!

Lichens are named for the fungal partner. Approximately 20 percent of all fungi are lichenized and of these, roughly 85 percent associate with a green alga, 10 percent with a cyanobacterium, and 3–4 percent with both. The discovery of basidiomycete yeasts that are regularly a part of some lichens added to the complexity of these unique organisms (Spribille et al. 2016). Much like the bacteria in human guts being vital to a healthy life, additional bacteria and fungi are likely vital for lichens—even modifying the form of their growth or their chemistry.

Lichens have no flowers, leaves, stems, roots. They have no tissues to filter or move food and water from one part to another. They have only rudimentary structures or barriers that control water intake and evaporation. Lichens exhibit *poikilohydry* in that they dry out during xeric periods and become wet during mesic periods. They receive nutrients (and contaminants) via *deposition* (when dust and airborne particulates land on them). Cyanolichens (cyanobacterial photo-partner) require liquid water to grow and they continue growing while fully saturated. Globally, cyanolichens are responsible for significant contributions to the nitrogen cycles in forests.

Substrate, Growth, and Reproduction

A trained eye will have a difficult time finding places without lichens. From the broadest perspective, lichens prefer cool, moist conditions. Worldwide, lichen diversity and biomass is greatest in boreal zones, particularly with a coastal climatic element. Substrates favored by lichens include bark, wood, rock, soil, moss, and almost any material that stays in place long enough for them to establish long-term colonies. While lichen symbionts interact strongly with each other, they have little impact on the substrate on which they grow.

Lichen Forms

Lichens are grouped into a few simple growth forms. Foliose lichens tend to have lobes that roughly resemble foliage: flat and leaf-like, with a lower surface that differs from the upper. Foliose lichens often lie more-or-less flat against the substrate. Fruticose lichens have narrow, round, or flattened branches—and can be droopy or erect—with no difference between the surfaces. Microlichens adhere tightly to their substrate, whether bark or rock, and include crustose, leprose, and other forms. The general rule is that if collecting a lichen the substrate is also extracted (without destroying the lichen) it's probably a crustose species. In North America, crustose species are less studied than the other forms, even though the majority of lichen species are crustose. In addition to those three most common forms, there are several variations. Squamulose lichens are distinctive but hard to describe with words: little flakes, small chips, scales (as in fish scales). Leprose lichens are macrolichens than grow tightly to the substrate and form a loose powder over it.

Lichens can be 1) Fruticose (shrubby or hairlike) 2) Foliose (leafy with a distinct upper and lower surface or 3) crustose (thin crust tightly attached to substrate). Photo by Shelly Benson.

Ecological studies of lichens in Oregon's temperate forests, just north of the Klamath Mountains, have demonstrated that lichen diversity varies along moisture gradients. These gradients include the species that are found from coastal to interior mountains and along moisture gradients within forests from the moist understory upward toward the sunnier, drier treetops (see Epiphytes sidebar this chapter). In the wet coastal lowland forests of the Klamath Mountains, the understory can be so damp that bryophytes (mosses and liverworts) cover most substrates and lichens are sparse. In these conditions, moisture-loving cyanolichens proliferate in the lower-canopy, the mid-canopy is dominated by a variety of generalist lichens, and the upper-canopy hosts a number of drier-habitat hair or beard-like lichens that disperse by wind and easily get caught on new branches.

Diversity changes with forest age as well. After a disturbance event, mid-canopy generalists are the first to colonize. Over time, moisture conditions stratify through the canopy, and gradually other lichens arrive in the appropriate microhabitats. Over long periods of time, pin-lichens can establish colonies. These are tiny lichens that avoid competition for substrate with other lichens by growing on bark sheltered from rain—surviving on humidity alone. Such microhabitats are slow to form in most forests—requiring textured bark, leaning trees, and sufficient girth on the lower side of the lean to be completely blocked from rain.

Lichen communities on rock tend to be more dominated by crustose species, which are difficult to identify. Therefore, ecological understanding of lichens on rock lags behind forest lichens. In contrast, there have been numerous studies of lichens on soil in cool deserts like the Great Basin. These are also primarily crustose species. Along with mosses and cyanobacteria, these form biological soil crusts (or simply "biocrusts") that retain fragile soil and enrich its quality. Biocrusts are less known in the Klamath Region although they exist commonly on soil in pockets around rock outcrops and sometimes in open oak savannas. Prior to annual grass invasions, they were likely quite abundant among oaks from Oregon southward throughout California.

As is the story with plants, the Klamath Mountains offer diverse habitats that meet the needs of a broad range of lichen ecologies (Table 2). Here, the Mediterranean-type climate of California meets the mesic Pacific Northwest. Species common further north like *Ramalina thrausta, Lobaria oregana, Usnea longissima, Peltigera pacifica, Umbilicaria lambii,* and *Umbilicaria rigida* reach their southern range extension in or near the Klamath Mountain region (McCune and Geiser 2009). A few species such as *Parmelina coleae* mark the Klamaths as the northern limit of their global range. A number are able to divert inland and trickle up the eastern side of the Cascades, such as *Esslingeriana idahoensis*. *Scytinium siskiyouensis*, *Sulcaria badia* (groovy-beard), and *Umbilicaria phaea* var. *coccinea* (also known as the lipstick lichen because of its red color)—all nearly Klamath endemics—found locally abundant within the region but rarely outside of it. The Klamath Mountains also represent a refugium for at least one Pleistocene glacial

Table 2. Select species and their distribution within the Klamath Mountains. Bear in mind that geographic distributions of lichens are poorly known, especially within this region.

Select Lichen Species	Distribution relative to the Klamath Mountains
Collema quadrifidum	Endemic
Cornicularia normoerica	Southwestern limit
Letharia gracilis	Endemic
Lobaria oregana	Southern limit
Peltigera pacifica	Southern limit
Ramalina thrausta	Southern limit
Scytinium singulare	Northern limit; coastal California endemic
Scytinium siskiyouensis	Near-endemic with one outlier south
Umbilicaria lambii	Southern limit
Umbilicaria phaea var. *coccinia*	Near-endemic with two outliers north
Umbilicaria rigida	Southern limit
Varicellaria rhodocarpa	Southern limit

relict. *Solorina spongiosa* is a calcium-loving lichen that was presumably as abundant 12,000 years ago in the Klamath Region as it is now in Canada.

Despite the attention to lichens and developments in lichenology in the West over the past half century, they remain mysterious and little known—especially when compared to popular groups of life such as birds or butterflies. A quick analysis of publicly accessible herbaria shows that even for common or widespread species (e.g., *Megaspora verrucosa, Phaeophyscia orbicularis, Dimelaena thysanota, Micarea prasina*), there is only one recorded instance from the Klamath Region. Under-exploration seems a likely explanation, given the abundant and diverse habitats encompassed by the ~8,300,000 acres.

It should be kept in mind that generalizations about lichens in the Klamath region are certainly based on insufficient data. Consider the 13 well-documented acres of the 30-year-old New Botanical Garden of the University of Zürich, where 149 species are reported. Compare this with the records for the Klamath region in publicly accessible herbaria: 660 species in 157 genera. This is based on 5,950 collections made within 8,300,000 acres; this averages to one collection per square mile. Of the 660 species, 256 (39 percent) are reported once, and 360 (55 percent) are reported only twice (Peterson 1998). Get out and document some lichens in the Klamath Mountains please! Use iNaturalist and join our project: Biota of the Klamath Mountain Geomorphic Province.

wila
Bryoria fremontii

no common name
Calicium viride

pixie cup lichen
Cladonia cenotea

powderhorn
Cladonia fimbriata

lipstick powderhorn
Cladonia macilenta

tube lichens
Hypogymnia sp.

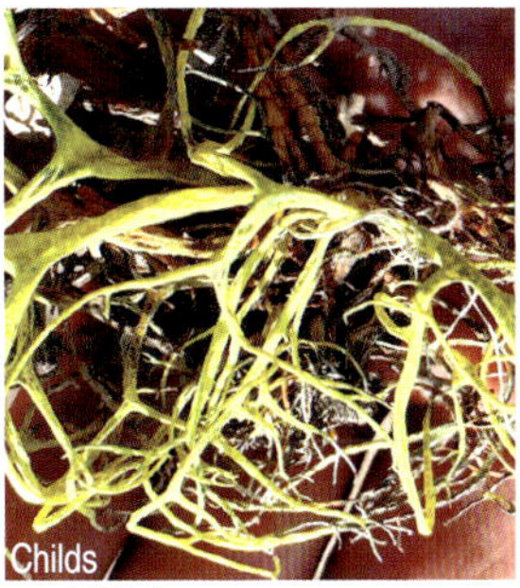

wolf lichen
Letharia gracilis

dimpled specklebelly
Lobaria anthraspis

pelt lichens
Peltigera sp.

green rock-posy
Rhizoplaca melanophthalma

snow lichens
Stereocaulon sp.

no common name
Tuckermanopsis merrillii

no common name
Verrucaria sp.

no common name
Vulpicida canadensis

sunburst lichen
Xanthomendoza hasseana

On a high-elevation serpentine rock in the eastern Klamath Mountains at least five species of lichens are seen including: 1) *Acarospora* sp., 2) *Candelariella* sp., 3) *Xanthomendoza* sp., (formerly *Caloplaca* sp.), 4) *Lecanora* sp., 5) *Lecidea sp.* (*L. atro-brunnea* group).

Epiphytes

Epiphytes are non-parasitic plants and fungi that grow on other plants. In the damp coastal forests of the Klamath Mountains, epiphytes like ferns, leafy liverworts, lichens, and mosses are common. These species perform invaluable ecological services for these forests including cyanolichens which fix nearly half of terrestrial nitrogen (Miller et al. 2017) and act as indicators for air quality and climate.

In the old-growth forests of the Pacific Northwest, epiphyte communities support high levels of biodiversity. Near the mouth of the Klamath River, 68 species of epiphytes were found in one charismatic Sitka spruce (*Picea sitchensis*) tree (Gorman et al. 2019). The synergistic interactions of epiphytes and big trees also support astounding animal diversity with mats of mosses and leafy liverworts creating habitat for a range of organisms including fungi, salamanders, mollusks, annelids, and arthropods. Birds, like the endangered marbled murrelet (*Brachyramphus marmoratus*), often use epiphytes for nesting material. Furthermore, decaying bryophyte mats in old-growth crowns can lead to soil accumulation, thereby promoting more biodiversity as plant and animal communities develop in these canopy soils (Ellyson and Sillett 2003). Finally, epiphytes provide shelter for arthropods, forage for birds, substrate for fungi, as well as nesting material for flying squirrels (*Glaucomys oregonensis*) and tree voles (*Arborimus pomo* and *A. longicaudus*) (Sillett and Antoine 2004).

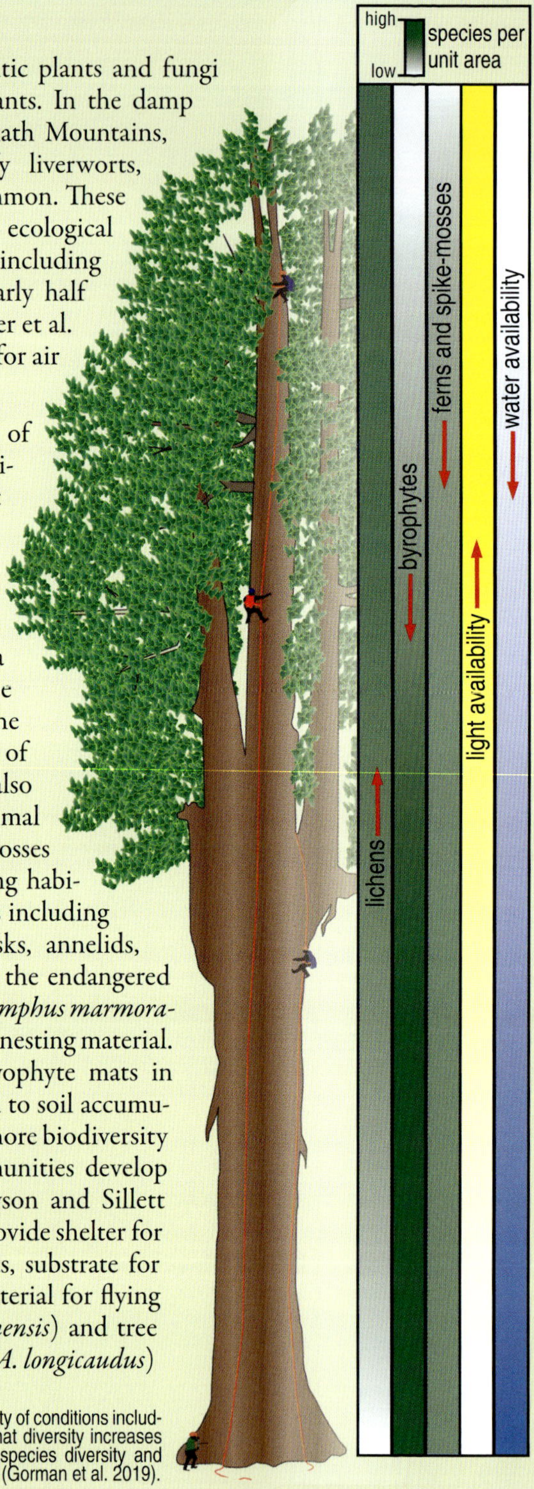

Researchers climb trees to study a variety of conditions including epiphytes. Some studies indicate that diversity increases with height–lichens show the highest species diversity and mosses have the highest percent cover (Gorman et al. 2019).

Bryophytes: mosses, liverworts, and hornworts

Bryophyte translates to "moss plant" and while mosses are bryophytes, so are liverworts and hornworts. For the past 400 million years or so, these three separate groups have been adapting and evolving to exploit a broad range of habitats where they avoid competition with vascular plants. The big picture is that the "moss plants" are composed of three separate phyla: Bryophyta (mosses), Marchantiophyta (liverworts), and Anthocerotophyta (hornworts).

Soon after plants colonized land the three bryophyte phyla began to diverge from one another. That deep-time evolutionary history has created differences but also unites these three groups. They are all non-vascular plants because they lack lignified vascular tissue. The lack of lignified vascular tissue is restrictive within the environments they live and therefore bryophytes achieve minimal size compared to vascular plants. All bryophytes lack roots and, instead, anchor to substrates with hair-like rhizoids. As with lichens and a few other vascular plants, all bryophytes are poikilohydric (unable to prevent desiccation). Most bryophytes are (more or less) able to be thoroughly dried out and, with water, resurrected. All bryophytes reproduce sexually via spores and some reproduce asexually via tiny clonal gemmae. Furthermore, bryophytes have swimming sperm that require liquid water for fertilization.

Bryophytes have life cycles dominated by the gametophyte phase. The green leafy part of a moss is the gametophyte, the plant that makes gametes (eggs and/or sperm). The phase that makes spores, the sporophyte, lives out its entire life attached to, and dependent on, its mother gametophyte. The sporophytes consist of a capsule usually on a long stalk called a seta.

Cryptogams

Seeds and sperm may be familiar concepts but other features uniting bryophytes involve cryptic aspects of their life cycle. All land plants have the same fundamental life cycle of alternating heteromorphic multicellular generations: the haploid generation (with one set of chromosomes) is the gametophyte that makes gametes (eggs or sperm) and the diploid generation (with two sets of chromosomes) is the sporophyte that makes spores. Bryophytes have a dominant gametophyte, and their sporophytes are short-lived, unbranched, and nutritionally dependent on the gametophyte. This situation is reversed in tracheophytes (vascular plants), where the sporophyte is dominant, relatively long-lived, and branched—while gametophytes are variously reduced and retained on the sporophyte. In other words, the green fuzzy moss is a gametophyte and the green leafy tree is a sporophyte.

Mosses

Mosses are the most familiar group of bryophytes, and with over 13,000 species they are also the most diverse. All mosses have leafy gametophytes and their leaves are spirally arranged upon the stem. Half of the 600+ mosses found in California are known from the Klamath Mountains. Like the vascular flora, the moss diversity is also associated with the Klamath Mountains' diverse geology, climate, topography, and plant communities. Vascular plants help provide shade, water retention, and soil stability for mosses. The favor is returned when mosses colonize rock outcrops and start a process of soil accumulation and moisture retention on shelves and in

antitrichia moss
Antitirichia sp.

Menzies' anacolia moss
Anacolia menziesii

plume moss
Dendroalsia abietina

grimmia
Grimmia sp.

mouse-tail moss
Isothecium stoloniferum

slender beaked moss
Kindbergia praelonga

bonfire moss
Funaria hygrometrica

Menzies' tree moss
Leucolepis acanthoneura

bladder-moss
Physcomitrium sp.

juniper haircap moss
Polytrichum juniperinum

bristle moss
Orthotrichum sp.

goose-necked moss
Rhytidiadelphus triquetrus

Table 3. Select mosses of the Klamath Mountains (Created by Dana York).

Soil Mosses: found mostly on soil, gravel, humus substrates or on rotting logs—many also grow on soil-covered rocks or boulders as well as tree trunks.	
Aulacomniaceae	drumsticks, lover's moss (*Aulacomnium androgynum*)
Bartramiaceae	swamp moss (*Philonotis fontana*) woolly apple moss (*Philonotis tomentella*)
Brachytheciaceae	whitish ragged moss (*Brachythecium albicans*)
Brachytheciaceae	Oregon beaked moss (*Kindbergia oregana*) slender beaked moss (*Kindbergia praelonga*)
Ditrichaceae	red roof moss, fire moss (*Ceratodon purpureus*)
Dicranaceae	Howell's broom moss (*Dicranum howellii*)
Hylocomiaceae	stair step moss, fern moss (*Hylocomium splendens*) bonfire moss (*Funaria hygrometrica*)
Hylocomiaceae	big shaggy moss, goose-necked moss (*Rhytidiadelphus triquetrus*)
Leucodontaceae	California antitrichia moss (*Antitrichia californica*)
Mniaceae	Menzies' tree moss (*Leucolepis acanthoneuron*) plagiomnium moss (*Plagiomnium venustum*)
Polytrichaceae	Lyall's haircap moss (*Meiotrichum lyallii*) juniper haircap moss (*Polytrichum juniperinum*)
Pottiaceae	Norway screw moss (*Syntrichia norvegica*) brown screw moss (*Syntrichia princeps*) great hairy screw moss (*Syntrichia ruralis*) controversial weissia moss (*Weissia controversa*)
Rock Mosses: found mostly on rock outcrops.	
Amblystegiaceae	claw brook moss (*Hygrohypnum ochraceum*)
Bartramiaceae	Menzies' anacolia moss (*Anacolia menziesii*)
Brachytheciaceae	Nevada homalothecium (*Homalothecium nevadense*) blunt-leaved moss (*Scleropodium obtusifolium*)
Orthotrichaceae	rock bristle moss (*Orthotrichum rupestre*) Grimmia (*Grimmia* spp.)
Polytrichaceae	hair-tipped haircap moss (*Polytrichum piliferum*)
Epiphytic Mosses: found on trunks or branches of vascular plants.	
Cryphaeaceae	plume moss (*Dendroalsia abietina*)
Hypnaceae	curly-leaf moss (*Hypnum circinale*) curly hypnum (*Hypnum subimponens*)
Lembophyllaceae	cat-tail moss (*Isothecium myosuroides*)
Neckeraceae	Menzies' neckera (*Metaneckera menziesii*) Douglas's neckera (*Neckera douglasii*)

crevices, especially along bedrock river gorges and moist cracks in subalpine rock outcrops. Creating soil that has some protection from the climate is the perfect nursery for many rock-loving vascular plants. Several rushes, sedges, and grasses are only found growing amongst mosses. The relationship between rock mosses and vascular plants is mutualistic—both benefit like partners in a good marriage. Mosses in California are generally divided into two main groups: peat mosses (*Sphagnum*) and mosses.

Peat Mosses (*Sphagnum*)

Sphagnum is not common in the Klamath Mountains but there are least seven species documented. *Sphagnum* species are mostly restricted to bogs, fens, and wet meadows. They differ from true mosses by having spreading branches that occur in fascicles or clusters. The *Sphagnum* sporophyte has a different capsule with spores distributed upward by an implosive mechanism that is triggered when moisture content drops below a certain level. *Sphagnum* has an incredible capacity to absorb large amounts of water. Because of this, indigenous Tribes use *Sphagnum* for such things as diapers, lining cradles, and wound dressing. Shaggy peat or shaggy sphagnum (*Sphagnum squarrosum*) is one of the more common peat mosses and is easily distinguished by its relatively large size and leaves spreading from the stem at right angles with protruding tips creating a shaggy appearance.

Mosses

For identification purposes, true mosses are further divided by two primary growth forms. Acrocarpous and pleurocarpous mosses have terminal (end of their stems) archegonia (female sex organs) and capsules. Archegonia and capsules on pleurocarpous mosses are born on lateral branches and not stem tips. They can also be grouped by the substrate a species is more likely to be found on. Identification to genus can often be done in the field, but most species require an examination of cell structure with a compound microscope to make an identification. See Table 3 for a list of common mosses of the Klamath Mountains.

Liverworts

Liverworts are the second most diverse group of bryophytes with approximately 5,200 species. Most liverworts are leafy and could be mistaken for mosses upon first impression. However, close examination of a leafy liverwort will reveal that, unlike mosses with their spirally arranged leaves, liverwort leaves are arranged in

great scented liverwort
Conocephalum conicum

common liverwort
Marchantia polymorpha

tree ruffle liverwort
Porella navicularis

two or three distinct rows. There are two lateral rows of leaves, and most leafy liverworts also have a third row of leaves on the ventral surface (not surprisingly, these are called underleaves). The non-leafy liverworts are considered thalloid. Eighty-one liverworts of the 135 known to occur in California are found in the Klamath Mountains.

Hornworts

Hornworts are the third and least common group of bryophytes. With fewer than 200 species worldwide, they are arguably relictual. Hornworts generally have a distinctive bluish-green thallus because of symbiotic cyanobacteria that live inside and provide a nitrogen source to the hornwort. They differ from a thalloid liverwort by having long needle-like capsules (horns). Five of the seven species known in California occur in the Klamath Mountains.

Lycophytes, Ferns, and Horsetails

These three groups have origins in the Devonian (420–360 Ma) with ancestors that were among the first vascular, spore-bearing land plants. All three groups are less highly derived than seed and woody tissue producing angiosperms and gymnosperms. However, the importance of the ascension of spores cannot be understated. Before the spore, distribution of offspring could only happen by water movement. These three groups are represented in the Klamath Mountain Region with highest diversity in the mesic coastal forests in the western part of the range. North of the Klamath Region, species richness and diversity continues to increase.

Lycophytes

In the Carboniferous (350-300 Ma) some lycophytes were forest-forming trees, reaching heights of 30 meters or more, but today, are a small and inconspicuous group. They are believed to be the oldest extant group of vascular plants having evolved about 410 Ma. Lycophytes have leaves called microphylls that contains one single, unbranched leaf vein. While a number of lycophytes are found in the fossil records, today the extant groups include spike-mosses, clubmosses, and quillworts (Table 4).

western quillwort
Isoëtes occidentalis

Oregon spikemoss
Selaginella oregana

Rocky mountain spikemoss
Selaginella scopulorum

Spike-mosses enjoy wet, shady environments but some are drought tolerant and can be found on rocky outcrops in the Klamath Mountains. Species here include Hansen's spikemoss (*Selaginella hansenii*), Rocky Mountain spikemoss (*S. scopulorum*), Oregon spikemoss (*S. oregana*), and Wallace's spikemoss (*S. wallacei*). A clubmoss has never been found in California's Klamath Mountains, but two species persist just to the west of the range in bogs and forests along the Pacific Coast. Running pine (*Lycopodium clavatum*) has been sporadically collected in Oregon along the low elevation river basins of the Siskiyou Mountains.

Quillworts grow in close association with water and four species are found in the Klamath Mountains where most prefer silt-bottom lakes. These species, from most to least common, are western quillwort (*Isoëtes occidentalis*), Bolander's quillwort (*I. bolanderi*), spiny quillwort (*I. echinospora*), and Howell's quillwort (*I. howellii*).

Ferns

Ferns evolved about 400 Ma and today have the most extant species of all vascular spore-bearing plants on Earth with over 150 genera and 12,000 species. All species have relatively large leaves divided into leaflets called pinnae. Stems, called rhizomes, grow horizontally at or below the soil surface.

Fern diversity in the region is concentrated, as one might expect, in the western coastal forests where as many as 15 species are found. But, somewhat surprisingly, a wide diversity is also found across the varied inland habitats depending on elevation, aspect, and the geological character of the landscape. A new species, Shasta maidenhair fern (*Adiantum shastense*) was recently described in Shasta County (Huiet et al. 2015). It was originally collected over 100 years ago and thought to be California maidenhair fern (*A. jordanii*). It is distinct in having dark brown to purplish brown rhizome scales and is persistent and green throughout the summer. *Adiantum jordanii* dies back during these dry months. It is currently known only from an area of the eastern Klamath Mountains, mostly on limestone.

Shield ferns (*Polystichum* spp.) have high diversity with eight species in the Klamath Mountains. Western sword fern (*Polystichum munitum*) is the most common fern in the western Klamath where it prefers the damp understory in coastal lowland forests, especially in the redwood belt. Surprisingly, relictual populations are found sporadically inland in the perfect mesic habitat. Narrowleaf sword fern (*Polystichum imbricans*) thrives on xeric, south-facing rocky slopes and is common across the lower to mid-elevations. Lemmon's holly fern (*Polystichum lemmonii*) is a common serpentine species in the mid- to upper-elevations of the Klamath Mountains with the only other populations occurring in the high Cascades of Washington. Mountain holly fern (*Polystichum scopulinum*) is a common species across the interior of western North America—reaching its western range in the Klamath where it occurs in the higher elevations of the Marble Mountains, Scott Mountains, and Trinity Alps. Northern holly fern (*Polystichum lonchitis*) is common across the Northern Hemisphere but reaches its southern range extension in the Klamath where it is an uncommon plant. Kruckeberg's holly fern (*Polystichum kruckebergii*) is a rarity that prefers rocky, subalpine habitats. It has only been reported in the Marble Mountains but is most likely found in other Klamath Mountain habitats—exclusively on carbonate rocks.

Table 4: Ferns, horsetails, and lycophytes of the Klamath Mountains. Adapted from Smith and Sawyer (2019).

LYCOPHYTES (8 TAXA)			
Family	**Common Name**	**Genus**	**Number of taxa**
Isoëtaceae	quillwort	*Isoëtes*	4
Selaginellaceae	spike-moss	*Selaginella*	4
FERNS AND HORSETAILS (46 TAXA)			
Family	**Common Name**	**Genus**	**Number of taxa**
Aspleniaceae	spleenwort	*Asplenium*	1
Athyriaceae	lady ferns	*Athyrium*	2
Azollaceae	mosquito fern	*Azolla*	1
Blechnaceae	deer ferns	*Struthiopteris*	1
	chain fern	*Woodwardia*	1
Cystopteridaceae	brittle fern	*Cystopteris*	1
Dennstaedtiaceae	bracken fern	*Pteridium*	1
Dryopteridaceae	wood fern	*Dryopteris*	1
	sword ferns	*Polystichum*	8
Equisetaceae	horsetails	*Equisetum*	4
Marsileaceae	water-clover ferns	*Marsilea*	2
	pillwort	*Pilularia*	1
Ophioglossaceae	moonworts	*Botrychium*	2
	rattlesnake fern	*Botrypus*	1
	grape-fern	*Sceptridium*	1
Polypodiaceae	polypody ferns	*Polypodium*	3
Pteridaceae	maidenhair ferns	*Adiantum*	3
	lacefern	*Aspidotis*	1
	parsley ferns	*Cryptogramma*	2
	lip ferns	*Myriopteris*	2
	cliffbrake ferns	*Pellaea*	3
	gold back fern	*Pentagramma*	1
Thelypteridaceae	Sierra marsh fern	*Thelypteris*	1
Woodsiaceae	cliff ferns	*Woodsia*	2

western maidenhair Fern
Adiantum aleuticum

California maidenhair fern
Adiantum jordanii

Shasta maidenhair fern
Adiantum shastense

serpentine fern
Aspidotis densa

alpine lady fern
Athyrium distentifolium

American parsley fern
Cryptogramma acrostichoides

brittle bladderfern
Cystopteris fragilis

lace lip fern
Myriopteris gracillima

Sierra cliffbrake fern
Pellaea brachyptera

Brewer's cliffbrake fern
Pellaea brewerii

goldback Fern
Pentagramma triangularis

western polypody
Polypodium hesperium

narrowleaf sword fern
Polystichum imbricans

Kruckeberg's holly fern
Polystichum kruckebergii

Lemmon's holly fern
Polystichum lemmonii

northern holly fern
Polystichum lonchitis

mountain holly fern
Polystichum scopulinum

leather grape-fern
Sceptridium multifidum

Cryptogams

Horsetails

Equisetaceae (horsetails) is a family of plants that evolved around 375 Ma. They have unique leaves, similar to lycophytes, called microphylls. Today extant members are concentrated in the non-tropical northern hemisphere. In the Klamath Mountains four species are found including common horsetail (*Equisetum arvense*) which is one of the ten most common plants in the world! Other species include rough horsetail (*E. hyemale*), smooth horsetail (*E. laevigatum*), and great horsetail (*E. telmateia*).

great horsetail
Equisetum telmateia

Acknowledgments

Many authors helped to make this difficult chapter a reality. Each subsection had a lead author or two. Jordan Mayor wrote the Forest Fungi subchapter; Tom Carlberg and Eric Peterson wrote the Lichen subchapter; Michael Kauffmann and Dana York wrote Bryophytes; and Michael Kauffmann and Dana York wrote the Lycophytes, Ferns, and Horsetails. Lucy Kerhoulas consulted throughout the writing. Thanks to all authors along with to Jim Agee and Marie Antoine for reviewing the final chapter.

Photo Credits

Photographic contributions by Shelly Benson, Julie Kierstead, Robert Childs, Thea Chesney, Ken DeCamp, John Game, Michael Kauffmann, Steve Matson, Eric Peterson, Christopher Lee, Christian Schwarz, Noah Siegel, and Dana York.

Literature Cited and Suggested Reading

Alexopoulos, C.J., C. Mims, M. Blackwell, eds., 1996. *Introductory Mycology, fourth edition*. John Wiley and Sons, Inc., U.S.A.

Brodo, I.M., S. Duran Sharnoff and S. Sharnoff. 2001. *Lichens of North America*. Yale University Press, New Haven and London. 795 pp.

Brundrett, M.C., L. Tedersoo. 2018. Evolutionary history of mycorrhizal symbioses and global host plant diversity. New *Phytologist* 220: 937-1336.

Castellano, M.A., Cázares, E., Fondrick, B., Dreisbach, T. 2003. *Handbook to additional fungal species of special concern in the Northwest Forest Plan*. United States Department of Agriculture, Forest Service, Pacific Northwest Research Station, Portland, OR.

Desjardin et al. 2019. Mykoweb. The Fungi of California. Available online at: https://www.mykoweb.com

Ellyson, W.J.T., S.C. Sillett. 2003. Epiphyte communities on Sitka spruce in an old growth redwood forest. *Bryology and Lichenology* 106(2): 197-211.

Gadd, G.M., ed., 2006. *Fungi in Biogeochemcial Cycles*. Cambridge University Press, U.K.

Gorman, AJ, L.P. Kerhoulas, W.T. Polda, N.J. Kerhoulas. 2019. Epiphyte diversity, abundance, and distribution in an old Sitka spruce crown. *Evansia*: 36: 12-22.

Hawksworth, D.L. 2001. The magnitude of fungal diversity: the 1.5 million species estimate revisited. *Mycological Research*, 105:1422-1432.

Henkel, T.W. 2005. Parakari, an indigenous fermented beverage using amylolytic Rhizopus in Guyana. *Mycologia* 97:1-11. Available online at: DOI:10.3852/mycologia.97.1.1

Horton T. R., R. Molina, and K. Hood. 2005. Douglas-fir ectomycorrhizae in 40- and 400-year-old stands: mycobiont availability to late successional western hemlock. *Mycorrhiza*, 15:393–403.

Honegger, R. 1996. Morphogenesis, pp. 65-87. In Nash, T.H. III (ed.) *Lichen Biology* (2nd. ed.). Cambridge University Press.

Huiet L, M. Lenz, J. K. Nelson, K.M. Pryer, A.R. Smith. 2015. *Adiantum shastense*, a new species of maidenhair fern from California. *PhytoKeys* 53: 73-81

Humphreys, C.P., Franks, P.J., Rees, M., Bidartondo, M.I., Leake, J.R., Beerling, D.J. 2010. Mutualistic mycorrhiza-like symbiosis in the most ancient group of land plants. Nature Communications, 1, https://doi.org/10.1038/ncomms1105

Kennedy, P.G., Izzo, A.D., Bruns, T.D. 2003. There is high potential for the formation of common mycorrhizal networks between understorey and canopy trees in a mixed evergreen forest. Journal of Ecology. 91:1071-1080. https://doi.org/10.1046/j.1365-2745.2003.00829.x

Cryptogams

Loron, C.C., François, C., Rainbird, R.H., Turner, E.C., Borensztajn, S., Javaux, E.J. 2019. Early fungi from the Proterozoic era in Arctic Canada. Nature. https://doi.org/10.1038/s41586-019-1217-0

McCune, B. and L. Geiser. 2009. *Macrolichens of the Pacific Northwest*. Oregon State University Press. 464 pp.

Miller, J., J. Villella, G. Carey, T. Carlberg, H.T. Root. 2017. Canopy distribution and survey detectability or a rare old-growth forest lichen. *Forest Ecology and Management* 392: 195-201.

Olson, D., DellaSala, D.A., Noss, R.F., Strittholt, J.R., Kass, J., Koopman, M.E., Allnutt, T.F. 2012. Climate change refugia for biodiversity in the Klamath-Siskiyou ecoregion. *Natural Areas Journal*, 32:65-74.

Peterson, E. B. 1998. Lichens of the Klamath Region: What do we know and why haven't we found rndemics? In: J. K. Beigel, E. S. Jules, and B. Snitkin (eds.), Proceedings of the First Conference on Siskiyou Ecology. Siskiyou Regional Education Project and The Nature Conservancy, Portland, Oregon.

Remy, W., T.N. Taylor, H. Hass, H. Kerp. 1994. Four hundred-million year-old vesicular arbuscular mycorrhizae. Proceedings of the National Academy of Sciences, USA, 91:11841–11843.

Saari S. K., C. D. Campbell, J. Russell, I. J. Alexander, and I. C. Anderson. 2005. Pine microsatellite markers allow roots and ectomycorrhizas to be linked to individual trees. *New Phytologist* 165:295–304.

Sawyer, J.O. 2007. Why are the Klamath Mountains and adjacent north coast floristically diverse? *Fremontia* 35:3-11.

Shah, F., et al. 2016. Ectomycorrhizal fungi decompose soil organic matter using oxidative mechanisms adapted from saprotrophic ancestors. *New Phytologist* 209: 1705-1719.

Siegel, N., Vellinga, E.C., Schwarz, C., Castellano, M.A., Ikeda, D. *A field guide to the rare fungi of California's national forests*. Bookmobile. Minneapolis, MN. 313 pp.

Sillett, S.C., M.E. Antoine. 2004. Lichens and Bryophytes in Forest Canopies. *Forest Canopies*, Margaret D. Lowman ed. 151-174.

Spribille, T., et al. 2016. Basidiomycete yeasts in the cortex of ascomycete macrolichens. *Science*. DOI: 10.1126/science.aaf8287.

Smith, J.P. and J.O. Sawyer. 2019. *The Vascular Plants of Northwest California*. California Native Plant Society. Sacramento, CA.

Smith, S.E., Read, D.J. (2008) *Mycorrhizal Symbiosis*, Third Edition. Academic Press, UK.

Steidinger, B.S., et al. 2019. Climatic controls of decomposition drive the global biogeography of forest-tree symbioses. *Nature* 569:404-408. https://doi.org/10.1038/s41586-019-1128-0

Tedersoo L., M. Bahram, S. Põlme, U. Kõljalg, N. S. Yorou, R. Wijesundera, et al. 2014. Global diversity and geography of soil fungi. *Science* 346:1256688.

Vitt, Marsh, Bovey. 1988. *Mosses, Lichens, and Ferns of Northwest North America*. Lone Pine Press.

van der Heijden M. G. A., F. M. Martin, M.A. Selosse, and I. R. Sanders. 2015. Mycorrhizal ecology and evolution: the past, the present, and the future. *New Phytologist*, 205:1406–1423.

Whittaker, R.H. 1960. Vegetation of the Siskiyou Mountains, Oregon and California. *Ecological Monographs* 30:279-338.

Whittaker, R.H. 1961. Vegetation history of the Pacific Coast states and the "central" significance of the Klamath Region. *Madroño* 16:5-23.

Chapter 8: Plant Communities

Michael Kauffmann, Julie Evens, Julie Kierstead, Michael Murray, and John O. Sawyer

An outcrop of ultramafic rock with complex plant assemblages in the high elevations of the Siskiyou Wilderness. Photo by Michael Kauffmann.

Introduction

The Klamath Mountains contain some of the most exceptional temperate plant communities in the world. Within its geographic boundaries, over 3,500 taxa (species, subspecies, and varieties) of vascular plants (Smith and Sawyer 2019), 35 conifer species (Kauffmann 2012), and 20 species of oak occur (Sawyer 2006). The broad vegetation patterns are similar to that in the Sierra Nevada—with a set of elevation belts ranging from foothill, montane, and subalpine. However, foothill woodlands and low-elevation chaparral are extensive only in the eastern Klamath, particularly in the southern foothills of the Trinity Mountains. The foothill forests of the western Klamath Mountains are classified as a temperate rainforest, receiving over 80 inches (200 cm) of precipitation per year. The main tree species of the higher elevation montane and subalpine forests are similar to nearby regions, but in the Klamath Mountains—due to heterogeneous topography, climatic complexity, and varied disturbance history—often exist as isolated patches rather than as a continuous forest. Even so, trees grow to the tops of the highest peaks and herbaceous plants that grow in other mountains in the subalpine occur in the Klamath Mountains at lower elevations.

A Deep Botanical History

Complex interactions between biotic and abiotic factors have encouraged and nurtured biodiversity in the Klamath Mountains over millions of years. The region hides relicts of epochs gone by called paleoendemics and also cradles other species—promoting the adaptive evolution of newer species called neoendemics. Complex climate and soils nurture microsites which, in essence, incubate biodiversity. The area also has a central location and continuity with other mountain ranges along the Pacific Slope. Across this landscape, a mosaic of habitats mix at a crossroads of five biotic regions—the Cascades, Coast Range, Great Basin, Central Valley, and Sierra Nevada—each helping to define the Klamath Mountains.

In the Tertiary, beginning around 65 million years ago (Ma), a temperate forest prevailed unlike any other in Earth's history. These forests existed on a landmass that would soon become North America, Europe, and Asia—a blending of conifers and broad-leaved trees dominated the landscape. With continental drift and climate change, the offspring of these great forests were fragmented. Over time, ice ages came and went, causing changes in floras as increasingly dry conditions became more common. The survivors became less extensive and more isolated, finding refuge in the higher and cooler regions which maintained a climate more similar to that of the early Tertiary—in what we now call northwest California and southwest Oregon. Today, we glimpse a forest that is similar to those of 65 Ma ago.

North America holds two of the most species-rich temperate forests in the world: those of the southern Appalachian and Klamath Mountains. What do these locations have in common? Glaciers and seas did not completely cover them during the Cenozoic and the mountains were monadnocks, or islands above the plains, offering refuge to plants and animals over time. Both locations have historically maintained a moderated climate and are beyond the southern terminus of the enormous continental ice sheets of the Pleistocene. Some plants remained through historic climatic change, while others repeatedly immigrated as climate cooled and glaciers advanced southward and then retreated following glaciers northward.

The current consequences of these historical patterns are that the Klamath Mountains and southern Appalachians have grand floristic diversity, a concentration of endemic plants, and a fundamental importance to the forest floras of nearby regions (Whittaker 1960, 1961). Per unit area, these two regions hold more plant taxa than any others in North America. Plant genera such as *Cornus* (dogwoods), *Asarum* (wild ginger), *Neviusia* (snow wreaths), *Clintonia* (blue-bead lily) and various conifers (*Pinus, Abies, Thuja, Chamaecyparis*) grow a continent apart while providing a comparative glimpse of an ancient flora.

It is also important to consider the effects Native People have and have had on vegetation communities across the Klamath Mountains. Everything we see today has been affected, to some degree, by long-term human habitation and land management practices. A major misconception of western science is the description of the assumed wild and the wilderness as absent of human impact—when this is far from the truth. Much of what we call wilderness today was, in part, sculpted by Native People's stewardship for its use as travel routes and crafting of goods including basketry, medicinal, and food resources to name only a few. For example, up north in British Columbia's coastal temperate rainforest Fisher et al. (2019) found that the plant communities around village sites had different plant assemblages than control sites and were dominated by plants with higher nutrient requirements and a cultural significance. Consider this next time you look at an oak woodland on a river bench.

This chapter will explore the region's diverse plant communities by starting in the lower elevations near the coast and moving inland and upward. The boundaries proposed are not observed strictly by all plants but instead offer a window into the ecology of species from the common to the rare. Plants will often be found in more than one of these vegetation delineations. It is important to remember that plant communities are a way to categorize patterns across the landscape. Human existence and ultimately our survival is, and always has been, dependent upon differentiating between, for instance, grasslands and forests and the plants that could be foraged there. The communities described herein are at the broadest of scales and a way to bring order out of the diverse landscapes of the Klamath Mountains.

The following treatments generalize vegetation into nine broad-reaching patterns. Within these patterns is a consistent overlap between the more than 3,500 regional plant taxa. For example, an important species to Native People, *Notholithocarpus densiflorus*, the tanoak, is found at the edge of chaparral, in the interior lowland forests, defining oak woodlands, and in the coastal lowland forests. Tanoak is highlighted here within the coastal lowland forest section but could also have been placed within the interior lowland forest or oak woodland sections. Because this is not comprehensive, we hope you are inspired to explore, read, and understand more about one of the most species-rich temperate floral regions on Earth and better interpret the many ways these patterns appear to us today.

Color Plate Legend

Tree

Shrub

Herb

Klamath mountain *Endemic species* (or near-endemic)

Figure 1: Sample vegetation zone distribution and exceptional conifers of the Klamath.

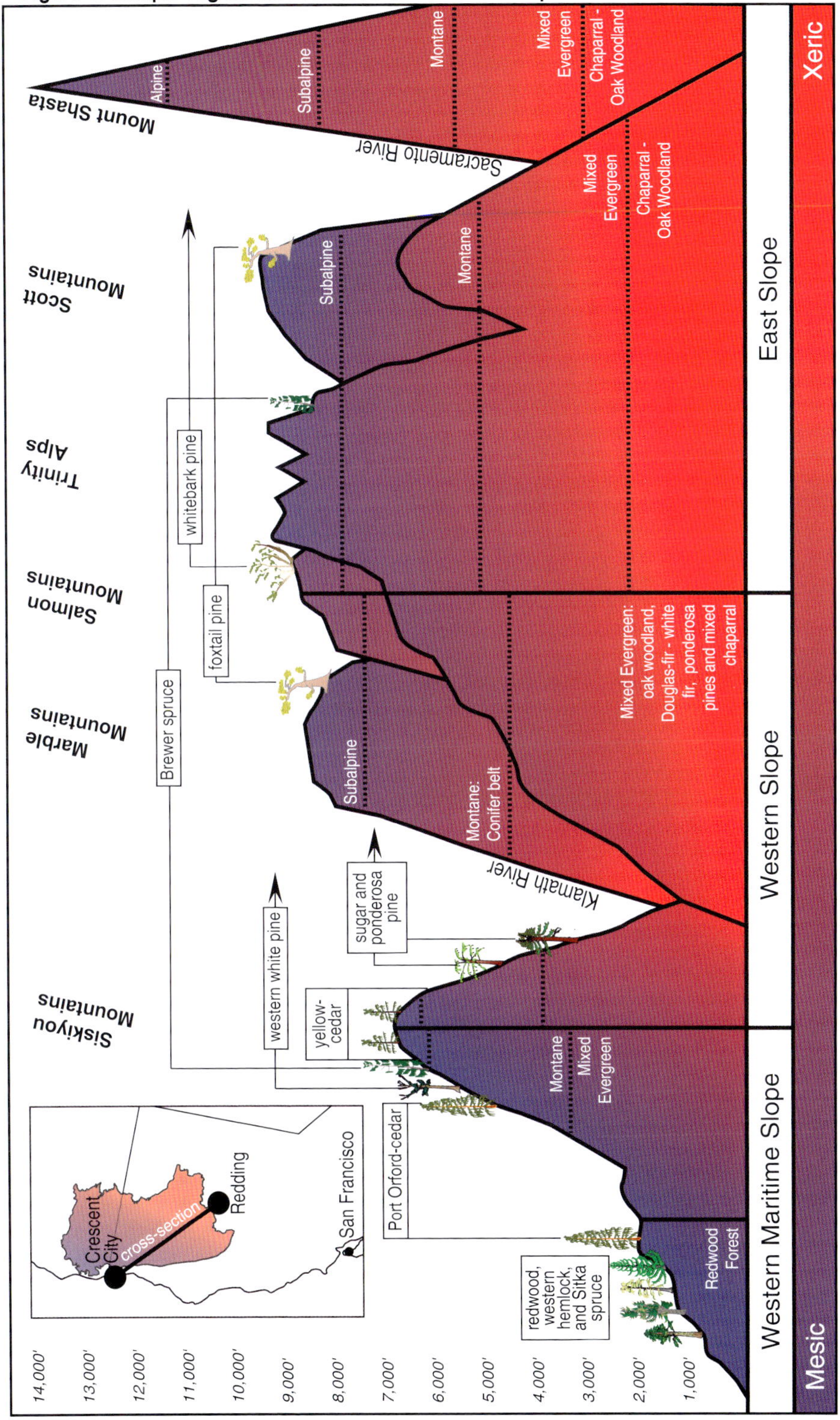

Plants

Coastal Forests

Coastal Forests of the Klamath Mountains.

This region receives the most annual rainfall of anywhere in the Klamath Mountains. From the north in Oregon, the Coquille, Rogue, and Illinois rivers drain southwest Oregon's rainforest. The Smith River country, in Del Norte County, holds California's only true temperate rainforest. The rich soils, year-round growing season, winter storms without hurricane force winds, and high rainfall with averages approaching 100 inches (250 cm) per year nurture these impressive forests with some of the tallest and most massive trees on Earth.

Klamath coastal forests are typified by dense, closed canopies on well-developed soils with occasional serpentine outcrops. The dominant tree species have high colonizing abilities (due to disturbance from flood or landslide), long life, and broad ecological tolerances to environmental conditions—including rain-free summers with persistent fog. These forests thrive below the snow belt on the extreme western slopes of the Siskiyou Mountains. Here, species like Douglas-fir (*Pseudotsuga menziesii*) and Port Orford-cedar (*Chamaecyparis lawsoniana*) secure the highest place in the canopy, approaching 265 feet (80 m) in height. A second canopy forms one hundred feet below and includes species like tanoak (*Notholithocarpus densiflorus*), giant chinquapin (*Chrysolepis chrysophylla*), and madrone (*Arbutus menziesii*). Any light that reaches the forest floor is used by shade-tolerant shrubs including salal (*Gaultheria shallon*) and evergreen huckleberry (*Vaccinium ovatum*). With the high rainfall each winter, coastal river banks and streams are frequently disturbed and repeatedly re-established by red alder (*Alnus rubra*) and grand fir (*Abies grandis*).

In addition to the deep, rich soils of the coastal forests, the largest serpentine outcrop in North America is found here. The Josephine Ophiolite (discussed further in the geology section) creates barrens of shrublands and stunted forests. Complex mosaics of lodgepole pine (*Pinus contorta*) and western white pine (*Pinus monticola*) occur at some of the lowest recorded elevations in California and Oregon. Poor soils nurture speciation events, highlighted by endemic shrubs like Howell's manzanita (*Arctostaphylos hispidula*) and Siskiyou kalmiopsis (*Kalmiopsis leachiana*).

Redwoods (*Sequoia sempervirens*), the world's tallest living plants at 380 feet (115 m), reach their northern range extension just north of the Oregon-California border. These impressive relic forests contain the most above-ground biomass anywhere on Earth. The majority of this biomass is accumulated by redwoods (including downed trees that are slow to decompose) and complemented by a number of other large conifers. Few redwood groves occur in the Klamath Mountains but many other tree species reach extraordinary size here.

red alder
Alnus rubra

bigleaf maple
Acer macrophyllum

Port Orford-cedar
Chamaecyparis lawsoniana

redwood
Sequoia sempervirens

western redcedar
Thuja plicata

western hemlock
Tsuga heterophylla

cascara buckthorn
Frangula purshiana

Pacific ninebark
Physocarpus capitatus

salal
Gaultheria shallon

Pacific rhododendron
Rhododendron macrophyllum

twinberry honeysuckle
Lonicera involucrata

evergreen huckleberry
Vaccinium ovatum

Select Coastal Forest Herbs

marbled wild ginger
Asarum marmoratum

Siskiyou Paintbrush
Castilleja elata

Oregon goldthread
Coptis laciniata

Oregon bleeding heart
Dicentra formosa ssp. *oregana*

Mahogany fawn lily
Erythronium revolutum

Siskiyou iris
Iris bracteata

Bolander's lily
Lilium bolanderi

Kellogg's lily
Lilium kelloggii

redwood sorrel
Oxalis oregana

western sweet coltsfoot
Petasites frigidus

thimbleberry
Rubus parviflorus

wedge leaf violet
Viola cuneata

Select Coastal Forest Vegetation Patterns

Along Oregon's Chetco River, the steep-walled canyons see frequent disturbance due to high winter flows. Red alder rapidly pioneer after these events.

Lower reaches of the South Fork Smith River nurture mixed-deciduous species along the river bank and mixed-evergreen on the slopes above the river.

Blue Creek, a tributary to the Klamath River, holds one of the most spectacular coastal forests in California including stands of massive Port Orford-cedar.

Coastal forest often have two canopy levels. Douglas-fir defines the upper canopy with a second tier including tanoak. Understory species include maples and dogwoods.

The dense understory in the coastal forests of southwest Oregon includes western sword fern (*Polystichum munitum*) and salal with an overstory of western hemlock.

Interior Lowland Forests

Interior Lowland Forests of the Klamath Mountains.

While several roads cut through the Klamath Mountains along the Trinity, Salmon, Klamath, and Sacramento rivers, the best way to get to know the interior lowland forests is by boat. Steep walled canyons, emerald blue pools, and perennial waterfalls are found at every bend (see the Water chapter). Throughout these lowland waterways, forest stands of Douglas-fir, ponderosa pine (*Pinus ponderosa*), California bay (*Umbellularia californica*) and tanoak intermix with a collection of oaks (treated in the next section). River banks are lined with white alder (*Alnus rhombifolia*), bigleaf maple (*Acer macrophyllum*), black cottonwood (*Populus trichocarpa*), Oregon ash (*Fraxinus latifolia*) and willows (*Salix* sp.). For the most part, these inland, lowland forests are a predictable mix of species that are often common in similar elevations across other regional mountain ranges.

The Trinity River tributary forms the grandest watershed in the region and is one of only a few places to find ghost pine (*Pinus sabiniana*), where it reaches its northern range extension along the Salmon River. Canyon live oak (*Quercus chrysolepis*) lines the canyon walls above riparian vegetation. In recently burned areas, watch for buckbrush ceanothus (*Ceanothus cuneatus*) and deer brush (*Ceanothus integerrimus*).

The South Fork of the Trinity River is highlighted by forests that are wild but not designated as wilderness. South from Willow Creek, the South Fork weaves it way toward Hyampom through fire-adapted landscapes with ghost pine, common manzanita (*Arctostaphylos manzanita*), western redbud (*Cercis occidentalis*), birch-leaf mountain mahogany (*Cercocarpus betuloides*), and a variety of other shrubs common in the Coast Ranges of central and southern California. As the South Fork winds its way upslope to the headwaters in the Yolla Bolly-Middle Eel Wilderness, mid-elevation conifer forests include incense-cedar (*Calocedrus decurrens*), Douglas-fir, sugar pine (*Pinus lambertiana*), and ponderosa pine.

The Sacramento River interior lowland forests are quite different because of the floral influence from the Great Basin and Cascades. Stands of ponderosa pine and Douglas-fir define the majority of the lower elevations in the watershed. Bigleaf maple and knobcone pine (*Pinus attenuata*) create a mosaic with chaparral. In the drier sites at the mid-elevations black oak (*Quercus kelloggii*) forms vast stands with ponderosa pine where bitter cherry (*Prunus emarginata*) and green leaf manzanita (*Arctostaphylos patula*) creep in the understory. The 2018 Delta, Carr, and Hirz fires stimulated massive germination of bush poppy (*Dendromecon rigida*) in the watershed where stand-replacing fire burned shrubby and forest overstory. The upper Rogue and Applegate rivers are similar to the Sacramento watershed.

Select Interior Lowland Forest Trees and shrubs

Pacific madrone
Arbutus menzeisii

Pacific dogwood
Cornus nuttallii

western redbud
Cercis occidentalis

Oregon ash
Fraxinus latifolia

tanoak
Notholithocarpus densiflorus

ponderosa pine
Pinus ponderosa

Fremont cottonwood
Populus fremontii

Pacific yew
Taxus brevifolia

birch-leaf mahogany
Cercocarpus betuloides

western burning bush
Euonymus occidentalis

Shasta snow-wreath
Neviusia cliftonii

California snowdrop bush
Styrax redivivus

Shasta huckleberry
Vaccinium shastense ssp. *shastense*

Hartweg's wild ginger
Asarum hartwegii

red baneberry
Actaea rubra

spotted coralroot
Corallorhiza maculata

umbrella plant
Darmera peltata

firecracker flower
Dichelostemma ida-maia

stream orchid
Epipactis gigantea

scarlet fritillary
Fritillaria recurva

warrior's plume
Pedicularis densiflora

Henderson's shooting star
Primula hendersonii

modesty
Whipplea modesta

narrowleaf mule's ears
Wyethia angustifolia

Select Interior Lowland Forest Vegetation Patterns

Black oak and ghost pine decorate the limestone outcrops above Lake Shasta.

Along Highway 299, ghost pine and canyon live oak line the Trinity River.

The Klamath River seen from Highway 96 near Somes Bar. Forest associations here includes Douglas-fir, ponderosa pine, and canyon live oak among other species.

Native People's Relationship with Geophytes

Mountain mahogany (*Cercocarpus ledifolius*) digging stick with elk (*Cervus canadensis*) antler handle. Photo by Frank Lake.

Geophytes are plants with underground storage organs. Energy and water are stored in a variety of structures including bulb, tubers, corms, or rhizomes. The global sweet-spots for geophytes are within the world's Mediterranean-type climates regions where—with summer drought and disturbance regimes including fire—the geophytic form of growth provides a successful survival strategy. The California Floristic Province includes a rich diversity of geophytes including nearly 240 species, or about five percent of the total native plant species in California. Forty percent of this diversity is in the lily family (97 species) that have their greatest importance in woodlands and grasslands that have been traditionally managed by Native People.

Indian potatoes or "root foods" are another name for geophytes because they are hidden under the ground. Indian potatoes were, and are, an important source of food for Native People throughout California because they are notably rich in carbohydrates. Some of the most important genera gathered by regional Tribes include *Allium, Brodiaea, Camassia, Chlorogalum, Calochortus, Dichelostemma, Lilium, Lomatium, Perideridia, Sanicula,* and *Triteleia.* The archeological record shows that geophyte harvesting is at least 10,000 years old on the Channel Islands and that bulbs, corms, and tubers were harvested across most of California during the last several thousand years (Andersen and Lake 2016). This storage organ even has its own month in many Tribal calendars indicating the most suitable harvest season. Before the ecological changes caused by European settlement, the geophytes of California were a staple food crop due to the fact they were abundant, widespread, storable, and nourishing. As little as 200 years ago, California was covered in geophyte flowers during the spring and summer—these were the gardens of the Native People.

Gathering the edible parts of geophytes, hidden below the surface of the soil, required extraction with a digging stick sharpened to a point. While digging time spanned the whole year, May to late October was the peak season. Among the Karuk, *Brodiaea, Calochortus, Dichelostemma,* and *Triteleia* are dug in the spring to summer and *Lilium* species in the fall. Traditional gathering sites are visited annually, and areas are specifically owned and maintained by particular families and passed down inter-generationally (Anderson and Lake 2016). Native People's harvest and management practices, including both consumption of large quantities of bulbs and corms and regular burning of geophyte habitats, allowed geophyte populations to thrive and created enormous bulb banks. Many of these still remained well after white settlement put an end to most of those practices. Next time you see a massive ground cover of lilies, thank the Native People for tending the wild and sculpting the vegetation communities we see today.

LEFT: Raw camas (*Camassia* sp.) bulbs and brodiaea (*Brodiaea* and *Triteleia* spp.) corms on a bed of bigleaf maple leaves prior to baking them in an earth oven. RIGHT: Cooked camas (*Camassia* sp.) bulbs and brodiaea (*Brodiaea* and *Triteleia* spp.) corms wrapped in a bed of bigleaf maple leaves after baking them in an earth oven. Photographs by Frank K. Lake.

Select Indian Potatoes with Karuk Names

imyuha

wavy-leaved soap plant
Chlorogalum pomeridianum

hanach'yu

scytheleaf onion
Allium falcifolium

pufish tayish

elegant brodiaea
Brodiaea elegans

xavin

elegant mariposa lily
Calochortus elegans

xiripipich

redwood lily
Lilium rubescens

axpaheekníkinach

cream fawn lily
Erythronium citrinum

Oak Woodlands

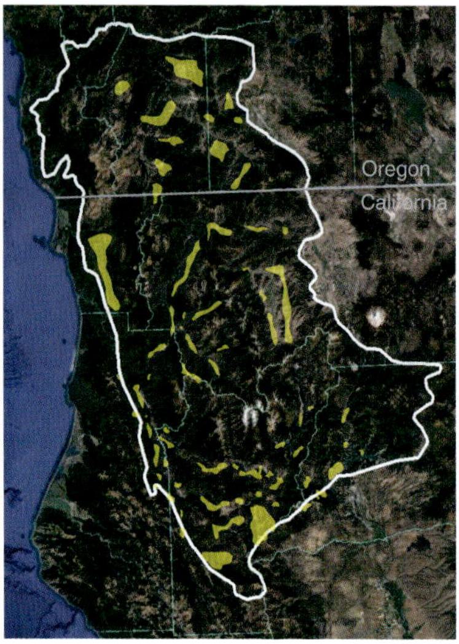

Mixed oak woodlands of the Klamath Mountains.

California is renowned for its diversity of oaks, a group that is a keystone in defining the legacy of our diverse landscape. Oak woodlands are one of the most important environments for Native American cultures and it is difficult to visit an oak woodland and not see the legacy of thousands of years of stewardship by them. Oak woodlands have been maintained in various ways—including frequent prescribed burns—in order to optimize food sources including access to acorns, geophytes, and game. Today, lack of fire and urbanization has led to the decline of many of the oak woodlands across the state.

Oaks are found across a wide range of ecological amplitudes in the Klamath Ranges. Stately canyon live oaks (*Q. chrysolepis*) and Oregon oaks (*Q. garryana*) occur commonly in the lowland forests. The Klamath Mountain endemic Sadler oaks (*Q. sadleriana*) enjoy the montane forests where dense stands fill the understory of the conifer-dominated canopy. The species is also quick to resprout after fire. But, again, these patterns do not always hold true. In the interior Klamath, canyon live oak are abundant on exposed, rocky sites above river and stream canyons. Oregon oaks populate riverside benches with thin soils or upland hillsides where frequent fire keeps Douglas-firs from overtopping them. On drier sites, California black oak associates with conifers and can even be found mixed with chaparral in the eastern Klamath Mountains.

In the southeastern Klamath Mountains, blue oaks (*Q. douglasii*) reach their northern range extension east of Shasta Lake around Montgomery Creek and in the Whiskeytown National Recreation Area. At Damnation Pass in Shasta County, *Q. sadleriana, Q. kelloggii, Q. vaccinifolia, Q. chrysolepis, Notholithocarpus densiflorus echinoides,* and *Chrysolepis sempervirens* are all found together. Along the Scott River, stands of interior live oak (*Q. wislizenii*) mix with those of canyon live oak.

In the higher elevations, montane chaparral is found on ridgetops, south-facing slopes, and areas with frequent fire return intervals. This habitat produces patches of huckleberry oak (*Q. vaccinifolia*), Brewer oak (*Q. garryana* var. *breweri*), and another species in the oak family (Fagaceae), bush chinquapin (*Chrysolepis sempervirens*). These species are foundational in montane chaparral communities.

Tanoak is endemic to the California Floristic Province and is a major component of the lower-elevation forests of the Klamath Mountains. It is also one of the species most seriously affected by sudden oak death (*Phytophthora ramorum*). Read more in the Forest Damagers chapter.

Select Oak Woodland Trees and Vegetation Patterns

white oak
Quercus garryana

Oregon white oak is commonly found in pure stands on benches with thin soils above riparian corridors.

Black oak
Quercus kelloggii

Canyon live oak, in the foreground, mix with black oak in the lower elevations of the eastern Klamath Mountains. In this environment, chaparral is often on south slopes with this mix on north slopes

blue oak
Quercus douglasii

Blue oak woodlands are only found in the eastern Klamath ranges, often mixing with chaparral.

canyon live oak
Quercus chrysolepis

Canyon live oak often forms pure stands on steep, rocky slopes along creeks and rivers.

Pacific Hound's Tongue
Adelinia grandis

Baby Blue Eyes
Nemophila menziesii

California brome
Bromus carinatus

Valley oak (*Quercus lobata*) is rare in the Klamath Mountains, reaching its northern range extension here.

poison oak
Toxicodendron diversilobum

ghost pine
Pinus sabiniana

Chaparral

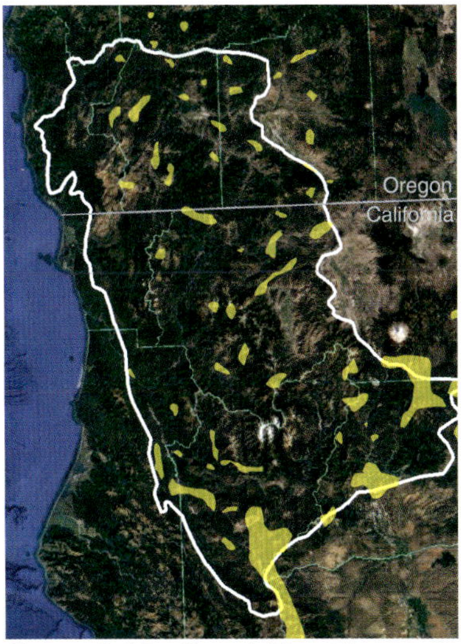

Chaparral of the Klamath Mountains.

It is impossible to understand the vegetation in California without discussing chaparral. Across the California Floristic Province, chaparral composes nine percent of the wildland vegetation and holds 20 percent of the plant species (Parker 2019). Mediterranean-type climates are rare on Earth today—occurring in only five areas around the world. These places share several commonalities, despite the vast distances between them. These include hot dry summers, cool wet winters, and frequent fire. When combined, these conditions favor shrublands as a dominant vegetation type. Mediterranean regions also share an unusually high number of locally endemic plants when compared to other vegetation types at similar latitudes.

In California, chaparral is characterized by evergreen shrubs and small trees with sclerophyllous (thick) leaves, shaped by summer drought and mild, wet winters. Fire is common and often of high intensity. The Klamath Mountains define the northern extent of the California Floristic Province and thus the northern extent of chaparral in North America. The largest and most contiguous stands of chaparral occur in central and southern California with less chaparral in the north state. However, in the xeric eastern Klamath Mountains chaparral covers vast areas.

Chaparral falls into a handful of broad categories with two main types occurring in the Klamath Mountains. The first is called mixed chaparral and is found in the low elevations of the eastern Klamath Mountains. The Whiskeytown National Recreation Area is a great place to find mixed chaparral with species like chamise (*Adenostoma fasciculatum*), whiteleaf manzanita (*Arctostaphylos viscida*), and several species of California lilac (*Ceanothus* spp.). A second type is montane chaparral. This typically occurs in south-facing openings and post-fire areas within the montane conifer belt and is defined by species like greenleaf manzanita (*Arctostaphylos patula*), pinemat manzanita (*A. nevadensis*), and huckleberry oak.

From lowland interior mixed chaparral upward into montane chaparral, knobcone pine feels right at home with a number of fire-dependent shrubs. These amazing trees need high-intensity, stand-replacing fires for two reasons. The first is to open the closed cones so seeds will fall to the ground. The second, also true for other chaparral species, is that fire will clear the ground of competitive species—thus opening habitat and facilitating rapid seedling recruitment. Because knobcone pine seeds are quick to germinate after a burn, seedlings establish first and will ultimately define the initial canopy in even-aged stands of saplings. Over time, fire will return and the cycles continue.

Select Chaparral Vegetation Patterns

Adenostoma-Ceanothus -Heteromeles mixed chaparral, typical across inland central and southern California, can be found in the lower elevations of the eastern Klamath Mountains.

High intensity fires reset this upland conifer forest to montane chaparral.

Chrysolepis-Quercus-Pinus attenuata chaparral is common on south slopes in the western Klamath.

Plants

Sawyer

In the higher elevations of the Trinity Alps Wilderness, montane chaparral is defined by the *Quercus vaccinifolia* alliance.

knobcone pine
Pinus attenuata

chamise
Adenostoma fasciculatum

pinemat manzanita
Arctostaphylos nevadensis

Del Norte manzanita
Arctostaphylos nortensis

greenleaf manzanita
Arctostaphylos patula

buckbrush
Ceanothus cuneatus

Lemmon's ceanothus
Ceanothus lemmonii

bush chinquapin
Chrysolepis sempervirens

yerba santa
Eriodictyon californicum

toyon
Heteromeles arbutifolia

huckleberry oak
Quercus vaccinifolia

Hollyleaf redberry
Rhamnus ilicifolia

Serpentines

One of the most interesting geologic features reflected in Klamath Mountain plant communities is told by the *ultramafic* (mega-igneous) rocks. Millions of years ago this rock formed deep in the ocean floor. As it cooled, heavy minerals concentrated (see Geology chapter). Today, these minerals constrain some plants from performing various metabolic functions; for example, high levels of magnesium restrict a plant's uptake of calcium. These restrictive soils interrupt the broader forest patterns across the Klamath Mountains, which hold some of the most extensive outcrops in North America.

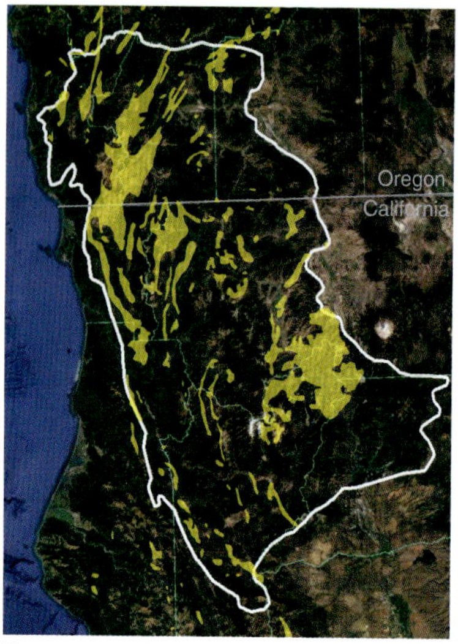

Serpentines of the Klamath Mountains.

Plants that do survive on these soils are slow-growing, even in areas like the Kalmiopsis Wilderness that receives 100 inches (250 cm) of rain per year. Many plant species, isolated on these outcrops, have become geographically and reproductively isolated from parent populations. In some cases, plants have evolved into new species. Sawyer (2006) recorded 200 neoendemic plants on the serpentine outcrops of northwestern California. On the Josephine ophiolite of the Smith River region, there are 70 endemic species—more endemics than any other serpentine outcrop in North America. This is considered the most diverse serpentine flora in western North America.

The transition from non-serpentine to serpentine is unmistakeable. The vegetation shifts are stark and dramatic—often with a dense wall of full-canopied forests abruptly ending and a sparse, red-rocked chaparral shrubland beginning. What makes serpentine habitats even more botanically interesting is that they occur from low elevations to mountain summits in both the mesic west and xeric east. The fact that they appear in such a broad range of habitats means that endemic species have arisen in just as many diverse places.

Because of the vast range of elevations, latitudes, and longitudes in which serpentine outcrops occur, vegetation can be classified in a variety of ways:

- Darlingtonia wetlands
- Foxtail pine subalpine woodland
- Jeffrey pine savanna and shrubland
- Montane mixed conifer forest
- Port Orford-cedar woodlands
- Rock outcrops and barrens
- Serpentine subalpine
- Serpentine riparian
- Serpentine meadows
- Serpentine lowland forest (coastal and inland)
- Western white pine subalpine woodland

Carnivorous Plants

Darlingtonia californica

Carnivorous plants have adapted to grow in places where nutrients are deficient. In the Klamath Mountains this often means peatlands (see Water chapter, page 144) and/or on serpentine soils. While carnivorous plants generate some energy from photosynthesis they supplement by trapping and consuming animals.

Cobra plant (*Darlingtonia californica*)

This insectivorous species is one of ten in the relict family Sarraceniaceae, and the only member of its genus remaining on Earth. The cobra plant supplements its diet by enticing insects inside its long, hollow stem where victims endure a slow death—by incarceration. The plant harbors bacteria and protozoa inside its "pitcher" and in a symbiotic relationship, the digestion is done by microbes and nutrients are shared. This species is restricted to fens.

Pinguicula macroceras

Horned butterwort (*Pinguicula macroceras*)

Restricted to serpentine soils where it grows on steep rocks or fissures in rocks near seeps, butterwort's roots must remain wet. The sticky light green leaves entice thirsty insects for a visit and the glands on the upper surface of the leaves initially trap the insect. On contact glands release more secretions, trapping the insect further, with more struggles, more stickiness, and the critter's fate is sealed—ensnared in sticky mucilage. Once the insect is trapped the leaves slowly curl over the carcass and release digestive enzymes to break down the "tasty" parts.

Drosera rotundifolia

Round-leaved sundew (*Drosera rotundifolia*)

One of the most widespread sundew species, it is found in lakes and fens from the coast to high mountain lakes. The hairy leaves extend from a basal rosette and hold sticky hairs on the rounded end. Insects are attracted to the sticky, sugary mucilage on the hairs and become stuck once in contact. The plant digests them from there. English sundew (*Drosera anglica*) also rarely occurs here.

Triantha occidentalis, photo by Morgan Stickrod.

Western false asphodel (*Triantha occidentalis*)

Among monocots, *Triantha* is the only known instance of a carnivory via a sticky-trap mechanism. For a short period every summer, flower stalks emerge for reproduction enlisting lucky invertebrates as pollinators and, at the same time, also trapping and digesting other invertebrates as prey along the stalk. This cryptic carnivore, only described as such in 2021, secretes a digestive enzyme to supplement upwards of 2/3 of its diet. Trapped spider circled in image.

Serpentine Vegetation Patterns

Windswept Jeffrey pine decorate Del Norte County's High Divide on the Josephine Ophiolite.

Deep in Oregon's Kalmiopsis Wilderness, western white pine and Port Orford-cedar woodlands rim a lake dotted with *Darlingtonia californica* in a classic serpentine-dominated landscape.

The western Siskiyous hold young ultramafic rock, so heavy in metals, that few plant species can survive on the soils.

Jeffrey pine savannas are common in the dry eastern half of the Klamath Mountains. Here in the Yolla Bolly-Middle Eel Wilderness.

Select Serpentine Associated Plants

incense-cedar
Calocedrus decurrens

Jeffrey pine
Pinus jeffreyi

leather oak
Quercus durata

California laurel
Umbellularia californica

whiteleaf manzanita
Arctostaphylos viscida

hoary manzanita
Arctostaphylos canescens

tobacco brush
Ceanothus velutinus

Siskiyou mat
Ceanothus pumilus

Serpentine milkweed
Asclepias solanoana

Waldo rockcress
Arabis aculeolata

Nuttall's toothwort
Cardamine nuttallii

Brandegee's spring beauty
Claytonia saxosa

Select Serpentine Associated Herbs

shorthorn steer's head
Dicentra pauciflora

Siskiyou fireweed
Epilobium siskiyouense

Siskiyou fritillary
Fritillaria glauca

Yolla Bolly bedstraw
Galium ambiguum

carrot-leaved horkelia
Horkelia daucifolia

Howell's lomatium
Lomatium howellii

Nuttall's sandwort
Minuartia nuttallii

flame ragwort
Packera greenei

brook wakerobin
Pseudotrillium rivale

Copeland's speedwell
Veronica copelandii

Tedoc stonecrop
Sedum rubiginosum

sickle fruit jewel flower
Streptanthus drepanoides

Siskiyou Kalmiopsis

In the 1930s, Lilla and John Leach were walking in the mountains west of Cave Junction when Lilla hesitated while gazing at a red patch of flowers—she got down on her knees and exclaimed "I have never seen anything like it before! Isn't it beautiful? I believe it is new." Her first impressions of the plant led her to conclude it might be a new form of black laurel (*Kalmia polifolia*) and previously not described by western science. She combined that belief with the Greek root "-opsis" meaning "seeing, like, sight, appearance" and coined a name that is forever in the heart and minds of plant lovers—*Kalmiopsis leachiana*. The two spent the following years searching for other locations throughout the rugged Siskiyou Mountains. Ultimately, a total of 20 individual populations are found within the 180,000 acre (72,840 hectare) Kalmiopsis Wilderness. A close relative, fragrant kalmiopsis (*Kalmiopsis fragrans*), is found outside the Klamath Mountains in the north Umpqua River region.

The Kalmiopsis Wilderness is a crossroads. Rainy season storms nourish the land with annual rainfall approaching 100 inches (250 cm) but this rain appears to have fallen in a virtual desert if one visits here in summer. Why is this? Serpentines! While the rich coastal soils foster the tallest trees in the world with the same amount of precipitation, the harsh soils of the Klamath Mountains nurture diminutive species just inland and upslope. Serpentine is not only restrictive to plant growth but also porous—so plants must be adept at water storage or retention to survive in the Kalmiopsis.

Beyond precipitation and geological character, this landscape is also sculpted by fire. The Biscuit Fire swept through the southwest corner of the state in 2002. It burned nearly 500,000 acres (202,340 hectares) across a landscape that receives significant rainfall yet is also exposed to extreme summer heat. The porous nature of the soil compounds the drying and in the summer of 2002, after a dry winter, the forest was ripe for a burn. The small amount of intact forest which survived the burn is not composed of large and/or old tree specimens. This evidence suggests that frequent fire is a common and necessary component here—and has been for thousands of years.

Montane Conifer Forests

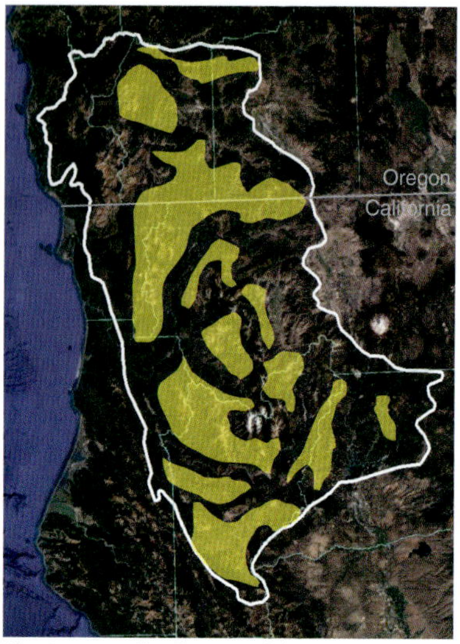

Montane conifer forests of the Klamath Mountains.

Conifers are ancient plants, and the places they survive today tell a story of migrations due to climatic changes over deep time. In this way, the montane conifer forests are the most compelling vegetation types in the region.

The white fir–Douglas-fir belt occurs at the lower extent (3,500–6,000 ft.) of the conifer belt and is the most common vegetation type in the Klamath Mountains. Mixing with these two dominant trees are ponderosa and sugar pines, at their upper elevational limits, along with another California Floristic Province specialist—the incense-cedar. These forests occur across a variety of soils and topographical features. At the mid-to-upper-elevations of this belt, white fir become the most common tree species. Both white fir and Douglas-fir's historic success across the region is due to their shade tolerance; they also have expanded their ranges due to the recent policy of fire suppression to grow timber over the last 100-150 years.

The appearance of Shasta fir (*Abies × shastensis*) south of the Klamath River and noble fir (*Abies procera*) to the north defines a gradient into the upper elevations of the montane belt (5,000–7,000 ft.) which eventually transitions to the even higher subalpine sky islands. While often quite extensive, this montane forest-type is broken up by pockets of montane chaparral, dwarf white oak woodlands, or serpentine outcrops. Other species associating with Shasta fir include mountain hemlock (*Tsuga mertensiana*), Pacific silver fir (*Abies amabilis*), and the endemic Brewer spruce (*Picea breweriana*). Near lakes and meadows, lodgepole pines (*Pinus contorta* ssp. *murrayana*), subalpine fir (*Abies lasiocarpa*), western white pine, and Engelmann spruce (*Picea engelmannii*) thrive.

Two eminent ecologists, Dale Thornburgh and John Sawyer, first documented the richness of conifers of this region in 1970, by defining a stand of 11 conifer species in the Horse Creek Range. In subsequent research trips their count jumped to 17 species surrounding the Duck and Sugar lakes basin in the Russian Wilderness. In 2014, western juniper (*Juniperus occidentalis*) was documented—establishing a temperate record of 18 species in one square mile. Dale and John named these diverse conifer assemblages the "enriched conifer stands" of the Klamath Mountains, and the one square mile in the Russian Wilderness is now known as the *Miracle Mile*. These forests occur for a variety of reasons including the frequency of small-scale disturbance, the interaction of complex soils, topographical diversity, and extreme climates (Sawyer 2006). Other notable enriched stands occur at Bear Basin Butte, Cedar Basin and Rock Creek Butte, Broken Rib, and China Mountain.

Select Montane Conifer Forest Trees

Pacific silver fir
Abies amabilis

white fir
Abies concolor

noble fir
Abies procera

Shasta fir
Abies × shastensis

common juniper
Juniperus communis

Engelmann spruce
Picea engelmannii

lodgepole pine
Pinus contorta ssp. *murrayana*

sugar pine
Pinus lambertiana

western white pine
Pinus monticola

Douglas-fir
Pseudotsuga menziesii

mountain hemlock
Tsuga mertensiana

Brewer spruce

At the beginning of the Cenozoic, about 65 Ma, a once-great flora existed across the Northern Hemisphere at a time when the Earth's climate was wet and warm. Species like magnolia and tulip trees flourished, co-habitating with conifers such as redwood and spruce. Gradually, the climate cooled when, for example, mountain-building events altered weather patterns. Populations of these once-vast forests became fragmented—members of the flora either went extinct, retracted southward, or became restricted within isolated pockets which retained moist, temperate climates. The Klamath Mountains offered, and continue to offer, these conditions; thus cradling several relicts from that nearly extinct floristic assemblage of the Tertiary, including the Klamath Mountain endemic Brewer spruce (*Picea breweriana*).

Brewer spruce currently favors the Klamath Mountains' cool, wet winters and warm, dry summers. With a relatively broad ecological amplitude within a narrow ecological range, it survives on numerous soils and across varied topographic conditions. Brewer spruce prefers unsaturated *terra firma* during the growing season and often establishes on rocky habitats (commonly ridgelines) that have remained fire-free. These rocky places are important in that they inhibit the encroachment of fire due to a lack of vegetation and ultimate fuel loading underneath, and around, trees. Fires have likely played a role in restricting populations to localized environs because the thin bark of Brewer spruce does not protect cambium well.

The branches of the species have a distinctive "weeping" habit. A multitude of pendulous branches hang like tinsel on a Christmas tree—often dangling eight feet in length. The weeping habit makes Brewer spruce easily identifiable at a distance. Crowns are often sparse and open with much shorter tendrils than are found lower on the tree.

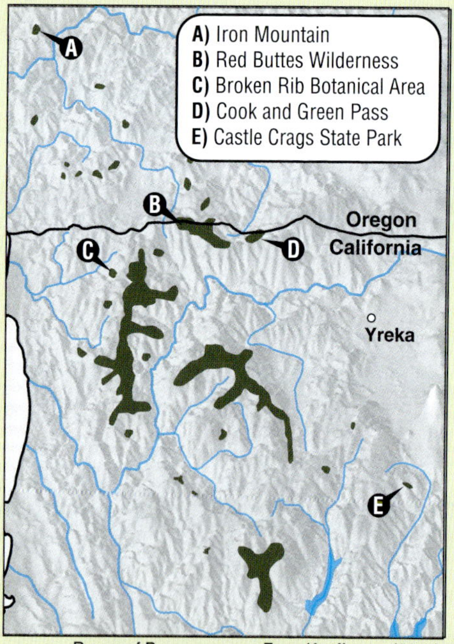

Range of Brewer spruce. From Kauffmann (2013).

Plants

Select Montane Conifer Forest Shrubs

vine maple
Acer circinatum

Rocky Mountain maple
Acer glabrum

serviceberry
Amelanchier alnifolia

Klamath manzanita
Arctostaphylos klamathensis

mountain grape
Berberis aquifolium

deerbrush ceanothus
Ceanothus integerrimus

mahala mat
Ceanothus prostratus

red osier dogwood
Cornus sericea

western wintergreen
Gaultheria ovatifolia

Sierra laurel
Leucothoe davisiae

Oregon boxwood
Paxistima myrsinites

Lewis's mock-orange
Philadelphus lewisii

whiteveined wintergreen
Pyrola picta

Sadler oak
Quercus sadleriana

Labrador tea
Rhododendron columbianum

prickly currant
Ribes lacustre

wood rose
Rosa gymnocarpa

mountain ash
Sorbus sp.

Siskiyou Mtn. huckleberry
Vaccinium coccineum

poke knotweed
Aconogonon phytolaccifolium

vanilla leaf
Achlys triphylla

American trailplant
Adenocaulon bicolor

windflower
Anemone deltoidea

goat's beard
Aruncus dioicus

Select Montane Conifer Forest Herbs

columbine
Aquilegia formosa

fairy slipper orchid
Calypso bulbosa

camas
Camassia quamash

pipsissewa
Chimaphila umbellata

bead lily
Clintonia uniflora

glacier lily
Erythronium grandiflorum

Rainier pleated gentian
Gentiana calycosa

narrow-leaved lotus
Hosackia oblongifolia

scarlet gilia
Ipomopsis aggregata

Siskiyou lewisia
Lewisia cotyledon

three-leaved lewisia
Lewisia triphylla

Cascade lily
Lilium washingtonianum

Heterotrophic Plants:
Parasites and Mycotrophs of the Klamath Mountains

Green plants are considered autotrophs because they photosynthesize—making sugar from water and carbon dioxide. The world of heterotrophic plants is complicated but all have moved from total energy production via photosynthesis. Some obtain organic carbon directly from other living beings while others have a parasitic relationship with fungi. Heterotrophic plants include parasitic and mycotrophic forms. The conifer forests of the western United States nurture an exceptional diversity and the Klamath Mountains might just hold the most heterotrophs in the West.

purple broomrape
Orobanche uniflora

woodland pinedrops
Pterospora andromedea

Parasitic plants include groundcones and broomrapes. These species contain no chlorophyll and obtain all nutrients by directly tapping the root system of host plants. Dwarf mistletoes (*Arceuthobium* spp.) are also parasitic while oak mistletoes (*Phoradendron* spp.) are hemiparasites. These species obtain water and some nutrients from its host tree, but also photosynthesize. See the Forest Pests chapter for more on mistletoes and the diversity in the Klamath Mountains.

snow plant
Sarcodes sanguinea

Mycotrophic plants are represented in our area by species in the heath (Ericaceae), orchid (Orchidaceae), and broomrape (Orobanchaceae) families. These plants all obtain nutrients through intermediary mycorrhizal fungi. Mycorrhizal relationships between fungi and plant are symbiotic—in that the fungi expand the root surface area and increases nutrient and water absorption for the plants while the fungi get nutrients in return. Mycotrophs exploit this symbiosis by parasitizing the hyphae and the fungi unwittingly feeds the mycotroph. Some species in the heath family have both leafless (heterotrophic) and leaf-bearing (autotrophic) forms.

California groundcone
Kopsiopsis strobilacea

Select Montane Conifer Forest Herbs

twinflower
Linnaea borealis

Pacific starflower
Lysimachia latifolia

woodland beardtongue
Nothochelone nemorosa

leafy lousewort
Pedicularis racemosa

short-spurred rein orchid
Piperia unalascensis

Woodbeauty
Drymocallis lactea

Nelson's catchfly
Silene nelsonii

giant white wakerobin
Trillium albidum

Oettinger's trillium
Trillium oettingeri

Pacific silver fir (*Abies amabilis*) is a common species of the montane belt in the Cascades. Pictured here is a nearly pure stand at the southern range extension for the species—in the Marble Mountain Wilderness.

Select Montane Conifer Vegetation Patterns

The Russian Wilderness holds the famed "Miracle Mile" of conifers. Eighteen species grow within one square mile in the Big and Little Duck lake basins. Russian Peak is at the center of the photo.

Pure stands of Shasta fir are the most common upper montane conifer forest type.

The Corey Peak Botanical Area in the Scott Mountains is home to the most extensive lodgepole pine - western white pine stands in the entire range. High elevation pines are much more common in the eastern Klamath Mountains.

Select Montane Conifer Vegetation Patterns

In Oregon's Siskiyou Mountains, south of the Applegate Valley near Miller Lake, a white fir - Douglas-fir forest shares space and time with the rare Baker cypress (*Hesperocyparis. bakeri*) in the left of the picture.

A pure stand of Port Orford-cedar in the upper reaches of the Siskiyou Wilderness.

John Sawyer shares his love of an enriched conifer stand in the Sugar Creek drainage, Russian Wilderness.

Meadows and Wetlands

Klamath meadows are biodiversity hotspots. On a per hectare basis, they support abundantly more native plant species than surrounding uplands based almost exclusively on water availability. They also provide habitat for a wide range of vertebrate species including reptiles, amphibians, mammals, and birds. Various meadow communities exist here, including sedge, dry meadow, sunken meadow, sedge-bunchgrass, lush-herbaceous, willow-alder, and corn-lily (Murray 2003). Besides being biodiversity hotspots, wet meadows create thousands of small reservoirs during spring snowmelt, recharging groundwater and releasing surface water in the dry season to support aquatic life downstream. As natural systems, meadows capture and hold surges of sediment from above and, over time, slowly meter it downslope—moderating impacts of erosion. As water seeps through rich meadow soils, decomposition of plant growth is slowed by high groundwater and low oxygen, tucking carbon away in peatlands.

Wet meadow habitats support a cover of emergent hydrophytes (need water) and mesophytes (require some water). Sedge meadows (dominated by *Carex* spp.) are found on soils at the upper extremes of moisture. Places such as lake margins, microbasins, and late-lying snowbanks commonly support sedge communities, but in the Klamath Mountains, these meadows don't reach the immense sizes of the sprawling Sierra sedgelands. Common species include shooting star (*Primula jeffreyi*), Merten's rush (*Juncus mertensianus*), alpine timothy (*Phleum alpinum*), and white marsh marigold (*Caltha leptosepala*). Shrublands are commonly found within larger herbaceous meadowlands. On valley bottoms willows (*Salix* spp.) thrive along stream margins, especially upon glacial gravel deposits (Van Sickle 1995). Slopes and higher elevations commonly support alder (*Alnus* spp.) along stream margins and seeps.

Dry meadows are also common on upper slopes and rocky hillocks—on up to ridge tops where low, matted plants resemble alpine tundra. Principal species are Idaho fescue (*Festuca idahoensis*), dwarf sagebrush (*Artemisia arbuscula*), and spreading phlox (*Phlox diffusa*). Sedge-bunchgrass meadows support columbia stipa (*Stipa columbiana*) and big sage (*Artemisia tridentata*). In contrast, the remarkable lush-herbaceous meadows are verdant jungles of towering flowers set on soils that remain damp throughout the summer. Hikers on the Pacific Crest Trail in the northern Marble Mountains encounter angelica (*Angelica tomentosa*), blue stickseed (*Hackelia micrantha*), California brome (*Bromus carinatus*), and Oregon checkermallow (*Sidalcea oregana*). Cornlily meadows are similar and found on valley bottoms and wetter slopes. Typical species are corn-lily (*Veratrum californicum*), yellow-staining collomia (*Collomia tinctoria*), and mountain thistle (*Cirsium callilepis*). Morris Meadow, in the Trinity Alps, is an expansive mesic grassland of California brome and needlegrass (*Stipa occidentalis* var. *californica*).

Non-native species are also found in Klamath Mountain meadows, including cows and a variety of plant species—particularly in and around the Scott Valley. Although meadows tend to be resistant to tree invasion, changes in the environment—such as decreased snowpack or livestock impacts—cause the prolific expansion of conifers including Shasta fir.

Select Meadow and Wetland Plants

subalpine fir
Abies lasiocarpa

thinleaf alder
Alnus viridis ssp. *sinuata*

monkshood
Aconitum columbianum

spikenard
Aralia californica

camas
Camassia quamash

Split hair paintbrush
Castilleja schizotrichia

Siskiyou sedge
Carex scabriuscula

fireweed
Chamerion angustifolium

sulfur buckwheat
Eriogonum umbellatum

Idaho fescue
Festuca idahoensis

monument plant
Frasera speciosa

Klamath gentian
Gentiana plurisetosa

Plants

Bigelow's sneezeweed
Helenium bigelovii

Davis' knotweed
Aconogonon davisiae

Vollmer's lily
Lilium pardalinum ssp. *vollmeri*

Wiggin's lily
Lilium pardalinum ssp. *wigginsii*

California bog asphodel
Narthecium californicum

marsh-grass of parnassus
Parnassia palustris

Sierra bog orchid
Platanthera dilatata

western coneflower
Rudbeckia occidentalis

Oregon checker mallow
Sidalcea oregana

yellow triteleia
Triteleia crocea var. *crocea*

Bridges' brodiaea
Triteleia bridgesii

California corn lily
Veratrum californicum

Meadow and Wetland Vegetation Patterns

Meadow mixes with western white pine in the upper elevations of the Red Mountain Meadows in the Trinity Alps Wilderness, Shasta-Trinity National Forest.

A meadow in late summer in the Siskiyou Wilderness is drying on the edges but still green where surface water flows.

A sunken meadow in the Red Buttes Wilderness supports year round water and a population of Rocky Mountain pond-lily (*Nuphar polysepala*).

A meadow on limestone in the Marble Mountains supports paintbrush (*Castilleja* sp.) and dwarf sagebrush (*Artemisia arbuscula*).

Meadowlands: A Disappearing Act?

Meadowlands and trees perform an ancient dance of shifting dominance. Meadow communities thrive only at sites supported by fundamentally tree-unfriendly edaphic properties such as heavy serpentines, permanent saturation, or extremely poor water retention. These traits assure the maintenance of meadows for centuries. However, the majority of vegetated habitats in the Klamath Mountains lack these site-specific hindrances to tree establishment. Thus, an array of natural agents work to secure, and even reclaim, meadowlands from forests. This is critical because many plant species cannot survive under shaded canopies including rare and endemic plants like Siskiyou lewisia (*Lewisia cotyledon* ssp. *cotyledon)*, Draba pterosperma *(Draba pterosperma)*, and crested cinquefoil (*Potentilla cristae)*. Further, a variety of wildlife and insects rely on meadow habitats.

Climate plays an important role in maintaining meadows. Where heavy snow accumulations occur at the tops of moderate to steep slopes, periodic avalanches will scour trees. Snowbanks that persist late into the growing season retard growth and establishment. Drought, coupled with high soil temperatures—most commonly on south-facing slopes—hinder survival of young seedlings. Frost damage on tender shoots further limits tree establishment.

Fires promote meadows. Common Klamath Mountain grasses and forbs (e.g., *Bromus carinatus, Chamerion angustifolium,* and *Achillea millefolium*) tend to be resilient to fire. Because invading trees are small, their thin bark renders vulnerability to lethal scorching. Many meadows are unable to carry fire due to high fuel moisture or sparse vegetation. Yet, burning embers are blown from surrounding forest to ignite tree patches far from forest edges. Additional agents that hinder tree cover in meadows include herbivory, frost heaving, and soil creep. A dense herbaceous canopy can prevent tree seeds from reaching the soil surface.

On occasion, forces overwhelm a meadow's natural defenses. Heavy grazing impacts meadow vegetation by compacting soils, bringing in non-native plants,

Disturbances like fire are needed to maintain meadow systems, like this one in the Siskiyou Wilderness after the 2017 Eclipse complex. While fire entered the meadow, some conifer seedlings remain.

and exposing more soil surface to falling tree seeds. If the grazing pressure is subsequently reduced, tree seedlings may proliferate. The suppression of lightning-caused fires further promotes the success of trees. Reduced snowpack, due to decreased precipitation and warming, also results in tree proliferation. Summer rainfall promotes tree establishment on well-drained meadows. Early seral forests invading meadows can happen because of a combination these factors.

During the mid-20th century, ecologists began to document young trees invading meadows in all the major western North American mountain ranges. At the same time, researchers were discovering how tree cover has expanded and contracted for millennia. Today, in the montane and subalpine zones of the Klamath Mountains, we are witnessing an overall loss of meadowlands. You may observe both trees and shrubs establishing along forest margins and well within large meadows as new woody patches. This recent invasion is related to a variety of causes—both natural and human.

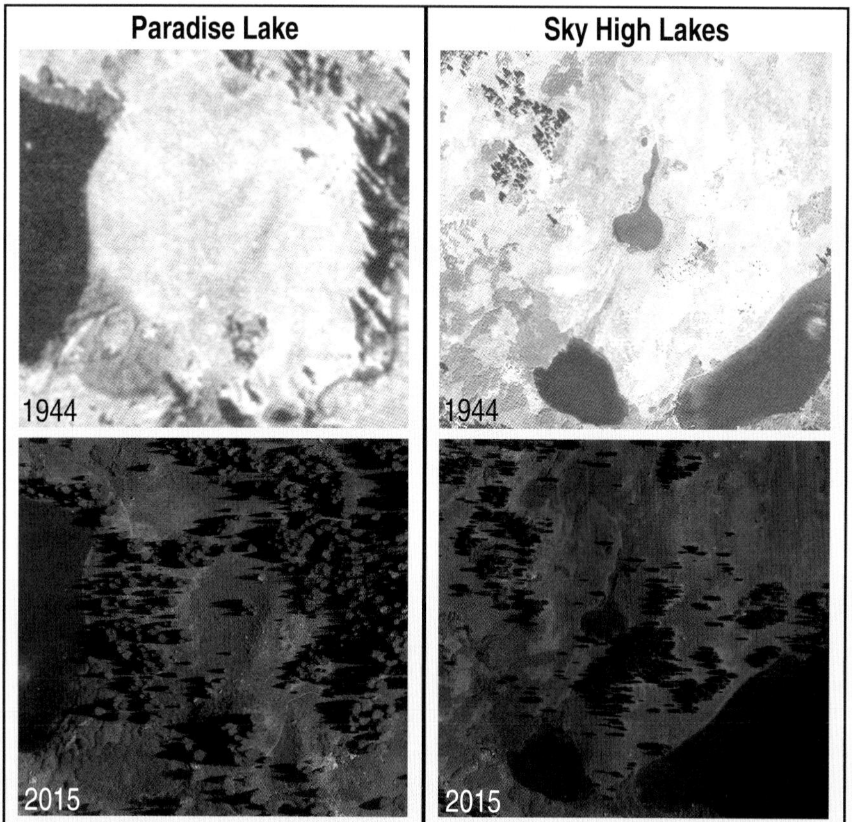

In the Marble Mountains, Russian Peaks, and Trinity Alps, meadow loss is evident by comparing historic and recent aerial photographs. Image analysis reveals, for instance, meadows surrounding Paradise Lake (elev. 6,080 ft) and Sky High Lakes (elev. 5,800 ft) were nearly treeless in 1944. The curtailment of heavy grazing between 1910 and 1935 likely enabled tree establishment. By 1999, each meadow lost approximately seven percent of its open area to an even mix of trees and shrubs. Overall, it is estimated that some meadows have experienced up to a 26 percent reduction due to tree and shrub establishment during this time period. In the late summer of 2014, the Happy Camp Complex Fire reached the invading trees at Paradise Lake. However, direct suppression actions, including pruning the lower branches of invading trees prevented fire from restoring this meadow.

Subalpine Sky Islands

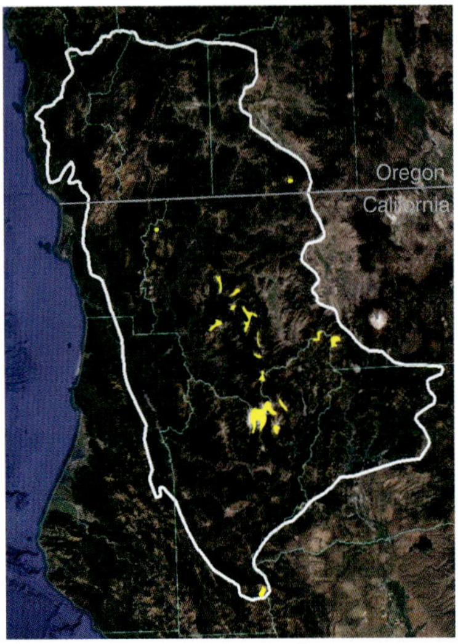

Sky Islands of the Klamath Mountains.

These communities are restricted to the highest peaks above 6,900 feet (2,100 m) in the Siskiyou, Marble, Trinity Alps, and Salmon Mountains. This zone has no upper limit since trees are able to grow to the tops of the highest peaks—hence no true alpine environment exists in the Klamath Mountains. Granitic plutons in the white Trinity Alps and Salmon Mountains complement 300 million year old serpentine outcrops in the eastern-most Scott Mountains. These regions, typified by shallow and/or ultramafic soils, create a subalpine forest character based on substrate, not low temperatures that prevent forest growth.

Subalpine forests are open, patchy woodlands with widely spaced trees with an understory of shrubs and herbs. These dimorphic rock types are crowned by regionally rare conifers including Klamath foxtail pine (*Pinus balfouriana* ssp. *balfouriana*) which is an endemic and whitebark pine (*Pinus albicaulis*) which is an outlier of a much broader range. These two light-loving species are generally found on ridgelines and south slopes while shade-tolerant species like mountain hemlock and Shasta fir occupy the north slopes. Dense stands of true firs (*Abies* spp.) and mountain hemlocks are found on more mesic sites with deeper soils. The number of plant species and associations in the subalpine sky islands has long been noted for its botanical importance (Whittaker 1960, Sawyer 2004). The preservation of this diversity relates to the persistence of small spatial-scale habitat. Scientists believe that high-elevation ecosystems are likely to be the first to register the impacts of global climate change and can thus catalog trends in vegetation and climatic shifts. In this discussion, it is important to remember that conifers are exclusively wind pollinated. If a mountain top is far enough away from another similar habitat, genetics of an isolated tree population may become restricted.

The long-term, dynamic story of the Klamath subalpine sky islands is best understood by interpreting climatic shifts and post-glaciation responses over the last 15,000 years. Briles et al. (2011) analyzed subalpine lake pollen in ultramafic and non-ultramafic settings spanning this time frame. Their data revealed that on non-ultramafic soils, dominant trees and shrubs shifted in elevation in response to climate variations. On ultramafic substrates changes were primarily structural—species composition did not change but tree and shrub densities did. Read more in the Climate Chapter and the Paleoclimate sidebar (pages 100-101).

What effects will our changing climate have on these ecologically restricted high elevation conifers of the Klamath Mountains? We hypothesize some scenarios in the Future chapter herein.

Klamath foxtail pine
P. balfouriana ssp. *balfouriana*

whitebark pine
Pinus albicaulis

Sawyer's pussy-toes
Antennaria sawyeri

balloon milkvetch
Astragalus whitneyi

dusty maidens
Chaenactis douglasii

winged-seed draba
Draba pterosperma

dwarf monkey flower
Diplacus nanus

prostrate buckwheat
Eriogonum lobbii

dwarf hulsea
Hulsea nana

quill leaf lewisia
Lewisia leeana

partridge foot
Luetkea pectinata

Mount Eddy sky pilot
Polemonium eddyense

Subalpine Sky Island Vegetation Patterns

Preston Peak, one of the few subalpine environments in the Siskiyou Mountains, is home to a variety of rare plants including yellow-cedar (*Callitropsis nootkatensis*) and noble fir (*Abies procera*).

Box Camp Mountain in the Marble Mountains, is one of the lowest elevation sky Islands in the range at only 7,267 feet (2,215 m). This mountain top is experiencing high levels of mortality in whitebark pine.

South China Mountain, like nearby China Mountain and Mount Eddy, hold sky island forests of whitebark pine and foxtail pine on ancient ultramafic rock.

The highest average elevation in the Klamath Mountains are along the granitic ridgelines of the Trinity Alps Wilderness where plants survive in microsites between decomposing rocks—here approaching 9,000 feet.

Plants

Juniper Woodlands

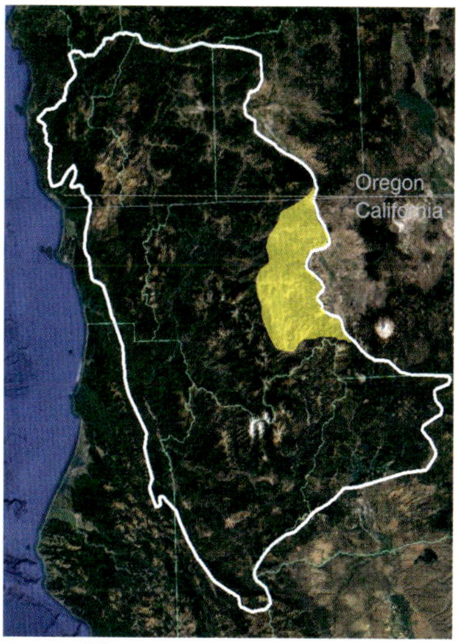

Juniper Woodlands of the Klamath Mountains.

The driest region in the Klamath Mountains is influenced by the flora of the Great Basin where xeric juniper communities grow on shallow, rocky, and heavy clay soils in widely spaced woodlands. Western juniper (*Juniperus occidentalis*) is the characteristic conifer, which is long-lived and, under optimal conditions, can reach 3,000 years old. This species was once considered to have two forms: var. *occidentalis* (western juniper) and var. *australis* (mountain juniper). Adams et al. (2006) analyzed the DNA and proved that these varieties are, in fact, two different species. The Sierra juniper (*Juniperus grandis*) occurs well south of the Klamath Mountains in the Sierra Nevada and south to the Transverse Ranges. In addition to stands of western juniper in the far eastern Klamath Mountians near Yreka and in the Scott Valley, outlier populations are found in the Salmon Mountains within the Miracle Mile as well as in the heart of the Yolla Bolly-Middle Eel Wilderness.

In the extreme eastern Klamath Mountains, the influence of the Great Basin is obvious. Common semi-arid desert species occuring in this region include mountain mahogany (*Cercocarpus ledifolius*), rabbitbrush (*Ericameria nauseosa*), and Great Basin sagebrush (*Artemisia tridentata*), while other more specific endemic species can also be found, particularly in the Scott Bar Mountains. Examples include the Yreka phlox (*Phlox hirsuta*), Scott Mountain phacelia (*Phacelia greenei*) and the Siskiyou mariposa lily (*Calochortus persistens*). A few species common elsewhere are also found here including Douglas-fir, which appears sporadically on north facing slopes.

A juniper woodland seen on the east side of the Scott Bar Mountains.

Select Juniper Woodland Plants

western juniper
Juniperus occidentalis

Great Basin sagebrush
Artemisia tridentata

water birch
Betula occidentalis

curl-leaf mountain mahogany
Cercocarpus ledifolius

rubber rabbitbrush
Ericameria nauseosa

antelope bush
Purshia tridentata

fragrant sumac
Rhus aromatica

Siskiyou mariposa lily
Calochortus persistens

cobwebby thistle
Cirsium occidentale

Scott Mountain bedstraw
Galium serpenticum

Scott Valley phacelia
Phacelia greenei

Yreka phlox
Phlox hirsuta

Acknowledgments

Parts this chapter were adapted from Sawyer and Kauffmann's Klamath Mountain chapter in *California's Botanical Landscapes* (Sawyer et al. 2016) after John Sawyer passed away. Thanks to Jim Agee, Leaf Hillman, and Lisa Hillman for their thoughtful reviews of the chapter.

Chapter Photographs

All photos by **Michael Kauffmann** unless noted. **Evens:** Julie Evens, **Kierstead:** Julie Kierstead, **Mergenthaler:** Kristi Mergenthaler, **Sawyer:** John Sawyer, **Stickrod:** Morgan Stickrod, **Whipple:** Jennifer Whipple, **York:** Dana York

Literature Cited and Suggested Reading

Adams, R.P., S. Nguyen, J.A. Morris and A.E. Schwarzbach. 2006. Re-examination of the taxonomy of the one-seeded, serrate leaf Juniperus of southwestern United States and northern Mexico (Cupressaceae). *Phytologia* 88(3):299-310.

Agee, J. K. 2007. *Steward's Fork*. University of California Press. Berkeley, CA.

Andersen, M. Kat, and F. Lake. 2016. Beauty, bounty, and biodiversity: the story of california indians' relationship with edible native geophytes. In *Fremontia*, California Native Plant Society publication. Sacramento, California.

Arno, S.F., and R.P. Hammerly. 2010. *Northwest Trees*. The Mountaineers Books. Seattle, WA.

Briles, C.E., C. Whitlock, C.N. Skinner, and J. Mohr. 2011. Holocene forest development and maintenance on different substrates in the Klamath Mountains, northern California, USA. *Ecology*, 92: 590-601. doi:10.1890/09-1772.1.

Fisher, J.A. et al. 2019. Indigenous peoples' habitation history drives present day forest biodiversity in British Columbia's coastal temperate rainforest. *People Nat.* 1: 103– 114.

Griffin, J.R. and W.B. Critchfield. 1976. *The Distribution of Forest Trees in California*. USDA Forest Service. Berkeley, CA.

Kauffmann, M. 2012. *Conifer Country*. Backcountry Press. Kneeland, CA.

Murray, M. 2003. Tree Encroachment on Klamath Mountain Meadows. *Fremontia*. California Native Plant Society. Sacramento, CA.

Parker, V.T. 2019. California chaparral. In: *Reference Module of Earth Systems and Environmental Sciences*, World Biomes.

Sawyer, J.O. 2006. *Northwest California*. University of California Press. Berkeley, CA.

Sawyer, J.O. and D.A. Thornburgh. 1969. "Ecological reconnaissance of relict conifers in the Klamath Region." USDA Forest Service, Pacific Southwest Coop-Aid Grant no. 7.

Sawyer, J.O. and D.A. Thornburgh. 1970. "The ecology of relict conifers in the Klamath Region, California." USDA Forest Service, Pacific Southwest Coop-Aid Grant no. 9, 21-52.

Sawyer, J.O. 2006 *Northwest California*. U.C. Press. Berkeley, California.

Sawyer, J.O. 2007. Forests of Northwest California. In: *Terrestrial Vegetation of California*. (editors) Barbour; Keeler-Wolf; and Schoenherr. University of California Press. Berkeley, CA.

Sawyer, J.O. and M.E. Kauffmann. 2016. Klamath Mountains. In, Barbour, M. G., et al. *California's Botanical Landscapes*. Pp. 67-87. California Native Plant Society. Sacramento, CA.

Skinner, C.N., A.H. Taylor, and J.K. Agee. 2006. Klamath Mountains In *Fire in California's Ecosystems*. Edited by N.G. Sugihara, J.W. van Wagtendonk, J. Fites-Kaufman, K.E Shaffer, A.E. Thode. 170-194. University of California Press. Berkeley, CA.

Smith, J.P. and J.O. Sawyer. 2019. *A Catalog of the Vascular Plants of Northwest California*. California Native Plant Society. Sacramento, California.

Stuart. J.D. and J.O. Sawyer. 2001. *Trees and Shrubs of California*. University of California Press. Berkeley, CA.

Sudworth, G.B. 1967. *Forest Trees of the Pacific Slope*. Dover Publications. New York, NY.

Van Sickle, V. 1995. Comparison of vegetation patch composition, biodiversity and wildlife, human and livestock use of Marble Mountain Wilderness Area Meadow Basins. Masters Thesis, University of California, Davis.

Wallace, D.R. 1983. *The Klamath Knot*. Sierra Club. San Francisco, CA.

Whittaker, R.H. 1965. Dominance and diversity in land plant communities. *Science* 147:250-260.

Whittaker, R.H. 1960. Vegetation of the Siskiyou Mountains, Oregon and California. *Ecological Monographs* 30:279-338.

Whittaker, R.H. 1961. Vegetation history of the Pacific Coast states and the central significance of the Klamath Region. *Madroño* 16:5–23.

Chapter 9: Forest Insects and Pathogens

Chris Lee

Ips bark beetle adults and larvae in a ponderosa pine.

Introduction

Forest pathogens and insects play important roles in ecosystems. For example, they may cull the growth of small, crowded, and suppressed trees; make openings in the forest canopy, releasing sunlight for new trees to grow; cause decay that creates living habitat for birds, small mammals, and other animals; and recycle nutrients back into the soil—among many other functions. The forest pathogens and insects in the Klamath Mountains, like those in other regions, help regulate individual tree life cycles as well as overall forest growth and succession. However, sometimes pathogens and insects cycle to epidemic levels through the forest, causing higher levels of tree damage and mortality than might otherwise be expected.

At least two scenarios can underlie tree mortality epidemics. First, when the forest undergoes extremes of environmental conditions—such as a prolonged, severe drought or a late freeze in spring after trees have begun leafing out or flowering—native insects and pathogens can take advantage of tree stress, increasing populations. Second, although much of the Klamath Mountain region is remote, it is not untouched by humans. Over the last few centuries we have introduced many non-native pathogens and insects to the region. These insects and pathogens have not co-evolved with the native trees and shrubs, so native plants have not developed effective defenses against them. Accordingly, the most important distinction in this chapter is that between native and non-native insects and pathogens.

Native forest damagers

Forest pathogens and insects can be divided into groups that attack various parts of trees. Beginning underground, some pests attack and decay the large roots that anchor plants in the ground, others attack smaller roots, and yet others attack the plant's vascular (water- and nutrient-conducting) system.

Heterobasidion irregulare and *Heterobasidion occidentale* are perhaps the most damaging and widespread root-attacking pathogens. The former kills pines, incense-cedar, and juniper, while the latter causes root rot and basal decay in true firs, spruces, hemlocks, and other conifers. These pathogens spread by spores that land on fresh tree wounds or on freshly created stump surfaces; they grow throughout the root system of the infected tree or stump and then spread to nearby trees with roots that touch the originally infected one.

Various species of *Armillaria* (including *A. mellea, A. gallica*, and *A. nabsnona*) also decay large roots, although they tend to cause less damage in the Klamath region, and in California as a whole, than they do in the Cascades of Oregon and other parts of the West. *Armillaria* is a fascinating fungus for many reasons. It produces shoestring-like structures called rhizomorphs that can grow through the soil and under tree bark in a quest for woody material, either living or nonliving, to digest. Certain parts of the fungus glow in the dark, i.e., bioluminesce. In addition, *Armillaria* individuals in Oregon and Michigan have been shown to be among the largest organisms in the world, with single clones of the fungus covering hundreds of hectares, aided by rhizomorphs.

The root-decaying cowpie fungus (*Phaeolus schweinitzii*), named for the distinc-

Overview of Select Native Forest Pests

Defoliators

- *Orgyia pseudotsugata* →
- *Choristoneura spp.*
- *Hyphantria cunea*
- *Neodiprion* spp.

Bark Beetles

- *Scolytus ventralis*
- *Dendroctonus brevicomis*
- *Dendroctonus ponderosae*
- *Dendroctonus valens*
- *Ips* spp. →

Foliar & Twig Pathogens

- *Lophodermium pinastri* →
- *Mycosphaerella arbuticola*
- *Rhabdocline pseudotsugae*
- *Tubakia californica*

Rusts

- *Gymnosporangium libocedri*
- *Endocronartium harknessii* →

Cankers

- *Atropellis piniphila* →
- *Cytospora abietis*
- *Diplodia pinea*
- *Diplodia corticola*

Aphids, adelgids, and scales

- *Cinara* spp.
- *Adelges piceae* →
- *Chionaspis pinifoliae*
- *Dynaspidiotus californica*
- *Matsucoccus bisetosus*

Wood Borers

- *Phaenops drummondi*
- *Phaenops californica*
- *Buprestis aurulenta*
- *Ergates spiculata* →

Root Rot Fungi

- *Phaeolus schweinitzii*
- *Heterobasidion irregulare* and *H. occidentale*
- *Armillaria* spp. →
- *Phellinus sulphurascens*
- *Pseudoinonotus dryadeus*

Select forest pathogen types of the Klamath Mountains.

Pathogen*	Description
Foliar Pathogens	Any number of fungi that infect the leaves of plants and cause leaf spots and needle loss. Often characterized by small fruiting bodies on leaves or needles which cause red to brown discoloration or may turn gray. In some cases, needles retain short green basal portions; in others, irregular discoloration occurs.
Rusts	These are fungal pathogens of the order Pucciniales with about 7000 species in 168 genera known worldwide. As a group, rusts affect many kinds of plants, but each rust species has a narrow range of hosts. They are obligate plant pathogens, meaning they only infect living plants. Rusts are so named because of their reddish spore-producing structures. The group is considered one of the most dangerous pathogens to agriculture.
Cankers	This is a general terms for a number of plant diseases caused by a wide range of organisms such as viruses, bacteria, and fungi and characterized by localized areas of dead tissue. These areas can lead to death of entire plant parts, such as branches. They are often sunken because the stem continues to get bigger elsewhere. Some are of only minor consequence, but others are lethal.
Root Rot Fungi	Wood decay is the process in which cellulose and lignin, the two most abundant organic compounds on Earth, are converted to carbon dioxide and water through the enzymatic activities of microorganisms. For practical purposes, fungi are the only agents of wood decay. Decay types can be broken into white, brown, and soft rots.

*Not included but also present are mistletoes, Oomycetes, and other wood decay fungi.

Insects*	Description
Defoliators	Most defoliators are moths or butterflies, although some are wasps, aphids, or grasshoppers. They eat leaves or needles and, like branch and tip insects, are most destructive when concentrated outbreaks occur, often encouraged by conducive environmental conditions such as extended or severe drought.
Bark beetles	Endemic throughout the West, these species are highly host-specific, and can be incredibly destructive when favorable environmental conditions and high levels of tree stress encourage the development of beetle populations. Bark beetles that cause the most tree mortality in California (including the Klamaths) are the western pine beetle (ponderosa and Coulter pines), the mountain pine beetle (five-needle, lodgepole, and ponderosa pines), the pine engraver beetles on various pine species, and the fir engraver beetle (true firs).
Aphids, adelgids, and scales	These insects from the order Hemiptera use piercing and sucking mouthparts to feed on tree sap, often causing strange host reactions such as needle and branch deformations or galls. As they feed, the sap is processed into honeydew, which leaves the insects and falls onto branches and leaves below, serving as a substrate for the growth of black, generally harmless sooty mold.
Wood Borers	Woodboring beetles differ from bark beetles in that the developing larvae generally feed not under the bark, but deep within the wood of already stressed, dying, or dead trees. Two exceptions in the Klamath Mountains are the flatheaded fir borer, which can kill healthy Douglas-firs, and the California flatheaded borer, which can kill healthy pines.

*Not included but also present are cone and seed insects and branch and tip insects.

tive look of its spore-producing structures when old and dried-up, is common throughout northwestern California. Its spores percolate through the soil, infecting Douglas-fir and pine roots causing undetectable but cumulatively severe root rot over decades. Laminated root rot (*Phellinus weirii* and *Phellinus sulphurascens*) is widespread and causes immense damage in Douglas-fir, grand fir, and mountain hemlock along the western slopes and crest of the Cascades but reaches its southern distribution in a few restricted areas sprinkled throughout the greater Klamath Mountain region. *Pseudoinonotus dryadeus*, also called the weeping conk or clinker conk, is a root-rotting fungus that infects primarily oaks in the eastern U.S. but true firs, hemlocks, and spruces in the West. It causes a white rot of the sapwood that is often quite severe in extent and that often decays the tree's large roots and butt wood, contributing to long-term decline and eventual windthrow.

Pseudoinonotus dryadeus on a white fir.

Other root pathogens work by decaying fine roots or interfering with tree physiology. The former category includes any number of pathogens in the genus *Phytophthora*, some of which will be mentioned within the non-native forest pathogens. These resemble fungi but are more closely related to brown algae and diatoms. They are primarily soil-borne and water-loving. Extremes of wet and dry weather promote the damage they cause: wet periods cause these organisms to produce excesses of spores that infect and destroy fine roots, which leaves trees unable to absorb enough water during ensuing dry periods, leading to even more stress on affected trees.

On the other hand, the pathogen *Leptographium wageneri*, which infects pines and Douglas-firs, works not by causing decay but by growing within the tree's vascular system—clogging it up and impeding water transport. This fungus is often brought into trees by root-feeding beetles and weevils.

Although these fungi work in different ways, the symptoms they cause are similar. Infected trees cannot transport enough water from roots to crown, and leaves or needles are slowly shed. As a result, live infected trees will usually have a thin tree crown, and pitch may exude from the base. Overall, root pathogens weaken the tree, opening the door for other pathogens or insects to join the attack.

Whereas many root pathogens can take decades to damage trees, insects and pathogens that attack stems cause extensive damage quickly. Several good examples exist within the various genera of bark and woodboring beetles. Although these beetles can attack both gymnosperms and angiosperms, the conifer-attacking bark beetles have attained the most notoriety. In some cases, they kill thousands of acres of susceptible forests in a single season.

Pitch tubes from *Dendroctonus brevicomis*.

In the Klamath Mountains, the bark beetles that cause the most damage are the western pine beetle (*Dendroctonus brevicomis*), which attacks ponderosa pine; mountain pine beetle (*Dendroctonus ponderosae*), which attacks ponderosa and lodgepole pines as well as the five-needle pines; and fir engraver beetle (*Scolytus ventralis*), which attacks all true fir species. These beetles are attracted to trees with compromised chemical defenses. When a beetle attempts to enter a tree, the tree will flood the attack site with resin (pitch) in an attempt to drown or eject the invader. If the tree cannot produce sufficient pitch and the beetle successfully enters, it will mine in the inner bark, laying eggs as it goes. Not only will the eggs hatch into larvae that eat more of the essential inner bark tissue but, also in many bark beetle species, the original invading adult will also emit a chemical signal attracting hundreds more beetles. Moreover, each adult inoculates the wood with fungal spores that it carries into the tree. These fungi can obstruct the tree's vascular system, helping the beetle overcome the tree's resistance and hastening the mortality process, although the fungus does not compromise the wood quality itself. After the tree dies, the fungal hyphae can be seen as a bluish stain throughout the wood, leading to the label "bluestain fungi" for this group of beetle symbionts. When attacked by these numbers, trees have little chance of survival.

Other beetles that invade trees use other tissue types for sustenance. Woodboring beetles, for example, generally make tunnels deep into the wood of trees rather than staying within bark tissues. For the most part, these beetles only take advantage of already dead or dying trees, with a couple of exceptions: the flatheaded fir borer (*Phaenops drummondi*) can kill apparently healthy Douglas-firs in the Klamath region, while the California flatheaded borer (*Phaenops californica*) can kill living pines. Ambrosia beetles are another kind of insect that bores deep into the wood of already dead or dying trees, but these beetles are noteworthy for another aspect of their lifestyle: they are farmers. They bring fungal spores, which adhere to their bodies, into trees with them. As they tunnel through the wood, the spores fall off the beetles, germinate in the wood, grow, and produce more spores. These fungal gardens provided by their parents give baby ambrosia beetles a reliable food source as they grow to adulthood, and when the young beetles emerge to fly to other dead or dying trees, they are covered with the sticky fungal spores, which they will take to those trees to inoculate them, continuing the remarkable symbiotic cycle.

One group of major stem- and branch-infecting pathogens are neither fungi nor insects. Dwarf mistletoes (*Arceuthobium* spp.) are parasitic vascular plants endemic to North America. Dwarf mistletoes are very different from the more familiar true mistletoes (*Phoradendron* spp.). The true mistletoes have well-developed leaves and large plants; many are deep green (most of the familiar mistletoe plants hung up at Christmas are true mistletoes). Dwarf mistletoe plants, on the other hand, are small, have leaves that are reduced to scales, and are yellowish- to brownish-green. The difference in color emerges from the ecology: true mistletoes parasitize their host trees mostly for water, whereas dwarf mistletoes draw both water and sugars from their host trees. In this process, dwarf mistletoes reduce growth and cause deformities—apparently by changing the distribution of nutrients and hormones within the tree system.

Dwarf mistletoes infect only conifers, while true mistletoes infect both hardwoods and conifers. Dwarf mistletoes are host-specific, with each mistletoe species infecting one conifer species—with some occasional secondary or "crossover" hosts

A Center of Dwarf Mistletoe Diversity

The parasitic plants called dwarf mistletoes infect only conifers. Their fascinating life cycle involves the production of sticky seeds that land on uninfected target trees, two years spent inside the host before emergence, and a highly aggressive parasitic lifestyle that causes great damage to the host as the plant shunts carbon, water, and minerals away from the tree's vascular system. Such functional complexity bespeaks a high level of host-parasite co-adaptation over millennia.

Brooming on a Douglas-fir caused by *Arceuthobium douglasii.*

This co-adaptation is also reflected in dwarf mistletoe's pinpoint specialization: each dwarf mistletoe species, and in some cases subspecies or special forms, infects one species of coniferous host tree. And in the Klamath Mountains, which contain the highest concentration of conifer species in the world's temperate regions, this diversity is reflected in a high number of dwarf mistletoe species, subspecies, and special forms. At least 12 dwarf mistletoe taxa exist in the Klamath Mountains to infect all but one of the regional conifer species.

Arceuthobium monticola on western white pine.

Although all dwarf mistletoes are very small and brownish- or yellowish-green, the careful observer will soon notice the subtle and interesting differences between them. Yellowish dwarf mistletoes on white firs often stand out sharply against the firs' glaucous foliage, while the tiny stubs of Douglas-fir dwarf mistletoe line the branches in single file and are impossible to see except at very close range. The primary symptom caused by the white fir dwarf mistletoe is a branch or trunk swelling at the site of infection, while Douglas-fir dwarf mistletoes cause huge, sweeping or circular branch clusters called "witch's brooms" that often take up a third or more of the entire tree crown. Meanwhile, the dwarf mistletoes that parasitize ghost pines cause hardly any visible symptoms at all.

All mistletoes are important to wildlife, providing an important source of berries, pollen, and nectar. Additionally, the changes in tree structure caused by witch's-broom-forming dwarf mistletoes provide an important habitat resource. Both small mammals and many species of birds—including threatened and endangered species such as the northern spotted owl—use witch's-brooms, especially the large, dense brooms of Douglas-fir, as nesting and hiding habitat.

infected. They tend to have severe effects on tree growth, another point of contrast with true mistletoes which cause less damage. One final difference between true and dwarf mistletoes involves their seed distribution. True mistletoes depend on birds to eat their seeds, transport them, and defecate them on host trees. Dwarf mistletoes produce seeds enclosed in sticky sheaths that are explosively propelled from the parent mistletoe plant to land on target trees up to dozens of feet away. Almost every conifer species within the Klamath Mountains has its particular dwarf mistletoe species, but some dwarf mistletoes are more pathogenic than others. Among the most heavily damaging dwarf mistletoes are the Douglas-fir dwarf mistletoe (*Arceuthobium douglasii*), which is among the tiniest of the genus but causes huge branching deformities called witch's-brooms in the host trees, lodgepole pine dwarf mistletoe (*A. americanum*), and western hemlock dwarf mistletoe (*A. tsugense* ssp. *tsugense*).

Another major stem and branch pathogen of Klamath conifers is western gall rust (*Endocronartium harknessii*), which infects most species of hard (non-5-needle) pines. This pathogen causes large swellings on branches and stems that can completely block water movement through the infected part. Depending on how prevalent western gall rust is in a particular stand of conifers, it can be a minor nuisance or a major cause of stand deterioration; it is more prevalent in areas with high humidity. It typically kills individual branches, only deforming or killing a whole tree if it establishes an infection on the stem. Rust fungi are the most advanced group of pathogens in the fungal kingdom. Most of them have five spore stages and require two different host plants to complete their life cycles. Western gall rust, however, has only two spore stages (with one rarely observed) and requires only the presence of pines to perpetuate its cycle.

A large number of other fungi infect branches and cause cankers (localized areas of dead tissue), small twig mortality, or entire branch mortality. Some of the best-known fungal genera containing these pathogens are *Diplodia, Fusarium, Seiridium, Neofusicoccum,* and *Cytospora*, but there are others. Most proliferate on trees that are suffering from other stresses like aridification, dwarf mistletoe infections, and insect attack. These branch pathogens rarely kill trees directly. Two exceptions are madrone pathogens *Botryosphaeria dothidea* and *Neofusicoccum arbuti.* These infect twigs and progress downwards to the trunks, killing branches and eventually the main stem. The affected tissue appears blackened as if charred by fire.

Another large group of opportunistic fungi, the decay fungi, can only colonize tissue that is already dead. These fungi are some of the most important nutrient recyclers in the forest ecosystem. Many of them produce the largest and most spectacular fungal fruiting bodies in the forest. Some of the better-known decay fungi in the Klamath Mountains include the Indian paint fungus (*Echinodontium tinctorium*), white pocket rot of pines and Douglas-fir (*Porodaedalia pini*), quinine conk (*Laricifomes officinalis*), popcorn fungus (*Cryptoporus volvatus*), incense-cedar pecky rot (*Oligoporus amarus*), sulfur shelf/chicken of the woods (*Laetiporus gilbertsonii* and *L. conifericola*), and lion's mane (*Hericium abietis* and *H. erinaceus*). Perhaps most magnificent of all is the noble conk or fuzzy Sandozi (*Bridgeoporus nobilissimus*), which features a 4- to 5-foot-wide perennial fruiting body and is known from only one location in northern California, in Redwood National Park

Decay Fungi: Pathogenic Pests or Ecological Essentials?

Many species we label pathogens also play vital ecological roles. They hasten successional pathways and processes; they break down the complex compounds inside trees into simpler chemical elements for soil enrichment. The mistletoes, for example, siphon water and nutrients from their host trees, but they also provide essential habitat for birds, squirrels, and other animals—not to mention food in the form of berries.

Decay fungi also partake of this double-edged nature. Most decay fungi break down dead tissue, including the wood of standing dead trees or the nonliving heartwood that helps support the tree. Some can both invade living tissue and decay dead material. Through this process, these pathogens slowly recycle entire forests, incrementally colonizing their hosts over decades and returning organic compounds and mineral nutrients to the soil.

The most pathogenic of these decay fungi can kill large swaths of host trees when environmental conditions disadvantage the trees (e.g., *Heterobasidion* spp.), and so they occasionally cause large-scale forest health problems. But throughout the history of the Klamath Mountains, they have also provided important benefits to both animals and humans.

Decay fungi are important to animals. Many of the wildlife habitat benefits we ascribe to old-growth trees accrue because of the sculpting action of these fungi over many years. This slow fungal artistry creates hollows, broken tops, dead branches, and eventually logs on the forest floor, all of which are used for hiding, roosting, nesting, and resting.

Cowpie fungus, velvet-top fungus, or Dyer's polypore (*Phaeolus schweinitzii*): This common fungus slowly destroys the root systems and interior bases of many conifer species over decades. It was also used by Native Americans for tinder. It produces excellent dyes for wool and paper in a wide range of colors from brown to yellow to green.

Kauffmann

Quinine conk (*Laricifomes officinalis*): This fungus produces gargantuan, perennially growing fruiting bodies on the trunks of old-growth conifers. The fruiting bodies only show up once the fungus has decayed a large portion of the tree heartwood. Pacific Northwest Tribes carved the fruiting bodies into sacred objects, and interior Tribes used them as medicine.

Jarvis

Indian paint fungus (*Echinodontium tinctorium*): Like the quinine conk, the fruiting bodies of this fungus show up on firs and hemlocks only after the fungus has decayed much of the stem heartwood. They were used to make paint and skin protectants to protect against snow blindness and insects.

Kauffmann

Pathogens

(all other known locations are from farther north in the Pacific Northwest).

At the top of trees, many pathogens and insect pests cause damage by consuming leaves and needles. Insects can mine in the leaf or needle tissue, much like bark beetles mining in the bark or they can eat the entirety of the leaf or needle. Such defoliators come from many insect orders, including Lepidoptera (butterflies and moths), Hymenoptera (wasps, bees, and ants), and Hemiptera (true bugs). Notable defoliators in the Klamath Mountains are the Douglas-fir tussock moth (*Orgyia pseudotsugata*) and fall webworm (*Hyphantria cunea*). Although both of these insects are known for periodic outbreaks that look quite damaging, they rarely impact forests at the landscape scale. Smaller-scale defoliators such as black pineleaf scale (*Dynaspidiotus californica*), pine needle scale (*Chionaspis pinifoliae*), California oak moth (*Phryganidia californica*), and pine sawflies (*Neodiprion* spp.) are commonly observed in Klamath Mountains forests.

Insects belonging to the family Hemiptera include pests with piercing/sucking mouthparts such as aphids, adelgids, and scales. These can damage various tree parts including leaves, needles, twigs, or tree boles. Some of the most notable in the Klamath Mountains are the various species of aphids such as the giant conifer aphids (*Cinara* spp.) and the oak pit scales (*Asterodiaspis* spp.). Many of these insects feed on tree sap and exude honeydew, which falls on branches, boles, and the ground and is quickly colonized by various species of black fungi known collectively as sooty molds. Although they look unsightly and are often assumed to harm the plant, the sooty molds cause relatively little damage.

Many native pests of Klamath Mountains trees are neither insects nor fungi. Several vertebrate animals damage or kill trees. Black bears, for example, strip the bark from Douglas-firs, Port Orford-cedars, redwoods, and other trees to reach the sugar-rich phloem and vascular cambium, mainly in the later winter and early spring when food resources are at a low ebb. If they strip the bark from around the entire circumference of the tree base, they will kill it. Squirrels, porcupines, and woodrats can do the same thing, generally to tree tops or branches rather than entire trees. Many other animals also kill trees or tree parts, from beavers to pocket gophers to deer to grouse to mountain beavers—some for direct use and others as incidental by-products of their home-building and hiding activities. In addition, many bacteria, viruses, nematodes, and other microorganisms parasitize woody plants. We know little about any of these organisms regionally, so they are ripe for study. However, the importance and harmful potential of invading exotic organisms is not in question, so it is to them that we turn next.

Non-native forest pathogens and insects

As is true around the world, exotic plant pests have found their ways to the Klamath Mountains. Introduction has occurred by various means from all parts of the world. The preeminent pathway has been by the importation of live plants for ornamental use, but others including wood packaging and other wood products have also introduced destructive insect pests. These pathways go both ways: the largest native bark beetle in the West, the red turpentine beetle (*Dendroctonus valens*), rarely kills trees in North America but is a virulent tree-killer in China, where it was introduced relatively recently.

Many exotic pests have been here for decades or even centuries and others have been introduced recently. Two of the oldest in our region are Port Orford-cedar root disease (*Phytophthora lateralis*) and white pine blister rust (*Cronartium ribicola*). Both were introduced to the Pacific Northwest in the early twentieth century and have been slowly making their way south since then. *Phytophthora lateralis* kills only Port Orford-cedars (*Chamaecyparis lawsoniana*) and a small number of Pacific yew (*Taxus brevifolia*) if they are growing close to cedars. The impact of this introduction is magnified by the limited number and range of the host tree: wild Port Orford-cedars are nearly limited to the Klamath Mountains. *Phytophthora lateralis* relies on water to move and reproduce, and like many other *Phytophthoras* it impacts its host by consuming fine roots, gradually moving up into the base of the tree and destroying the tissue (cambium) that forms the vascular system. In the case of *P. lateralis,* mobile spores move through waterways and saturated soil in search of the fine roots of the water-loving Port Orford-cedar. These spores persist in mud allowing vehicles, tools, and even animals to transport them from one watershed to another. When *Phytophthora lateralis* is transported into a new watercourse, it spreads downstream from that point, killing trees as it goes.

White pine blister rust infects five-needle pines. In the Klamath Mountains these include sugar pine, western white pine, whitebark pine, and foxtail pine. Unlike western gall rust, white pine blister rust exhibits all five spore stages characteristic of rust fungi and requires two hosts to complete its life cycle: the five-needle pine plus any one of a number of currant or gooseberry (*Ribes*) species. Theoretically, the absence of *Ribes* should spell the end of a white pine blister rust epidemic in an area, since the fungus would not be able to complete its life cycle. This hypothesis initiated intensive government-sponsored efforts in the mid-twentieth century to remove *Ribes* in areas with five-needle pines in both the eastern and western United States. In the East, due to more gentle topography and fewer species of *Ribes* and white pine, the project was successful in some areas. In the West, however, more rugged topography ensured that the effort was largely unsuccessful. Moreover, researchers more recently determined that some other plant genera, such as *Castilleja* (paintbrushes) and *Pedicularis* (louseworts), can also function as alternate hosts. Since then, blister rust has slowly spread from the Pacific Northwest to the mountains of southern New Mexico and Arizona. It resembles western gall rust in that it does not kill trees until an infection becomes established on the trunk—but for an infected tree in an area where the disease is established, this is only a matter of time. This rust is especially concerning on whitebark pine because of the tree's keystone role in providing food and habitat in subalpine ecosystems and because of its vulnerability to other stressors including mountain pine beetles and climate change.

The invasion of Port Orford-cedar root disease and white pine blister rust illustrate the impacts of invasive plant pathogens and insects. If introduced pests encounter plant species related to native hosts they are able to infect them but these new hosts have not evolved defenses against the invaders. However, the good news is that pathologists have found individual trees that are resistant to the invaders. These discoveries have led to the establishment of breeding programs designed to evaluate these resistant trees and cross them to produce viable seed for reforestation. The Forest Service began the blister rust resistance program in the 1950s and the POC root disease resistance program in the 1990s; both programs have progressed and test trees have been replanted and genetics shared with public and private land managers.

Two more recently introduced non-native pests also need addressing. The first, *Phytophthora ramorum,* which causes sudden oak death, is loosely tied to the distribution of redwood within the fog belt along the California and Oregon coast. Unlike most *Phytophthora* pathogens, *P. ramorum* infects hosts above the soil line, attacking trunks and leaves. This species prefers humid areas where moisture persists year-round, so far moving only into margins of arid inland regions. Although *P. ramorum* attacks over one hundred plant species, both woody and herbaceous, only a few suffer mortality. In the Klamath region, in descending order of susceptibility, these are tanoak (*Notholithocarpus densiflorus*), California black oak (*Quercus kelloggii*), and canyon live oak (*Quercus chrysolepis*).

One intriguing, and troublesome, aspect of the occurrence of *P. ramorum* concerns the existence of several plant species that suffer limited leaf damage but on which the pathogen reproduces prolifically enough that these hosts help the disease spread. The most important of these in California is California bay laurel (*Umbellularia californica*), but tanoak itself also contributes to the spread in this way. Several ornamental plant hosts, including cultivars of rhododendron (*Rhododendron* spp.), viburnum (*Viburnum* spp.), andromeda or fetterbush (*Pieris* spp.), and lilac (*Syringa* spp.) also act as hosts. As sudden oak death spreads, new hosts continue to be discovered. In England in 2009, it was found on Japanese larch (*Larix kaempferi*) and has killed thousands of larch since. The pathogen has also shown up on larch in Ireland and France. More recently, it has been found infecting native manzanitas around San Francisco Bay. Although this pathogen has only appeared as a serious threat at the fringes of the Klamath Mountains, if it were to establish itself in the region there is high damage potential because of locally high moisture levels, and it presents a threat of unknown magnitude to other species.

Another major non-native pest has appeared in the Klamath region even more recently. The balsam woolly adelgid (*Adelges piceae*) is a tiny sucking insect that infests true firs and causes major damage on subalpine fir (*Abies lasiocarpa*), Pacific silver fir (*Abies amabilis*), and grand fir (*Abies grandis*). Adelgids begin life as tiny crawlers that are blown by wind from tree to tree. When they find a suitable place to settle down and insert their sucking mouthparts, their lifecycle becomes sedentary. As they grow, a coat of white "wool" protects them from predators. On infected trees, the cotton-ball-like insects infest trunks and branches. When the trunks are infested tree mortality is likely while branch infestations usually lead to strange, twisted branch symptoms (called "branch gouting") which leads to slower crown decline.

Subalpine fir and Pacific silver fir only exist in California in limited distributions within the Klamath Mountains. Unfortunately, in 2018 the balsam woolly adelgid was found infesting subalpine firs in the southern part of the Russian Wilderness. With heavy damage previously recorded to subalpine fir along the eastern slope of the Cascades in Oregon and Washington, it is likely this pest will pose

Forest and Canopy Decline

In the Klamath Mountains, canopy decline is greatest in the higher elevation subalpine forests, particularly in Shasta fir (*Abies magnifica* var. *shastensis*).

Extended, severe drought from 2011-2017 caused unprecedented tree mortality in the Sierra Nevada and other parts of California, with an estimated 130 million trees dying during that time. Occasional sudden die-offs happen in the Klamath Mountains ecoregion too, typically at smaller scales, since the Klamath Mountains, unlike the spatially homogeneous Sierra Nevada, are characterized by small-scale topographic complexity, spatial heterogeneity, and species diversity.

But more gradual tree mortality events also occur and observers usually call them forest or tree "declines." These declines are characterized by slow but steady mortality of particular tree species over a period of years, often with an accompanying suite of biotic and abiotic causal factors. Forest pathologist Paul Manion hypothesized that most tree declines occur as a gradual "death spiral" and that the causal factors can be sorted into those that are predisposing (i.e., underlying tree stress leading to physiological weakness, such as a long-term water deficit), inciting (more sudden events that cause severe physiological disruption, such as an unseasonal freeze), and contributing (the attackers that take advantage of physiological dysfunction and ultimately kill the tree, such as a normally weak bark beetle).

As average temperatures warm and precipitation or long-term water storage decreases in some regions, these declines are becoming more evident in the Klamath Mountains, indicating trees under continuing anomalous levels of stress. Mortality events, whether sudden or gradual, cause shifts in species composition, changed interactions between species, unknown consequences for wildlife and future wildfires, and potential loss of stored carbon.

One example of increased tree mortality was recently documented by Bost et al. (2019) in the Russian Wilderness. While mortality varied by species, the highest proportions occurred in subalpine fir (*Abies lasiocarpa*, 35.3 percent), Shasta fir (*Abies* × *shastensis*, 28.6 percent), and lodgepole pine (*Pinus contorta* ssp. *murrayana*, 22 percent). DeSiervo et al. (2018) found that most Shasta red fir mortality was related to biotic agents including dwarf mistletoe (*Arceuthobium abietinum* ssp. *wiensii*) infection, Cytospora canker (*Cytospora abietis*), and fir engraver (*Scolytus ventralis*) activity.

Whether the recent red fir mortality is classed as a decline or a die-off, these studies support the conclusion that there is an active increase in tree mortality in the Klamath Mountains. It also demonstrates the vulnerability of high-elevation dense old-growth forests to increasing temperatures, decreasing water availability, and invasive species—as is the case for the arrival of the balsam woolly adelgid and increased mortality of subalpine fir.

an existential threat to these rare conifers. A similar story applies along the coast, where balsam woolly adelgid has been found causing the decline of mature grand fir. The hemlock woolly adelgid (*Adelges tsugae*) has caused catastrophic declines in high-elevation conifers in the southern Appalachians and other parts of the East; however, strangely enough, although the hemlock woolly adelgid is present in northwest California, it has yet to cause harm to its principal host, western hemlock (*Tsuga heterophylla*).

These four pests, along with other potential non-native threats to ecosystem integrity in California (including other soilborne *Phytophthora* species, pitch canker, introduced ambrosia beetle-fungus combinations, other invasive boring beetles, and exotic weed species) illustrate how non-native species spread once established in a novel environment. They also highlight the importance of vigilance, education, and early intervention in reducing human-facilitated pest movement from region to region. Unlike the relatively predictable cycles of damage caused by native pests acting in response to climatic perturbations and other disturbances, non-natives have the potential to change the appearance of the Klamath landscape in ways we can, at present, only dimly perceive.

Chapter Photographs

All photos by Chris Lee unless noted. Other photos contributed by Kauffmann: Michael Kauffmann, McComb: David McComb, Jarvis: Steph Jarvis. Ellis and Everhart: Ellis, J.B.; Everhart, B.M.

Chapter Reviewers:

Thanks to David Shaw, Jeffrey Kane, Jim Agee, and Michael Murray for their thoughtful reviews.

Literature Cited and Suggested Reading

Bost, D. et al. (2019). Assessing spatial and temporal patterns of canopy decline across a diverse montane landscape in the Klamath Mountains, CA, USA using a 30-year Landsat time series. *Landscape Ecology*. 34.

DeSiervo, M., E.S. Jules, D.S. Bost, E.L. De Stigter , and R.J. Butz. 2018. Patterns and drivers of recent tree mortality in diverse conifer forests of the Klamath Mountains, California. *Forest Science* 64(4): 371-382.

Furniss, R.L., and V.M. Carolin. 1977. Western Forest Insects. Washington, D.C.: USDA Forest Service Misc. Pub. No. 1339. 654 pp.

Goheen, D.J. and E.M. Hansen. 1993. Effects of pathogens and bark beetles on forests. in Schowalter, T.D., and G.M. Filip, eds. *Beetle-Pathogen Interactions in Conifer Forests*. London: Academic Press. pp. 175-196.

Hansen, E.M. 2008. Alien forest pathogens: *Phytophthora* species are changing world forests. *Boreal Environment Research* 13: 33-41.

Hansen, E.M. and E.M Goheen. 2004. *Phellinus weirii* and other native root pathogens as determinants of forest structure and process in western North America. *Annual Review of Phytopathology* 38: 515-539.

Hawkins, A.E. and T.W. Henkel. 2011. Native forest pathogens facilitate persistence of Douglas-fir in old-growth forests of northwestern California. *Canadian Journal of Forest Research* 41(6): 1256-1266.

Maloney, P.E. 2011. Incidence and distribution of white pine blister rust in the high-elevation forests of California. *Forest Pathology* 41(4): 308-316.

Mathiasen, R. and K. Marshall. 1999. Dwarf mistletoe diversity in the Siskiyou-Klamath Mountain region. *Natural Areas Journal* 19(4): 379-385.

Mathiasen, R.L. 2021. *Mistletoes of the Continental United States and Canada*. Fort Worth, TX: BRIT Press. 220 pp.

Swiecki, T.J. and E.A. Bernhardt. 2006. A Field Guide to Insects and Diseases of California Oaks. Gen. Tech. Rep. PSW-GTR-197. Albany, CA: Pacific Southwest Research Station, Forest Service, U.S. Department of Agriculture. 151 pp.

Wood, D.L, T.W. Koerber, R.F. Scharpf, and A.J. Storer. 2003. *Pests of the Native California Conifers*. Berkeley: UC Press. 233 pp.

Chapter 10: Invertebrates

Michael Kauffmann, Justin Garwood, Len Mazur, Brian Dykstra, Rod Crawford, Dana Ross, and Paul Hammond

A goldenrod crab spider (*Misumena vatia*) on bear grass (*Xerophyllum tenax*). Photo by Michael Kauffmann.

Introduction

Invertebrates are animals without a backbone or bony skeleton. They are, by far, the largest group in the animal kingdom including somewhere between 95 and 97 percent of all animals. Around the world, 1.25 million invertebrate species have been described with many more waiting for science to give them a name. There may be as many as five million invertebrate species while only 60,000 vertebrates are known to science. Invertebrates find global success in a variety of ways—including a rapid rate of reproduction with methods including ants and bees whose eggs develop without fertilization (workers) and sponges that are hermaphroditic. Insects are also adaptable by way of opportunistic foraging, surviving in extreme environments, and flight. Here we treat arthropods, mollusks, sponges, and tardigrades as they relate to the Klamath Mountains and beyond.

Arthropods

These are invertebrate animals with an exoskeleton, a segmented body, and paired jointed appendages. Herein we treat four subphyla (Chelicerata, Crustacea, Myriapoda, and Hexapoda)

Subphylum Chelicerata

Arachnida are joint-legged invertebrates including spiders, scorpions, ticks, mites, harvestmen, and solifuges. Adult arachnids have four pairs of legs and are terrestrial—though some inhabit aquatic environments. They comprise over 100,000 named species.

The spider fauna of North America is rich—with at least 68 families, 569 genera, and more than 3,700 species (Ubick et al. 2005). Globally, spiders are one of the most important consumers of insects in temperate regions. The Klamath Mountains have more than 500 known species with likely twice that being undescribed or undocumented (Crawford 2015a). Distribution depends on whether the species disperse by air (ballooning) or not. Endemism is more common in the non-ballooning spiders due to increased genetic isolation. Of the 524 species documented in the Klamath Mountains, approximately one-quarter are non-ballooning. Of the 131 documented non-ballooning species, 36 are considered endemic and 43 reach their northern or southern range limit in the Klamath region. Of the 391 ballooning spiders, only ten percent are considered endemic to the Klamath. The Klamath Mountains are one of the centers of spider endemism for non-ballooning spiders (Crawford 2015a).

Probably the most distinctive and unique Klamath Mountain spider story is told by a newly described family (Trogloraptoridae) from a cave near Grants Pass (see Obligate Subterranean sidebar, 274-275). Other interesting non-ballooning spiders include the sheet- and funnel-like web-making Amaurobiidae. For this family, it appears the Klamath Mountains are a region of speciation including a variety of endemic species like *Callobius angelus, C. manzanita, C. panther, C. paskenta,* and *C. paynei*. Speaking of cryptic speciation, the *Cybaeus* make webs that are barely

California turret spider burrow (*Atypoides riversi*) by Garwood.

Invertebrates

visible and include Klamath endemics like *Cybaeus somesbar, C. gidneyi,* and *C. multnoma*.

Taracus fluvipileus is a member of Harvestmen (Order Opiliones). Photo by Tom Rickman.

Harvestmen are a group related to spiders but, unlike spiders, lack venom or silk glands, have two eyes, and have a segmented abdomen. The most familiar species have globular bodies and long legs. Most harvestmen can secrete a defensive scent to repel predators. Like the non-ballooning spiders, harvestmen have no means of rapid, long-distance dispersal. Of the 30 named species documented in the Klamath, eleven are believed to be endemic, with ten more reaching their northern or southern range extension (Crawford 2015b).

Western forest scorpion *(Uroctonus mordax)*. Photo by Michael Kauffmann.

Scorpions and Pseudoscorpions are cryptic groups that occur across the Klamath Mountains. Scorpions are considered the oldest predators on land—almost twice as old as the first dinosaurs. Western forest scorpions (*Uroctonus mordax*) are found under the bark of standing or fallen dead trees. Bryson et al. (2016) found that Klamath scorpions have their own distinct clade due to isolation. Graham's cave scorpion (*Uroctonus grahami*) is restricted to underground habitats in the eastern Klamath—read more in the obligate subterranean sidebar. The pseudoscorpions of the Klamath Range are less known but at least three species have been recorded (Roth 2015).

Mites are different from spiders in having one body part compared to the two body parts of spiders. This distinction is easy to see in the largest mites—ticks. Mite bodies tend to be flat, though an exception locally is the oribatids which have globular bodies. While many people are innately fearful of spiders, mites—specifically ticks—are much more troublesome due to the diseases they carry. The highest rates of Lyme disease in California occur in and around the Klamath Mountain region where one percent of western black-legged ticks (*Ixodes pacificus*) are carriers in Trinity County (CDPH Website 2018). Another noteworthy mite is *Sarraceniopus darlingtonia*—an obligate inhabitant of the California pitcher plant (*Darlingtonia californica*). It feeds on the rich microbial growth associated with decomposing arthropods captured in the pitcher and lays its eggs in the pitcher wall above the fluid line (Fashing 2004). For more on the pitcher plant, see the Carnivorous Plant sidebar, page 213.

Subphylum Crustacea

A diverse group of primarily marine invertebrates is represented by over 60,000 species worldwide. When compared to other areas in North America, the diversity of crustaceans is lower in the Klamath Mountains. This is due to the Mediterranean-type climate that offers dry summer months but also the abundance of ultramafic rocks deficient in calcium—a vital mineral used to build crustacean's exoskeletons (Rogers 2015). Crustaceans occurring in the Klamath region maintain diverse ecological roles as carnivores, herbivores, detritivores, scavengers, and parasites in wetlands and forest leaf litter communities.

Regional diversity is mainly microscopic and concentrated in lentic (lake and pond) zooplankton and benthic communities along with the phylum's Rotifera (rotifers) and Nematoda (roundworms). Larger macrocrustacean members, such as crayfish, can exceed six inches (150 mm) in length. Most small (< 1mm) crustaceans filter feed on phytoplankton using modified antennae and thoracic limbs to target diatoms, flagellates, and small green algae. Larger (>1mm) crustacean zooplankters, such as *Daphnia*, are filter feeders, where others are opportunistic omnivores. Some members, such as *Leptodora,* are predators having modified legs for seizing prey. Collectively, crustacean zooplankters highlight the profound trophic complexities occurring at the base of aquatic food webs by transferring energy from phytoplankton and algae to larger predatory consumers. Crustaceans also play a vital role in maintaining the water quality of freshwater ecosystems by filter feeding and recycling nutrients. Four classes of Crustaceans occur in the Klamath Mountains: Branchiopoda, Maxillopoda, Ostracoda, and Malacostraca.

Class Branchiopoda—are the least highly derived of all living Crustacea and members found in the Klamath Mountains include fairy shrimp (Anostraca) and water fleas (Cladocera). Unlike other classes of Crustaceans dominated by marine forms, branchiopod diversity is concentrated in inland fresh and saline waters. Only two species of fairy shrimp have been documented in the Klamath region (see sidebar next page) where they require unique, montane to subalpine, temporary ponds to complete their intricate lifecycle. Except for Decapods, fairy shrimp are far larger than other classes of Klamath crustaceans, measuring up to 1.5 inches (40 mm). In contrast, water fleas in the Klamath tend to be microscopic, occurring in open water and vegetated shoreline zooplankton communities during warmer months—while having a resting state in sediments during the winter. Here they compete with other zooplankton, such as rotifers and copepods, by occupying a range of foraging niches as small herbivores to large predators.

Class Maxillopoda (copepods and barnacles)—most are marine, with barnacles being exclusively marine. Copepods are nearly ubiquitous in marine and freshwater systems and are thought to represent the greatest animal biomass on Earth. Copepods are often the most abundant freshwater class found in zooplankton, benthic, and vegetated shoreline communities—in addition to wet, semi-terrestrial habitats such as leaf litter. Copepods also occur in subterranean waters (groundwater aquifers and cave streams) although practically all such species in our region are undescribed. Members of the genus *Diaptomus* are perhaps the most impressive to observe with a mask and snorkel in high-elevation lakes of the Klamath Mountains. These giant sesame seed sized Copepods can reach up to 3 mm across and are bright scarlet red. Some Copepods are parasitic. The most notable in the Klamath region being sea lice that parasitize salmonids in the ocean—later hitching a brief ride up the swollen rivers on adult salmon and trout migrating to spawn. The lice fall off and perish within days after their host enters freshwater.

The Fairy Shrimps of Montane Vernal Ponds

Fairy shrimp sightings are inimitable experiences. Their transparent, glacier blue color and elegant, wave-like metachronal locomotion—powered by a series of undulating leaf-like appendages—reveal a tiny charismatic creature strikingly unique among other freshwater aquatic life. Fairy shrimp belong to the Order Anostraca and are among the most primitive of living crustaceans—a diverse group that also includes crabs, crayfish, barnacles, and krill. Their Class name, Branchiopoda (branchio= gill; poda= feet), refers to their swimming paddles which also serve as gills, salt regulation structures, and appendages for food collection.

Fairy shrimp require temporary ponds to complete their complex lifecycles. Like other animals that live in ephemeral waters, they benefit in having fewer predators compared to perennial waters. Ponds where spiny-tail fairy shrimp (*Streptocephalus coloradensis*) are found in the Klamath Mountains are shallow, rocky, hard-pan, or clay-lined depressions in meadows surrounded by conifers. The water is often stained brown with tannins derived from water-soluble de-

The dramatic hydrological cycle of a shallow vernal pond in the Siskiyou Mountains that hosts breeding populations of spiny-tail fairy shrimp and Pacific treefrogs (*Pseudacris regilla*). Top photo from July 3rd, 2019 and the bottom photo is from August 4th, 2016. Photographs by Justin Garwood.

Invertebrates

Female spinytail fairy shrimp (left) from the Siskiyou Mountains, Del Norte County, CA. Female Oregon fairy shrimp (right) from the Hogback Ridge area, South Fork Trinity River, Humboldt County, CA. Note brood pouch and cyst eggs in each image. Photos by Justin Garwood.

caying vegetation. Spiny-tail fairy shrimp reproduce sexually. The female carries eggs in an ovisac where they form a thick, multilayered shell around each tiny embryo (~0.3mm) able to withstand extreme environmental conditions brought by desiccation and freezing. Once the eggs are ejected from the female they become cysts, or resting eggs, that persist for decades until key environmental signals induce hatching. In the Klamath Mountains, spiny-tail fairy shrimp cysts hatch into a new generation of larvae when snowmelt water refills temporary ponds each spring. Larvae then proceed through a series of molts to complete the annual lifecycle as adults.

The major dispersal mechanism for fairy shrimp is the cyst. This rugged dust-like capsule sticks to an animal's fur or a bird's feathers. It also passes, unharmed, through the digestive tract of birds that eat female shrimps. Until 2020, spinytail fairy shrimp were the only species of Anostracans documented in the Klamath Mountains, despite the Oregon fairy shrimp (*Eubranchipus oregonus*) and vernal pool fairy shrimp (*Branchinecta lynchi*) occurring just outside the border in the prairie-vernal pools of the Agate Desert in the Rogue Valley, Oregon. While regional spiny-tailed fairy shrimp distribution remains poorly understood, they are generally restricted to the montane conifer belt between 4,000–7,200 ft (1,200–2,200 m). Only two high-elevation ponds in the Siskiyou Mountains were identified as spiny-tail fairy shrimp locations by Eriksen and Belk (1999) who mapped the distributions of all 25 Anostraca species known to California. However, a subsequent survey of high-elevation Klamath wetlands in the early 2000s found over 50 ponds occupied by spiny-tailed fairy shrimp across the Yolla Bollys (1), Trinity Alps (20), Trinity Mountains (7), and the Marble Mountains (39) suggesting a broader distribution. The discovery of Oregon fairy shrimp by Justin Garwood and Jamie Bettaso in 2020 at an inconspicuous pond (2,263 ft, 690 m) along the South Fork Trinity River marks only the third known location for the species in California (others near Weed and Rio Dell). There is much more to learn about fairy shrimps in the Klamath Mountains.

Class Ostracoda (seed shrimps)—are tiny crustaceans, usually less than 1mm across, with flattened bodies protected by a hinged chitinous carapace similar in function to the shell of bivalve mollusks. Seed shrimps are found in standing and running waters where they are most commonly associated with substrates including algae, rooted aquatic plants, decaying vegetation, silt, mud, and gravel (Pennak 1989). Although seed shrimps represent an abundant group in benthic freshwater communities, few are planktonic in contrast to other Crustacean zooplankters from Cladocera and Copepoda. Another unique character is from the family Entocytheridae. This group of tiny seed shrimps live as ectosymbionts on the gills of other host crustaceans including the signal crayfish (*Pacifastacus leniusculus*) from the order Decapoda. Ostrocods also occur in subterranean waters (groundwater aquifers and cave streams) although practically all are undescribed in the Klamath.

Class Malacostraca—are represented by Amphipoda (scuds, sideswimmers), Decapoda (shrimp, crayfish, crawdads, crabs), Isopoda (pill bugs, sow bugs, woodlice) and Mysida (opossum shrimp) in the Klamath Mountains. A brief discussion of each follows.

Amphipods, such as *Hyalella azteca*, are laterally compressed (< 6 mm), nocturnal scavengers common along vegetated zones of ponds and lakes as well as caves and aquifers. Subterranean amphipods from the genus *Stygobromus* (< 6 mm) occur in groundwater aquifers and caves and, due to their inherent isolations, have experienced rapid diversification across North America (Ethridge et al. 2013). For example, *Stygobromus mysticus* was described from a subterranean aquifer in the Scott Valley of Siskiyou County and *S. trinus* is known only to a single cave in Trinity County (Wang and Holsinger 2001). More Klamath species are likely to be discovered, especially in isolated cave systems (see Obligate Subterranean sidebar).

Decapods are represented in the Klamath by one native species, the signal crayfish. Today, its native distribution is thought to be restricted to coastal watersheds from the Klamath River to the Rogue River (Larson and Olden 2011). The species has been widely introduced throughout California, where it has had severe impacts on aquatic ecosystem structure and native biodiversity. For example, east of the Klamath Mountains in the Pit River, populations of the endangered Shasta crayfish (*P. fortis*) have been extirpated from areas with introduced populations of its larger, aggressive cousin the signal crayfish. Introduced populations appear to cause ample confusion regarding the status of the species in the Klamath Mountains, especially in California. Current knowledge supports the species being native to most of the Klamath Mountain watersheds where it is an important member of stream food webs. Crayfish are opportunistic foragers, eating detritus, carrion, plants, and animals. Juvenile crayfish are an abundant prey for salmonid

Signal crayfish (*Pacifastacus leniusculus*). Photo by Justin Garwood.

fishes while adult crayfish are targeted by aquatic mustelids—including river otters and mink. Where crayfish are abundant, they can act as a major habitat disturber by way of excavating and maintaining small caves and interstitial areas.

Brackenridgia heroldi is a member of woodlice and pillbugs (Suborder Oniscidea). Photo by Ken-ichi Ueda.

Isopods are familiar to us because of the garden varietal pill bug *Armadillium*, which was introduced to North America. Native Klamath Mountain isopods include woodlice (*Ligidium* sp.). Woodlice make there living as detritovores in damp soil, leaf litter, and decomposing organic matter. Freshwater isopods like *Caecidotea occidentalis* are common in cool, clean, and well oxygenated streams and rivers. Other isopod species are restricted to springs and cave streams.

Mysida, or opossum shrimp, resemble true shrimp and regionally inhabit large rivers and deep lakes. They are usually white to grey and range from 4-12 mm.

Subphylum Myrapoda

Centipedes

Although not as diverse as many arthropods, myrapods are fascinating nonetheless. Species have anywhere from ten to nearly 200 pairs of appendages and range in size from microscopic to 12 inches (30 cm) long. Myriapods prefer humid environments where they are found in soils, leaf litters, and under stones and wood. Many have a repugnatorial gland that emits foul-tasting compounds for defense. While the fossil record is sparse, evidence suggests the group is quite old—living on land as early as 400 million years ago.

Over 50 centipede species are known from California or the Pacific Northwest but one of them (*Stylopauropus californicus*) also has isolated populations in Siberia and the Appalachians—suggesting that this species was once more widespread (Roth 2015). The Klamath Mountain endemic, *Stenophilus californicus,* is found only in a cave in Shasta County (see Obligate Subterranean Invertebrate sidebar).

Millipedes

Millipedes are easy to identify as a group because their body partitions are fused in pairs with two legs on each side. They are particularly diverse in the moist forests near the coast from central California north to Alaska—and the Klamath Mountains hold evolutionary significance for the continent's millipedes (Shelley 1994). Several species in the orders Julida and Polydesmida are only found in the central Klamath Mountains. Oregon Caves National Monument is the type (and only known) locality of *Aprosphylosoma darcenae*—the sole member of the family Aprosphylosomatidae—one of the rarest millipedes in the world.

Yellow-spotted millipede (*Harpaphe haydeniana*). Photo by Michael Kauffmann.

Two other species, *Opiona catorycha* and *Taiyutyla caseophila,* were described from Oregon Caves in 2019 (Shear 2019). Other Klamath Mountain endemics include *Californiulus dorsovittatus,* and *C. chamberlini*. The southern Appalachian Mountains in the eastern United States have been documented as a prime global center of millipede evolution, and similar significance should be given to the California/Oregon border centered around the Klamath Mountains (Roth 2015).

Subphylum Hexapoda

Insecta

This largest group within the arthropod phylum have an exoskeleton, a three-part body (head, thorax and abdomen), three pairs of jointed legs, compound eyes, and one pair of antennae. More than a million described species represent more than half of all known living organisms. Insects are found in nearly all environments but are less common in oceans where another arthropod, the crustaceans, dominate. The irony is that they are the least understood group of all animals both globally and locally. What follows is a synopsis of available information from the common to the rare for select orders found across the Klamath Mountain region.

Coleoptera

It is estimated that half of all animal species alive today are beetles, making this the largest order in the animal kingdom. Conservatively, this means there are at least three million beetle species on the Earth—including some of the largest insects (Pakaluk and Slipinski 1995). At least one-third of all insects in the Pacific Northwest are beetles (Haggard and Haggard 2006). Beetles inhabit terrestrial and freshwater habitats across the region and have mouthparts, as both adults and larvae, to feed on a variety of foods. Coleoptera means "sheathed wing" to describe the two pairs of wings. The first pair (elytra) is enlarged and thickened into sheaths which cover the more delicate hind wings. Because the elytra are hard, there are better fossil records for beetles than other insect group. The oldest fossil beetles are from the Permian (about 265 million years old; Ponomarenko, 1995).

Spined woodborer (*Trichocnemis spiculatus*). Photo by Eric Peterson.

The Klamath Mountains and Appalachians have some of the highest diversity in the predatory Carabinae (tribe Carabini) beetle family. Degenerative hind-wing selection in this family, over time, has played a role in speciation because the inability to fly has increased geographic isolation (Su et al. 2004). The Klamath Mountains have exceptional diversity in land snails (*Monadenia*) on which snail-eating beetles (Genus *Scaphinotus)* prey. There are about 60 species of *Scaphinotus,* all native to North America. About 20 percent of these species are presumed to live in the Klamath Mountains (Roth 2015b).

California glowworm (*Ellychnia californica*). Photo by Justin Garwood.

Nebria Praedicta and Nebria turmaduodecima

There are five *Nebria* species in the Klamath Mountains. *Nebria siskiyouensis* and *Nebria triad* survive along montane stream margins (between 2,700–5,300 feet) while *Nebria eschscholtzii* occur in similar habitats from 4,600 feet down to the coast. *Nebria praedicta* and *Nebria turmaduodecima* live amongst the high elevation snowfields and ice in the Trinity Alps and are commonly referred to as ice beetles. All *Nebria* species are strictly nocturnal—avoiding both diurnal predators and warmer daytime temperatures.

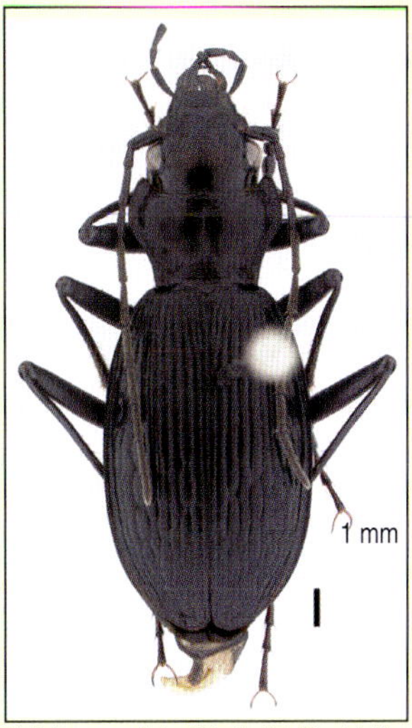

Nebria praedicta. Photo by David Kavanaugh.

N. praedicta and *N. turmaduodecima* are perfectly adapted for life along the margins of snow and ice in the extreme subalpine zones of the Klamath Mountains. With the help of special enzymes, these beetles have evolved to function at low temperatures—staying active below freezing when other animals can't. In another exceptional evolutionary leap, they produce a chemical that works like antifreeze—lowering the freezing point of water in their bodies. These two species avoid daylight at all costs—hiding from both light and predators under rocks. Under cover of the night, they emerge to feed on other snow-loving (or snow-stranded) insects. The same insect foraging strategy on snow and ice is commonly employed during daylight hours by the american pipit (*Anthus rubescens*), a small passerine bird that would likely forage on *Nebria* if it were not for their nocturnal habits.

In 1980, *Nebria turmaduodecima*, was first collected and described by Dave Kavanaugh on a large snowfield near Upper Caribou Lake (between 6,900–7,500 feet). In 2008 Kavanaugh found the snowfield above Caribou Lake disappeared and it was thought *N. turmaduodecima* could have gone extinct. That fall, Sean Scoville, a graduate student of Kavanaugh's, rediscovered *N. turmaduodecima* on the margin of Grizzly Glacier below Thompson Peak in the Trinity Alps. However, the expedition was first motivated by Kavanaugh who predicted Scoville would find a new species of *Nebria* at the small isolated glacier. On that trip, Scoville also collected the first specimens of *Nebria praedicta*—described as a new species in 2009—with perhaps the smallest distribution of any animal occurring in the Klamath Mountains.

The recent discovery of these two endemic *Nebria* beetles highlights their total existence is dependent on just a few acres of available habitat. There are 85 described species of *Nebria* beetles in North America. With the decline in snowpack across the West, and ultimately glaciers, it is only a matter of time before some of these species disappear forever (see Glacier sidebar, page 130-131).

Diptera

True flies, in the order Diptera, are some of the most successful and common insects in the Klamath Mountains. With at least 20,000 species in North America alone, it is difficult to quantify the diversity here. The diversity in species, and thus habitats, is due to the complexity in mouthparts which are used to eat plants, prey on other insects, and even suck blood—to name only a few.

Net-winged midges (Blephariceridae; about 320 species) are a small family important to stream ecosystems where they are usually restricted to cool, clear, well-oxygenated mountain streams. *Blepharicera kalmiopsis* is endemic to several rivers in southwestern Oregon and northwestern California and appears to have one of the most restricted ranges in the genus (Jacobson and Courtney 2008). Based on the high level of plant diversity in the region it is easy to speculate that host-specific plant-eating families of flies like gall-midges (Cecidomyiidae) and leafminers (Agromyzidae) are likely high in endemism in the Klamath Mountains. Like many other invertebrates, more research is needed.

Hymenoptera

The unique and rich assemblage of plant communities in Klamath Mountains is often thought of as originating primarily from geological processes, but Hymenopterous insects also play a fundamental role in shaping regional botanical diversity. Hymenoptera (bees, wasps, sawflies, and ants) have been sculpting plant diversity through pollination, seed dispersal, nutrient cycling, predation, parasitism, herbivory, and more since the Triassic—over 280 Ma. Hymenoptera are characterized by microscopic distinctions and are diverse in shape, size, color, phenology, diet, and behavior. Most Hymenopterans live solitary lives with no adult inter-generational overlap, no castes, and no cooperative care. Only a small proportion live in social groups.

Nearctic bumble bee (*Bombus vancouverensis* ssp. *nearcticus*) on Klamath coneflower (*Rudbeckia klamathensis*). Photo by Michael Kauffmann.

Globally there are over 150,000 described species but estimates predict there could be as many as 2.5 million species. In the Klamath Mountains while the total number of species is unknown, predictive diversity is high based on the documentation of bees and hunting wasps. Particularly diverse habitats are open, xeric, serpentine, and fire-prone serpentines. Shadier, conifer and oak dominated forests harbor herbivorous and xylophagous sawflies, carpenter ants, and wasp parasitoids of herbivorous insects.

Bees

The renowned floral diversity of the Klamath Mountains gives rise to an amazing diversity of native bee species. In fact, northwest California and southwest Oregon are exceptionally rich in bumble bees (*Bombus* spp.) with ~25 of the world's ~250 bumble bee species. Bees originated around 140 Ma and are separated from closely related wasps by the presence of branched (plumose) hairs, which facilitate pollen collection. It is not known how many of the world's approximately 20,000 bee species live in the Klamath Mountains but with over 1,600 species in California and at least 700 species in Oregon—the Klamath surely holds many. One study in the diverse plant communities of the Horse and Grouse Mountain area of Humboldt County found 229 bee species in less than 0.07 mi² (17.6 hectares) of survey area (Lopez 2017)!

Andrena sp.on *Balsamorhiza* sp. Photo by Brian Dykstra.

Most bees species are 'solitary' in that one mated mother bee will facilitate nest building, foraging, provisioning, and egg laying for their progeny. Intergenerational contact for these species is non-existent, or limited. A smaller number of bee species are subsocial or social, meaning that there are multiple generations in a nest. The only native social bees living in a colony are bumble bees (*Bombus* spp.). The introduced and propagated, non-native European honey bee (*Apis mellifera*) is another social colony-forming bee species. Most bee species make a nest in the ground, although some nest in existing cavities such as beetle burrows, snail shells, and pine cones—as well a variety of other substrates. Nesting material can be glandular secretions, or collected items such as mud or plants. Larvae feed on pollen/nectar provisions and, over time, metamorphose into adult bees—usually emerging in the next season. The timing of emergence, and length of adult activity, can vary widely between species—which is related to the phenology of floral resources utilized. Most bee species provision pollen and nectar for their young.

Female bees transport pollen in dense areas of hair (scopae) located on the legs or abdomen—or packed within concave areas fringed with long hairs (corbiculae) on the hind legs. Hair characteristics of bees (e.g., spacing, length, type of branching) may vary according to the morphology of pollen grains they prefer to collect, just as tongue length may be related to the floral morphology of preferred nectar sources. Bees vary in size and this affects foraging abilities. Some plant groups only allow certain bee groups to efficiently collect pollen. This is done in a variety of ways, such as corolla size and shape,

A carpenter bee (*Xylocopa* sp.) in the Trinity Alps. Photo by Kyle Keegan.

Invertebrates

Decline of the world's rarest bumble bee

Bumble bees (*Bombus* spp.) are an important group of wild North American pollinators. However, in the latter half of the 20th century, concerns have arisen because of noticeable declines in many species' populations—due to a variety of factors including pesticides and diseases from managed bees.

The Klamath Mountain endemic Franklin's bumble bee (*Bombus franklini*) was named in 1921 for Henry J. Franklin, who authored the *Bumble Bees of North and South America* (1913). The Franklin's is even more vulnerable than other *Bombus* species due to its restricted range—confined to Siskiyou and Trinity counties in California; and Jackson, Douglas and Josephine counties in Oregon. This is considered the smallest range of any *Bombus* species in the world.

Robbin Thorp, the late professor of entomology at U.C. Davis, worked tirelessly to better understand the species and the decline he witnessed. Until 1998, Franklin's bumble bee was prevalent throughout its range—and even in yards around Ashland, Oregon. Surveys since 2004 have been almost entirely incapable of observing a single specimen. The last time the species was seen was by Thorp, on Mount Ashland, Oregon, in 2006. In 2021, the species was listed as endangered by the U.S. Fish and Wildlife Service. But, is it too late?

Colla et al. (2012) found that bumblebee species with a narrow climatic niche, species at the edge of their climatic niche, and species restricted to certain habitat types are more vulnerable. This checks all the boxes for the Franklin's bumblebee and its relationship to the Klamath Mountains. As lovers and explorers of the region, it is our responsibility to diligently watch for this species—which is possible to identify with good photos but approaching impossible to find.

Franklin's bumblebees have yellow on the anterior top of the thorax that extends behind the wing bases and forms an upside-down U around the central patch of black. There is no yellow on the abdomen, which has white at its tip, and the face is mostly black. Males and females are similar. They are sustained by lupines (*Lupinus* spp.), poppies (*Eschscholzia* spp.), hummingbird mints (*Agastache* spp.), coyote mint (*Monardella* spp.), and vetches (*Vicia* spp.). They prefer abandoned rodent burrows for colonies.

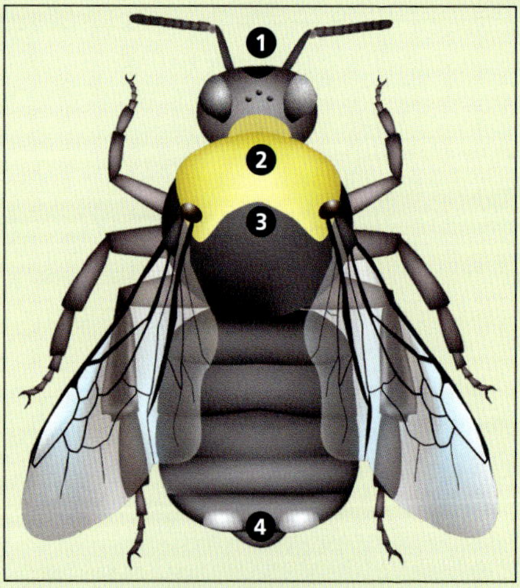

Distinct characteristics of the Franklin's bumble bee including (1) predominantly black face with yellow on the vertex (top of the head), (2) yellow on the anterior top of the thorax extends well behind the wing bases and (3) forms an inverted U-shape around the central patch of black, (4) white at the tip of the abdomen. Illustration by Elaine Evans, courtesy of the Xerces Society.

Invertebrates

and stamen type. Some plants (*Primula* spp., *Pyrola* spp., others) require a specific vibration, achieved by a limited set of species, for pollen to effectively release.

The connection between the Klamath Mountain botanical diversity and bee diversity is seen in specialist bees, which are adapted to collect pollen from a specific plant or group of plants. *Osmia penstemonis* and *O. brevis* may rely solely on *Penstemon* species for pollen needs. Other examples include *Duforea dentipes* (*Calochortus* spp.*)*, *Osmia pikei* (*Ribes* spp.), and *Anthidium placidum* (*Cordylanthus* spp.). *Agoseris, Arctostpahylos, Fragaria, Phacelia, Potentilla, Ranunculus* and *Sildacea* are a just a few of the plant genera with specialist bee relationships.

Wasps

Wasps play important and varied ecological roles as predators, parasitoids, and pollinators. Over 100,000 wasp species are described, but the global total may exceed 1 million species. The number of species in Klamath Mountains is unknown, but likely high. As in bees, social behavior is much less common than a solitary life cycle. The majority of wasps rely on insect protein for larval development and very few are vegetarian (e.g., *Pseudomasaris vespoides* provisions its young a mixture of nectar and pollen from *Penstemon* spp.). Wasps may seek specific animals (flies, caterpillars, leafhoppers, tree crickets/ katydids, grasshoppers/cicadas, aphids, bees, flies, spiders, other wasps, etc.) to provision nests with and lay an egg on or into as a host.

The amazing and nuanced specificity is evidenced by wasps that only hunt specific spiders (e.g., *Calopompilus pyrrhomelas* hunts only turret spiders in the genus *Antrodiaetus*) or wasps that hunt specific bee groups (*Philanthus* species hunt Halictid bees). Another example of a solitary parasitoid wasp is the velvet "ant." Wingless females are often covered with bright scarlet or orange hair. They lay an egg within an underground bee or wasp nest, and the larva feeds on larvae of that bee or wasp host. Parasitoid wasps may specialize on hosts such as certain butterflies and moths, beetles, flies, and true bugs. Hyperparasitoid wasps are parasitoids of parasitoids (there can be multiple tiers). Kleptoparasitoid wasps take advantage of existing wasp or bee nests and lay eggs on the provisions of others.

Gall-forming wasps lay eggs in plant tissue, inducing formation of uniquely shaped galls. Larvae feed on the galls' nutritive tissue. Oaks (Fagaceae) are the primary host plant, and among the estimated 1,000 species of the oak gall wasp tribe (Cynipini) over 99 percent use only *Quercus*. Unique gall wasps are found on the wide diversity of Klamath Mountain oaks including chinquapin (*Chrysolepis chrysophylla, C. sempervirens*) and tanoak (*Notholithocarpus densiflorus*).

Golden oak apple gall wasp galls (*Andricus vacciniifoliae*) on a huckleberry oak (*Quercus vaccinifolia*). Photo by Kyle Keegan.

Invertebrates

Sawflies

Sawflies (wood wasps, parasitic wood wasps, stem sawflies, and horntails) are in the Suborder Symphyta. There are approximately 8,200 known species in the world and it is, again, unknown how many reside in the Klamath Mountains. In the short-lived adults, the thorax and abdomen are fused such that they appear as one—unlike ants, bees, and wasps which typically have an observably constricted waist. Microscopic examination of a sawfly ovipositor reveals a "saw-like" morphology to facilitate deposition of eggs into plants. Sawflies do not have stingers.

Insect communities in the veiled polypore

The wood-rotting *Cryptoporus volvatus* is nearly impossible to confuse with other fungi. What is confusing is that, while it is a polypore, the pore surface is cryptically veiled by a tissue flap giving it a puffball-like appearance. The veil is an evolutionary advancement wherein the fungus traps and holds mature spores to entice a variety of insects to move in, consume various parts of the fruiting body, and then carry the spores away. When insects leave, particularly beetles, and then bore into new wood, *Cryptoporus* spores enter the novel habitat as well where they eventually germinate—and the process begins again.

Veiled polypore on a burned Douglas-fir. Photo by Michael Kauffmann.

The species thrives in the recently-dead wood of conifers across North America which makes it entirely dependent on cyclical disturbance regimes caused primarily by wildfire and bark beetle infestations. In the Klamath Mountains, *Cryptoporus* most commonly grows on Douglas-fir (*Pseudotsuga menziesii*) but they are also found on pines (*Pinus* spp.) and true fir (*Abies* spp.).

In a niche partitioning study, *Cryptoporus volvatus* was used to test for the competitive coexistence of insect species. Niche partitioning posits that different insect species use different resources, developmental and/or life stages, or parts of a given resource to reduce competition and increase survival rates. A total of 17 insect species—including 11 species of *Coleoptera*, three species of *Lepidoptera*, and three species of *Diptera*—emerged from *Cryptoporus vulvatus* fruiting bodies between mid-May and mid-July (Kadowaki 2010). The temporal distribution of the emerging insects not only ensures their survival but allows the spores of the veiled polypore to be transported by insects to new trees over a longer time period. This symbiotic relationship also guarantees the veiled polypore will continue to thrive, feasting on the remaining sugars left in the recently deceased conifers of the Klamath Mountains.

Larvae resemble brownish or yellowish green caterpillars, often with dark heads. The larvae of most species are plant feeders, and common names indicate their preferred host plants: alder sawfly, willow gall sawfly, raspberry horntail (raspberry, salmonberry, and thimbleberry), conifer sawflies, and so on. The cedar wood wasp *Syntexis libocedrii* lays its eggs in fire-burned, still-smouldering incense-cedars.

Ants

Ants live in social colonies composed of hundreds to thousands of individual ants in extensive underground nests. The building and habitation of these nests creates bioturbation effects (cycling nutrients, distributing and decomposing organic materials, aerating soil, and facilitating water and gas exchange) with cascading effects on local ecosystem functioning and biodiversity.

Carpenter ants in the genus *Camponotus* are common across the range. Photo by Kauffmann.

Numerous plants in the Klamath Mountains have seed dispersal strategies involving ants. These plants have seeds with an attachment, known as an eliosome, a specifically attractive reward (rich in specific lipids, amino acids, and other nutrients) to ants that don't eat seeds. When collected by the right ants, these seeds have enhanced germination and growth after being deposited in an ant nest environment rich in phosphorous and nitrates. These underground locations also offer protection from fire and from seed-eating ants and rodents.

Ants also have a variety of mutualisms with butterflies. Some ants protect gossamer-winged butterfly (Lycanidae) caterpillars in exchange for nutritive honey dew excreted from the caterpillar's special gland. Some caterpillars are even brought into ant nests where they are attended to, and where they pupate and emerge.

Lepidoptera

Butterflies and moths (lepidopterans) comprise about 180,000 described species making them the second largest order of insects. Several characteristics generalize this order including scales on the body, wings, a proboscis, membranous wings, and complete metamorphosis. The larvae are commonly called caterpillars and are completely different from adults—having a cylindrical body, mandibles, three pairs of thoracic legs and a series of prolegs. Approximately 118 species of butterflies and over 600 species of macromoths have been documented in the Klamath Mountains (Shapiro 1990). Many of these, like our rich diversity of plants, have extant relatives in Mexico, the Appalachian Mountains, or eastern Asia. Several endemic species call the range home as well.

Most butterflies and moths are host-specific in their use of larval food plants and regional plant diversity plays a role in supporting butterfly diversity. Because many plants are at their geographic limit in the Klamath Mountains, downsized moth and butterfly populations are reflected in isolation from larger populations as well. Ten butterfly genera with endemic Californian species reach their northern limit here. A dozen major groups of larval food plants and serpentine habitats are primarily responsible for supporting most of the Lepidoptera diversity in the Klamath Mountains. What follows is a summary of that diversity based on plants and habitats.

Pinaceae (pines, spruces, hemlocks, and firs) hosts feeders like *Neophasia menapia* and *Callophrys eryphon*. Butterflies that feed on mistletoes growing on conifers include two hairstreaks *Callophrys spinetorum* and the increasingly rare and localized *C. johnsoni*. *Lithophane jeffreyi* is a moth that reaches its northernmost range limit in the Klamath, as does its host, Jeffrey pine (*Pinus jeffreyi*). Other conifer feeders include *Egira variabilis*, the rare *Panthea gigantea*, *Phaeoura mexicanaria*, and *Cochisea sonomensis*. The latter moth is at its northernmost limit in the Klamath, although its Douglas-fir host extends much farther north.

Cupressaceae (cedars, junipers, or cypresses) hosts such butterflies as *Callophrys gryneus.* Examples of very rare and endemic moths that feed exclusively on cedars and cypress include *Sphinx sequoiae*, *Lithophane gausapata*, *Abagrotis mirabilis*, and *Abagrotis rubricundis*.

Evergreen Fagaceae (canyon oak, tanoak and chinquapin) host feeding butterflies like the golden hairstreak (*Habrodais grunus*) and the California sister (*Adelpha californica*). Moths that feed exclusively on evergreen oaks include *Catocala ophelia*, *Perigonica pectinata*, *Perigonica angulata*, and *Zale termina*.

Deciduous Fagaceae (Oregon white oak, Brewer's oak, and California black oak) host feeding butterflies like the goldhunter's hairstreak (*Satyrium auretorum*). Mistletoes on these oaks (*Phoradendron* spp.) are host to the brilliant blue hairstreak butterfly *Atlides halesus*. Examples of moths that feed exclusively on deciduous oaks including Edwards' glassy-wing moth (*Pseudohemihyalea edwardsii*), marble dagger moth (*Acronicta marmorata*), *Abagrotis baueri*, *Feralia februalis*, *Lithophane contenta*, *Mesogona subcuprea*, *Orthosia ferrigera*, *Perigonica tertia*, and the extremely rare *Pseudocopivaleria sonoma*. Underwing moths include *Catocala aholibah*, *C. ilia*, and *C. verrilliana*.

Pine white (*Neophasia menapia*) on western false asphodel (*Triantha occidentalis*). Photo by Michael Kauffmann.

Evergreen Ericaceae (manzanitas and madrones) host feeder moths on madrone, manzanita species, salal, and rhododendron include the ten-spotted rhododendron moth (*Dysstroma sobria*), *Orthosia transparens, O. pulchella, O. mys, Cochisea sinuaria, Saturnia mendocino, Mesogona rubra,* and *Pleromella opter* on manzanitas.

California sister (*Adelpha californica*). Photo by Ken DeCamp.

western sulphur
Colias occidentalis

snowberry clearwing moth
Hemaris thetis

pink sheep moth
Hemileuca eglanterina shastensis

Lindsey's skipper
Hesperia lindseyi

common buckeye
Junonia coenia

clodius parnassian
Parnassius clodius

Rocky Mountain parnassian (larva)
Parnassius smintheus

Rocky Mountain parnassian (adult)
Parnassius smintheus

Sierra blue
Plebejus podarce klamathensis

western cloudywing
Thorybes diversus

mountain-mahogany hairstreak
Satyrium tetra

callippe fritillary
Speyeria callippe

Deciduous Ericaceae (huckleberries) host butterflies like a locally unique form of *Lycaena mariposa* (typically found in or near *Darlingtonia californica* fens), and again, *Callophrys augustinus*. Rare moths that feed either exclusively or primarily on huckleberries include *Platypolia contadina*, *Lithomoia germana*, *Abagrotis trigona*, *Eulithis destinata,* and *Syngrapha orophila*. *S. epigaea* is one of the few regional moths to reach its southern limit in the region. This may reflect the unusual nature of heaths, which are more common in temperate rather than tropical regions. Bailey's pinion moth (*Lithophane baileyi*) lives only in the Pacific Northwest and in eastern North America—a disjunction likely caused by the drying and cooling of the last 15 million years. All these moths are most common on moist north-facing slopes at high elevations where huckleberries are particularly abundant. Also in the fens is *Euros osticollis,* a moth first described in 2006. It probably feeds on sedges growing in the fens. Some of these genera and families may be part of the boreotropic fauna that began to decline in the Northwest 34 million years ago as increasing drought separated *Darlingtonia californica* from other pitcher plant genera.

Euros osticollis is a moth of the family Noctuidae. It is endemic to streams and seeps in southwestern Oregon. Photo by Merrill Peterson/PNW Moths.

Rhamnaceae (wild lilacs or snowbrush–*Ceanothus* spp.) host butterflies like pacuvius duskywing (*Erynnis pacuvius*), pale swallowtail (*Papilio eurymedon*), hedgerow hairstreak (*Satyrium saepium*), echo azure (*Celastrina echo*), and California tortoiseshell (*Nymphalis californica*). Moths that feed exclusively on *Ceanothus* include a dozen species of the geometrid *Drepanulatrix*, as well as bordered fawn moth (*Sericosema juturnaria*) and elegant sphinx moth (*Sphinx perelegans*). A smaller number of moths on *Frangula* and *Rhamnus* include tawny eupithecia moth (*E. ravocostaliata*) and tissue moth (*Triphosa haesitata*)—which particularly favor cascara. *Speranza guenearia* is near its northern limit in the region, as it feeds exclusively on California coffeeberry (*Frangula californica*).

Salicaceae (willows and cottonwoods) include riparian butterflies like western tiger swallowtail (*Papilio rutulus*), Sylvan hairstreak (*Satyrium sylvinus*), green comma (*Polygonia faunus*), Lorquin's admiral (*Limenitis lorquini*), mourning cloak (*Nymphalis antiopa*), and dreamy duskywing (*Erynnis icelus*). Moths on willows and poplars include one-eyed sphinx (*Smerinthus cerisyi*), and the royal moth (*Scoliopteryx libatrix*).

Betulaceae (birches, alders, and hazelnuts) host butterflies like the pale swallowtail (*Papilio eurymedon*). Moths include fingered dagger moth (*Acronicta dactylina*), stormy arches moth (*Polia nimbosa*), *Andropolia aedon*, and *Drepana arcuata*. Like the heaths, members of Betulaceae tend to be common in boreal and temperate latitudes—a pattern that helps to place *Autographa corusca* near its southern limit. *Dysstroma ochrofuscaria* only feeds on hazelnuts and is also near its southern limit in the region.

Rosaceae (roses, blackberries, hawthorns, serviceberry, ocean-spray, cherry, and mountain mahogany) host butterflies like *Papilio multicaudata* (on chokecherry), *Celastrina echo,* and *Satyrium tetra* (on mountain mahogany). Moths include elegant sheepmoth (*Hemileuca eglanterina*), great ash sphinx moth (*Sphinx chersis*), *Ceranemota fasciata*, *Abagrotis forbesi*, and fragile (*Acronicta fragilis*) and impressed (*A. impressa*) dagger moths. Enigmatic dart moth (*Cerastis enigmatica*) and *Lacinipolia patalis* are specialists feeding on blackberries (*Rubus* spp.) *Sympistis columbia* and *S. dunbari* feed on ocean-spray (*Holodiscus discolor*). *Drepanulatrix bifilata* and *Stamnodes coenonymphata* only feed on mountain mahogany (*Cercocarpus* spp.).

Western tent caterpillar (*Malacosoma californicum*) on *Ceanothus cordulatus*. Photo by Michael Kauffmann.

Caprifoliaceae (honeysuckle) hosts *Euphydryas colon,* a butterfly that feeds on snowberry. Moths on snowberry and honeysuckle include *Sphinx vashti*, snowberry clearwing moth (*Hemaris diffinis*), *Abagrotis erratica*, *Pleromelloida cinerea*, *Behrensia conchiformis*, *Pleromelloida obliquata,* and *Sympistis ragani.*

Western Tiger Swallowtail (*Papilio rutulus*). Photo by Ken DeCamp.

Grossulariaceae (currants and gooseberries–*Ribes* spp.) host butterflies like the tailed copper (*Lycaena arota*) and hoary comma (*Polygonia gracilis*). Moths include *Dysstroma formosa*, *D. brunneata*, *Speranza bitactata*, *Neoterpes trianguliferata*, *Stretchia plusiaeformis*, *S. muricina*, *S. pictipennis*, and *Platypolia loda*.

Serpentine Meadow Habitats have certain moths tied to them including *Protorthodes alfkeni* and *Properigea niveirena*. *Pseudobryomima muscosa* feeds on rock ferns and *Cryphia cuerva* feeds on rock lichens. This special habitat with its regional endemics includes Edith's copper (*Lycaena editha pseudonexa*) on broomrapes and *Speyeria zerene* ssp. *gloriosa* and Great Basin fritillary (*S. egleis mattooni*) on violets (*Viola* spp.). Bunchgrass is fed upon by *Apamea siskiyou*, a rare species largely endemic to this habitat and ecoregion. Other species include *Apamea albina, Apamea centralis* and *Apamea cinefacta*. *Melanapamea mixta* is primarily an Appalachian species of eastern North America, but a disjunct population occurs on rocky serpentine habitats of the Klamath Mountains. It appears to be an ancient relict that previously had a wider distribution across North America.

Obligate Subterranean Invertebrates

California's complex geography and physiography support a diversity of karst (carbonate rocks), volcanic, littoral (sea), and other cave types. While over 4,000 caves are documented in California, approximately 1,000 have been sampled for biological diversity. Of those, 197 are in California's Klamath Mountains. The Klamath Mountains hold five distinct cave regions, all of which are carbonate limestone and marble karst from the Permian Period. Those areas are Oregon's central Siskiyou Mountains, Marble Mountain Karst, Shasta Limestone, Trinity Alps, and the Hayfork region (see Cavern sidebar, pages 74-75).

Of the 1,366 taxa documented in California's caves, 102 are terrestrial troglobites (live only underground) with 44 more taxa classified as aquatic cave obligates or groundwater dependent—for a total of 146 obligate subterranean species (Elliott et al. 2017). An astounding 494 taxa (37 percent of those documented) remain unclassified! Obligate subterranean invertebrates are of interest for a variety of reasons including their high rates of localized endemism as well as their unique morphology—including divergent physical characteristics like diminutive eye size and reduced pigmentation compared to closely related terrestrial species. In the Klamath Mountains, 25 endemic obligate subterranean species have been documented. Documentation in Klamath Mountain caves, especially for invertebrates, began only in the 1970s so there is surely much more to discover.

Terrestrial Cave Obligates

Arachnids: Scorpions, pseudoscorpions, harvestmen, spiders

Arachnids, which prefer moist, dark habitats both above and below ground, contribute significantly to the overall cavernicole fauna of the Klamath Mountains. In California, 250 arachnid species have been found in caves with 24 percent (or 57 species) being troglobites. In the Klamath Mountains, Graham's cave scorpion (*Uroctonus grahami*) is one of only a few discovered scorpion troglobites and is currently only known from a single cave in the Shasta Limestone of the eastern Klamath.

Uroctonus grahami. Photo by Warren E. Savary.

The unique hooked claw of *Trogloraptor marchingtoni*. Photo by Charles E. Griswold.

In 2010, a completely new family, genus, and species of spiders was discovered by spelunkers in caves around Grants Pass, Oregon. Geo Graening, Neil Marchington, Ron Davis, and Daniel Snyder found *Trogloraptor marchingtoni* (family Trogloraptoridae) and sent specimens to the California Academy of

Sciences where the discovery was confirmed (Griswold et al. 2013). The species is exceptional in that they have only six eyes and remarkable hook-like (raptorial) claws on the last segments of their legs.

Myriapods: Millipedes and Centipedes

This ancient group, dating to 428 Ma, is considered the first terrestrial arthropod. They are common in moist forest caves where millipedes are found decomposing organic matter and centipedes fill the niche of nocturnal predator. Many myriapod troglobites are endemic to a single cave or localized cave system. However, *Opiona graeningi* is found in 11 caves across the Klamath Region.

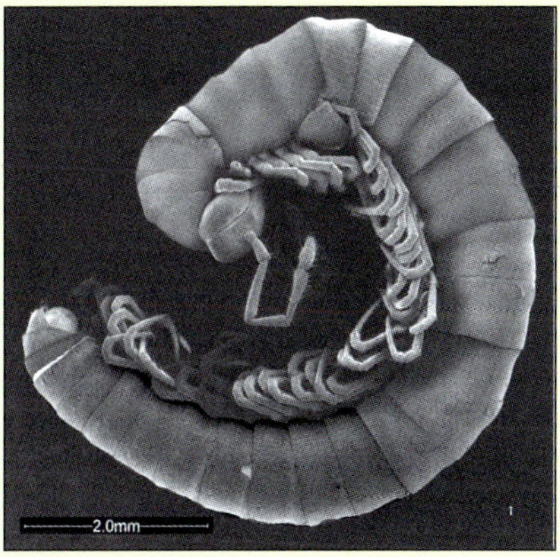

Opiona graeningi. Photo by William A. Shear.

Collembolans: Springtails

These primitive arthropods are like insects but lack wings, among other traits. They are generally small (< 6 mm) and are cave detritivores. California has 43 species, with more yet to be described. Little is known about this group in the Klamath Mountains.

Insects: Ice Crawlers

Surprisingly, this group does not appear to be significant in defining cave fauna in California (Elliott et al. 2017). Grylloblattids, or ice crawlers, live on mountains and near glaciers—and also in caves. Somewhere between a cricket and a cockroach, there are 32 species worldwide, 13 in California, with at least three species endemic to the Klamath Mountains. *Grylloblatta oregonensis* and *Grylloblatta siskiyouensis* are endemic to Oregon Caves with the former preferring surface and twilight zones with the latter inhabiting the cave's dark zones. *Grylloblatta marmoreus* is restricted to caves and snow fields in the Marble Mountains. Additional populations are likely to be found in the Klamath Mountain region, particularly the Trinity Alps. Continued efforts to understand the diversity and evolutionary history of the genus is important for understanding historical climate changes (Schoville 2012).

Grylloblatta marmoreus. Photo by Sean Schoville.

Aquatic Cave Obligates

Amphipoda

With 148 described species worldwide, crustaceans in the genus *Stygobromus* are important components of subterranean habitats across North America and Asia. The secret amphipod (*Stygobromus mysticus*) is the only representative yet documented in Klamath Mountain caves.

Notoptera

Grylloblattids, or the ice crawlers, are patchily distributed on glaciers in North America, China, Siberia, Korea and Japan. Several species are distinctive to the Klamath Mountains and discussed in the Obligate Subterranean Invertebrate sidebar on previous pages.

Odonata

Dragonflies and damselflies are an ancient group of carnivorous insects that have existed since the Permian (275 Ma). They are characterized by rounded heads with large, compound eyes; legs for catching prey in flight; two pairs of long, transparent wings; and elongated abdomens. Odonata in the Klamath include Pacific clubtail (*Phanogomphus kurilis*), grappletail (*Octogomphus specularis*), eight-spotted skimmer (*Libellula forensis*), flame skimmer (*Libellula saturata*), widow skimmer (*Libellula luctuosa*), Pacific spiketail *(Cordulegaster dorsalis*), boreal bluet (*Enallagma boreale*).

Widow skimmer (*Libellula luctuosa*). Photo by Ken DeCamp.

Boreal Bluet (*Enallagma boreale*). Photo By Ken DeCamp.

Orthoptera

Includes grasshoppers, locusts, and crickets represented by over 20,000 species worldwide. They are characterized by cylindrical bodies and elongated hindlegs for jumping. They generally consume plants and are easily found in meadows, grasslands, and exposed upland habitats across the region. A variety of species from sooty grouse (*Dendragapus fuliginosus*) to gray wolves (*Canis lupus*) target them as a high-quality food source in mid- to late-summer. Species in the Klamath include Jerusalem crickets (*Stenopelmatus* spp.), devastating grasshopper (*Melanoplus devastator*), diabolic shieldback (*Neduba diabolica*), Siskiyou shieldback (*Idionotus siskiyou*), and Square-legged camel cricket (*Tropidischia xanthostoma*). The Podismine grasshopper tribe, an ancient lineage distributed mainly in Eurasia—along with the western and eastern regions of North America—are relicts of glacial times. They are represented here by the slow mountain grasshopper (*Bradynotes obesa*) in the higher elevations around Mount Eddy and in the Trinity Alps.

Slow mountain grasshopper (*Bradynotes obesa*). Photo by Justin Garwood.

Siskiyou shieldback (*Idionotus siskiyou*). Photo by Jeff Cole.

An Overview: Freshwater Aquatic Invertebrates

While we explore fairy shrimp and freshwater mussels and clams, few other aquatic invertebrates are discussed herein. This brief discussion of other aquatic invertebrates will impart a cursory context for this group in the Klamath Mountains. Aquatic invertebrates play an important role in the energy and nutrient systems of rivers, streams, and lakes. From the littoral zones near the surface of the water to the benthic zones at the bottom of lakes, aquatic invertebrates are the primary driver of decomposition and the foundation of aquatic food webs.

A caddisfly nymph with a rock-casing in the Smith River. Photo by Justin Garwood.

An adult stonefly (Plecoptera). Photo by Justin Garwood.

It is estimated that there are well over 11,000 species of freshwater invertebrates in the United States alone with some unknown fraction of that occurring in the Klamath Mountain region. Compared with marine

Isonychia velma, an adult mayfly in the Ephemeroptera order. Photo by Nathan McCanne.

Invertebrates

environments, the amount of freshwater environments is seemingly incidental, but in the Klamath, there are no salt water environments. Freshwater invertebrates are an immigrant fauna because they moved from salt water to fresh. In doing so, the groups had to make leaps in order to change environments including osmoregulation which balances internal water and salt concentrations.

Three common orders of aquatic invertebrates are Ephemeroptera (mayflies), Plecoptera (stoneflies), and Tricoptera (caddisflies). The aquatic larvae of all three are found in a wide variety of habitats such as streams, rivers, lakes, ponds, spring seeps, and temporary waters (vernal pools). Tricoptera larvae are distinct in that many use silk to make protective cases, which are often strengthened with gravel, sand, twigs, or other debris. All three groups have larvae that exhibit various feeding strategies including predation, leaf shredding, algal grazing. Most adults emerge from the water and have short lives during which they do not feed and focus on reproduction.

Mollusks

From sea to summit, snails, slugs, abalone, limpets, nudibranchs, clams, oysters, scallops, mussels, and octopus represent the second-largest phylum of invertebrates—with 85,000 extant species. Mollusks of the ocean are the most speciose marine phylum, representing over 20 percent of described marine organisms. While not as diverse as in the ocean, freshwater and terrestrial mollusks are found across a range of aquatic and terrestrial habitats where they are important nutrient recyclers—maintaining a balance between production and decomposition. They are also an abundant food item, and shelled mollusks provide a source of calcium for many predators. Major groups of mollusks represented in the Klamath Mountains include gastropods (slugs and snails) and bivalves (mussels and clams). Slugs are terrestrial; mussels and clams are fully aquatic; and snails can range from being fully aquatic. to semi-aquatic. to fully terrestrial.

Like other invertebrate groups, native mollusks of the Klamath Mountains, and across the Pacific Northwest, have been understudied. It is believed that about half of the freshwater and terrestrial mollusks in the Pacific Northwest have not been formally described (Roth 2015). This is due to remoteness but also the overwhelming cryptic diversity—especially in both aquatic and terrestrial snails. Individuals within the same species display a wide variety of morphological characteristics across their range. Identification often requires genetic analysis or dissection and careful examination of reproductive organs. However, a relatively accurate assumption to the species level can be made based on location. Perhaps the best approach is to enjoy each encounter with reverence for individual beauty, and to appreciate another story of regional biodiversity unfolding in the Klamath Knot.

Across the Klamath region, there are 80 species and subspecies of native land snails and slugs, dozens of aquatic snails from at least five families, at least three species of freshwater mussels, and a family of miniature freshwater clams that are easily overlooked. Many taxa, especially land and aquatic snails, are listed as sensitive or endangered. For example, some gastropods are dependent on late seral forests and are official Survey and Manage Species in the Northwest Forest Plan.

The plan requires surveys for specific species prior to logging or other disturbances and the establishment of Federal forest reserves for sensitive species largely under the umbrella of spotted owl (*Strix occidentalis*) habitats.

Gastropoda – Snails and Slugs

Terrestrial Snails and Slugs

With no comprehensive list available, and limited research across the Klamath Mountains (Dunk et al. 2004), all described terrestrial gastropods were cataloged using the most definitive keys and range descriptions found in Roth and Sadeghian (2006) and Burke (2013). With these data, at least three slug and 17 terrestrial snail families are represented—which includes 66 species and another 14 subspecies, totaling 80 (71 snails and nine slugs). Many taxa on this list are disputed—especially the subspecies. Unique snail specimens mentioned in scientific literature, that lacked formal descriptions, were omitted.

Like salamanders, terrestrial gastropods occupy a wide range of moist environments to avoid desiccation. These areas include entire old-growth forest stands down to edges of small springs in arid landscapes. Many endemic and near-endemic taxa in the upper Sacramento River of Shasta County appear to be associated with cavernous limestone formations and are commonly found in caves. At a local scale, talus (rock) slopes, stream riparian zones, thick leaf litter, and decaying downed wood are common habitats—providing moisture refuges when the surrounding landscape is dry. The snail shell is a habitat feature because it assists in maintaining body moisture while providing defense against predators. Terrestrial gastropods can be scavengers, herbivores, predators, or omnivores. Taildropper slugs also consume mycorrhizal fungi and are likely important mushroom spore dispersal agents. For most species, reproduction is hermaphroditic, with the individual carrying both male and female reproductive organs. Miniature snail offspring hatch from egg masses that are deposited in dark and moist locations.

Although native land snails and slugs are cryptic, they are Klamath Range biodiversity ambassadors—alongside salamanders, butterflies, and conifers. Land snail diversity here is similar to the Appalachian Mountains where over 120 species have been described (Ricketts et al. 1999). This rich evolutionary story is the product of geographic location, climate, geology, and mountain building over the eons. Like the southern Appalachians, the Klamath Mountains escaped the massive ice sheet during the Ice Age resulting in isolated refuges. Land snails are poor dispersers and thus commonly isolated in mesic patches within a xeric landscape. This fragmentation is amplified by other barriers including streams and ridgelines. Geological factors, such as calcium deficient serpentine barrens are also thought to limit terrestrial snail distributions.

Seventy out of the 71 land snails had enough distribution information to categorize taxa into geographic categories relative to the Klamath Mountains (Figure 1). A total of 24 percent are endemic with the greatest number found in Siskiyou, Trinity, and Shasta counties. An additional 17 percent are near endemics, having almost their entire distribution restricted to the Klamath Mountains. Some of these local taxa are more coastal, others more inland, while some are isolated in the central core. Eleven percent are regionally distributed in northern California

Mollusk diversity in the Klamath Mountains (red text are endemic)

California lancetooth *Haplotrema minimum*

northwest hesperian
Vespericola columbianus

Pacific sideband
Monadenia fidelis ssp. smithiana

Trinity bristle snail
Monadenia setosa

robust lancetooth snail
Haplotrema vancouverense

Shasta sideband
Monadenia troglodytes ssp. *troglodytes*

reticulate taildropper
Prophysaon andersoni

and southern Oregon, whereas 47 percent are widespread relative to the western United States. Pilsbry (1948) divided North America into ten terrestrial mollusk provinces. The Oregonian Province spans from western British Columbia through Oregon west of the Cascades crest and into northwest California to the southern tip of the Klamath Mountains. This abrupt southern boundary marks a clear break in regional diversity indicating episodic periods of species isolation driving local evolution. For example, 24 percent of regional and widespread snails reach their southern terminus here while 39 percent of widespread California species reach their northern terminus here. Only 36 percent of widespread snail taxa, and 21 percent of all snail taxa, show no Klamath Mountain influence on distribution.

The most common helicoid (spiral shelled) snails across the Klamath Mountains are the genus *Monadenia*, with over eighteen species and subspecies. *Monadenia* are relatively slow growing and long lived, with an average lifespan of 10–15 years. They are ecologically diverse, with species that inhabit talus rock crevices and rockslides, and others that inhabit deep forest duff. There are some semi-arboreal forms and others that crawl out on low plant cover. Across their range, various assemblages of climate, geology, and terrestrial habitat form what can be understood as conceptual islands—fragments of broader populations of organisms. Given enough time, these isolated groups change and exhibit unique, localized characteristics.

Take, for example, the rare and endemic Trinity bristle snail (*Monadenia setosa*) that was first described from a small area around three tributary streams near Big Bar along the Trinity River. This cryptic species requires a complex assemblage of soil composition, soil drainage, depth of leaf litter, climate, elevation, slope and aspect, the behavior of nearby creeks and streams, vegetation composition, canopy cover, forest decadence, disturbance regime, surface moisture, and proximity to other ideal habitats. When taken together the perfect, unique habitat is generated for the species. Interestingly, other optimal environments remain uninhabited—likely a result of the snail's limited dispersal abilities.

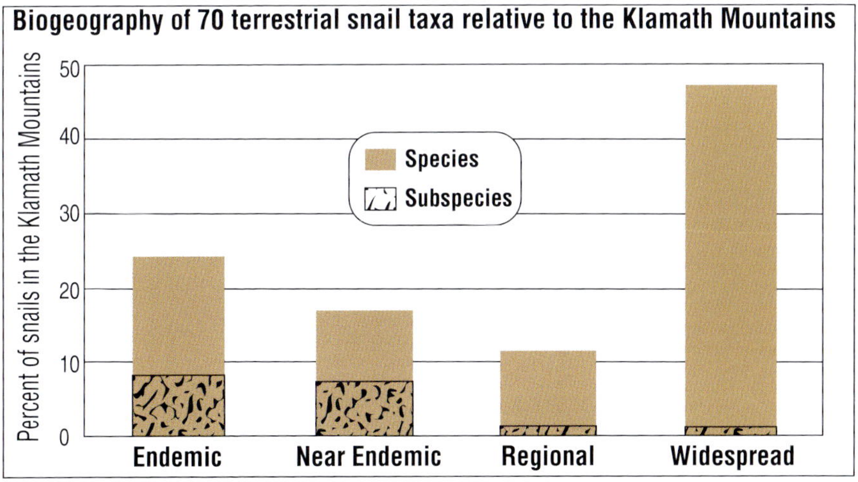

Figure 1. Biogeographic distribution of land snails found in the Klamath Mountains.

Despite many species being commonly observed, native slugs only represent nine species. Many (if not all) slugs appear to be habitat generalists, especially when compared to their shelled counterparts. The reticulate taildropper (*Prophysaon andersoni*), for example, is found throughout the coast ranges as well as Button's banana slug (*Ariolimax buttoni*) and the black westernslug (*Hesperarion niger*). The Klamath Mountains represent the southern range extents for of the blue-grey taildropper (*Prophysaon coeruleum*), papillose taildropper (*Prophysaon dubium*), and yellow-bordered taildropper (*Prophysaon foliolatum*)—all of which extend northward through Oregon and Washington.

Aquatic Snails

Although their taxonomy is in constant revision, California has at least 101 and Oregon has at least 79 described species of aquatic snails (143 for both states combined) making California the third and Oregon the seventh most speciose states for aquatic snails in North America (Johnson et al. 2013). Like terrestrial snails, little is known about the taxonomy, ecology, and distributions of aquatic snails in the Klamath Mountains making them one of the largest remaining regional mysteries. Generally, five families turn up with regularity including two air breathing families: bladder snails (Physidae) and ramshorn snails (Planorbidae); two gill breathing families: *Juga* snails (Semisulcospiridae) and minute pebblesnails (Lithoglyphidae); and one family that breathes through the mantle cavity, freshwater limpets (Lymnaeidae).

Highcap lanx (*Lanx alta*) in the Smith River. Photo by Justin Garwood.

Black juga (*Juga nigrina*) in the Smith River. Photo by Justin Garwood.

Juga species in the Smith River. Photo by Justin Garwood.

Most aquatic snails here have a spiral discoidal shell. The limpets are an exception by having a flattened cone-shaped shell. Most aquatic snails are grazers and use a rough tongue to scrape organic material—like algae and detritus—but some are known to also shred plant material, including leaves, from riparian trees. Depending on species, reproduction can be sexual or hermaphroditic, and miniature snail offspring hatch from aquatic egg masses. They are prey for a variety of ducks, spotted sandpipers (*Actitis macularius*), amphibians, and aquatic insect larvae. Some snails, such as genus *Juga*, serve as intermediate hosts for parasites including the parasitic fluke *Nanophyetus salmincola*. This fluke is responsible for salmon poisoning in dogs and requires three alternative hosts to complete its life cycle: *Juga* snails, salmonid fishes, and canines.

Aquatic snails are found in a variety of habitats ranging from seeps and springs to major rivers and ephemeral pools to lakes. Ramshorn and bladder snails are found in aquatic habitats with emergent vegetation, especially river backwaters, lower elevation lakes, and perennial ponds. Pebblesnails of the upper Sacramento

River watershed show the highest degree of local diversification and many species are confined to isolated springs. With most measuring less than 4 mm, many of these minute snails, such as the Potem Creek pebblesnail (*Fluminicola potemicus*), are crenophiles—found only in cold oxygen-rich spring headwaters (Frest and Johannes 1999). Other crenophiles are only found in large springs with cobbles, while the nugget pebblesnail (*Fluminicola seminalis*) is found in large streams and rivers lined with cobbles and boulders.

The limpets found in the region include the highcap lanx (*Lanx alta*) and the kneecap lanx (*L. patelloides*). Both species require well-oxygenated, clear rivers with fast currents that have boulders and bedrock—including the Rogue, Smith, Trinity, and McCloud rivers. *Juga* snails are also found in large streams and rivers with stable cobble and boulder beds where their densities can reach dozens per square meter. The robust walker (*Pomatiopsis binneyi*), a minute semi-aquatic species, has only been found in perennial seeps of the Chetco and Illinois rivers that are protected from seasonal flooding.

Bivalvia – Mussels and Clams

Bivalves are filter feeding aquatic mollusks enclosed in a laterally compressed two-part hinged shell and are commonly referred to as mussels or clams.

Freshwater Mussels

With over 300 species, North America hosts the highest diversity of freshwater mussels in the world. Two-thirds of these species live in the Mississippi River basin (Haag 2012). At least three, and potentially up to five, species of mussels inhabit major rivers throughout the Klamath Mountains. These include the western ridged mussel (*Gonidea angulata*), the western pearlshell (*Margaritifera falcata*), and floaters (*Anodonta* spp.). Regional taxonomy of the *Anodonta* genus is unresolved with genetic analysis needed to identify species. Potential *Anodonta* species include the Oregon floater (*A. oregonensis*), western floater (*A. kennerlyi*), and winged or California floater (*A. nuttalliana*). Although species ranges are not well described, it appears the western pearlshell has the broadest distribution—occurring in all major rivers throughout the Klamath Mountains. Even so, the western ridged mussel appears to be the most abundant species in the Klamath River (Davis et al. 2013) and is commonly found together with western pearlshells. Remarkably, some individual mussel beds in the Klamath River basin have all three genera living together (Davis et al. 2013, Johannes and Clark 2016).

Western pearlshell (*Margaritifera falcata*) bed in the Smith River. Photo by Justin Garwood.

With shells having detailed, tree-like annual growth rings, we know some mussels are among the longest-lived animals on Earth (Haag 2012). The western pearlshell is a longevity champion—becoming

Invertebrates

mature in 9–12 years and living well over a century (Nedeau et al. 2009). Although dams have exerted profoundly negative impacts on native mussel populations, large pearlshells alive today could easily pre-date the dam building era of the mid-20th century. In contrast, floaters reach maturity in four to five years and have a short life span of around 20 years. Western ridged mussels fall in the middle, reaching maturity around seven years and living over 30 years to perhaps even 60 years (E. Blevins, pers. comm.).

Due to their longevity, sensitivity to high water temperatures, low oxygen levels, and dewatering, freshwater mussels are important indicators of a river's long-term health. Similar to other regions, the three genera in the Klamath Mountains have experienced substantial population declines over the past century (Howard et al. 2015, Blevins et al. 2017). Healthy populations contain thousands of individuals, densely lining interstitial cavities of stream banks. As they move and burrow into sediments using a retractable muscular foot, they aerate stream sediments and create habitat for other stream invertebrates. Individual mussels can filter up to 20 gallons of water a day making a large mussel colony a living water purification system. By filtering particulate matter, plankton, and algae, mussels capture and transfer suspended nutrients to the stream bed that would otherwise be unavailable. Mussels are food for larger wildlife including river otters and raccoons. Indigenous communities of the region use them as a natural and cultural resource (Davis et al. 2013).

Perhaps the most fascinating feature of the mussel life cycle is a bizarre parasitic relationship between tiny mussel larvae, called glochidia, and their fish hosts. Female mussels brood eggs in their gills that develop into mature glochidias. These are then released into the water column. Although mussels can live independently for decades, glochidias survival is entirely dependent on finding a specific host fish species and clamping to their gills as a hitchhiking parasite for a few weeks. Eventually the glochidia releases from the fish and burrows into sediments. Here they remain through maturity and the cycle begins again. Many mussels require specific hosts. The western pearlshell only parasitize salmonids, the western ridged mussel parasitizes sculpin and dace, while floaters are generalists (Nedeau et al. 2009).

Freshwater Clams

Tiny freshwater bivalves, from the family Sphaeriidae, called fingernail clams are easily overlooked. Often mistaken for small gravel or a droplet of pine sap, adults are typically less than 6 mm across. Clams may be among the least expected taxa to see in high-elevation ponds and lakes of the Klamath Range, but members from the genus *Pisidium* are surprisingly common in lentic habitats above 6,000 feet (1,800 m). These clams are filter-feeders, living in sediment and detritus layers. Similar to fairy shrimp cysts, clams disperse among isolated waters

Tiny pea clams (*Pisidium* sp.) are common in lakes and ponds of the Klamath Mountains—even those occurring above 6,000 feet. Photo by Justin Garwood.

through passive means by sticking to animals' fur and feathers in mud, and when ingested by water birds, passing unharmed through the digestive tract.

Sponges

Though it is not apparent, sponges (Porifera) are animals—but not single individuals. Instead, they function as a collection of individuals working in unison. The body consists of microscopic individuals connected by living tissue held rigid by tiny silica rods. There are about 30 freshwater species in North America and all are indicators of clean, clear water. While not all are green, the color comes from algae living symbiotically within the tissue of the sponge. Similar to lichens, the algae photosynthesize and offer food to the sponge in return for a place to grow. Sponge colonies are found on bedrock walls in the region's rivers.

A freshwater sponge from the South Fork Trinity River in the family Spongillidae. Photo by Justin Garwood.

Tardigrades

Tardigrades (aka: water bears, moss piglets) are understudied aquatic micro-metazoans that occur globally across a range of marine, freshwater, and terrestrial habitats. In the Klamath Mountains they are reliably encountered in the forest canopy on epiphytic vegetation such as in lichens and bryophytes. Villella et al. (2020) found 12 species in 43 sampled red tree vole (*Arborimus longicaudus*) nests along an east-west transect in southern Oregon's Klamath Mountains. Of the 12 species of tardigrades, seven had not been previously reported in Oregon. This study suggests tardigrades prefer forests with large trees in regions with higher precipitation regimes. They also found that vole nests occurring higher in the canopy had greater tardigrade abundance than in lower nests —nest building by small mammals may facilitate tardigrade dispersal in forest canopies.

Tardigrade as seen under a compound microscope. Photo by Alexander Young.

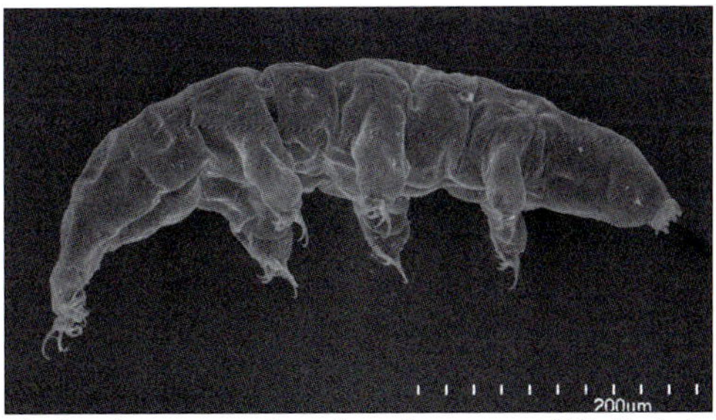

Tardigrade as seen under a dissecting scope at 30X. Photo by Alexander Young.

Invertebrates

Acknowledgements

Many authors made this chapter a reality. Dana Ross and Paul Hammond worked only on butterflies, Rod Crawford on arachnids, Len Mazur contributed to our mollusk section, and Brian Dykstra led the way for hymenoptera.

Thanks to Jamie Bettaso, Candace Fallon, and Emilie Blevins for reviewing Mollusks. Subterranean invertebrates was created with input from William Shear, Sean Schoville, Warren Savary, and Charles Griswold. David Kavanaugh provided feedback on the *Nebria* sidebar.

Photographic Contributions

Photos were granted from many sources including iNaturalist and from scientific papers. We thank all the additional photographers who contributed to this chapter including Rob Santry, Warren E. Savary, Charles E. Griswold, Jeff Cole, Nathan McCanne, Ryan Elliot, Sean Scoville, William A. Shear, Merrill Peterson, Ken DeCamp, David Kavanaugh, Eric Peterson, Ken-ichi Ueda, Tom Rickman, Rob Santry, Alexander Young, and Brenna Green.

Literature Cited and Suggested Reading

Blevins. E., S. Jepson, J. B. Box, D. Nez, J. Howard, A. Manine, C. O'Brien. 2017. Extinction risk of western North American freshwater mussels: Anodonta nuttalliana, the Anodonta oregonensis/ kennerlyi clade, Gonidea angulate, and Margaritifera falcata. Freshwater Mollosk Biology and Conservation 20:71-88.

Burke, T.E. 2013. *Land snails and slugs of the Pacific Northwest*. Oregon State University Press.

Bryson, R.W. et al. 2016. Genomic data reveal ancient microendemism in forest scorpions across the California Floristic Province. *Molecular Ecolology*, 25: 3731-3751.

California Department of Public Health (CDPH). 2018. Tick Borne Diseases. Accessed 2019. https://www.cdph.ca.gov/Programs/CID/DCDC/Pages/Tick-Borne-Diseases

Colla, S.R., F. Gadallah, L. Richardson, et al. 2012. Assessing declines of North American bumble bees (*Bombus* spp.) using museum specimens. *Biodiverse Conserve* 21, 3585–3595.

Crawford, R. 2015a. Spiders. In *The Klamath-Siskiyous: Timely Treasures of an Iconic Bioregion*.Unpublished Report.

Crawford, R. 2015b. Harvestmen. In *The Klamath-Siskiyous: Timely Treasures of an Iconic Bioregion*. Unpublished Report.

Davis, E.A., A.T. David, K.M. Norgaard, T.H. Parker, K. McKay, C. Tennant, T. Soto, K. Rowe, and R. Reed. 2013. Distribution and abundance of freshwater mussels in the mid Klamath subbasin, California. *Northwest Science* 87:189-206.

Dunk, J. R., W. J.Zielinski, and H.K. Preisler. 2004. Predicting the occurrence of rare mollusks in Northern California forests. Ecological Applications 14:713-729.

Elliott, W.R. et al., 2017. The cave fauna of California. Proceedings of the California Academy of Sciences, 64(4), pp.1-311.

Eriksen, C. and D. Belk. 1999. Fairy shrimps of California's puddles, pools, and playas. Mad River Press, Eureka, CA. 196 pages.

Ethridge,J., J.R. Gibson, and C. Nice. 2013. Cryptic diversity within and amongst spring-associated Stygobromus amphipods (Amphipoda: Crangonyctidae). Zoological Journal of the Linnean Society 167:227-242.

Fashing, N. J. 2004. Biology of *Sarraceniopus darlingtoniae* (Histiostomatidae: Astigmata), an obligatory inhabitant of the fluid-filled pitchers of *Darlingtonia californica* (Sarraceniaceae). Phytophaga, XIV: 299-305.

Forbes, A.A., R.K. Bagley, M.A. Beer, et al. 2018. Quantifying the unquantifiable: why Hymenoptera, not Coleoptera, is the most speciose animal order. BMC Ecol 18, 21.

Frest, T. J. and E. J. Johannes. 1999. Field guide to survey and manage freshwater mollusk species. 117p.

Griswold, C. E., T. Audisio, and J. M. Ledford. 2012. An extraordinary new family of spiders from caves in the Pacific Northwest (Araneae, Trogloraptoridae, new family). *ZooKeys*, (215), 77–102. doi:10.3897/zookeys.215.3547

Haag, W.R. 2012. *North American freshwater mussels: natural history, ecology, and conservation*. First edition. Cambridge University Press.

Haggard, P. and J. Haggard. 2006. *Insects of the Pacific Northwest.* Timber Press. Portland, Oregon.

Howard, J. K., J. L. Furnish, J. Brim Box, and S. Jepsen. 2015. The decline of native freshwater mussels (Bivalvia: Unionoida) in California as determined from historical and current surveys. California Fish and Game 101:8-23.

Jacobson, A.J. and G.W. Courtney. 2008. A New Species of Blephariceră Macquart (Diptera: Blephariceridae) from Western North America. Proceedings of the Entomological Society of Washington. 110(4), 978-987.

Johannes, E. J., and S. A. Clark. 2016. Freshwater mollusc declines. Local extinctions and introductions in five northern California streams. Tentacle 24:22-25.

Johnson, P. D. et al. 2013. Conservation Status of Freshwater Gastropods of Canada and the United States. Fisheries 38:247-282.

Jules, E.S. 1996. Yellow Jackets (*Vespula vulgaris*) as a Second Seed Disperser for the Myrmecochorous Plant, *Trillium ovatum*. The American Midland Naturalist Vol. 135, No. 2.

Kadowaki, K. 2010. Behavioral observation of two fungivorous beetles (Coleoptera: Tenebrionidae) on wood-decaying bracket fungus *Cryptoporus volvatus. Entomological Science*, 13: 159-161.

Larson, E., and J. Olden. 2011. The state of crayfish in the Pacific Northwest. Fisheries 36:60-73.

Lopez, C. 2017. The bee fauna of the Horse Mountain and Grouse Mountain Region, Humboldt County, California. Cal Poly Humboldt theses and projects.

Pennak, R. W. 1989. Fresh-water invertebrates of the United States: Protozoa to Mullusca. Third Edition. John Wiley and Sons, Inc. 628p.

Pakaluk, J. and S.A. Slipinski (eds.). 1995. Biology, Phylogeny, and Classification of Coleoptera. Papers Celebrating the 80th Birthday of Roy A. Crowson. Muzeum i Instytut Zoologii PAN, Warszawa.

Pilsbry, H.A. 1948. *Land mollusca of North America.* (north of Mexico. Academy of Natural Sciences of Philadelphia, monograph no. 3, II:575-994.

Ponomarenko, A.G. 1995. The geological history of beetles. Pp. 155-171 in Pakaluk and Slipinski.

Rogers, D.C. 2015. Crustaceans. In *The Klamath-Siskiyous: Timely Treasures of an Iconic Bioregion*. Unpublished Report.

Shear, W.A. 2011. Cave millipeds of the United States. XI. *Opiona graeningi*, n. sp., a troglomorphic milliped from Siskiyou County, California. *Zootaxa* 3114: 50-56.

Shear, W.A., and R. Crawford. 2019. Cave millipedes of the United States. XVI. Two new species from Oregon Caves. Zootaxa 4674 (5): 571–580.

Nedeau, E., A.K. Smith, J. Stone, and S. Jepsen. 2009. Freshwater Mussels of the Pacific Northwest, 2nd Edition. The Xerces Society, Portland.

Ricketts, T.H., E. Dinerstein, D. Olson, C. Loucks, W. Eichbaum, D. DellaSala, K. Kavanagh, P. Hedao, P. Hurley, K. Carney, R. Abell, and S. Walters. 1999. Terrestrial Ecoregions of North America: A Conservation Assessment. Island Press, Washington, DC.

Roth, B. and P. Sadeghian. 2006. Checklist of land snails and slugs of California. Second edition. Santa Barbara Museum of Natural History Contributions in Science, no 3.

Ross, D. and P. Hammond. 2015. Butterflies and Moths. In *The Klamath-Siskiyous: Timely Treasures of an Iconic Bioregion*. Unpublished Report.

Roth, J. 2015. Editor. In *The Klamath-Siskiyous: Timely Treasures of an Iconic Bioregion*. Unpublished Report.

Schoville, S. 2012. Three new species of *Grylloblatta* Walker (Insecta: Grylloblattodea: Grylloblattidae), from southern Oregon and northern California. Zootaxa 3412: 42–52.

Shapiro, A. 1990. Significant Additions to the Butterflies of the Trinity Alps and Mount Eddy, Northern California. Journal of the Lepidoptera 29(1-2):33-36.

Shear, W.A. and R. L. Crawford (2019). Cave millipedes of theUnited States. XVI. Two new species from Oregon Caves National Monument, Oregon (Chordumatida, Conotylidae and Caseyidae). Zootaxa 4674 (5): 571-580.

Shelley, R.M. 1994. Revision of the milliped family Paeromopodidae, and elevation of the Aprosphylosomatinae to family status (Julida: Paeromopodoidea). Entomologica Scandinavica 25: 169-214.

Su, Zhi-Hui. 2004. Pattern of phylogenetic diversification of the Cychrini ground beetles in the world as deduced mainly from sequence comparisons of the mitochondrial genes. Gene 326 (4): 43-57.

Ubick D, P. Paquin, P.E. Cushing, V. Roth (Eds) (2005) *The Spider Genera of North America: An Identification Manual.* The American Arachnological Society, 377 pages.

Villella, J. C. et al. 2020. Tardigrades in the Forest Canopy: Associations with Red Tree Vole Nests in Southwest Oregon. *Northwest Science*, Vol. 94, No. 1.

Wang, D., and J.R. Holsinger. 2001. Systematics of the subterranean amphipod genus Stygobromus (Crangonyctidae) in western North America, with emphasis on species of the hubbsi group. Amphipacifica 3:39-147.

Chapter 11: Vertebrates

Justin Garwood, Chris Feldman, Michael Kauffmann, Karen Reiss
Greg Gray, Dan Barton, Tiana Williams-Clausen

Vertebrates of the Klamath Mountains

An Introduction

Animals within the subphylum Vertebrata all have backbones and are thus vertebrates. About 3 percent of Earth's animals, or about 75,000 species, are vertebrates with the remaining 97 percent being invertebrates. Vertebrates either absorb heat from their environment (ectotherms including fishes, amphibians, and reptiles) or make their own heat (endotherms including birds and mammals). Herein, we roughly follow the phylogeny and fossil record, but also the thermal physiology of these charismatic and familiar animal groups. Justin Garwood and Andrew Kinziger begin our treatments with the fishes whose oldest regional representative, based on the arrival of the jaw, is the Pacific lamprey (*Entosphenus tridentatus*). Justin then presents the epic tale of amphibians with the coastal tailed frog (*Ascaphus truei*) being part of the most ancient amphibian family regionally and globally. Justin and Chris Feldman tell the story of a reptile crossroads with the western pond turtle (*Actinemys marmorata*) representing the most ancient reptile lineage while also sharing the greatest vertebrate longevity in the region alongside our two sturgeon species. Greg Gray, Michael Kauffmann, Daniel Barton, and Tiana Williams-Claussen take us up the rivers, into the mountains, and across the sky for an exploration of the birds, which are nested within the reptiles (Figure 1) with one of the oldest representative, the California condor, on its way back to the Klamath Mountains. Last, Karen Reiss, Michael Kauffmann, and Chris Feldman wrap up the treatments with mammals with the mountain beaver (*Aplodontia rufa*) having been on Earth longer than any other regional furbearer.

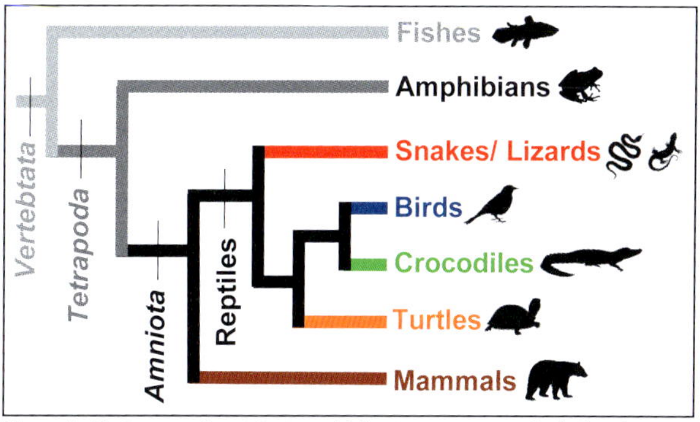

Figure 1: Cladogram of vertebrates with best-consensus arrival of each group based on the fossil record.

Table 1: Estimated number of vertebrate families, genera, and species within the Klamath Mountains.

Vertebrate Group	Klamath Families	Klamath Genera	Klamath Species	World's Species
Fishes	9	15	31	~34,500
Amphibians	9	13	26	~8,400
Reptiles	8	15	23	~11,600
Birds	51	154	244	~10,400
Mammals	22	56	81	~6,500
Total	**99**	**253**	**405**	**~75,000**

Fishes of the Klamath Mountains

Justin Garwood and Andrew Kinziger

Introduction

Fishes are ectothermic vertebrates that obtain dissolved oxygen in water through gills. They are the primary vertebrate inhabitants of aquatic ecosystems throughout the world and, based on the fossil record, have existed for roughly half a billion years. They are the most species rich vertebrate group in the world with over 34,500 described species, which is more than all other vertebrate groups combined (i.e., birds, mammals, amphibians, and reptiles). The total number of described fish species increases by about 200 per year, revealing that a large number of fish species remain to be formally recognized (Eschmeyer 1998).

While only about 0.01 percent of global surface water is fresh, nearly half of the world's fish species live in fresh water (Lundberg et al. 2000, Nelson et al. 2016). This freshwater fish diversity has been driven, in part, by speciation through isolation brought on by long-term and episodic natural forces including climatic extremes, sea-level fluctuations, volcanic activity, Pleistocene glaciers, floods, mountain building, landslides, and more. These same forces have been on wide display in the Klamath Mountains and surrounding regions over millions of years.

Natural forces have continuously created, connected, and severed rivers across the Klamath region (and the broader west), resulting in the distinct local evolutionary lineages of fishes we see today. California has around 65 described native freshwater fish species (Moyle 2002, McGinnis 2006) while Oregon has around 75 (Markle 2016). Still, California and Oregon contain less than 30 percent the diversity of many southeastern states (Abell et al. 2000). Fish diversity in the Klamath Mountains is modest, with 31 native fishes identified from nine families (Table 1). Twenty-four species are persistent inhabitants while another six species occasionally disperse from neighboring bioregions, in the case of rare anadromous species. The bull trout is the only modern native species known to be entirely extirpated from the Klamath Mountains, where it once maintained the only known population in California (Moyle 2002). Species such as Chinook salmon and coho salmon have suffered long-term declines in many Klamath Mountain streams, largely due to dams, disease, and habitat destruction from historic mining, logging, road building, and water extraction for agriculture.

Fishes have significant commercial, recreational, and cultural importance to peoples in the Klamath Mountains. Many species are cultural keystones for native peoples—contributing to cultural identity, subsistence, and ceremonies. The most iconic cultural fishes include salmon and trout (salmonids), lampreys, and sturgeons, but native languages also have names and uses for many other species. A variety of regional fishes in the upper Sacramento-Pit river portion of the Klamath Mountains are important to Winnemem, Wintu, and Achumawi Tribes including hardhead, Sacramento pikeminnow, and Sacramento sucker (Gobalet et al. 2004). This importance is also reflected in the archeological record. Chinook salmon, steelhead, rainbow trout, and coastal cutthroat trout are fished throughout the rivers of the Klamath Mountains by recreational anglers. To many anglers,

Table 1. List of 31 native fish species historically found in the Klamath Mountains. See watershed maps in the Water Chapter for boundaries. Species abundance and distribution ranks are weighted by available habitat and historic distributions in the Klamath Mountains.

● = Abundant, widespread	◔ = Rare or localized		Rogue	Chetco	Smith	Klamath	Upper Sac
◕ = Moderate, widespread	○ = Extirpated						
◑ = Rare, moderate distribution	† = Sporadic	**?** = Unknown					
Lampreys (PETROMYZONTIDAE)							
ˇ Pit-Klamath Brook Lamprey	*Entosphenus lethophagus*					◑	◑
ˇ Klamath River Lamprey	*Entosphenus similis*					◕	
►Pacific Lamprey	*Entosphenus tridentatus*		◕	◕	◕	◕	○
Sturgeons (ACIPENSERIDAE)							
►Green Sturgeon	*Acipenser medirostris*		◑	?	◔	◑	
►White Sturgeon	*Acipenser transmontanus*		◔	?	?	◔	◔[1]
Minnows (LEUCISCIDAE)							
ˇ Tui Chub	*Siphateles bicolor*					†	†
ˇ Sacramento Hitch	*Lavinia exilicauda*						◔
ˇ California Roach	*Hesperoleucus symmetricus*						◑
ˇ Hardhead	*Mylopharodon conocephalus*						◕
ˇ Klamath Speckled Dace	*Rhinichthys klamathensis kl.*		◑			●	
ˇ Sacramento Speckled Dace	*Rhinichthys kl. achomawi*						◑
ˇ Sacramento Pikeminnow	*Ptychocheilus grandis*						◑
Suckers (CATOSTOMIDAE)							
ˇ Sacramento Sucker	*Catostomus occidentalis*						◑
ˇ Klamath Smallscale Sucker	*Catostomus rimiculus*		◕		◑[2]	◕	
Smelts (OSMERIDAE)							
►Eulachon[4]	*Thaleichthys pacificus*		?			†	
Salmonids (SALMONIDAE)							
►Pink Salmon	*Oncorhynchus gorbuscha*		†	†	†	†	
►Chum Salmon	*Oncorhynchus keta*		†	†	†	†	
►Sockeye Salmon	*Oncorhynchus nerka*		†	†	†	†	
►Coho Salmon[4]	*Oncorhynchus kisutch*		◑	◔	◔	◕	
►Chinook Salmon (fall/spring/winter)	*Oncorhynchus tshawytscha*		◕/◑	◕/?	◕/†	◕/◔[4]	○/○[4]/○[4]
►Rainbow Trout (steelhead) (fall-winter/summer/resident[3])	*Oncorhynchus mykiss*		◕/◑/◕	◕/†/●	◕/†/●	◕/◔/◕	○[4]/?/◕
▷Coastal Cutthroat Trout[3]	*Oncorhynchus clarkii*		◕	●	●	◔	
ˇ Bull Trout[4]	*Salvelinus confluentus*						○
Sticklebacks (GASTEROSTEIDAE)							
▷Threespine Stickleback	*Gasterosteus aculeatus*		◕	◔	◔	◕	
Sculpins (COTTIDAE)							
‡ Coast Range Sculpin	*Cottus aleuticus*		◑	◔	◔	◑	
‡ Prickly Sculpin	*Cottus asper*		◑	◔	◔	◑	
ˇ Riffle Sculpin	*Cottus gulosus*						◑
ˇ Pit Sculpin	*Cottus pitensis*						◔
ˇ Reticulate Sculpin	*Cottus perplexus*		◕				
ˇ Rough Sculpin[4]	*Cottus asperrimus*						†
ˇ Bigeye Marbled Sculpin	*Cottus klamathensis macrops*						◔
ˇ Klamath Marbled Sculpin	*Cottus klamathensis polyporus*					◕	
Perches (EMBIOTOCIDAE)							
ˇ Sacramento Tule Perch	*Hysterocarpus traskii*						◔
Total Number of Extant Species by Basin (● + ◕ + ◑ + ◔):			13	8	9	15	13

ˇ Strictly fresh water, ►Strictly anadromous, ▷Occasionally anadromous, ‡Occasional marine migrations, [1]Relic population occurs in Shasta Reservoir, [2]Smith River population introduced (Kinziger et al. 2021), [3]Resident populations often found above anadromous migration barriers, [4]Protected under State/Federal ESA.

steelhead offer the greatest and most challenging allure. The strength and chrome-bright iridescence of steelhead represent resilience and the untamed wild beauty of the Klamath Mountains. Tenuous commercial ocean fisheries target Chinook salmon stocks originating from rivers of the Klamath Mountains that become mixed with fish from multiple other rivers along the Pacific Coast, complicating population management in individual rivers.

Traditional salmon bake with a wood fire pit and Chinook salmon threaded on redwood sticks. Photo by Thomas Dunklin.

Seeing fishes up close in their natural habitats reveals just how difficult it is to grasp, as terrestrial creatures, what an aquatic lifestyle means. One of the best ways to understand regional fishes is to watch them underwater. After carefully investigating different habitats, such as pools, riffles, and edge water, it becomes apparent how complex fish communities are. For example, in spring, one may see hundreds of fishes in a single pool, organized by species, age groups, and distinct habitat zones. Two- and three-year-old juvenile steelhead parr aggressively take the swift food-rich currents at the head of the pool. These are flanked by yearlings working the lower velocities at the side. Tiny steelhead, sucker, and stickleback fry are found sheltering from predators along shallow margins, where they benefit metabolically from the warmer edge waters. Chinook salmon parr prefer the upper pelagic zone in the middle of the pool while coho salmon parr are tucked along shadowed undercut banks and surrounding woody cover. Adult suckers and sculpins are found in the deepest zone where they forage on the bottom while lamprey larvae filter-feed detritus, buried tail first, in sandy river margins. If one were to survey the same pool at night, this community organization would be much different, again highlighting ecological complexity.

This chapter enters the underwater world to explore the region's native fish, their unique natural histories, and their importance to aquatic and terrestrial landscapes. Fish diversity is usually defined at the watershed level so our geographical boundaries are somewhat arbitrary. Two great rivers, the Sacramento's Pit, and the Klamath, originate in the eastern flanks of the volcanically complex Cascade Range where high arid tablelands and fault-block topography hold a fish fauna and landscape that have little resemblance to those found in the Klamath Mountains. These two rivers eventually pierce through the Cascades abrupt western geological boundary and into the forested canyons of the Klamath Mountains. Although some fishes indeed migrate between these divergent landscapes, this chapter highlights fishes that live and reproduce within the Klamath Mountains.

Biogeographic Patterns and Deep Time

The composition of the fish community in the Klamath Mountains is dynamic. Some species have come and gone, like the giant (~7.5 ft, 2.3 m) anadromous spike tooth (sabertooth) salmon (*Oncorhynchus rastrosus*) that had a set of enlarged lateral fangs in the upper jaw and ate plankton (Cavender and Miller 1972). Based on fossil locations in Oregon and California, this giant salmon migrated through ancient Klamath Mountain streams to spawn before going extinct around 5 million years ago (Ma). Conspicuous evidence of ancient, abandoned river channels—perhaps used at one time by spike tooth salmon—can be seen within the Klamath Mountains along roadcuts where river bedload deposits are elevated high above current drainages due to faulting. Aside from many human translocations of fishes, the modern geographic native fish distributions are likely a complex product of geological and climatic forces. To add to this complexity, fish communities have also been shaped by adjoining ecoregions by ancient river and lake systems, drainage development, barriers, and stream capture.

Today there are two general types of fishes occurring in the Klamath Mountains: marine and freshwater dispersants. Marine dispersants, such as anadromous fishes, can move through the ocean and thus have distributions explained by marine migrations. Marine dispersants are flexible in a changing landscape without becoming isolated. However, a few examples of marine and anadromous fishes have evolved regionally into obligate freshwater forms, such as the Sacramento tule perch and freshwater resident lampreys (closely related to anadromous Pacific

Spike tooth salmon (extinct). Drawing by Ray Troll, printed with permission.

lamprey). Freshwater dispersants are fishes that cannot tolerate salt water and thus have distributions that are best explained via migration through freshwater routes. Given that ocean and land serve as impervious barriers to freshwater dispersants migration, these fish are candidates for long periods of isolation. Some freshwater fishes of the Klamath Mountains hold fascinating biogeographic mysteries. For example, how does the Klamath River have a freshwater dispersant sculpin that is closely related to another found in the Pit River of the Sacramento basin where there are no watershed connections? Because geological events and fish dispersal events have not been independently verified for the region, it remains difficult to piece together just how fishes have landed where they are with certainty. What follows are some hypotheses—advanced through various geological, archaeological, and genomic investigations.

To appreciate the historical biogeography of regional freshwater fishes, we need to look outside the Klamath Mountains for potential ancient watershed connections influenced by faulting, volcanos, climate, glaciers, and ancient rivers. During the wetter episodic stages of the Pleistocene Epoch (2.5 million to 11,000 years ago), and perhaps earlier, vast internal lake systems, known as pluvial lakes,

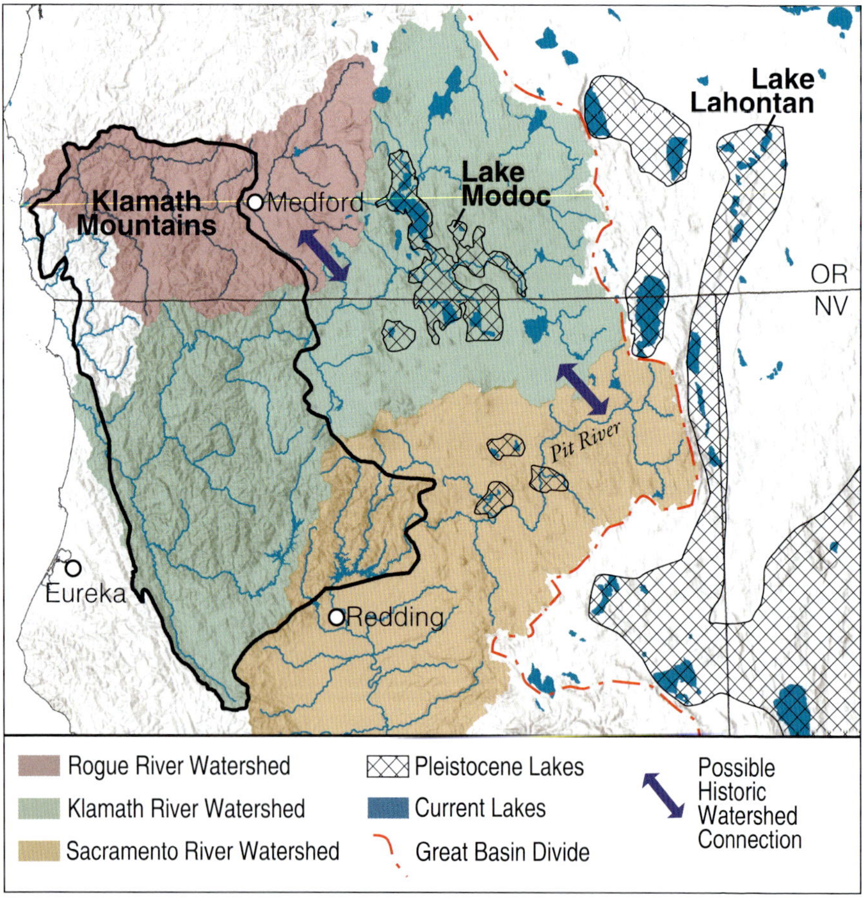

Pleistocene pluvial lakes of the upper Rogue, Klamath, and Sacramento watersheds.

inundated and connected large areas in the high tablelands east of the Cascade Range and throughout the Great Basin (Orme 2008, Ibarra et al. 2014). Ancient lakes just east of the Klamath Mountains include Alturas Lake, Goose Lake, and Lake Modoc which encompasses today's Upper Klamath, Lower Klamath, and Tule lakes (Pease 1965, Dicken 1980). Lake Modoc was a giant waterbody that filled and overflowed west, forming the present Klamath River drainage (Adam et al. 1990). Lake Modoc's formation may have severed an ancient connection between the Klamath and Pit rivers, when large volcanic flows creating the expansive Medicine Lake Highlands likely blocked an ancient northbound Pit River that was connected to the Klamath River drainage (Pease, 1965, Adam et al. 1990).

The size and depth of Pleistocene lakes varied dramatically based on climate shifts (Ibarra et al. 2014). Transient connections among lakes most certainly caused intermittent isolation and dispersal of aquatic species. For example, freshwater clam fossils are shared among many of these ancient lake basins (Taylor 1960). Closely related Klamath Mountain fish species that are shared between the Pit and Klamath rivers include the marbled sculpin and the Pit-Klamath brook lamprey—possibly a dwarf successor of ancient landlocked Pacific lampreys (Moyle 2002) (Table 1). In addition, the slender sculpin (*Cottus tenuis*), a species found in the upper Klamath River in Oregon, and the rough sculpin (*C. asperrimus*) endemic to the Pit River, share a close ancestry (Baumsteiger et al. 2012) that is also likely a result of this past connection. The severed ancient Pit River was likely captured by the Sacramento River and invaded by the Sacramento fish fauna. Different fish species assemblages occur above and below Pit River Falls (above the eastern edge of the Klamath Mountains) resulting in a mixture of lower Sacramento River species (such as the Sacramento tule perch) and others sharing ancestries with the Klamath basin fishes that occur today (Robins and Miller 1957, Taylor 1960).

Other freshwater dispersant fishes of the Klamath Mountains that are shared among isolated Klamath rivers, but have much broader distributions across the west, include the tui chub (a rare Klamath Mountain visitor, more common in the upper Klamath and Pit rivers) and the 2 subspecies speckled dace (widely distributed in the Pit and Klamath rivers but poorly documented in the Rogue River) (Table 1). Given their broad distributions, these fishes likely spread across the west throughout complex geological stages by oscillating river and pluvial lake connections. One long-standing proposal is that the Klamath and Sacramento rivers were influenced by an ancient Snake River that flowed west from Idaho before being captured by the Columbia River around 3.1 Ma (Smith et al. 2000). Various mollusk and fish fossils from the Snake River in Idaho are aligned with those found in the Klamath and Sacramento river regions (Taylor 1960, Stearley and Smith 2016). In addition, Idaho's Snake River sediments found in the Klamath Mountains were deposited during the Miocene Epoch >5 Ma (Aalto et al. 1998) when the Klamath region was near sea-level and a flat plain (Aalto 2006)—prior to the rapid mountain building that created today's Klamath Mountains. Valley glaciers also caused stream capture that could have influenced fish distributions. A great example is in the headwaters of the South Fork Salmon River, near Big Flat, where a massive terminal glacier moraine dammed Coffee Creek, a tributary of the Trinity River, causing the stream's headwaters to get captured by the Salmon River

The Immense Journeys of Anadromous Fishes

Most streams of the Klamath Mountains host a variety of migratory fishes that display different life history patterns to accommodate a common strategy—gain most off their size in the ocean and reproduce in streams. Anadromous fishes are born and rear in fresh water from a few months, to over five years, depending on the species. They then go through dramatic physiological transformations allowing them to crossover between fresh water and salt water environments. Anadromous fishes of the Klamath Mountains share the food rich waters of the eastern Pacific Ocean with millions of other fishes migrating from thousands of streams spread across western North America from Alaska to Baja California,

Chinook Salmon
Oncorhynchus tshawytscha
♂ 35 in (89 cm)

Coho Salmon
Oncorhynchus kisutch
♂ 27 in (69 cm)

Coastal Cutthroat Trout
Oncorhynchus clarkii
♀ 14 in (36 cm)

Mexico. They usually spend at least a year—and often several years—maturing in the Pacific before returning to their freshwater streams to spawn. Anadromous fishes grow rapidly and much larger in the ocean than they would in streams alone, allowing for far greater reproductive output. Pacific lamprey and some salmonids die after spawning—leaving behind ocean derived nutrients that subsidize stream and forest ecosystems. Others, such as steelhead and green sturgeon, return to the ocean after spawning. The Klamath Mountains host at least twelve species of anadromous fishes (Table 1), with the six species pictured below being the most common and widespread in the range. Migrations of these fishes can cover hundreds of miles in streams and thousands of miles in the Pacific Ocean. **Illustrations by Michael Zontos**

Steelhead
Oncorhynchus mykiss
♀ 28 in (71 cm)

5th year adult

3rd year smolt

2nd year parr

1st year parr

Green Sturgeon
Acipenser medirostris
72 in (183 cm)

Pacific Lamprey
Entosphenus tridentatus
23 in (58 cm)

(Hershey 1900).

Modern Klamath Mountain Fish Faunas

The fish faunas of the Klamath Mountains are generally divided into four regions: the Rogue River, the smaller Smith and Chetco coastal rivers, the Klamath River, and the upper Sacramento River—including the Sacramento headwaters, the Pit River, and the McCloud River (Table 1). This section provides a brief description of each region and its native fishes; for more on the rivers see the Water Chapter.

Rogue River Region

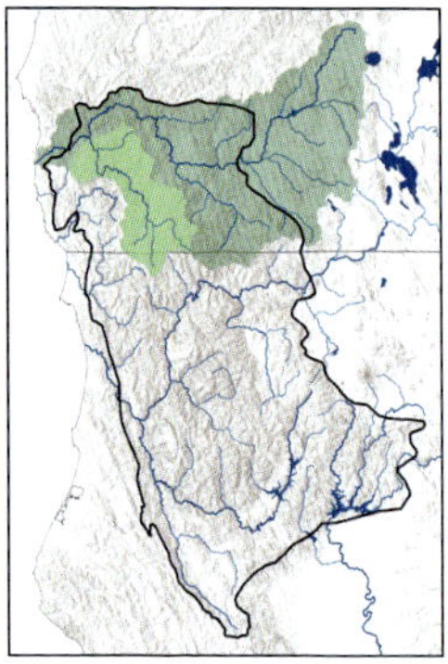

The Rogue is a deep, powerful river originating on the western slopes of the Cascade Range that then meanders through some of the wildest expanses of the Klamath Mountains while absorbing the Applegate and Illinois tributaries draining the north slopes of the Siskiyou Crest. Because the Rogue River has not cut through the Cascade Range, its fish diversity is sparse (at least 13 species) relative to the Klamath and Pit rivers that bisect coastal and inland bioregions (Table 1). The fish assemblage is also distinct from the adjacent Umpqua River to the north, which has greater fish diversity and is closely aligned with the Columbia River fish fauna to the north (Markle 2019). Rogue fishes, and their general life history patterns, align most with the Klamath River fauna in the Klamath Mountains and the fishes of these watersheds are lumped by biogeographers into the Lower Klamath Province (Moyle 2002, Markle 2019). For example, Chinook salmon from the Rogue River and the Klamath River, above the confluence of the Klamath and Trinity rivers, form a sister group and are more closely related to each other than they are to Chinook salmon from the Smith and Chetco rivers (Clemento et al. 2014).

The Rogue River only has three freshwater dispersant species (Table 1). The Klamath smallscale sucker is shared with the Klamath River and the Klamath speckled dace is shared with the Klamath River. How these species made it to the isolated Rogue Basin remains a puzzle. Jenny Creek is a small Klamath River tributary in Oregon that has both smallscale suckers and Klamath speckled dace occurring above a 33 ft (10 m) waterfall (Hohler 1981, Wiesenfeld et al. 2018). Gentle topography is all that separates the headwaters of the Klamath (Jenny Creek) and Rogue (Little Butte Creek) basins. A recent genetics study on dace found the Jenny Creek population was distinct from other Klamath River populations and most similar to the Little Butte Creek population which supports either a historic connection or a recent translocation between the basins (see Wiesenfeld et al. 2018). The freshwater obligate reticulate sculpin is the only fish from the Klamath Mountains unique to the Rogue River and is also found in Oregon coastal basins north of the Klamath Mountains (Markle 2016). The Rogue River is one of the few critical spawning streams for sturgeon, especially green sturgeon (Erickson and Webb 2007). The basin also has maintained promising runs of spring and fall Chinook salmon and summer and winter steelhead—with recent dam removals providing a major catalyst for largescale population recoveries.

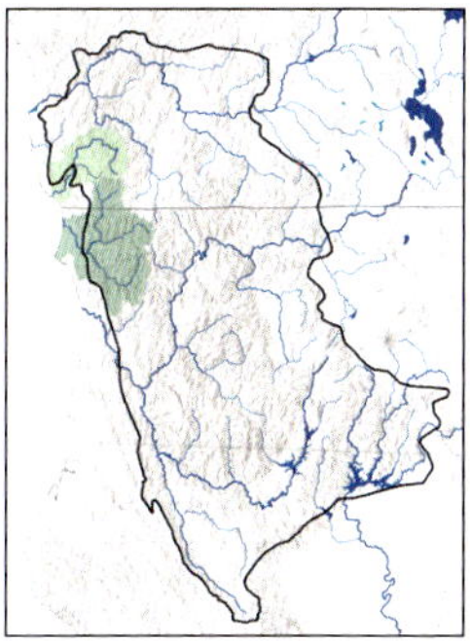

Smith-Chetco Region

Unlike the larger basins of the Klamath Mountains, the Smith and Chetco rivers are small, isolated coastal drainages resulting in lower fish biodiversity (Table 1) due to the lack of freshwater dispersants. However, exceptional water quality and fish habitats make these rivers important wild anadromous fish reserves—especially in the face of climate change. Both streams are free flowing within the rainiest region of the Klamath Mountains. They also have minimal fine sediment loads that impair other coastal streams. Winter flows are flashy, which constantly modifies and maintains habitats. Robust summer base flows are maintained by springs, resulting in excellent water quality during warm summer months. For these reasons, the Smith and Chetco rivers support among the most resilient salmonid habitats in the Klamath Mountains with strong populations of resident coastal cutthroat and rainbow trout in addition to moderate runs of fall Chinook salmon and winter steelhead. They also support rare regional life histories, including stream-type Chinook salmon parr which often over-summer in the river (Walkley and Garwood 2017) before transforming into silvery smolts and migrating to the ocean. Despite having among the highest quality stream habitats in the Pacific Northwest, the estuaries of these streams have been substantially altered by people, limiting productivity for juvenile salmonids. The threat of catastrophic wildfires is also always looming—as seen in the 2002 Biscuit Fire and the 2017 Chetco Bar Fire.

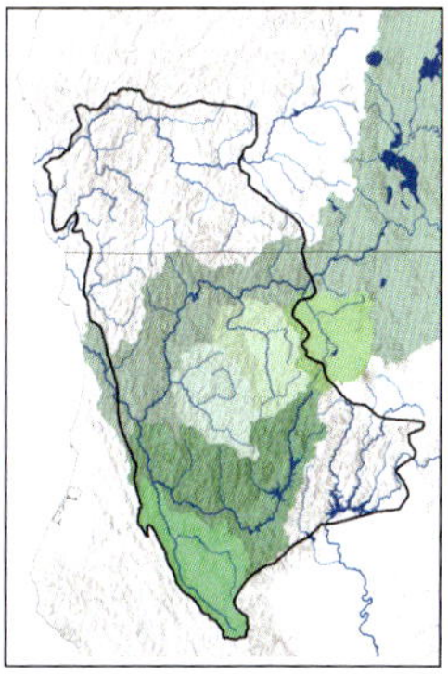

Klamath River Region

The upper Klamath River and its numerous tributaries drain through a starkly different bioregion than where it flows downstream through the Klamath Mountains. The upper Klamath includes many relict pluvial lake basins and slow meandering streams that are flanked by the Cascade Range and an arid volcanic plateau that adjoins the Great Basin Province. The connection between the upper and lower Klamath River is recent in geological time, and because of its unique geological history and long-term isolation, the upper basin hosts many endemic fishes. At least 12 obligate freshwater fish species occur in the upper basin (Moyle 2002)—twice as many as found west in the Klamath Mountains portion of the watershed. The western half of the Klamath watershed drains roughly half of the Klamath Mountains as the mainstem Klamath River absorbs many large tributaries as it flows through lush, forested canyons before exiting into the lower coastal redwood belt and ultimately terminating at a coastal estuary. The Klamath portion of the watershed hosts more native fishes than any river basin in the Klamath Mountains, with at least 15 species (Table 1). Six of these are freshwater dispersants, including the Pit-Klamath brook lamprey, Klamath River lamprey, speckled dace, Klamath smallscale sucker, Klamath marbled sculpin, and the tui chub.

Even with the Klamath River having only six freshwater dispersants in the Klamath Mountains, some of these fishes hold fascinating cryptic genetic diversity that is not obvious based on the drainage patterns we see today. This diversity is partially a product of the complicated history in drainage patterns and ancient dispersal barriers that isolated some fish populations over long periods. For example, the Klamath River lamprey and Pit-Klamath lamprey are both freshwater species that are closely related to the anadromous Pacific lamprey (Lang et al. 2009) and likely evolved into different forms based on long-term isolation in freshwater habitats. The speckled dace is found throughout the Klamath River Basin, however a genetic analysis by Weisenfeld et al. (2018) found that dace in the Trinity River exhibited species-level differentiation from populations found in the mainstem Klamath River. There are no apparent geographic barriers separating these populations and hybrids between them have been identified in the lower Trinity and Salmon rivers. A third genetic group was discovered above a waterfall in Jenny Creek, an Oregon tributary of the Klamath River, that is most similar to dace found in a nearby Rogue River stream.

The Klamath River is an important drainage for anadromous fishes—hosting Pacific lamprey, green and white sturgeon, numerous salmonids, and the rarely encountered eulachon (Table 1). Pacific lamprey, Chinook salmon, coho salmon, and steelhead had far greater abundance and distributions in the Klamath River basin before the arrival of long-term watershed devastation and unsustainable resource exploitation. Destructive gold mining and land conversion began in the 1850s, followed by extensive ocean and in-river salmonid fisheries by the early 1900s, and then the dam building era from the early 1900s to the 1960s. Rapid salmonid population declines were obvious by the early 1900s as noted by fish biologist John O. Snyder in 1934: "Depletion of Klamath salmon is not only apparent, but it seems to be progressing at an alarming rate."

Among the most devastating for Klamath fishes were the eight significant dams built in the upper Klamath River for irrigation, hydroelectricity, and intermittently maintaining wildlife marshlands. The dams have had dramatic impacts on the Klamath River by blocking hundreds of stream miles formerly used by anadromous fishes (Hamilton et al. 2016). The reservoirs behind some dams on the mainstem Klamath River are hyper-eutrophic, with toxic algal blooms greatly reducing the river's water quality. The dams also starve the mainstem Klamath River from receiving sediments during high flows. This altered flow and sediment regime is linked to an increase in disease by creating conditions that favor a polychaete worm that is the intermediate host for *Ceratonova shasta*, a parasite that infects and kills salmonids (Bartholomew et al. 1997), especially Chinook salmon smolts. The Trinity River, the Klamath's largest tributary, had its high-elevation headwaters blocked in 1962 by two dams that continue to sever anadromous fishes from ~109 miles (175 km) of former stream habitats (USDI 2000). Critical historic salmon and steelhead alluvial spawning reaches are now submerged under the smaller Lewiston Reservoir and the massive Trinity Reservoir. Trinity Reservoir produces hydroelectricity while diverting ~50 percent of Trinity River's water (nearly 75 percent before 1997) up through local mountains and down to

Whiskeytown Reservoir and the Sacramento River drainage for the Central Valley Project. To mitigate for losses in salmon and steelhead populations, both the Klamath and Trinity rivers have large-scale fish hatcheries that raise, and annually release, juvenile Chinook and coho salmon while the Trinity River Hatchery also raises and releases steelhead.

Despite many chronic issues and anadromous fish population declines caused by the region's dams, the Klamath River region still collectively hosts a remarkable variety of anadromous fish life histories. Fall Chinook salmon and winter steelhead maintain the largest populations. Wild populations of imperiled coho salmon, spring Chinook salmon, summer steelhead, and coastal cutthroat trout are considered rare as suitable habitats have become more isolated. Many tributaries of the Klamath River, such as the Scott River, Salmon River, Wooley Creek, Blue Creek, Clear Creek, Dillon Creek, Indian Creek, Seiad Creek, and tributaries of the Trinity River, such as the South Fork, New River, North Fork, Canyon Creek, and Rush Creek, are the fragile lifelines supporting current diversity. These tributaries are free-flowing and snow-fed, with most providing high-quality habitats and cool water temperatures throughout the summer months. During this writing, four of the Klamath River dams are slated for removal, which will become the largest dam removal project in history—worldwide. Removing these dams and restoring large-scale habitats in undammed tributaries, such as the Scott and South Fork Trinity rivers, will be a big step towards improving water quality, reducing disease, and greatly expanding available spawning and rearing capacity for lamprey, salmonids, and other native fishes of the Klamath River basin.

Upper Sacramento-Pit River Region

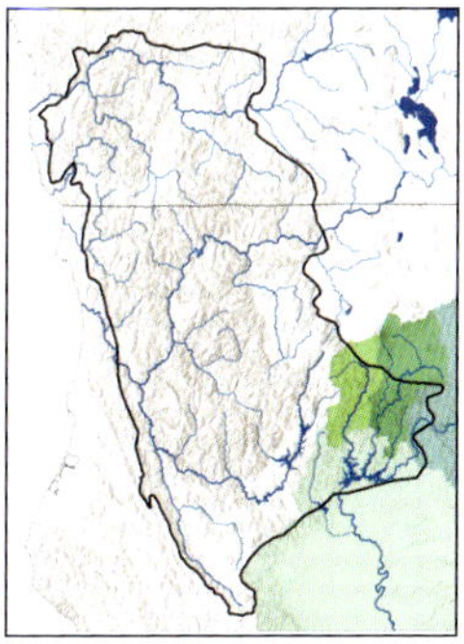

The Klamath Mountains contain a tiny portion of the Sacramento River Basin, and this small section has been highly modified by dams. The Whiskeytown Reservoir is an impassable barrier preventing anadromous fishes from accessing the upper Clear Creek watershed. The upper Sacramento River, the majority of the McCloud River, and the lowermost reach of the extensive Pit River all flow through and converge in the Klamath Mountains. Until nearly 80 years ago, these rivers were free-flowing and combined to form the headwaters of the immense Sacramento River Basin. Because the Pit and McCloud rivers originate within the volcanic lands of the Cascade Range, massive lava tube springs and glacier runoff from Mount Shasta contribute cold water to these streams year-round—an irreplaceable natural resource in the face of climate change. This little-known eastern expanse of the Klamath Mountains hosted at least 17 native fishes before it was dammed. The massive 602 ft (183 m) tall Shasta Dam was completed by 1945 near the southern border of the Klamath Mountains and severely altered the region's biological and cultural resources. The dam severed some of the most important cold-water spawning and rearing habitats in the entire Sacramento River watershed for anadromous Pacific lamprey, white sturgeon, steelhead, and Chinook salmon. The dams also contributed to the extirpation of California's only bull trout population in the McCloud River by the late 1970s. Until then, this isolated population

Genetic and run-timing diversity in anadromous fishes

Chinook salmon, steelhead, and Pacific lamprey typify the anadromous ichthyofauna of the Klamath Mountains. These species exhibit ecotypic diversity with genetically distinct varieties adapted to respond to seasonal cues. Like investors who diversify portfolios to protect against financial losses, anadromous fishes of the Klamath Mountains exhibit ecotypic variation allowing them to protect against reproductive failure (Schindler et al. 2010). Thus, while a disturbance event may negatively impact one ecotype the others may flourish. This differentiation helps maintain species abundance despite environmental conditions that often fluctuate through space and time.

Chinook salmon are a vital cultural resource to peoples of the Pacific Northwest. In the Klamath and Rogue rivers, which today support important recreational, commercial, and Tribal subsistence fisheries, Chinook salmon are characterized by when adults make their terminal spawning migrations into fresh water. Spring-run Chinook salmon move upriver when streams are swollen with spring snowmelt runoff in a reproductively immature condition then over-summer in deep, cold pools where they cease feeding, and ultimately spawn in the early fall. Fall-run Chinook salmon initiate their adult freshwater migration in late summer to early fall in a more advanced state of maturation, and spawn later in the fall.

The migration timing difference between spring and fall Chinook salmon in the Klamath River has been traced to a small region on chromosome 28, known as the Region of Strongest Association or RoSA, that determines when fish migrate upriver (Thompson et al. 2020). While spring and fall are otherwise genetically similar, the RoSA predicts migration timing of Klamath River Chinook salmon, with a cut-off date of about August 1st. Chinook salmon that possess two cop-

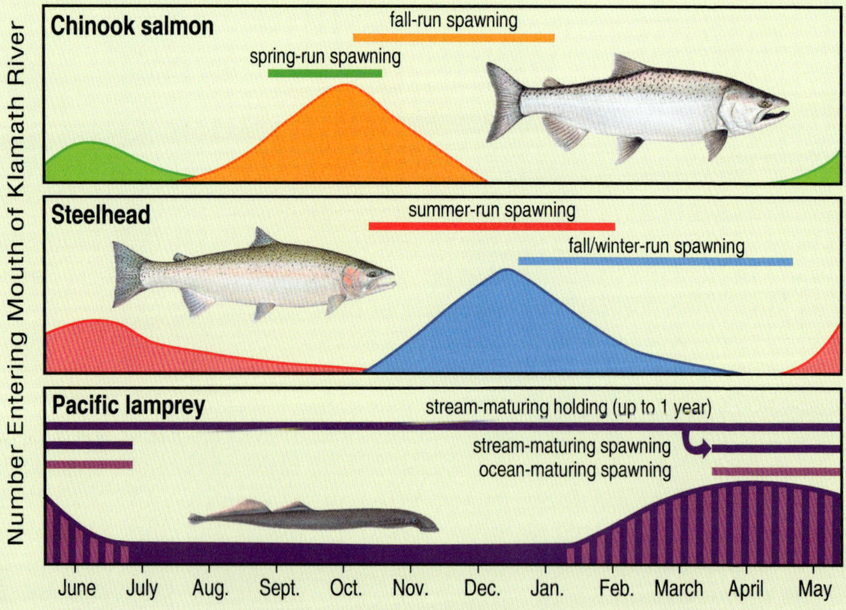

ies of the early allele (EE homozygous) enter the Klamath River before August 1, whereas those with two copies of the late allele (LL homozygous) enter after this date. Individuals that are a mix of both alleles (EL heterozygous) have intermediate migration times. Interestingly, these findings suggest that interbreeding between a pair of individuals that are both heterozygous EL, could produce offspring with a mix of all three genotypes, EE, EL, and LL. Or, put another way, brothers and sisters may exhibit completely opposite migration timing, some having the spring-run migration timing and others with the fall-run migration timing. Ultimately, the genetic diversity of the RoSA region underpins ecotypic diversity in Chinook salmon.

Steelhead are an anadromous form of rainbow trout and prized by recreational fishers. They display ecotypic variation parallel to Chinook salmon. In steelhead, two migration phenotypes are exhibited in the Rogue and Klamath rivers, a summer-run and fall/winter-run. Fall/winter-run steelhead enter freshwater rivers from December–April in a more mature state and spawn shortly after entry. Summer-run steelhead enter natal rivers from May–October in an immature state and hold in cold pools until spawning in early winter. This variation is also traced to a similar genetic region impacting run timing in Chinook salmon (Hess et al. 2016). While the precise genetic mechanisms remain to be elucidated, genetic diversity is critical for maintaining fall/winter- and summer-run steelhead.

Pacific lamprey is a species that serves as a vital protein source for Native Americans of the Pacific Northwest during times of year when salmon are not available. The Pacific lamprey of the Klamath River, and likely other regional rivers, exhibit two ecotypes—ocean-maturing and river-maturing. The two ecotypes differ in maturity levels at the time they enter fresh water to begin their spawning migrations, with the ocean-maturing entering in a more mature state and possibly spawning shortly thereafter. In contrast, the river-maturing enter in a less mature condition and are thought to spawn up to year or more after arrival. The ecotypic variation in Pacific lamprey has also been found to be under genetic control, predicted by a gene interaction, and appears more complicated than the genetic underpinnings of ecotypic variation in salmon (Parker et al. 2019). While Pacific lamprey ecotypic variation is less understood than that of salmonids, it is clear this variation is vital to the species' survival.

Collectively, we see that evolutionary forces have acted in parallel to drive ecotypic diversification among regional anadromous fishes. Chinook salmon, steelhead, and Pacific lamprey, which represent highly divergent evolutionary lineages, all share an anadromous life history and exhibit a diverse portfolio of ecotype partitioning and reproductive strategies. This diversity can be viewed as a legacy of unique natural forces within the Klamath Mountains which has hosted these species for eons. Today, Pacific lamprey are in decline throughout much of their geographic range and spring-run Chinook salmon and summer-run steelhead are the most imperiled salmonid populations in the Klamath River basin. The presence of multiple ecotypes within anadromous fishes appears to produce a diverse portfolio that ensures these species can persist into the future.

persisted in the spring-fed McCloud but perhaps had a much greater distribution in California during earlier times with cooler climates (Moyle 2002).

Even with the extensive losses of anadromous fishes and bull trout, at least 13 native freshwater fishes of the Sacramento basin persist, including nine found nowhere else in the Klamath Mountains (Table 1). Moreover, at least 13 obligate freshwater species have been documented here—a fascinating evolutionary legacy of the massive Sacramento Basin. Against the odds, many of these unique fishes now persist in the fluctuating Shasta and Pit reservoirs and inlet streams, but also with novel reservoir fish communities dominated by introduced fishes. Minnows are an especially diverse group of the region's native fishes which includes the Sacramento pikeminnow, hardhead, California roach, Sacramento speckled dace, and the Sacramento hitch. The region also hosts four sculpin species in the combined upper Sacramento, lower McCloud, and lower Pit rivers—all endemic to the Sacramento basin. Last, and perhaps the most fascinating, the lower Pit River and its chain of hydroelectric reservoirs still maintain a population of Sacramento tule perch, the only freshwater member of the surfperches (*Embiotocidae*), a family otherwise only found in coastal waters of the Pacific.

Family Accounts

This section provides the general biology and ecology of the nine fish families that are native to the Klamath Mountains.

Lampreys (Petromyzontidae)

There are over 40 described lamprey species in the world. Lampreys are the most ancient lineage of extant vertebrates found in the Klamath Mountains with fossils extending over 450 million years. They are from the primitive superclass of jawless ancient fishes known as Agnathans. Lampreys are eel-like but are dramatically different from eels and other modern fishes. Besides lacking jaws, paired fins, and scales, they have a cartilaginous skeleton and a notochord—a precursor to the vertebral column found in modern fishes. Their most notable feature is a specialized sucker mouth with radial horny teeth-like plates that are used for attaching to substrates and foraging. Parasitic species latch on with their sharp tongue and teeth to suck blood and fluids from the host's body. The Klamath Mountains host at least three lampreys including the anadromous Pacific Lamprey, and two resident freshwater species: the Klamath River lamprey and the Pit-Klamath brook lamprey (Table 1). Remarkably, the greater Klamath River basin, including the portions above and below the Klamath Mountains, hosts nine species of lamprey—more than any other river basin in the world (Markle 2016).

Lampreys have one of the most complex and least understood lifecycles among regional fishes. Adults range in length from under 7 in (18 cm) for the Pit-Klamath brook lamprey to more than 28 in (72 cm) for the Pacific lamprey. Lampreys spawn in fresh water during spring and early summer by excavating compact circular nests (redds) often resembling small, submerged rock-lined fire pits. They spawn in similar stream habitats to salmon but, in contrast, both sexes excavate the nest of pebbles, gravels, or cobbles by grasping and moving rocks with their mouths instead of their tails. All three lampreys of the Klamath Mountains most likely die

Three Pacific lamprey constructing a redd for spawning. Photo by Michael van Hattem. Inset of Pacific lamprey mouth by Thomas Dunklin.

after a single spawning event. Embryos hatch into free swimming eyeless larvae (ammocoetes) that get swept downstream and burrow into sandy depositional habitats. Here ammocetes filter feed mainly on algae and detritus (Moyle 2002). The ammocoete duration for Pacific lamprey ranges from three to seven years (Dawson et al. 2015). Ammocoetes then metamorphose into a juvenile form—developing large eyes and a suction disc mouth. The anadromous Pacific lamprey then migrate to the ocean, where they spend multiple years maturing before migrating up rivers to spawn. Remarkably, the Pacific lamprey can take a decade to become reproductively fertile. This is more than twice that of salmon.

The Pacific lamprey and the Klamath River lamprey are parasitic over the adult portion of their life. Hosts include various fishes including salmonids, suckers, rock fishes, cods, flatfishes, and at least five whale species (Moyle 2002, Clemens et al. 2019). Adult Pacific lampreys are commonly preyed on by fishes, seals, and sea lions in marine and estuarine waters. River otters and ospreys often prey on adults in rivers and are also likely predators of adult Klamath River lamprey. In contrast, and like other small brook lampreys, the Pit-Klamath brook lamprey ammocoetes likely filter-feed exclusively on detritus, algae, and microbes (Moyle 2002) and do not feed once they transform into adults. Lamprey die after spawning and many scavengers, from invertebrates to mammals, feed on their nutrient-rich carcasses.

Like salmonids, Pacific lamprey express multiple life history strategies that are still not well understood. For example, many migrate into streams as fully mature adults, with some likely spawning within several weeks of leaving the ocean. Others migrate into streams a year early, maturing in fresh water, before spawning the following spring (Beamish 1980, Parker et al. 2019) (see run-timing sidebar).

Sturgeons (Acipenseridae)

There are fewer than 30 species of sturgeon in the world and all are restricted to the northern hemisphere. The green sturgeon and white sturgeon are two anadromous species found in the Klamath Mountains. Sturgeons have fossil origins extending to 200 Ma. Like lampreys, sturgeon evolved independently from other modern fishes for millions of years. They possess many less highly derived features, such as a notochord, offering a rare portal into the deep history of fishes. Like coelacanths and other ancient fishes, sturgeon possess nasal cavities below the eyes (spiracles). They have scaleless sandpaper-rough skin with five rows of large bony plates (scutes). Sturgeon also have a heterocercal caudal fin, as seen in sharks, where the dorsal lobe is much longer than the ventral lobe. Sturgeon also have many specialized features. Their skeletons are made almost entirely of cartilage. Their unique, flexible sucker-like mouths, and series of four sensory taste barbels located on the underside of the snout, make them highly specialized for benthic feeding. Sturgeons in the Klamath Mountains forage on a variety of fishes and invertebrates including crustaceans and mollusks, with diets becoming more diverse as they get older.

Both green and white sturgeon are found along the Pacific coast of North America from Alaska to Mexico—with the Sacramento River hosting the southernmost spawning habitats for both species. With their affinity for deep pools and strong currents, sturgeons are among the most rarely seen and enigmatic vertebrates of the Klamath Mountains. Major regional green sturgeon spawning populations are found in the Rogue River up to Rainie Falls (Erickson and Webb 2007) and the Klamath River up to Ishi Pishi Falls near Somes Bar (Benson et al. 2007). The Trinity River provides spawning habitats up to Gray's Falls (Moyle 2002). Although multiple adult observations have occurred in the Smith River (Garwood, unpublished data) and Salmon River up to Wooley Creek (Moyle et al. 1994), reproduction has not been verified in these waters. White sturgeon are often found in estuaries of most Klamath Mountains rivers. Small numbers have been recorded in the Rogue and Klamath rivers, though evidence indicating consistent reproduction is lacking (Moyle 2002, Erickson and Webb 2007). The Sacramento River basin is an important region for sturgeons and the lower Pit River in the Klamath Mountains historically supported a spawning run of white sturgeon before being severed and flooded by the construction of Shasta Dam.

Sturgeons are the largest fishes in the Klamath Mountains and among the longest-living creatures of the region. In fact, the white sturgeon is the largest freshwater fish in North America with a maximum recorded length exceeding 20 ft (6 m) and weight exceeding 1,400 lb (630 kg) (Moyle 2002). However, due to long-term overexploitation, individuals rarely reach 10 ft (3 m) today. Green sturgeon are smaller and rarely exceed 6.5 ft (2 m). The maximum recorded size is from the Klamath River at nearly 9 ft (2.7 m) and weighing 350 lb (159 kg) (Moyle 2002). The oldest recorded white sturgeon is 104 years (Rein and Beamesderfer 1994) while the oldest known green sturgeon is 53 years (Farr et al. 2002). Both species are slow-growing and late to mature relative to other Klamath fishes with females beginning to reproduce at 12–16 years (Moyle 2002, Beamesderfer et al.

The Ghost Sturgeons of Shasta Reservoir

In the spring of 2020, regional media broadly reported a 3.4 meter-long (11 ft) sturgeon was caught by a fisher in Shasta Reservoir. Large white sturgeon have been seen and captured in the reservoir over the years, but this fish was a monster. Based on its length, regional fisheries biologists estimated the sturgeon could be up to 90 years old! The official white sturgeon record for the reservoir was a dead fish discovered in 1963 that measured 2.9 meters (9.5 ft), weighing 225 kg (496 lbs), and aged to 67 years (Moyle 2002). Prior to the construction of Shasta Dam, anadromous white sturgeon once migrated from the Pacific to above the dam site and most likely spawned in the lower Pit River. The Pit River is the largest of the upper Sacramento tributaries now flooded by the reservoir. It receives a massive amount of spring water from volcanic groundwater basins connected to Mount Shasta and the broader southern Cascades and Warner Mountains.

Shasta Dam was built between 1937 and 1945, and based on images of the dam's construction, the original river channel was free flowing at least through 1943. Various age classes of white sturgeon became trapped in the reservoir and later lost their spawning grounds when two hydroelectric reservoirs were built in the lower Pit River during the 1960s. Because of their great longevity, the giant sturgeons in the Shasta Reservoir today are a ghost population from reclamations past because they are doomed to live out there lives without the ability to reproduce. The birth of the monster fish captured (and released) in 2020 most likely predates the moment Shasta Dam destroyed many unique anadromous fish populations of the Upper Sacramento. It's anyone's guess how long the population of ghost sturgeon will remain in Shasta Reservoir before disappearing. Given the sizes of some of these fish, it's easy to imagine many living well beyond 100 years. However, it should be expected that fewer and fewer white sturgeon will be found in future years as individuals die from old age.

Snorkeling with green sturgeon and observing these prehistoric creatures circle placidly in deep sun-dappled pools is a stunning experience. Photo by Alex Rose.

2007). Sturgeons are anadromous and iteroparous, meaning they can have repeat spawning migrations throughout their lives. They enter Klamath streams in late winter, with peak spawning between April and June (Erickson and Webb 2007, Benson et al. 2007). Sturgeons are broadcast spawners, releasing many thousands of tiny, sticky eggs. Juveniles grow for one to three years in rivers and estuaries before entering the ocean. Many spawning and non-spawning adults hold in deep pools throughout the summer with outmigration to the ocean at the onset of increased winter stream flows.

Minnows (Leuciscidae, formerly Cyprinidae)

Minnows represent one of the most diverse families of fishes in the world. Early fossil records from the Eocene Epoch place their origins in Asia. Except for a few marine species, they are a freshwater group with species distributed across Africa, Eurasia, and North America. The group is taxonomically challenging due to morphological similarity among species, hybridization potential among species, as well as individual species having substantial genetic divergence that has resulted in further sub-specific divisions. California and Oregon together host over 30 native species (Moyle 2002, Markle 2016, Baumsteiger and Moyle 2019) and the Klamath Mountains contain six native minnows (Table 1). Within the Klamath Mountains, the upper Sacramento River watershed hosts all six species. The Klamath River only has two native minnows—the Klamath speckled dace and the rarely observed tui chub that occasionally migrates downstream from upper Klamath basin strongholds. The Rogue River only hosts the Klamath speckled dace, while the Smith and Chetco are devoid of minnows. The absence of minnows from small coastal rivers in the region contrasts sharply with other parts of North America where minnows are the dominant fish.

Despite what their common name implies, minnows of the Klamath region range from small (<4.5 in [11 cm] long), such as in the speckled dace and California roach; to medium (<14 in [35 cm] long) as in the Sacramento hitch; to

Klamath speckled dace in the South Fork Trinity River. Photo by Justin Garwood.

large (up to 39 in, [100 cm] long) as in the hardhead and Sacramento pikeminnow. Minnows are closely related to suckers (Catostomidae) and similarly have pharyngeal teeth located in the throat that are specialized in each species based on its specific feeding habits: from grinding, to filtering, to chopping. Minnows also have advanced hearing and release chemical scents when injured or threatened, providing an escape warning to others in their shoal.

Most minnows of the Klamath Mountains prefer warmer habitat zones with lower flow velocities including pools and backwaters. Many even thrive in reservoirs. However, speckled dace and roach are found in fast moving riffles and a broader range of water temperatures including habitats often shared by salmonids. All minnows of the Klamath Mountains are omnivores, except for the Sacramento pikeminnow. This species is a formidable nocturnal predator that primarily targets other fishes and crustaceans with its large mouth and knife-like pharyngeal teeth. Most minnows eat algae, small crustaceans, and insects on a seasonal basis. The Sacramento hardhead forages for zooplankton in the water column (Moyle 2002). Minnows breed in the spring and early summer during pronounced water temperature increases. Most minnows are communal broadcast spawners with high fecundities for their size, producing many thousands of eggs. Fertilized eggs drop onto the stream bottom and become lodged into the gravel. Some species, including the Sacramento pikeminnow and the tui chub, have sticky eggs that adhere to substrates. California roach breed in large communal groups splashing along the shallows like marine grunions.

Suckers (Catostomidae)

Suckers are closely related to minnows (*Leuciscidae*) and are from an ancient lineage of freshwater fishes with fossils dating to the early Eocene Epoch nearly 50 Ma. There are about 80 extant species found almost exclusively in North America. California and Oregon have 13 species combined (Moyle 2002, Markle 2016). Two species are found in the Klamath Mountains, including the Sacramento sucker and the Klamath smallscale sucker. The Sacramento sucker has a widespread distribution in California. In the Klamath Mountains, they occur in streams throughout the upper Sacramento River above the Shasta Reservoir. They also occur in the Shasta and lower Pit River reservoirs. The Klamath smallscale sucker is nearly endemic to the Klamath Mountains with native populations occurring in the Klamath River and Trinity River basins as well as the Rogue River basin. New genetic evidence indicates a third smaller population found in the Smith River is nonnative and was likely introduced from the Klamath River (Kinziger et al. 2021). The Klamath smallscale sucker is related to three other endemic suckers found in the upper Klamath River basin upstream of the Klamath Mountains.

Both sucker species are relatively long-lived (>15 years) and in rivers rarely exceed 16 in (40 cm) in length (Moyle 2002). They have stout muscular bodies and strong swimming abilities in turbulent waters. With ventral facing mouths and extendable fleshy lips they browse on stream bottoms where they suck up food and sift out debris with comb-like pharyngeal teeth. These unique anatomical features allow suckers to exploit specific stream resources that are generally unavailable to other fishes. They are generalist omnivores that largely eat detritus, algae, and

School of Sacramento suckers. Photo by Thomas Dunklin.

invertebrates. Their long-coiled intestines process foods that are difficult to digest such as algae. Similar to sculpins, suckers are often blamed for salmonid declines through competition and egg predation, despite a lack of evidence (Moyle 2002). They have a broad thermal tolerance and are found in various habitats including cool turbulent streams, deep pools, backwaters, and some reservoirs. Both suckers commonly undertake upstream movements for spawning. Like other catostomid species, it is thought that individuals return to the same spawning areas across multiple reproductive seasons where they often produce over 10,000 eggs. They spawn in the spring over gravel where their eggs drift and drop into interstitial cavities or settle out in slow water areas. Hatchlings settle out along the slower, warmer margins of streams.

Smelts (Osmeridae)

There are 15 smelt species in the world, and most are marine restricted to the Pacific and Atlantic oceans in the Northern Hemisphere. One smelt species, the eulachon, occurs in the Klamath Mountains. The eulachon is a large (<10 in [26 cm]) anadromous smelt occurring in Pacific coastal waters from northern California to Alaska where they feed on plankton (Gustafson et al. 2010). Eulachon are sometimes referred to by the common name "candlefish" because of fatty flesh that burns like a candle when dried. Adult eulachon enter the mouths of spawning streams from January–May and peak in the spring (March–April). They spawn

Eulachon. Photo by Connie Taylor.

Fishes

in slow moving waters, typically within lower river reaches. Eulachon are not known to exhibit natal homing as precisely as anadromous salmonids, but instead return to suitable spawning habitats in river systems where they were born. Female eulachon produce as many as 30,000 eggs, which are adhesive to sand or gravel substrates. Following spawning, most adult eulachon die. After an incubation period of 21–40 days eulachon larvae migrate to the sea where they spend two to six years in the ocean prior to returning to fresh water to spawn (Moyle 2002).

The number of eulachon returning to spawn in a particular river varies dramatically from year to year depending upon stream conditions and ocean survival rates. The lower Klamath River is the only stream in the Klamath Mountains that has historically supported large runs of eulachon (Gustafson et al. 2010). They most likely use Klamath River reaches that are west of Pecwan, but they have also historically been found between Pecwan and Weitchpec (Larson and Belchik 1998) (Table 1). Because of range-wide declines, eulachon populations from Washington state to California were listed as Threatened in 2010 under the Federal Endangered Species Act. As with Pacific lamprey, eulachon are recognized as a “salvation” fish for the Yurok Tribe and other Native Americans throughout the Pacific Northwest because they migrate and spawn in late winter and early spring when other subsistence food supplies are scarce.

Salmon and Trout (Salmonidae)

There are around 226 salmon, trout, and char species in the world with native distributions restricted to the Northern Hemisphere (Eschmeyer 1998). No fishes define Pacific Northwest stream ecosystems more than salmon and trout, and they are the most abundant and widespread group of fishes in the Klamath Mountains. Salmonids dominate stream food webs of the Klamath Mountains and serve as a keystone group for maintaining stream and riparian productivity (see Marine Nutrients sidebar) while serving as indicators of ecosystem health. Salmon are foundational elements in religion, culture, and subsistence for fishing Tribes of the Pacific Northwest. Many species are anadromous, leaving tiny, steep, conifer shaded tributaries deep in the heart of the range as parr and migrating to the immense Pacific Ocean as smolts only to return one day to breed at their birthplace by smell (see Anadromous Fishes sidebar) (Quinn 2018). Others permanently reside in isolated streams—sometimes even above insurmountable waterfalls. The diversity of salmonids in the Klamath Mountains lies in their numerous life histories that are a product of their own genes and regional factors such as snowmelt runoff, local precipitation regimes, and habitat distributions.

Eight salmonid species from two genera have been documented in the Klamath Mountains including seven from *Oncorhynchus* (Pacific salmon and trout) and the bull trout from *Salvelinus* (chars) (Table 1). Five Pacific salmon are strictly anadromous and die after spawning. Chinook salmon are the most abundant and widely distributed of the Pacific salmon species, followed by coho salmon. Of the less common species, marginal spawning runs of chum salmon are documented sporadically in the Smith and Klamath rivers while sockeye and pink salmon are rarely seen. Sockeye salmon have been observed spawning during summer on a few occasions in the Smith River (Walkley and Garwood 2017). Small numbers of juvenile sockeye salmon are documented annually in fish outmigrant traps on the Trinity River, but these may come from the Trinity reservoirs where kokanee (resident landlocked sockeye salmon) have been introduced to support a sport fishery. Pink salmon are so rarely encountered in the Klamath Mountains that only a few records exist. Adult pink salmon were reported at a fish weir in the McCloud River in 1891 (Moyle 2002) and juveniles were found in the Klamath River above Interstate 5 bridge in 2003 (Carter and Kirk 2008). The Klamath Mountains also host two species of trout including the rainbow trout (steelhead) and the coastal cutthroat trout. Rainbow trout are, by far, the most abundant and widespread regional salmonid in the Klamath Mountains,. Both steelhead and coastal cutthroat have broad life history patterns including freshwater resident and anadromous forms.

The following species ecology treatments provide a brief look into Klamath Mountain salmonids. Due to their rarity, we omit further discussion of chum, sockeye, and pink salmon.

Bull Trout (Extirpated from the Klamath Mountains)

The only population of bull trout known to California was in the McCloud River prior to going extinct by the 1970s (Moyle 2002). How they arrived in the McCloud River system is a mystery as the nearest population is found in the upper Klamath River in Oregon (Markle 2016). It's possible the McCloud River bull trout population was the last remnant of much broader regional bull trout distributions during cooler Pleistocene climates. Bull trout are fish predators that prefer cold streams and it's thought the Shasta and McCloud dams cut off their abundant anadromous fish prey while also eliminating their breeding habitat. Furthermore, the introduction of non-native eastern brook trout and brown trout likely increased competition (Moyle 2002).

Chinook Salmon

North American Chinook (king) salmon populations extend from California's San Joaquin River to Alaska. They are the largest-bodied species among the five North American Pacific Salmon. Spawning adults are 27–31 in (70–80 cm) and sometimes exceed 51 in (130 cm) and weigh more than 50 lb (>23 kg)! Adult Chinook salmon enter streams a bright silver-gold and typically darken to olive-green, brown, and reddish during spawning. They have densely clustered spots on the upper half of the body that are usually much larger than those found on coho salmon. Chinook salmon have black mouths and gums which also distinguishes them from coho salmon and steelhead regardless of body color.

The Klamath Mountains host two federally recognized Chinook salmon groups or Evolutionarily Significant Units (ESUs). The Southern Oregon/ Northern California Coastal Chinook ESU includes the Rogue, Chetco, and Smith rivers and the lower Klamath River up to the confluence with the Trinity River. The Upper-Klamath-Trinity-River Chinook ESU includes the Trinity River and all other Klamath River populations found above the Klamath-Trinity confluence. Each of these two ESU groups also have fall-run and spring-run ecotypes that are largely separated by run timing (see run-timing sidebar). In 2021, the Upper-Klamath-Trinity-River spring Chinook salmon was listed as Threatened under the

Anadromy and Marine Nutrients in Klamath Watersheds

Anadromous fishes of the Pacific Rim take advantage of the productive waters along the Pacific Ocean to attain much larger body sizes than their closely related resident freshwater forms. Larger sizes, gained in the ocean due to food availability, provides greater fitness and agility for long and tricky upstream migrations to spawning and rearing grounds. Anadromous fishes produce far more offspring than their freshwater residents. For example, Chinook salmon produce > 4,000 eggs while a resident trout may produce only 1,000 eggs (Moyle 2002).

The Klamath Mountains have five anadromous fishes that historically had a broad influence on subsidizing the region's intricate stream and riparian ecosystems through a massive annual pulses of marine-derived biomass. Pacific lamprey, Chinook salmon, and coho salmon are semelparous—dying soon after spawning. Steelhead and coastal cutthroat trout are iteroparous—having the ability to return to the ocean after spawning to attempt future spawning migrations. Semelparous fishes contribute more marine biomass to stream ecosystems through death as adult iteroparous fishes often survive spawning and return to the Ocean. However, iteroparous fishes contribute nutrients and biomass to streams through waste excretion, eggs and sperm from spawning, and as occasional carrion.

A high ratio of the nitrogen isotope known as ^{15}N provides a stealthy tracer that reveals marine nutrient inputs from anadromous fishes throughout fresh water and terrestrial forest ecosystems. It has been found in a variety of organisms including algae, aquatic invertebrates, and even conifers. Salmon and lamprey are conveyor belts, transporting ocean energy and nutrients deep into watersheds. This biomass is delivered to stream and riparian ecosystems through pathways that include hundreds of consumers (including humans!) and decomposers. Predators like river otters prey on live fish while scavengers like black bears or bald eagles feed on carcasses and move fish biomass to the land. Macro-nutrients from the fish, including nitrogen but also phosphorus and potassium, fertilize forest plants. These nutrients can also kick-start algal growth and invertebrate productivity in streams that ultimately provides food for young lampreys and salmonids.

Lamprey and salmonid runs are a fraction of their historic levels resulting in far less biomass and nutrients being delivered to the Klamath Mountains. Restoring these species to their former abundance also leads to broader ecosystem health and its promising capacity to support more fish.

A female Chinook salmon carcass, Smith River, California. Chinook salmon cease feeding during their in-river spawning migrations and it is common to find spawning adults, dead on the shores or that are still living, with highly eroded fins and decaying bodies. Photo by Zach Wenderott.

California state Endangered Species Act due to low abundance of wild populations with fewer than 1,000 individuals in the Salmon and Trinity rivers. Available scientific evidence indicates the spring-run does not actually represent a distinct evolutionary lineage from fall-run (Thompson et al. 2020) so the listing was based on the low numbers of spring-run. The listing of spring-run Chinook salmon by the State of California represents a unique management strategy that departs from most other state and federal agencies approaches that typically require formal species recognition as prerequisite for listing. The upper Sacramento River and its tributaries in the Klamath Mountains also supported similar fall and spring ecotypes, in addition to a unique winter-run before the construction of Shasta and Whiskeytown dams. These dams extirpated salmon and steelhead from upper Clear Creek and the Pit, McCloud, and upper Sacramento rivers (Table 1). Wild fish from these historic runs persist in the Sacramento River below dams and in a few tributaries but they are all imperiled.

Despite varying abundance of local populations, wild Chinook salmon are broadly distributed in coastal streams, with adults spawning in mainstem rivers and small tributaries (see Chinook spawning sidebar). The spring and fall ecotypes also add more temporal variability in Chinook salmon distributions. Most Chinook salmon in the Klamath Mountains return to spawn as 3- and 4-year-olds, but ages range from 2-year-old jacks to 5-year-old adults. The distinctively small male "jack" salmon are passive and sneak fertilizations rather than fight for access to spawning like older and larger males. Juvenile Chinook salmon of the Klamath Mountains commonly rear in fresh water for three to seven months and move downstream at a variety of sizes. Some temperature-impaired streams, such as the Klamath and Rogue, have shorter juvenile freshwater rearing periods (ocean-type rearing life-history) due to elevated summer stream temperatures while the Smith River, and cooler tributaries in the Klamath and Rogue rivers, can host juvenile Chinook salmon through September (stream-type rearing life-history). Juvenile Chinook salmon feed in a variety of stream habitats and spring emigrants tend to spend more time in estuaries prior to ocean entry than other anadromous salmonids.

Coho Salmon

In North America, coho (silver) salmon range from central California (Santa Cruz) to Alaska where they are found in coastal streams and rivers. Spawning adults average 26 in (65 cm) in length, making them medium-sized among Pacific salmon. Adults entering fresh water are bright silver and quickly transform into shades of reds, greens, and pinks with small black spots along the upper body. Males often take on bright scarlet-red colors and develop hooked, toothy jaws for sparring with other males during spawning. In the Klamath Mountains they are found in the Rogue, Chetco, Smith, and Klamath rivers (Table 1). These populations are experiencing dramatic declines, and all are listed as Threatened under both State and Federal endangered species acts.

The number of coho salmon spawning in natural areas within the Klamath Mountains is generally low. Coho salmon spawning locations are isolated and are mainly associated with forested, low-gradient tributaries with medium-sized spawning gravels (Quinn 2018). Adults migrate into interior streams, such as the

The Structure and Chronology of a Chinook Salmon Redd

All salmonids and lampreys of the Klamath Mountains build conspicuous gravel spawning nests called "redds" where they deposit their eggs in rivers and streams. Chinook salmon create the largest redds sometimes exceeding 6 meters (20 ft) in length. **Series illustrated by Michael Zontos.** Adapted from Burner (1951).

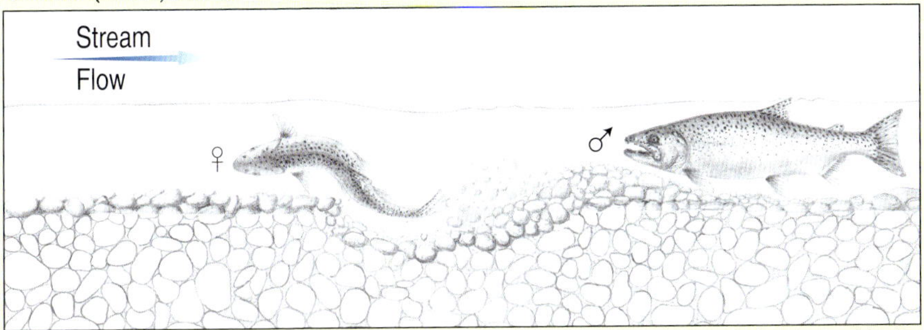

(A) Prespawning: 1 to 5 days. During the fall, a female Chinook salmon finds a suitable location and begins excavating the streambed using strong lateral strokes of the body and tail. This powerful action creates a boiling current that lifts gravels and silts into the water column where they are swept by stream currents a short distance downstream, forming a loose pile called a tailspill. A depression upstream of the tailspill, known as a pot, forms as the alluvium is excavated. Males begin to assemble near the female and aggressively spar for the opportunity to mate. The female chases off males until she is ready to spawn.

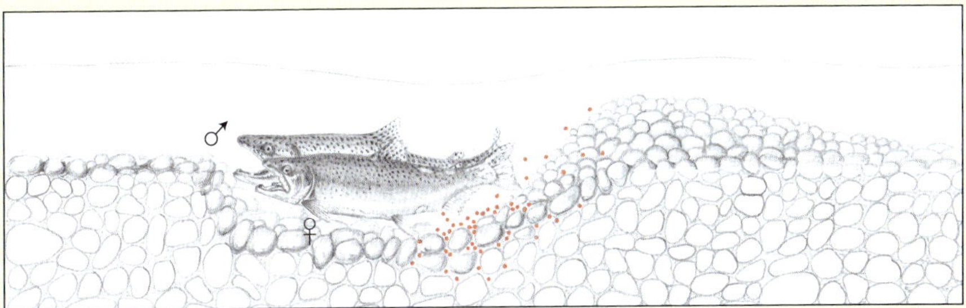

(B) Spawning: 1 to 5 days. After the pot is excavated and the tailspill becomes more substantial, the female begins depositing her eggs throughout the downstream end of the pot and the base of the tailspill. Eggs are simultaneously fertilized by a male. This process continues in a series—with the female depositing eggs between new layers of sorted alluvium while continuing to bury fertilized eggs and excavating the pot further upstream.

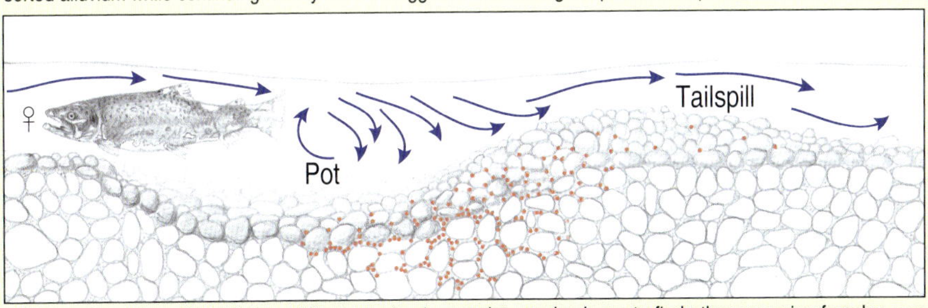

(C) Post Spawning: 5 to 14 days. After spawning is complete, males leave to find other spawning females—or die in the process. The resulting oval-shaped streambed nest, known as a redd, is a marvel of careful engineering. Fragile eggs develop in the streambed often through many powerful winter flow events with egg survival influenced by the severity of flood events. The profile of the redd also promotes local hydraulic circulation throughout the tailspill—delivering oxygen to the eggs while removing carbon dioxide and other wastes from the egg pockets. The female stays and guards the redd while she rapidly deteriorates and eventually dies. Even after spawning, female salmon will continue ineffective digging as they weaken until their tails are almost completely worn off. Eggs, and later alevins with yolk sacs, incubate in the redd for around 18 weeks before emerging as fry in the spring.

Juvenile coho salmon parr in headwaters of Baldface Creek, Smith River Watershed. Photo by Justin Garwood.

Klamath River, from September through November and spawn between November and December. They generally enter coastal streams, such as the Smith River, from November- January and spawn through February. Juvenile coho salmon spend slightly more than a year rearing in fresh water before smolt transformation and seaward migration. Survival during the summer in the Klamath Mountains is largely controlled by the availability of cold-water summer-rearing habitats found in tributaries and colder, thermal niches along mainstem rivers. Winter survival is influenced by the availability of complex overwintering habitats that offer stable refugia from winter high water events. In general, they prefer rearing in off-channel backwater habitats, and deep pools lined with undercut banks and downed logs. Throughout the Pacific Northwest, juvenile coho salmon are also commonly associated with habitats created and maintained by beavers—which have also suffered severe historic declines in the Klamath Mountains (see Life of Beavers sidebar page 134–135).

Many coho salmon that are born in Klamath Mountain streams migrate downriver, west of the Klamath Mountains, to coastal plains where they overwinter in small coastal streams and fresh water portions of estuaries. After smolt transformation and marine migration, coho salmon spend less than two years at sea (two summers) before returning to their natal river basins at age three to spawn and die. Smaller male jack coho salmon also return to streams after spending only one summer at sea. Imperiled natural populations are persisting in the Klamath, Trinity, and Rogue rivers and are supplemented with larger hatchery stocks as a mitigation measure for habitat loss above various dams.

Coastal Rainbow Trout and Steelhead

Coastal rainbow trout and steelhead are the same species, but each has a very different life history. Rainbow trout complete their entire lifecycle as residents in fresh water while steelhead are anadromous. To add to the complexity, resident forms can produce steelhead offspring, steelhead can produce resident offspring, and both forms can breed together where distribution overlaps . Furthermore, population age structure is much more complex and variable than other salmonids because juvenile anadromous forms remain in fresh water from 1- 3 years and adults have the capacity for repeat spawning. This life history diversity has allowed rainbow trout to adapt to a variety of habitats and become among the most successful freshwater fishes in western North America—with native distributions from Alaska to Baja California, Mexico. In the Klamath Mountains they are widespread and clearly the region's most common native fish. Rainbow trout are variable in appearance with most heavily spotted, with whitish bellies, an olive-green back, and a pink-to-purple band along the sides. Steelhead have similar colorations, especially as adults darken after holding for long periods in streams. Steelhead entering streams from the ocean are stunning—with chrome-bright silver flanks, a steel-blue back and head, and transparent fins. Rainbow trout and steelhead have white mouths and gums distinguishing them from both Chinook and coho salmon. Because of their ocean migrations, steelhead average around 28–30 in (70–75 cm) in length while native resident rainbows rarely exceed 16 in (40 cm).

Extant Klamath Mountain steelhead populations are grouped into the discrete Klamath Mountains Province population segment . Nested within this larger group are three main migration life histories: half-pounder, summer-run, and fall- winter-run. The term "half-pounder" was coined by J. O. Snyder (1925), to describe runs of juvenile steelhead (usually < 14 in [35 cm]) that return to rivers roughly 2–4 months after ocean entry to feed (and rarely spawn) before returning to the ocean the following spring. Half-pounders are well documented in the Eel, Klamath, and Rogue rivers and are a popular annual river fishery before the onset of the winter steelhead season. Summer-run steelhead enter streams during the spring and summer in an immature condition and over-summer in deep cool pools. Fall-winter-run steelhead enter streams in the late fall and over-winter in a mature condition (see run-timing sidebar). Like other anadromous fishes blocked by dams, steelhead have been extirpated from the Sacramento River portion of the Klamath Mountains that historically had a fall-winter-run and possibly a summer-run (Table 1). Discerning experts subdivide these ecotypes further in individual river systems to capture local life history variation. Steelhead spawning ranges from October- April with summer-run spawning earlier than the fall-winter-run.

Juvenile rainbow trout and steelhead emerge from redds throughout the late winter and spring and typically forage in protected edge waters of streams into their first summer. As juveniles get older, they take up positions in habitats with stronger currents where larger and older fish defend feeding areas in riffles, runs, and at the head of pools. Typical Klamath Mountain steelhead transform into smolts after they are two to three years old, where they are around 7 in (18 cm) in length. Steelhead spend 1–4 years at sea before making their first migration up rivers to spawn (Moyle 2002).

Coastal Cutthroat Trout

Coastal cutthroat trout and rainbow trout diverged over 10 Ma (Trotter et al. 2018) though the two species still occasionally hybridize (Rizza 2015). They are adorned with multiple heavy black spots and red slashes or "cuts" under each side of their lower jaw and gills. They have distinct basibranchial teeth at the base of their tongue that can be felt with your finger. Coastal cutthroat trout are found in streams throughout the coastal Pacific Northwest from Prince William Sound, Alaska to the Eel River, California. Their distributions largely coincide with the coastal temperate rainforest belt west of the Cascade Range (Trotter et al. 2018).

Coastal Cutthroat Trout in the clear waters of Blue Creek, Klamath River, California. Photo by Thomas Dunklin.

In the Klamath Mountains they are found in streams of the Rogue, Chetco, and Smith rivers and extend their distributions up to the smallest headwater tributaries, even above waterfall barriers. They are rare in the Klamath River basin but appear to be common in Blue Creek, a cold-water coastal Klamath River tributary located on the western edge of the Klamath Mountains (Gerstung 1997). Larger piscivorous individuals are common in mainstem sections of the Smith and Chetco rivers making them the top resident fish predator and a flagship conservation species in these resilient coastal streams. They have a variety of life history patterns, including freshwater residents, freshwater migrants, and anadromous populations. However, marine migrations are rare in Klamath Mountain streams relative to populations in northern latitudes (Gerstung 1997, Moyle 2002). Like rainbow trout, coastal cutthroat trout spawn in late winter and spring.

Sticklebacks (Gasterosteidae)

The threespine stickleback is the only native stickleback in the Klamath Mountains. They originally radiated into freshwater habitats from marine environments, and despite their widespread freshwater distributions, can tolerate a broad range of salinity. Some populations are marine, estuarine, or anadromous while others are strictly freshwater residents. Because of their small size and marine origins, threespine stickleback originally had limited distributions below high-velocity barriers along coastal drainages of the western Klamath Mountains. For example, they barely enter the Smith River's turbulent, cascading South and Middle Fork gorges. Some populations have been introduced inadvertently into reservoirs of the region through sportfish stocking, where they have expanded into stream reaches occurring above historic coastal barriers. The brook stickleback (*Culaea inconstans*) was

Threespine stickleback displaying its sharp dorsal spines. Photo by Justin Garwood.

introduced into the Scott River watershed prior to the 1990s where it has become widespread (Moyle 2002).

Threespine sticklebacks are small (usually <2.5 in [6 cm] long), laterally compressed, with a distinctly narrow caudal peduncle. They lack scales and are armored with bony plates on their sides. They have three sharp spines above their dorsal fin and each of their pelvic fins are largely reduced to ventral spines. When fully erect, the combined dorsal and ventral spines arm them with a formidable defense from some predators. They are a schooling fish and seen in large groups along stream edges, backwaters, and in reservoirs where they forage around aquatic vegetation and bottom sediments for small invertebrates. They are also found in mixed assemblages with young salmonids and other small fishes.

Sculpins (Cottidae)

The genus *Cottus* comprises the freshwater sculpins with at least 68 species across North America, Eurasia, and Japan. They are a complex group and often difficult to identify to species because of a wide variation of features. Two or more species are often found in the same habitats, adding to the confusion. At least seven species are found in the Klamath Mountains making them the region's most speciose group of fishes alongside the salmonids. They occupy a range of streams in the Klamath Mountains and have intriguing regional distribution patterns. A few species can handle warm streams during the summer and most prefer turbulent and rocky streams with good water quality and cool temperatures. Many species occur in riffle portions of streams with abundant boulders and cobbles with interstitial cavities but some also use deep pools as well. Vegetated areas are used, especially at locations offering cooler water temperatures during summer.

The coast range sculpin and prickly sculpin are both found in the western portions of the Klamath, Smith, Chetco, and Rogue rivers. These two are the only marine dispersant sculpins for the region—also using estuaries and nearshore marine habitats. The five remaining species, including the rarely encountered rough sculpin, are strictly freshwater fishes and only disperse through inland streams. Three species are only found in the Pit River basin, which hosts four species in total—the most of any watershed in the Klamath Mountains. The Rogue is the only

Prickly sculpin displaying its large, fan-shaped pectoral fins. Photo by Justin Garwood.

drainage in the range to host the reticulate sculpin, a common species throughout Oregon's coastal streams (Markle 2016). Last, the Klamath and Pit rivers each have separate subspecies of the marbled sculpin (Table 1) adding to evidence suggesting an ancient connection between these drainages (see biogeography section).

Sculpins are small, rarely exceeding 5 in (12 cm) in length that specialize as benthic invertebrate predators on stream bottoms. Bodies are flattened with a broad head and large mouth that tapers toward the tale. Large eyes are pointed upward. They also have exceptionally large fan-shaped pectoral fins with ribbed spiny-tipped fin rays on the ventral surface. These specialized fins help sculpin live on stream bottoms with strong currents. Sculpins of the Klamath Mountains are mottled and banded with Earth tone colors allowing them to blend in with the stream bottom while stalking invertebrates, small fish prey, and intercepting stray eggs when salmonids are spawning. Their cryptic appearance also helps conceal them from predatory fish and piscivorous birds, including the common merganser (*Mergus merganser*)—a major sculpin predator. Female sculpins deposit their eggs on the underside of cobbles and boulders that are accessed by a narrow tunnel. Males fertilize the eggs and actively guard the nest from egg predation.

Surfperches (Embiotocidae)

Around 20 species of surfperch are found along the Pacific Coast of North America making them a diverse group of near-shore marine fishes (Love 1996). The Sacramento tule perch is the only entirely freshwater surfperch from the group and is endemic to inland waters of California. They are widespread throughout the Sacramento River basin and were historically distributed up to Pit River Falls, which is a natural barrier along the lower Pit River roughly 300 miles (483 km) from salt water of San Francisco Bay and 31 stream miles (50 km) upstream of the eastern Klamath Mountain boundary. Today, they are found throughout the Pit River portion of the Klamath Mountains and have adapted to living in the

Sacramento tule perch. Photo by Vince Golder.

various Pit River hydroelectric reservoirs, the Shasta Reservoir, and short riverine sections between. Sacramento tule perch are small compared to many marine surfperch species, with lengths rarely exceeding 6 in (15 cm). All surfperches are viviparous—giving birth to live young. The genus for tule perch (*Hysterocarpus*) means "womb-fruit" and describes their reproductive biology of giving birth to fully developed young. They eat a variety of small invertebrates in addition to large zooplankton.

Nonnative fishes

Many nonnative fishes have been introduced both intentionally and unintentionally into waters throughout the Klamath Mountains. The diversity of exotic fishes in the Klamath Mountains is not tabulated here but is likely equal to or exceeds the number of native fish species in the Klamath Mountains (see Moyle 2002 for a detailed summary). Some nonnative sportfish populations are directly managed through periodic hatchery stocking, but most introduced populations successfully reproduce in the wild and have restructured many of the region's aquatic communities. High elevation lakes and streams were historically fishless due to the barriers created by steep mountain slopes and a lack of riverine connections. Beginning in the early-1900s, these waters were stocked to provide sportfishing opportunities with brook trout (*Salvelinus fontinalus*), rainbow trout (*Oncorhynchus mykiss*), brown trout (*Salmo trutta*), and rarely golden trout (*O. aquabonita*). Rogue stocking of baitfish in high elevation lakes, such as golden shiners (*Notemigonus crysoleucas*) and speckled dace (*Rhinichthys* spp.), is an increasing problem in recent decades. Stocking activities can be detrimental to native amphibians including Cascades frogs (*Rana cascadae*) and long-toed salamanders (*Ambystoma macrodactylum*) (Welsh et al. 2006). To help these declining native species, beginning in the early 2000s, the California Department of Fish and Wildlife dramatically reduced their mountain lakes fish stocking program to allow for some ecologically sensitive, high-elevation waters to become fishless again.

At lower elevations, warmwater non-native species like catfish, bass, and sunfish thrive in artificial habitats including water storage ponds and reservoirs. They also occur in tailwater streams below reservoirs and river reaches with altered flow regimes. Notable nonnative species in lower elevations include smallmouth bass (*Micropterus dolomieu*), largemouth bass (*M. salmoides*), spotted bass (*M. punctulatus*), black crappie (*Pomoxis nigromaculatus*), bluegill (*Lepomis macrochirus*), green sunfish (*L. cyanellus*), pumpkinseed (*L. gibbosus*), yellow perch (*Perca flavescens*), mosquitofish (*Gambusia affinis*), brown bullhead (*Ameiurus nebulosus*), channel catfish (*Ictalurus punctatus*), goldfish (*Carassius auratus*), golden shiner, fathead minnow (*Pimephales promelas*), brook stickleback (*Culaea inconstans*) in the Scott River, and both redside shiner (*Richardsonius balteatus*) and Umpqua pikeminnow (*Ptychocheilus umpquae*) in the Rogue River. Artificial populations of landlocked salmonids, including sockeye (kokanee) salmon (*O. nerka*) and Chinook salmon (*O. tshawytscha*), are periodically stocked in reservoirs. Previously introduced brown trout persist and reproduce in some reservoirs and rivers. American shad (*Alosa sapidissima*), an anadromous species introduced to the western U.S. in the 1800s, has intermittent spawning runs in the Klamath, Smith, and Rogue rivers. Exotic species are commonly cited as the second greatest threat to biodiversity following habitat loss. Little is known regarding potential impacts most introduced species are having on native species in the Klamath Mountains, but the largest impact appears to be predation.

Acknowledgements

Thank you to Darren Ward, Mark Clifford, Stewart Reid, and Ken Lindke, who reviewed the chapter and provided excellent feedback. Michael Zontos created the beautiful color plates of Klamath Mountain anadromous fishes. In addition to being an accomplished artist, Michael is a fish biologist with a keen eye for accurate representation, including the forms and color phases one would see naturally in the Klamath region. Many thanks to the Smith River Alliance who partially sponsored the artwork for this chapter. Thanks to Thomas Dunklin and Alexandra Rose for exceptional fish photos in their natural habitats and Ray Troll for providing us his wonderful spike tooth salmon drawing. Sam Plemons, Ken Lindke, Stewart Reid, Samantha Kannry, Terry Roelofs, Koen Breedveld, and Morgan Knechtle provided detailed local information from across the region that enriched the chapter.

Literature Cited and Suggested Reading

Aalto, K. 2006. The Klamath peneplain: a review of J.S. Diller's classic erosion surface. Pp. 451-463 *In*: A. Snoke and C. Barnes, eds. Geological Studies in the Klamath Mountains Province, California and Oregon: Geological Society of America Special Paper 410.

Aalto, K., W. Sharp, and P. Renne. 1998. ^{40}AR/^{39}Ar dating of detrital micas from Oligocene-Pleistocene sandstones of the Olympic Peninsula, Klamath Mountains and northern California Coast Ranges: Provenance and paleodrainage patterns. Canadian Journal of Earth Sciences 35:735-745.

Abell R.A., et al. 2000. Freshwater ecoregions of North America: a conservation assessment. Washington, DC: Island Press.

Adam, D., J. Bradbury, H. Rieck, and A. Sarna-Wojcicki 1990. Environmental changes in the Tule Lake basin, Siskiyou and Modoc Counties, California, from 3 to 2 million years before present. US Geol. Surv. Bull.13pp.

Bartholomew, J.L., M.J. Whipple, D.G. Stevens, and J.L. Fryer, J.L. 1997. The life cycle of *Ceratomyxa shasta*, a myxosporean parasite of salmonids, requires a freshwater polychaete as an alternate host. The Journal of Parasitology 83:859-868.

Baumsteiger, J., A.P. Kinziger, and A. Aguilar. 2012. Life history and biogeographic diversification of an endemic western North American freshwater fish clade using a comparative species tree approach. Mol. Phylogenetic Evol. 65:940-952.

Baumsteiger, J. and P. B. Moyle. 2019. A reappraisal of the California Roach/Hitch (Cypriniformes, Cyprinidae, *Hesperoleucus/Lavinia*) species complex. Zootaxa 4543:2221-240.

Beamesderfer, R., M. Simpson, and G. Kopp 2007. Use of life history information in a population model for Sacramento green sturgeon. Environmental Biology of Fishes 79:315-337.

Beamish, R. 1980. Adult biology of the river lamprey (*Lampetra ayresi*) and the Pacific lamprey (*Lampetra tridentata*) from the Pacific coast of Canada. Canadian Journal of Fisheries and Aquatic Sciences 37:1906-1923.

Benson, R., S. Turo, and B. McCovey. 2007. Migration and movement patterns of green sturgeon (*Acipenser medirostris*) in the Klamath and Trinity rivers, California, USA. Environmental Biology of Fishes 79:269-279.

Burner, C.J. 1951. Characteristics of spawning nests of Columbia River salmon. U. S. Fish and Wildlife Service Fishery Bulletin 61:97-110.

Carter, K., and S. Kirk. 2008. Fish and fishery resources of the Klamath River Basin. North Coast Regional Water Quality Control Board, Oregon Department of Environmental Quality. 32p.

Cavender, T. and R. Miller. 1972. *Smilodonichthys rastrotus*: A new Pliocene salmonid fish from western United States. University of Oregon, Museum of Natural History, Bulletin 18, Eugene.

Clemens, B., L. Weitkamp, K. Siwicke, J. Wade, J. Harris, J. Hess, L. Porter, K. Parker, T. Sutton, and A. Orlov. 2019. Marine biology of the pacific lamprey *Entosphenus tridentatus*. Reviews in Fish Biology and Fisheries 29:767-788.

Clemento, A.J., E.D. Crandall, J.C. Garza, and E.C. Anderson. 2014. Evaluation of a single nucleotide polymorphism baseline for genetic stock identification of Chinook Salmon (*Oncorhynchus tshawytscha*) in the California Current Large Marine Ecosystem. Fishery Bulletin 112:112-130.

Dawson, H., B. Quintella, P. Almeida, A. Treble, and C. Jolley. 2015. The ecology of larval and metamorphosing lampreys. *In*: Docker MF, ed. Lampreys: biology, conservation, and control. Springer, Dordrecht, pp. 75-137.

Dicken, S.N. 1980. Pluvial Lake Modoc, Klamath County, Oregon, and Modoc and Siskiyou Counties, California. Oregon Geology 42:179-187.

Erickson, D. and M. Webb. 2007. Spawning periodicity, spawning migration, and size at maturity of green sturgeon, *Acipenser medirostris*, in the Rogue River, Oregon. Environmental Biology of Fishes 79:255-268.

Eschmeyer, W.N. 1998. Catalog of fishes. California Academy of Sciences, San Francisco.

Farr, R., M. Hughes, and T. Rien. 2002. Green sturgeon population characteristics in Oregon. Annual Progress Report. Sport Fish Restoration Project F-178-R. 27 p.

Gerstung, E.R. 1997. Status of coastal cutthroat trout in California. Pp. 43-56 *In* J.D. Hall, P.A. Bisson, and R.E. Gresswell, eds. Sea-run cutthroat trout: biology, management, and future conservation. Corvallis, OR: Oregon Chapter, American Fisheries Society.

Gobalet, K., P. Schulz, T. Wake, and N. Siefkin. 2004. Archaeological perspectives on native American fisheries of California, with emphasis on steelhead and salmon. Trans. Am. Fish. Soc. 133:801-833.

Gustafson, R., M. Ford, D. Teel, and J. Drake. 2010. Status review of eulachon (*Thaleichthys pacificus*) in Washington, Oregon, and California. NOAA Tech. Memo. NMFS-NWFSC-105.

Hamilton, J.B. et al. 2016. The persistence and characteristics of Chinook salmon migrations to the upper Klamath river prior to exclusion by dams. Oregon Historical Quarterly. 117(3):326-377.

Hershey, O.H. 1900. Ancient alpine glaciers of the Sierra Costa Mountains in California. Journal of Geology 8:42-57.

Hess, J., J. Zendt, A. Matala, and S. Narum. 2016. Genetic basis of adult migration timing in anadromous steelhead discovered through multivariate association testing. Proc. R. Soc. B Biol. Sci. 283(1830):20153064.

Hohler, D. 1981. A dwarfed population of *Catostomus rimiculus* (Catostomidae: Pisces) in Jenny Creek, Jackson, County, Oregon. MS Thesis, Oregon State University.

Ibarra, D., A. Egger, K. Weaver, C. Harris, and K. Maher. 2014. Rise and fall of late Pleistocene pluvial lakes in response to reduced evaporation and precipitation: evidence from Lake Surprise, California. Geol. Soc. Amer. Bull. 126:1387-1415.

Kinziger, A.P., J.L. White, R. Nakamoto, and B.C. Harvey. 2021. Recent, small beginnings: genetic analysis suggests *Catostomus rimiculus* (Klamath smallscale sucker) in the Smith River, California, are introduced. Journal of Fish Biology 98:1321-1328.

Lang, N.J. et al. 2009. Novel relationships among lampreys (Petromyzontiformes) revealed by a taxonomically comprehensive molecular data set. *In* Biology, Management, and Conservation of Lampreys in North America (Brown, L., S. Chase, M. Mesa, R. Beamish, and P. Moyle, eds), pp. 41 – 55. American Fisheries Society Symposium 72.

Larson, Z., and M. Belchik. 1998. A preliminary status review of eulachon and Pacific lamprey in the Klamath River Basin. Yurok Tribal Fisheries Program, Klamath, CA.

Love, M. 1996. Probably more than you want to know about the fishes of the Pacific Coast. Really Big Press, Santa Barbara.

Lundberg, J., M. Kottelat, G. Smith, M. Stiassny, and A. Gill. 2000. So many fishes, so little time: An overview of recent ichthyological discovery in continental waters. Annals of the Missouri Botanical Garden 87:1. pp. 26-62.

Markle, D.F. 2016. A guide to freshwater fishes of Oregon. Oregon State University Press, Corvallis, Oregon.

Markle, D.F. 2019. Drainage evolution and freshwater fish zoogeography in coastal Oregon and Washington. Northwestern Naturalist 100: 71-89.

McGinnis, S.M. 2006. Field guide to freshwater fishes of California. University of California Press, Berkeley.

Moyle, P., P. Foley, and R. Yoshiyama. 1994. Status and biology of the green sturgeon, *Acipenser medirostris*. Sturgeon Quarterly 2:7.

Moyle, P.B. 2002. Inland fishes of California. University of California Press, Berkeley.

Nelson, S., T. Grande, and M. Wilson. 2016. Fishes of the world. Fifth edition. John Wiley and Sons, New Jersey.

Orme, A. 2008. Pleistocene pluvial lakes of the American West: a short history of research. In: R. Grapes, D. Oldroyd, and A. Grigelis, eds. History of geomorphology and Quaternary geology. Geological Society, London, Special Publications 301:51-78.

Parker, K., J. Hess, S. Narum, and A. Kinziger. 2019. Evidence for the genetic basis and epistatic interactions underlying ocean- and river-maturing ecotypes of Pacific Lamprey (*Entosphenus tridentatus*) returning to the Klamath River, California. Molecular Ecology 28:3171-3185.

Pease, R.W. 1965. Modoc County, a geographic time continuum on the California volcanic tableland-pt. I. Geography of the pre-historic past: University of California Publications in Geography 17:7-45.

Quinn, T.P. 2018. The behavior and ecology of Pacific salmon and trout. Second edition. University of Washington Press, Seattle.

Rien, T. and R. Beamesderfer. 1994. Accuracy and precision of white sturgeon age estimates from pectoral fin rays. Trans. Am. Fish. Soc.123:255-265.

Rizza, S. 2015. Asymmetric Introgression between Coastal Cutthroat Trout and Steelhead: Variable Introgression by Linkage Group. Thesis, Cal Poly Humboldt, Arcata, CA.

Robins, C. and R. Miller 1957. Classification, variation, and distribution of the sculpins, genus *cottus*, inhabiting Pacific slope waters of California and southern Oregon, with a key to the species. California Department of Fish and Game 43:213-233.

Schindler, D., R. Hilborn, B. Chasco, C. Boatright, T. Quinn, L. Rodgers, and M. Webster. 2010. Population diversity and the portfolio effect in an exploited species. Nature 465:609-612.

Smith, G., N. Morgan, and E. Gustafson. 2000. Fishes of the Mio-Pliocene Ringold Formation, Washington: Pliocene capture of the Snake River by the Columbia River. University of Michigan Papers on Paleontology No. 32.

Snyder, J.O. 1925. The half-pounder of Eel River, a steelhead trout. California Fish and Game 11:49-55.

Stearley, R. and G. Smith. 2016. Salmonid fishes from Mio-Pliocene lake sediments in the Western Snake River Plain and the Great Basin. *In*: W. Fink and N. Carpenter, eds. Fishes of the Mio-Pliocene western Snake River Plain and vicinity. Misc. Pub. Museum Zoology, University of Michigan 204:1-45.

Taylor, D.W. 1960. Distribution of the freshwater clam, *Pisidium ultramontanum*; a zoogeographic inquiry. American Journal of Science. 258A:325-334.

Thompson, N., E. Anderson, A. Clemento, M. Campbell, D. Pearse, J. Hearsey, A. Kinziger, and J.C. Garza. 2020. A complex phenotype in salmon controlled by a simple change in migratory timing. Science 80:609-613.

Trotter, P., et al. 2018. A special workshop on the taxonomy and evolutionary biology of cutthroat trout. *In*: P. Trotter, P. Bisson, L. Schultz, and B. Roper, eds. Cutthroat trout: evolutionary biology and taxonomy. American Fisheries Society, Bethesda, MD.

U.S. Department of the Interior. 2000. Record of decision, Trinity River mainstem fishery restoration final environmental impact statement/environmental impact report. U.S. Department of Interior, Washington D.C.

Walkley, J. and J. Garwood. 2017. 2011-2016 Salmonid Redd Abundance and Juvenile Salmonid Spatial Structure in the Smith River Basin, California and Oregon." Final Report to the California Department of Fish and Wildlife, FRGP Grant: P1210524. 88pp.

Welsh Jr., H., K. Pope, and D. Boiano. 2006. Sub-alpine amphibian distributions related to species palatability to non-native salmonids in the Klamath Mountains of northern California. Diversity and Distributions 12:298-309.

Wiesenfeld, J., D. Goodman, and A. Kinziger. 2018. Riverscape genetics identifies speckled dace (*Rhinichthys osculus*) cryptic diversity in the Klamath-Trinity basin. Conservation Genetics 19:111-127.

Amphibians of the Klamath Mountains

Justin Garwood

Introduction

Foothill yellow-legged frog (Rana boylii) by Spencer Riffle, who took all photos in this chapter unless otherwise noted.

The roots *amphi* (double) and *bios* (life) refer to the complex lifecycle of most amphibians, living at first an aquatic life and then metamorphosing into a terrestrial form. Like reptiles, amphibians are ectotherms and do not generate metabolic heat to regulate their body temperatures as do endotherms, such as birds and mammals. Instead, amphibians either stay close to favorable air, water, and substrate temperatures, or respond with behavioral and physiological adaptations to buffer against exposures to physiologically stressful (hot, cold, or dry) conditions. All amphibians have glandular skin, whether smooth or roughened that is semi-permeable to air and water, allowing for cutaneous respiration (skin-breathing) and maintaining water balance. Because their skin functions as an essential gas exchange and water balance organ, all amphibians require moist environments.

Amphibians are an ancient group, and were the first vertebrates to conquer land, evolving from fishes during the Devonian (~380 million years ago [Ma]) and dominating terrestrial and freshwater systems for 60 million years. The three modern lineages arose in the late Permian (~250 Ma), and include salamanders (*Caudata*), frogs and toads (*Anura*), and caecilians (*Gymnophiona*). Over 8,400 amphibian species have been described, of which 88 percent are frogs and toads, 9 percent are salamanders, and 3 percent are caecilians; the latter is a poorly known group of wormlike species that live primarily underground. The moist tropical forests of the Amazon basin in South America dominate global frog and caecilian biodiversity whereas salamanders are mostly distributed throughout the Northern Hemisphere with the temperate Appalachian Mountains of the southeastern United States holding the greatest global concentration of salamander species.

While the Pacific Northwest has fewer amphibians than the southeastern United States, it contains some unique lineages and a variety of regionally endemic species, especially salamanders. Of the 49 native species (34 salamanders and 15 frogs), 32 (65 percent) only occur in the Pacific Northwest, most of which are salamanders (88 percent). This diversity is explained, in part, by a rich biogeographical history shaped by geology, tectonic processes, and shifting climate regimes. Many regions acted as refuges during periods of past glaciation, including much of the Klamath. The Klamath Mountains hold the greatest amphibian biodiversity for any region of its size across the Pacific Northwest including at least 26 native species from 9 families (Table 1) and one invasive species, the American bullfrog.

This chapter touches on the exceptional story of Klamath Mountain amphibian diversity through historical biogeography, species distributions, modern genetic analysis, and the broad suite of habitats that make this region a world-renowned amphibian hotspot.

Table 1. Amphibians of the Klamath Mountains Geological Province (KMP) and their biogeographic associations. Most associations adopted from Bury and Pearl (1999). See species range maps for detailed distributions.

Common Name	Scientific Name	HABITATS	BIOGEOGRAPHY
FROGS AND TOADS (ANURA)			
True Toads	**Bufonidae**		
Western Toad	*Anaxyrus boreas*	RV, LP, ST	Northern
Treefrogs	**Hylidae**		
Pacific Treefrog	*Pseudacris regilla*	LP, LT, RV	Widespread
Tailed Frogs	**Ascaphidae**		
Coastal Tailed Frog	*Ascaphus truei*	SE, ST, RI	Northern
True Frogs	**Ranidae**		
Northern Red-legged Frog	*Rana aurora*	LP, LT, RI, ST	Northern
Foothill Yellow-legged Frog	*Rana boylii*	RV, ST, RI, SE	Southern
Cascades Frog	*Rana cascadae*	LP, LT, ST, SE	Northern
American Bullfrog	*Rana catesbeiana*	LP, RV, ST	Introduced
SALAMANDERS (CAUDATA)			
Mole Salamanders	**Ambystomatidae**		
Northwestern Salamander	*Ambystoma gracile*	LP, ST, RI	Northern
Long-toed Salamander	*Ambystoma macrodactylum*	LP, TA, UG	Widespread
Giant Salamanders	**Dicamptodontidae**		
Coastal Giant Salamander	*Dicamptodon tenebrosus*	ST, RI, LP, SE, RV	Northern
Lungless Salamanders	**Plethodontidae**		
Clouded Salamander	*Aneides ferreus*	DW, AR, TA	Northern
Wandering Salamander	*Aneides vagrans*	DW, AR, TA	Regional
Klamath Black Salamander	*Aneides klamathensis*	TA, UG, DW	Near-endemic
Shasta Black Salamander	*Aneides iecanus*	TA, UG, DW, RI	Near-endemic
California Slender Salamander	*Batrachoseps attenuatus*	DW, LL, TA	Southern
Oregon Ensatina	*Ensatina eschscholtzii oregonensis*	DW, LL, TA	Widespread
Painted Ensatina	*Ensatina eschscholtzii picta*	DW, LL, TA	Regional
Shasta Salamander	*Hydromantes shastae*	UG, DW, LL, TA	Endemic
Samwel Shasta Salamander	*Hydromantes samweli*	UG, DW, LL, TA	Endemic
Wintu Shasta Salamander	*Hydromantes wintu*	UG, DW, LL, TA	Endemic
Scott Bar Salamander	*Plethodon asupak*	TA, DW	Endemic
Dunn's Salamander	*Plethodon dunni*	RI, TA, DW, LL	Northern
Del Norte Salamander	*Plethodon elongatus*	TA, DW, RI, LL	Near-endemic
Siskiyou Mountains Salamander	*Plethodon stormi*	TA, DW	Endemic
Western Red-backed Salamander	*Plethodon vehiculum*	TA, LL, DW, RI	Northern
Torrent Salamanders	**Rhyacotritonidae**		
Southern Torrent Salamander	*Rhyacotriton variegatus*	SE, ST, RI	Northern
Newts	**Salamandridae**		
Rough-skinned Newt	*Taricha granulosa*	LP, ST, RV	Northern
Sierra Newt	*Taricha sierra*	ST, RV, LP	Southern

HABITATS—Amphibians use a variety of habitats. Primary habitats for each species are listed in order of importance. Aquatic: Seeps (SE), Streams (ST), Rivers (RV), Lentic Permanent (LP), Lentic Temporary (LT). Terrestrial Rock: Talus and Loose Rock (TA), Underground including Burrows, Caves, and Cracks (UG), Terrestrial Forest: Downed Wood (DW), Leaf Litter (LL), Moist Riparian (RI), Arboreal (AR).

BIOGEOGRAPHY—Rangewide distribution relative to Klamath Mountain Province (KMP). Endemic= only occurs in the KMP; Near-endemic= Majority of range in KMP; Regional= Portion of range in KMP and regionally endemic; Northern= Most of range north of KMP; Southern= Most of range south of KMP; Widespread= Range broadly spans north and south of KMP; Introduced= Widespread invasive species.

Salamanders (Caudata)

The order *Caudata* means "with tail." Salamanders are tailed amphibians with elongate bodies that appear lizard-like. In fact, many salamanders even have the same ability as lizards to "shed" their tails (called caudal autotomy) when grasped by a predator. However, this overall resemblance to lizards is merely superficial. Besides being only distantly related to lizards and other reptiles, salamanders lack scales, claws, and external ear openings (canals) to name a few major differences.

Klamath salamanders vary greatly in body form and size. The California slender salamander is miniature and elongated due to its thin trunk and vestigial limbs, and is among the smallest salamanders in North America, while the wide-bodied coastal giant salamander is one of the largest salamanders in the world, approaching a foot in length and supported by large muscular limbs. Aquatic salamanders in lake and pond (lentic) habitats have long tails and pronounced dorsal fins extending up their backs to aid in swimming. Stream dwelling (lotic) salamanders have much shorter dorsal fins with vertically compressed bodies for living in high-velocity water. Most terrestrial salamanders in the Klamath Mountains are thin resulting in a large body surface area as they obtain most of their oxygen through skin-breathing. Terrestrial salamanders also have a variety of limb and toe shapes to assist in climbing. The Klamath Mountains have at least 20 salamander species from five families including Ambystomatidae (mole salamanders), Dicamptodontidae (giant salamanders), Plethodontidae (lungless salamanders), Rhyacotritonidae (torrent salamanders), and Salamandridae (newts) (Table 1). This diversity grows to 21 if the painted ensatina, a small bodied form with highly variable color patterns from the Smith-Chetco river region, is included.

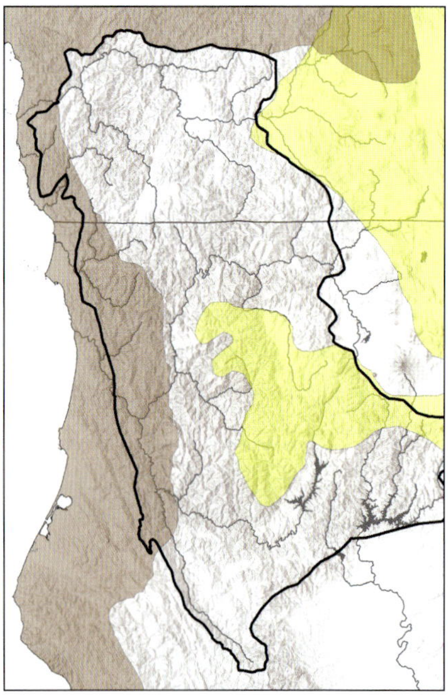

Amphibians

Shasta black salamander
Aneides iecanus

wandering salamander
Aneides vagrans

Klamath black salamander
Aneides klamathensis

clouded salamander
Aneides ferreus

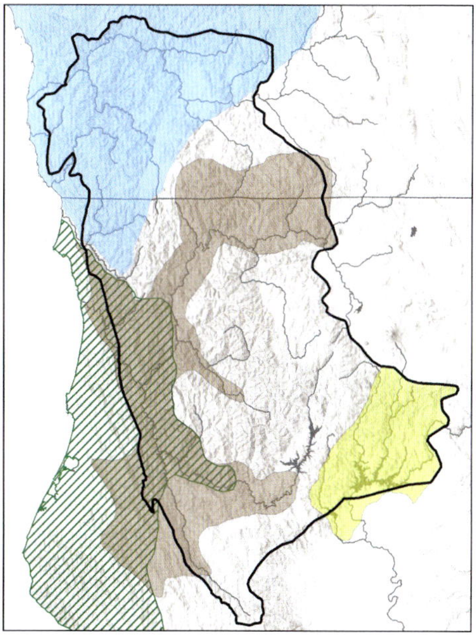

Samwel Shasta salamander
Hydromantes samweli

Marisa Ishimatsu

Wintu Shasta salamander
Hydromantes wintu

Shasta salamander
Hydromantes shasta

ensatina
Ensatina eschscholtzii

California slender salamander
Batrachoseps attenuatus

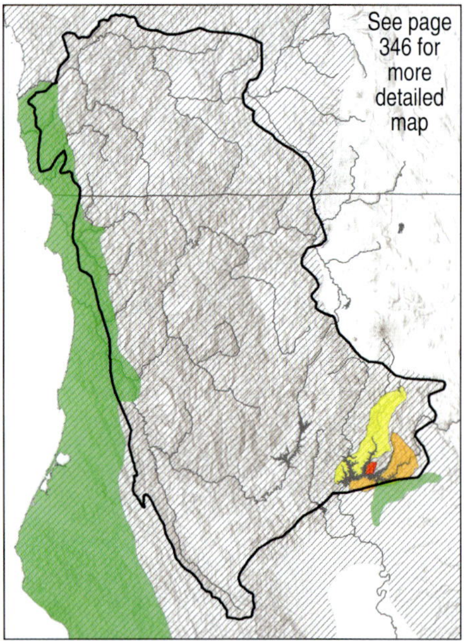

Del Norte salamander
Plethodon elongatus

Siskiyou Mountains salamander
Plethodon stormi

Western red-backed salamander
Plethodon vehiculum

Scott Bar salamander
Plethodon asupak

Dunn's salamander
Plethodon dunni

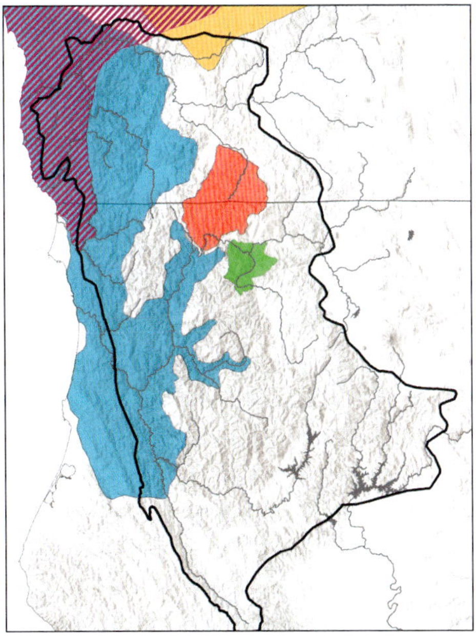

southern torrent salamander
Rhyacotriton variegatus

rough-skinned newt
Taricha granulosa

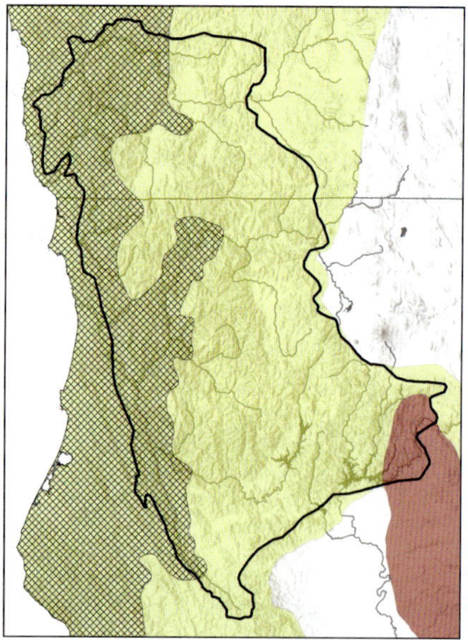

Sierra newt
Taricha sierra

Frogs and Toads (Anura)

The order Anura means "without tail", because unlike salamanders, anurans lack tails, except during the larval stage. Anurans have compact bodies with a shortened and stout vertebral column compared to salamanders. Most anurans also have long muscularized hind limbs with webbed clawless toes and shorter front limbs with unwebbed toes. These anatomical features assist in jumping, climbing, and swimming. With large eyes and large mouths, anurans are insatiable predators of terrestrial and aquatic insects and spiders. Anurans occur on every continent except Antarctica.

The Klamath Mountains contain six species from four families including Ascaphidae (tailed frogs), Hylidae (treefrogs), Ranidae (true frogs), and Bufonidae (true toads) (Table 1). Three of the species—including the Pacific treefrog, western toad, and coastal tailed frog, are widespread across all elevations in isolated streams and wetlands throughout the Klamath Mountains. The three ranid frogs have little overlap in distribution and habitats. The foothill yellow-legged frog is largely restricted to streams and rivers below 4,000 ft (1,220 m) but has been observed higher in large subalpine basins. The Cascades frog prefers montane and subalpine habitats above 4,000 ft in the Klamath and upper Sacramento River basins where it breeds in glacial formed lakes and ponds. The northern red-legged frog is found only in ponds and associated streams along the western edge of the range but is more widespread in the lower mountains and coastal plains to the west. The voracious American bullfrog is the only introduced amphibian in the Klamath Mountains and is among the greatest threats to regional wetland ecosystems. They are most abundant in lower elevation reservoirs and ponds, as well as slow-water habitats in rivers and streams..

Pacific treefrog (*Pseudacris regilla*) have highly variable dorsal patterns and colors.

western toad
Anaxyrus boreas

coastal tailed frog
Ascaphus truei

Cascades frog
Rana cascadae
Photo by J. Garwood

foothill yellow-legged frog
Rana boylii

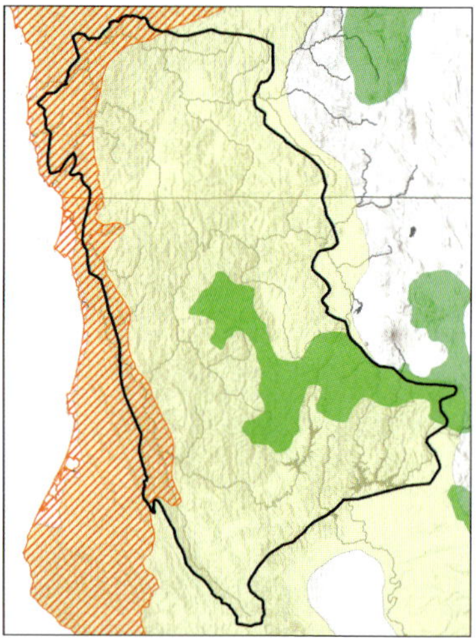

Northern red-legged frog
Rana aurora

Amphibians

Ecology

Due to their small size, secretive behaviors, nocturnal activities, and often inconspicuous habitats, amphibians are among the most overlooked land vertebrates. And yet in some places amphibians are the dominant vertebrate group by abundance and biomass. Indeed, amphibians are often the most abundant vertebrates in Klamath ecosystems leading to an underestimate of their ecological roles and importance in forest and aquatic ecosystems. For example, predatory coastal giant salamanders can be so abundant in streams their biomass often exceeds all other aquatic vertebrates combined—including salmonid fishes. Terrestrial salamanders are key predators of small soil invertebrates which influences nutrient cycling, soil building, and carbon storage. In turn, amphibians are a common prey for larger vertebrate carnivores including fishes, reptiles, birds, and mammals.

Because they occur in both aquatic and terrestrial habitats, amphibians are considered key links in the flow of nutrients from aquatic to terrestrial food webs. Aquatic breeding amphibians typically have high fecundities, with breeding sites sometimes containing thousands of larvae. These locations can support a large predator assemblage often including unsuspected species, especially as larvae grow larger and water surfaces dry down making them vulnerable and easy prey. One fall afternoon I recorded four common garter snakes, over 40 American robins (*Turdus migratorius*), two American dippers (*Cinclus mexicanus*), two Clark's nutcrackers (*Nucifraga columbiana*), and even chipmunks (Sciuridae) preying on tadpoles and recently metamorphosed Cascades frogs and Pacific treefrogs at a drying pond in the Trinity Alps. Although amphibians are often an abundant prey for other animals, many in the Klamath region can live upwards of 20 years—especially those occurring in higher elevations with shorter summer activity periods.

Habitats

The Klamath Mountains contain a rich assortment of natural wetlands used by aquatic breeding amphibians, along with cool moist forest and rocky interstitial microclimates used by fully terrestrial amphibians. In Mediterranean-like climates such as that in the Klamath, the environmental conditions of most amphibian habitats change dramatically across seasons from cold, wet winter to dry, hot summers. In response, many species move between and within habitat types based on life-history needs, including reproduction, foraging, torpor, or dispersal. Further, amphibian habitats are often isolated and ephemeral—especially as summer brings drier conditions forcing amphibians to retreat to stable moist habitats.

Seeps, Streams, and Rivers—Lotic Habitats

Thirteen species of Klamath Mountain amphibians (7 salamanders, 6 anurans) are associated with lotic habitats extending from sources at cold headwater seeps down to the broad reaches of warmer alluvial rivers. Besides providing water, lotic environments also maintain cool and moist riparian zones that buffer stream environments from the much drier and warmer terrestrial landscape—especially in smaller streams and seeps. Southern torrent salamanders frequent seep habitats on steep slopes where cool spring water emerges from the Earth. Further downstream, cascades, riffles, and pools form small streams. Here, coastal tailed frogs

Coastal Tailed Frog: A Living Fossil Hiding on Ancient Rock

Male coastal tailed frog, *Ascaphus truei.* Photo by Justin Garwood.

Perhaps the most spectacular amphibian that calls the Klamath Mountains home is the coastal tailed frog. Though it is widespread across the Pacific Northwest, it is rarely seen because of its nocturnal habits, affinity for cold high-gradient streams, and patchy distributions. The coastal tailed frog is one of only two living species in the ancient family *Ascaphidae*. The coastal tailed frog and the Rocky Mountain tailed frog (*A. montanus*) both live in the Pacific Northwest and likely diverged into separate species with the uplift of the Cascade Range between 28 and 10 million years ago in the late Miocene. Habitats diverged with this uplift and the creation of a rain shadow dispersal barrier (Nielson et al. 2006).

Tailed frogs represent the basal lineage of all 7,400+ living frogs and toads (anurans) on the planet. The closest living frog relatives are three species of *Leiopelmatidae* which are endemic to New Zealand. The term 'close' is relative in both space and time as it appears ancestors of the two families diverged around 180 million years ago during the Jurassic Period when the supercontinent *Pangaea* was splitting apart (Roelants and Bussuyt 2005). Unlike modern anurans, tailed frogs lack eardrums and do not vocalize for communication—perhaps unnecessary features in a loud, cascading stream environment. Evidence suggests

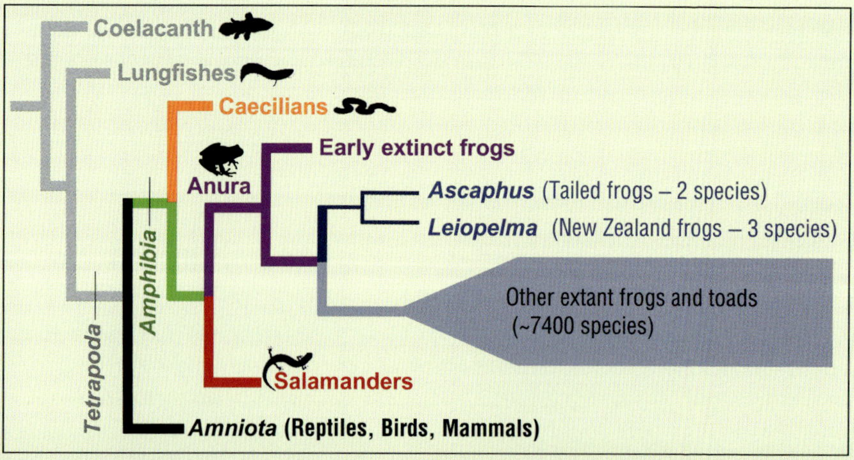

Cladogram of *Ascaphus* as they relate to other amphibians.

Amphibians

that males and females find each other by chemical cues instead (Asay et al. 2005). Tailed frogs have retained free ribs, a trait shared with fossils of extinct anurans but absent in modern species. Vertebrae loss has been a clear pattern throughout anuran evolution and tailed frogs have more vertebrae than all modern species. Unlike modern anurans, tailed frogs have also retained tail-wagging muscles (Ritland 1955), a likely holdover from ancestors prior to anuran and salamander divergence.

The coastal tailed frog offers a rare living portal into ancient anuran evolution. Yet, notwithstanding their ancient origin, the specialized adaptations tailed frogs have developed for life in streams have their own evolutionary merit. Because tailed frogs live in isolated cold-water streams, they have much slower growth rates, longer generation times, and lower fecundity relative to most modern anurans in warmer habitats. Depending on the inhabited elevation larval periods can last to 4 years, age to sexual maturity can exceed 7 years, and longevity can approach 20 years. Tailed frogs have reduced lungs that decrease buoyancy while crawling on stream bottoms and hardened fingertips for gripping and climbing rocks. Male frogs are reproductively unique among living frogs. The copulatory organ, (penis) or "tail", is an extension of the cloaca and used for internal fertilization while mating in flowing water. Females typically mate in the fall and can store sperm for months through the winter, until stream conditions are suitable to oviposit their sticky eggs under large submerged rocks the following spring. Emerging tadpoles have streamlined muscular bodies and sucker-like mouths used for gripping rocks in fast water, climbing waterfall rock faces, and grazing on algae. Collectively, these specialized adaptations are absent in other North American frogs, showing this ancient species is far from an evolutionary dead end.

Sucker-like mouths are used for gripping rocks in fast water, climbing waterfall rock faces, and grazing on algae. Photos by Spencer Riffle.

and coastal giant salamanders occur in cool water with rocky substrates. Coastal giant salamanders can be found much further down the stream network, but eventually become limited by warmer water temperatures in rivers. In contrast to cold-water specialists, the foothill yellow-legged frogs and western toads prefer larger open streams and rivers. Many other widespread species use lotic habitats facultatively (Table 1), such as western toads and Pacific treefrogs that breed along shallow river margins and backwaters. Some fully terrestrial salamanders can also be found in moist riparian zones, especially the Dunn's salamander and Klamath black salamander.

Lakes and Ponds—Lentic Habitats

Lentic habitats are largely restricted to glacial-formed subalpine lake districts and western Klamath mid-elevation lakes (see Water chapter). Ten species of amphibians (5 salamanders and 5 anurans) use lentic habitats in the Klamath Mountains. Most lentic amphibians in the region prefer sites with shallow vegetated shorelines that are used for egg deposition, foraging, and cover. Deeper areas (or deeper sites nearby) are used for overwintering when wetlands are covered in ice. The amphibian community at a given lake or pond is largely based on its hydroperiod, elevation, and if other supporting wetlands are nearby. Some temporary ponds only last a month or two before drying, while deep glacial-formed lakes persist for thousands of years.

Other wetlands occur in between—drying only in drought years or once every decade or two in severe droughts. High-elevation waters are colder and have relatively short ice-free seasons compared to lower waters in milder climates and that slows amphibian development and limits activity periods. For example, the larvae of long-toed and northwestern salamanders usually overwinter in perennial lentic habitats to survive before transforming into adults the following year. Others, such as the Cascades frog and Pacific treefrog do not overwinter as larvae and metamorphose by the fall. These frogs will also breed in temporary ponds causing boom-and-bust recruitment cycles dictated by individual water years.

Terrestrial Forest Amphibian Habitats

Fourteen species of salamanders in the Klamath Mountains from the family Plethodontidae are fully terrestrial. Because these terrestrial salamanders require moisture to breathe through their skin, surface activity periods are strongly associated with moist and humid conditions generally occurring between November and April. Different regions of the Klamath Mountains have unique precipitation and moisture regimes, further influencing the duration of surface activity for salamanders. All terrestrial salamanders in the Klamath Mountains require moist thermally-buffered habitats in which to retreat for depositing eggs and when surface conditions are too dry, warm, or cold. Thus, the availability of such microhabitats largely defines the distribution of land salamanders across multiple spatial scales.

Retreat habitats come in three general forms: rocky substrates with interstitial cavities, downed and decaying wood, and the organic duff layer on forest floors. All these habitats maintain moisture and can remain at remarkably stable temperatures throughout the year. Talus is the most widespread rocky habitat used by terrestrial salamanders in the Klamath, providing interstitial cavities that penetrate deep underground. Salamanders respond to changing environmental conditions in talus slopes by migrating vertically within the rock layers to maintain their ideal climactic envelope as conditions change. The Shasta County region of the Klamath Mountains has large limestone formations with deep cracks and caves that are used by three species of cave salamanders (genus *Hydromantes*). Heavily forested regions can provide decaying wood and deep forest litter. A single large decaying Douglas-fir log can provide substantial salamander habitat for decades while also hosting salamander prey from diverse invertebrate communities. Lastly, clouded salamanders and wandering salamanders possess substantial climbing abilities and often take up residence high in tree canopies by burrowing into cavities and shelves with mosses and epiphytes—some have been found well over 100 ft (>30 m) in tree canopies!

Reproduction

Amphibian reproductive biology is the most diverse among vertebrates and is on exemplary display across the 26 native Klamath species. Twelve (46 percent) species, including six anurans and six salamanders, have aquatic egg and larval stages which are the most common and less highly derived form of amphibian reproduction. The Klamath Mountains offer a broad range of aquatic habitats and almost all are used for reproduction. Cold high-gradient seeps and streams are used exclusively by coastal tailed frogs and southern torrent salamanders, while foothill yellow-legged frogs breed exclusively in open canopy streams with cobbles and boulders. Coastal giant salamanders breed in creeks and streams with high or low gradients. The Cascades frog, northern red-legged frog, and northwestern salamander deposit their eggs in ponds and lakes and rarely use stream backwaters. The newts (Salamandridae), Pacific treefrog, and western toad are generalists that reproduce in a wide range of streams, lakes, and other wetlands.

PLATE A. Reproductive cycle of an aquatic breeding amphibian. The Cascades frog (*Rana cascadae*) occurs in rugged high elevations of the eastern Klamath Mountains in California. Breeding occurs as soon as ponds and lakes become free of snow and ice, signaling males to start calling and defending their positions as females arrive to the breeding site. The male grips the female (as shown above), known as amplexus, and fertilizes eggs as they are spawned. Eggs hatch weeks later into gilled larvae that forage on algae and grow throughout the summer. Metamorphosis into juvenile frogs occurs in the fall before the onset of winter dormancy. Cascades frogs hibernate up to seven months each year in ponds, lakes, and springs, under deep snow and ice. Males require at least three years to maturity while females require at least four years. Illustration by Michael Zontos

Antipredator Chemical Defense in Klamath Amphibians

The skin of all amphibians is covered in multi-purpose dermal glands that keep the body surface moist for respiration while also delivering antimicrobial peptides to combat pathenogenic bacteria, fungi, and other potentially hazardous microbes. Skin secretions also allow for the rapid delivery of various irritating, distasteful, and even lethal toxins from specialized poison glands. If you have ever taken a swim in a lake in the Marbles or Siskiyous, you have likely marveled at the abundance of rough-skinned newts. Despite their small size, these air gulping swimmers seem to accept your presence and often even prod you with curiosity.

Tetrodotoxin (TTX) is thought to be created with the help of special bacteria and is among the most lethal animal poisons on Earth (roughly 1,000 times deadlier than cyanide!). This neurotoxin is found in both rough-skinned and Sierra newts that forms a potent chemical defense. When ingested by a predator, TTX blocks the flow of sodium into nerve and muscle cells by plugging sodium channels. Without sodium, nerve and muscle cells fail to generate electrical signals for nerve impulses and muscular contractions, causing paralysis and even death by asphyxiation. TTX is concentrated in the skin glands of newts but is also found in ovaries and eggs, making it a more complex defensive strategy for reproducing females and protecting subsequent early life stages. One odd thing about Pacific newts is why some populations are so deadly poisonous (where a single newt contains enough TTX to kill 30–50 people!) while others are barely toxic at all. It turns out that some predatory garter snakes (*Thamnophis*) can eat newts and have co-evolved various levels of resistance to TTX. This unique predator-prey relationship has been well studied and considered a classic example of a 'co-evolutionary arms race' (*see* Brodie and Brodie 1999)—the likely reason for many newt populations having extreme concentrations of TTX beyond those necessary to defend themselves against most predators. Nearly all other animals are vulnerable to TTX, including humans. The same compound is found in, and named after, the fish family Tetradontidae which includes many species of puffer fishes including the dangerous-delicacy prepared by master chefs in Japan called fugu.

Ensatina eschscholtzii shown in defensive posture while exuding toxic white secretions from glands on its head and tail. Photo by Spencer Riffle.

As with many animals armed with strong chemical defense compounds, rough-skinned and Sierra newts have bright color patterns that serve as a warning to potential predators. When threatened or attacked, newts

Rough-skinned newt in the 'Unken reflex' defensive posture showing bright orange undersides while exuding the powerful neurotoxin TTX. Photo by Justin Garwood.

will excrete TTX from skin glands and often display their bright orange undersides by arching their back and curling their head and tail backward in an extreme defensive posture known as 'Unken reflex'. Similar to the newts found in the Klamath Mountains, the color of the co-occurring ensatina salamander is a mix of dark contrasted with brighter orange patterns. In what may be a stunt of deceptive Batesian mimicry, ensatinas also present a defensive posture to predators by raising their body and tail. Although ensatinas lack potent TTX, they can secrete a distasteful sticky white substance that is far less toxic to most predators but acts as a sticky glue that can immobilize the jaws of some predators.

Many other species of amphibians have distinct glands and warts concentrating chemical defenses to certain regions of the body. Two Klamath Mountain species who have large parotid glands on their heads include the western toad and the northwestern salamander. Both assume defensive postures by lifting their hind legs and lowering their heads while secreting toxic substances from their parotid glands. Western toad tadpoles are also distasteful and avoided by predators such as salmonid fishes in streams and lakes.

All anurans except the coastal tailed frog (see sidebar) have external egg fertilization. On the other hand, all the region's salamanders have internal fertilization. In aquatic breeding salamanders, males produce sperm packets (spermatophores) that are picked up by the female's vent and stored until environmental conditions are favorable for fertilization and egg deposition. Eggs are then deposited in water and hatch into free-swimming gill-breathing larvae that later metamorphose into juveniles (PLATES A and B). Anuran larvae largely feed on algae (periphyton) while salamander larvae are predators primarily on aquatic invertebrates.

Amphibian metamorphosis involves many physiological and morphological changes with gill and fin resorption being the most notable. However, not all species strictly follow this model. Three Klamath species can retain gills, fins, and an aquatic life as sexually mature adults including the coastal giant salamander, northwestern salamander, and rough-skinned newt (PLATE B)—though it's extremely rare in newts. This adaptive strategy, called paedomorphism, likely evolved as a genetic and physiological mechanism for populations to adapt to local environmental conditions. Adults living in dryer inland habitats, or in subalpine elevations, remain aquatic paedomorphs, whereas adults in mesic coastal landscapes of the western Klamath Mountains have examples of both aquatic and terrestrial forms. Paedomorphic coastal giant salamanders are widespread and under appreciated in high-elevation lakes of the eastern Klamath Mountains where they serve as formidable predators in naturally fishless lakes.

Aquatic breeding amphibians have well-defined breeding seasons closely tied to precipitation accumulation, air temperature, and landscape factors such as elevation and wetland hydrology. Regional aquatic amphibians tend to have short breeding seasons during late winter to early summer depending on elevations and snowpack. One exception is the prolonged breeding season of the Pacific treefrog in lower elevations. All male anurans, except the coastal tailed frog, use vocalization

PLATE B. Reproductive cycle of aquatic breeding amphibian with a paedomorphic gilled adult (guarding eggs) and terrestrial adult lacking gills (on stream edge). The coastal giant salamander (*Dicamptodon tenebrosus*) occurs in streams and lakes throughout the Klamath Mountains often being the most abundant vertebrate predator in small streams. Males produce sperm capsules (spermatophores) that females pick up and store in their cloaca until environmental conditions are optimal for egg deposition. In streams, females oviposit their eggs during spring in interstitial nest chambers, by hanging them singly from a rock or log, and then guard the chamber until hatching. Multiple year classes of predatory gilled larvae co-occur, taking one to three years to reach sexual maturity. Depending on environmental and genetic cues, the coastal giant salamander can either remain as an aquatic adult with gills (paedomorph) or it can permanently transform into a terrestrial adult. These physiological changes are controlled by the thyroid gland, allowing greater flexibility in life history between aquatic and terrestrial habitats. Illustration by Michael Zontos (calls) to entice females to the breeding site and to synchronize breeding activity. The intense commotion and aggressive posturing by males at an active breeding site is easy to observe.

Male foothill yellow-legged frogs, Cascades frogs, and northern red-legged frogs defend territories while calling—a comparable mating strategy of the sage grouse where a small territory (lek) is used for enticing courtship displays. Aquatic breeding salamanders of the Klamath do not call like frogs but have a variety of courtship modes. The most bizarre breeding behavior is seen when dozens of rough-skinned newts gather underwater in a tight courtship mass known as a "newt ball." The coastal giant salamander is the only documented aquatic breeding amphibian in our area to show parental care, with adult females guarding her eggs until they hatch (PLATE B).

The remaining 14 (54 percent) species of amphibians in the Klamath Mountains are direct-developing terrestrial salamanders from the family Plethodonidae. Using elaborate courtship behaviors, males emit pheromones and display seductive body positions and tail wagging, to ensure the female finds and picks up his deposited spermatophore with her vent (PLATE C). The female then stores the sperm packet after mating until environmental conditions are ideal for egg deposition. At this point, eggs are laid in moist microhabitats underground or in narrow cracks of rocks, downed logs, or cavities in live trees. Terrestrial Klamath salamanders show advanced parental care where the female keeps the clutch moist and free of parasites, and even guards her eggs from predators until hatching. This advanced form of amphibian reproduction skips the free-swimming larval period altogether, with juvenile hatchings emerging from terrestrial eggs as miniature versions of adults (PLATE C). Despite our understanding of these general stages, many aspects of the reproductive ecology of this group largely remain a mystery.

PLATE C. Reproductive cycle of a terrestrial direct-developing amphibian. The ensatina (*Ensatina eschscholtzii*) is from the family Plethodontidae, the lungless salamanders. The ensatina is widespread throughout the Klamath Mountains in conifer and hardwood forests having rocky substrates and decaying logs. This salamander family is entirely terrestrial, skipping the larval stage with eggs hatching into fully formed miniature versions of the adults. All plethodontids have elaborate courtship behaviors. The scene above shows the 'tail-straddling walk' where the female places her chin on the males' tail and is led to walk forward as the female straddles his tail. The male eventually deposits white sperm packets (spermatophores) on the substrate that the female picks up with her vent after the male entices the female by secreting courtship pheromones from skin glands. Females can store sperm for months until environmental conditions are right for fertilization and egg deposition. Eggs are laid in deep moist cracks in live and downed trees or underground where they are guarded by the mother through hatching. Illustration by Michael Zontos

Biogeography

The best way to grasp amphibian diversity and distributions in the Klamath Mountains is to study range maps. While some species show no refined spatial patterns, others are greatly restricted. For example, the Wintu Shasta salamander is currently only known from a three-square mile limestone mountain slope above Shasta Reservoir. The Western red-backed salamander also barely enters the range from the north. In contrast the Pacific treefrog is ubiquitous, breeding in just about every type of natural and artificial wetland imaginable, including cattle troughs and roadside ditches. Widespread species—those occurring throughout the entire range and elevational gradients—include the ensatina salamander, rough-skinned newt, coastal giant salamander, coastal tailed frog (only in cold waters), Pacific treefrog, and western toad. The foothill yellow-legged frog is also widespread but is only found in streams. Many species that are widespread in the moisture-rich redwood and Oregon coast regions only enter the mesic western edge of the Klamath Mountains including the northwestern salamander, Dunn's salamander, California slender salamander, and northern red-legged frog. The clouded salamander, wandering salamander, Del Norte salamander, and southern torrent salamander are also widespread along the coastal mountains and extend somewhat further inland along major river canyons of the Rogue, Klamath, and Trinity rivers.

Nineteen species of native amphibians were known to occur in the Klamath Mountains at the end of the 20th century, which made the Klamath Province the most species-rich mountain range across the Pacific Northwest (Bury and Pearl 1999). The past few decades have brought intense scientific interest in Klamath salamander speciation motivated by the broad application of advanced molecular genetic techniques. Since the 1999 Bury and Pearl checklist, an astonishing five new endemic salamanders have been described—all from the family Plethodontidae, the speciose group of lungless terrestrial salamanders. Further, the Sierra newt extends into the eastern Klamath Mountains in Shasta County and was described as a separate species from the California newt (*Taricha torosa*) in the early 2000s.

To put recent genetic discoveries in the Klamath Mountains into perspective, only three new amphibians have been described over the same timeframe across the Pacific Northwest from Oregon to Alaska and east to Montana. Most newly described Klamath salamanders are part of cryptic species complexes, meaning the

Coastal giant salamander *(Dicamptodon tenebrosus)*. Photo by Spencer Riffle.

animals are difficult to distinguish using appearance alone, but contain substantial genetic differences indicating long-term isolation and no inter-breeding. With ongoing research, the evolutionary story of Klamath amphibians is still unfolding, and more discoveries certainly await.

The origin of the amphibian diversity in the Klamath Mountains is similar to the evolutionary histories of other lifeforms here such as conifers, ferns, butterflies, and mollusks. Like amphibians, these Klamath endemics often have poor dispersal abilities coupled with narrow habitat and microclimate preferences, which can result in very patchy distributions. Substantial dispersal barriers such as streams, ridgelines, and isolated habitats also result in genetic differences that can accrue over long periods. In part, marked differences reflect the Klamath Range as an ancient rock island that has persisted longer than the surrounding landscapes. Its complex geology, tectonic processes, topography, and climate have allowed for species to persist and evolve in shifting refugial habitats during extreme climate cycles such as the Pleistocene Ice ages that completely wiped out taxa from entire regions to the north. As conditions improve, species hiding in isolated niches become more widespread creating a cyclical pattern of expansion and contraction across millions of years. See Figure 1 on page 441 showcasing this local amphibian diversity.

Four regions with exceptional amphibian biodiversity are highlighted below to illustrate the complexity of Klamath amphibian distributions and related evolutionary histories.

The Great Meeting Place: The Smith River

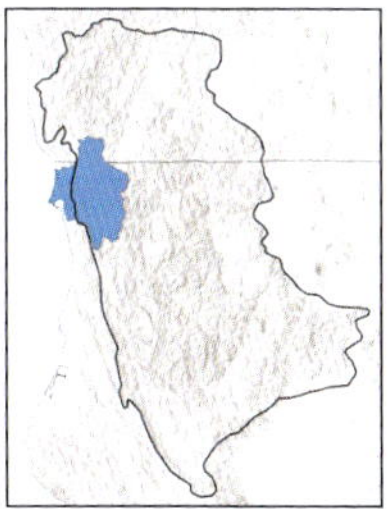

The Smith River watershed is the epicenter of amphibian diversity in the Klamath Mountains with 16 of the 26 (62 percent) species converging here. Draining west along the western slopes of the Siskiyou Mountains, the Smith River bisects the region near the California and Oregon border. What makes the area remarkable for biodiversity is the serpentine-rich rock of the Siskiyous abruptly meeting the western edge of North America within a few miles of the Pacific Ocean. The maritime climate elicits extensive orographic precipitation in the Smith and surrounding

California slender salamander (*Batrachoseps attenuatus*). Photo by Spencer Riffle.

watersheds, creating a lush temperate rainforest that is among the wettest locations in North America (see Climate chapter). Due to the rocky soils, runoff is swift, resulting in a powerful bedrock-lined river with deep pools. Based on the exceptional amphibian diversity, the Smith River region has likely been a persistently moist coastal refugium for amphibians during colder glacial and drier climate periods when temperature and moisture regimes of other regions were too extreme.

The mainstem and south fork region of the Smith River defines the range limits of four terrestrial salamander species with the river serving as major barrier and potential contact zone disrupting range expansions and gene flow. The Dunn's and clouded salamanders extend their Oregon distributions south to the Smith's northern banks. The wandering and Klamath black salamanders extend their more southerly and easterly ranges north to the Smith's southern banks. The Smith River has also been identified as a break in the genetic structure of the highly aquatic southern torrent salamander with populations north of the river currently recognized as a separate genetic lineage from those to the south (Miller et al. 2006). Torrent salamanders thrive in isolated cold-water seeps and have low dispersal capabilities so a large, warmer river like the Smith can represent a formidable barrier for such a habitat specialist. Refined genomic studies are still needed to understand torrent salamander divergence here.

While the Smith River is a barrier for some salamanders, it appears to have had little effect on the distribution and genetic breaks of others—including the Del Norte salamander (Mahoney 2004) and the California slender salamander (Martinez-Solano et al. 2007) which extends north to the south side of the Rogue River. This contrast suggests the regions salamander fauna have a complex and episodic evolutionary history where individual species have likely expanded and retracted ranges on different timelines.

The Klamath Refuge: Heart of the Range

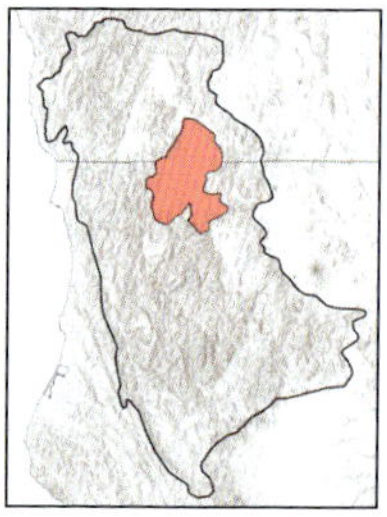

Steep forested mountains rise from the valley floor along the Klamath River from the mouth of the Scott River, through the Seiad Valley amphitheater, and down to Happy Camp. This remote setting is about as close to the center of the Klamath Mountains one can get without leaving a paved road. Here the deepest reach of the Klamath River canyon divides major glacially carved subranges—including the Marble and Scott Mountains to the south and the Siskiyou Crest to the north. Many powerful snow-fed streams, such as Seiad Creek and Grider Creek, drain these mountains with lush narrow riparian zones along their lengths. Given the extraordinary amphibian diversity on display throughout the Klamath Mountains, it is fitting that here in the center of the range is an evolutionary nucleus for endemic terrestrial woodland salamanders from the genus *Plethodon*.

The mid-Klamath corridor contains at least eight species of salamanders including the easternmost populations of southern torrent salamanders. The region is most known for a mysterious species contact zone for *Plethodon* salamanders that has perplexed herpetologists and geneticists for decades (Degross and Bury 2007, Mead et al. 2005). The Del Norte salamander, Siskiyou Mountains salamander,

Siskiyou mountains salamander (*Plethodon stormi*). Photo by Spencer Riffle.

and Scott Bar salamander converge here despite the region being among the driest for terrestrial salamanders in the Klamath Mountains. All three species rely on cool moist areas and deep rocky talus slopes. These habitats are also commonly associated with mature forests on north-facing slopes. In general, habitats are patchy resulting in fragmented salamander distributions. Both the Siskiyou Mountains and Scott Bar salamanders are endemic to the Klamath Mountains and have the most restricted ranges among the eight *Plethodon* species found in the Pacific Northwest.

The Del Norte salamander was first described in 1916 and is the most widespread among the three species—extending from the lower Trinity River up the Klamath River canyon to Seiad Valley. It also extends north along the western flank of the Klamath Mountains and coastal drainages from the Humboldt Bay region to the Rogue River. A genetic study across the range of Del Norte salamanders found three distinct genetic groups have experienced long periods of isolation and restricted gene flow (Mahoney 2004). The southern group centered on the Trinity River appears to be the most distinct while the two northern groups show evidence of recent range expansions.

The Siskiyou Mountains salamander was described in 1965 and has a compact range largely north of the Klamath River from Indian Creek to Seiad Creek and north over the Siskiyou crest into the Applegate River in Oregon. Two distinct genetic lineages have been identified within the range indicating populations have experienced long periods of isolation (Mead et al. 2005). In the past, populations of Siskiyou Mountains salamander may have been separated for long periods by the high Siskiyou crest near the border of California and Oregon. The Scott Bar Salamander was described in 2005 and occurs in a small region south of the Klamath River in the lower Scott River basin and surrounding mountains. Despite its extremely small range, the Scott Bar salamander displays a high degree of divergence from Del Norte and Siskiyou Mountains salamanders and is thought to represent the most ancient lineage of the three species (Mead et al. 2005).

The Shasta Refuge: Where Rivers and Limestone Collide

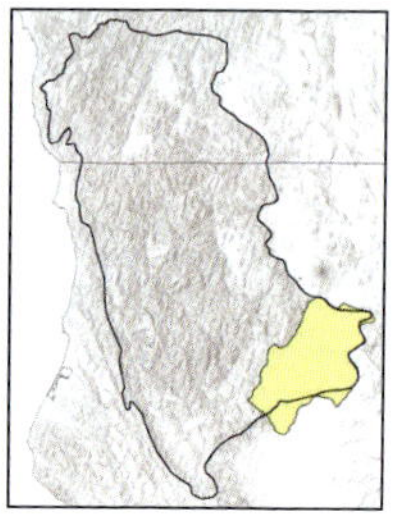

The extreme eastern edge of the Klamath Mountains in Shasta County hosts some of the oldest rock formations in the entire range, including massive limestone outcrops with caves and deep fissures. The western edge of the subregion is also flanked by the Trinity Mountains—a major north trending sub-range giving rise to the higher glacial characterized mountains of Trinity County. The Shasta region has a mild climate due to its lower elevations (~1,500 ft, 450 m) and receives substantially more rainfall than the arid Central Valley meeting its southern border near Redding (see Climate chapter). The region is dissected by major rivers—including the McCloud and Pit rivers that join the headwaters of the Sacramento River. Taken together, these factors have created an isolated refugial hotspot for amphibian evolution unmatched along the arid eastern flanks of the Klamath Mountains to the north and south.

With many recent discoveries, the mystery of amphibian evolution in the Shasta region is still unfolding. The region harbors at least 13 species of native amphibians, including an isolated population of California slender salamanders roughly 70 miles (113 km) east of the closest Klamath population (Stebbins 2003). This is also the only location where the recently described Sierra newt (Kuchta and Tan 2006) extends into the Klamath Mountains. Coastal tailed frog populations in Shasta are rare and appear to be genetically and morphometrically distinct—though this work is still in progress (Bury et al. 2020). The Trinity Mountains to the west are a barrier separating the Shasta black salamander from the Klamath black salamander—which was recently recognized as a new species (Reilly and Wake 2019). However, no amphibian group displays more dramatic patterns of local evolution in the Shasta region than the cryptic cave salamander (*Hydromantes*) species complex. Prior to 2018 only three species were known to North America—all being endemic to California. The Shasta salamander had the most northern distribution centered on Shasta Reservoir in Shasta County. After an in-

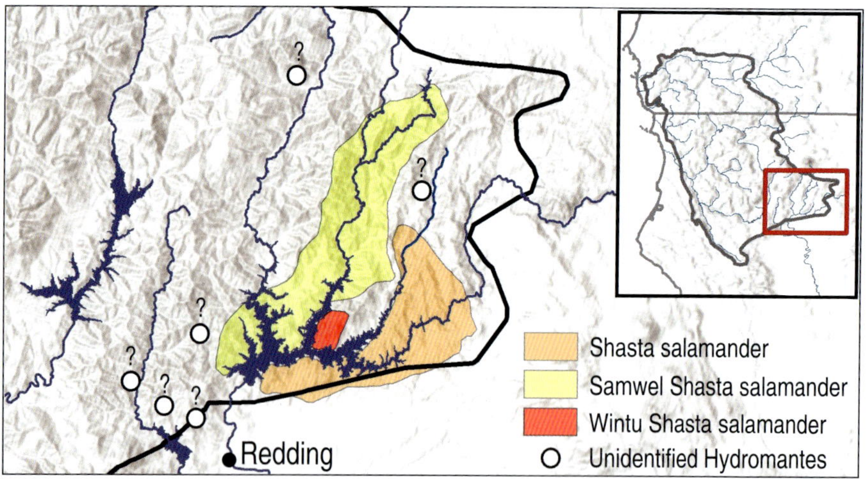

tensive genetic and morphometric study, two additional genetic lineages originally thought to be *Shasta salamanders* were described as two new species: the Wintu Shasta salamander and the Samwel Shasta salamander (Bingham et al. 2018, see Table 1). This discovery shows this *Hydromantes* complex has experienced little dispersal and gene flow across tiny spatial scales relative to most other vertebrates.

In October of 2016, Tim Laughlin snapped a photo of a uniquely colored *Hydromantes* with a broad lateral stripe while climbing at Castle Crags State Park in the upper Sacramento watershed. This was the second time Tim spotted a salamander here that year, with the earlier observation noted July 6th while he was suspended around 180 feet (55 m) up a granite wall. These observations are at least 10 miles (16 km) northwest of the nearest known locality in the McCloud watershed and 26 miles (42 km) north of the nearest location in the same Sacramento watershed. While such small distances and isolation seems trivial for most vertebrates, these discoveries remain an exciting mystery from which another new species may yet emerge.

Rock Islands: The Subalpine

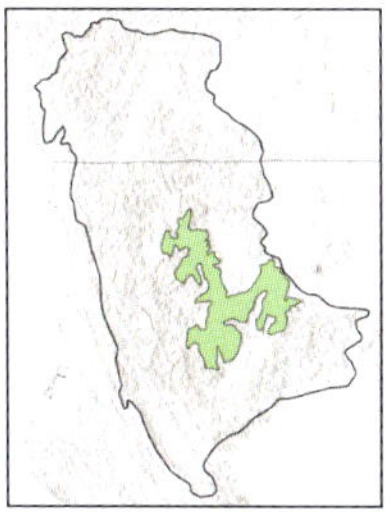

The subalpine zone of the Trinity Alps, Salmon, Marbles, and eastward through the Scott-Mount Eddy and Shasta-Trinity mountains contains a remarkable assemblage of glacial-formed lakes, wetlands, and streams isolated by an arid rocky landscape. No other region of the Klamath has such a concentration of natural wetlands available for aquatic breeding amphibians. In contrast, most lower elevation amphibian habitats of the range are restricted to streams, moist forests, and rocky talus slopes. Much of this subalpine region was covered with large glaciers during the Pleistocene ice ages, so amphibian colonization here has occurred within the past 10,000 years. These mountains accumulate deep winter snowpack and ice resulting in short seasons of amphibian activity. Despite these factors, this rugged topography hosts five species of salamanders and five species of anurans.

The Cascades frog and long-toed salamander are the only true 'sky island' amphibians of the Klamath Mountains, reproducing almost exclusively in the upper montane and subalpine zones. There are only a few iNaturalist records of adult long-toed salamanders occurring at lower elevations in Etna Valley (~2,900 ft, 884

The yellow striping on the back of the long-toed salamander (*Ambystoma macrodactylum*) matches the yellow lichen on its high-elevation rocky perch. Photo by Steve Johnson.

m) and the lowest Cascades frog breeding population that I have found occurs at 4,280 ft (1,305 m). Both have dramatic appearances with black, yellow, and golden colors contrasted in sharp patterns making them striking vertebrate ambassadors of high elevation wetlands. Southern, northern, and coastal areas were likely not Pleistocene refugia for these species, as their present-day distributions are restricted to the interior Klamath River basin east of the North Fork Trinity River and west of the upper Sacramento River. In contrast to many other Klamath amphibians adapting to local climate refuges, Cascades frogs and long-toed salamanders likely colonized the Klamath region from the southern Cascades range around Mount Shasta once the Klamath glaciers receded. Both Cascades frogs and long-toed salamanders radiated from the southern Cascades far north into the Pacific Northwest.

Seven widespread Klamath Mountain amphibian species also occur in the subalpine including the coastal tailed frog, Pacific treefrog, western toad, coastal giant salamander, rough-skinned newt, northwestern salamander, and ensatina. The northwestern salamander makes a narrow appearance in subalpine lakes and ponds of the western Trinity Alps, and are likely in the western Siskiyou Mountains. The foothill yellow-legged frog is also rarely encountered in subalpine streams but has been found to overlap with the closely related Cascades frog in some streams during the summer. Unlike lower elevations of the Klamath region where terrestrial salamanders predominate in forests, most amphibians in subalpine live here exclusively in water with only short terrestrial forays during summer rains. One exception is the thick-skinned western toad, which is often found skittering along trails at night or resting in shallow burrows in loamy soils. The only documented terrestrial breeding amphibian found in the subalpine zone is the ensatina. Because forests are so sparse in the subalpine, the ensatina and adults of the aquatic breeding long-toed salamander, lead cryptic lives in deep rocky talus slopes and small mammal burrows. These subterranean habitats provide a hidden niche by maintaining moisture and cool stable temperatures throughout the year in an otherwise exposed arid summer and frozen winter environment. The ensatina was only recently discovered in the subalpine zone after unidentified salamander remains were found in three separate stomach samples from garter snakes that were later confirmed through genetic analysis as ensatina (Reilly et al. 2010). After surveying this region for 20 years, I finally stumbled across an adult ensatina during a damp fall evening in 2019 at a 7,400 ft (2,256 m) mountain pass.

Acknowledgements

Thanks to Bruce Bury, Hartwell Welsh, Jr., Chris Feldman, and Jim Agee for chapter review. John Reiss reviewed the sidebar on *Ascaphus* and Dave Wake reviewed the species list. Michael Zontos applied his drawing skills and biological knowledge to create the amphibian reproductive cycle plates while Spencer Riffle provided exceptional photographs of all species unless noted. Additional photos were provided by Marisa Ishimatsu, Michael van Hattem, and Steve Johnson. Field excursions with Jamie Bettaso, Ryan Bourque, Monty Larson, Clara Wheeler, and Michael van Hattem over the years have sharpened my knowledge on Klamath amphibians. Thanks to Bruce Bury and Jason Reilly for helping clarify Oregon biogeography data. California Department of Fish and Wildlife, VertNet, HerpMapper, and iNaturalist provided useful datasets for developing species range maps.

Literature Cited and Suggested Reading

Asay, M. J., P. G. Harowicz, and L. Su. 2005. Chemically mediated mate recognition in the tailed frog (*Ascaphus truei*). *In* R. T. Mason, M. P. LeMaster, and D. Müller-Schwarze (Eds.), Chemical Signals in Vertebrates 10 (pp. 24–31). New York: Springer.

Bingham, R. E., T. J. Papenfuss, L. Lindstrand III, and D. B. Wake. 2018. Phylogeography and species boundaries in the *Hydromantes shastae* complex, with description of two new species (Amphibia: Caudata: Plethodontidae). Bulletin of the Museum of Comparative Zoology 161(10):403-427.

Brodie III, E. D., and E. D. Brodie Jr. 1999. Predator-prey arms races: asymmetrical selection on predators and prey may be reduced when prey are dangerous. BioScience 49:557-568.

Bury, R. B. and C. A. Pearl. 1999. Klamath-Siskiyou herpetofauna: biogeographic patterns and conservation strategies. Natural Areas Journal 19:341-350.

Bury, G. W., R. B. Bury, N. Waters, M. Leppin, T.R. Wahbe and H.H. Welsh. 2020. Phylogenetics and morphological analyses of tailed frogs (*Ascaphus truei*) in the Pacific Northwest reveal distinct species or clades in need of conservation attention. Abstract. World Congress of Herpetology, New Zealand.

Degross, D., and R.B. Bury. 2007. Science Review for the Scott Bar salamander (*Plethodon asupak*) and the Siskiyou Mountains salamanders (*P. stormi*): Biology, Taxonomy, Habitat, and Detection Probabilities/Occupancy. U.S. Geological Survey Open-File Report 2007-1352. 14pp.

Kuchta, S. R. and An-Ming Tan. 2006. Lineage diversification on an evolving landscape: phylogeography of the California newt, *Taricha torosa* (caudata: Salamandridae). Biological Journal of the Linneann Society 89:213-239.

Mahoney, M. 2004. Molecular systematics and phylogeography of the *Plethodon elongatus* species group: combining phylogenetic and population genetic methods to investigate species history. Molecular Ecology 13:149-166.

Martınez-Solano I, E. L. Jockusch, and D. B. Wake. 2007. Extreme population subdivision throughout a continuous range: phylogeography of *Batrachoseps attenuatus* (Caudata: Plethodontidae) in western North America. Molecular Ecology 16:4335-4355.

Mead, L., D. Clayton, R. Nauman, D. Olson, and M. Pfrender. 2005. Newly discovered populations of salamanders from Siskiyou County California represent a species distinct from *Plethodon stormi*. Herpetologica 61:158-177.

Miller, M. P., S. M. Haig, and R. S. Wagner. 2006. Phylogeography and spatial genetic structure of the southern torrent salamander: implications for conservation and management. Journal of Heredity 97:561-570.

Nielson, M., K. Lohman, C.H. Daugherty, F.W. Allendorf, K.L. Knudsen, and J. Sullivan. 2006. Allozyme and mitochondrial DNA variation in the Tailed Frog (Anura: *Ascaphus*): The influence of geography and gene flow. Herpetologica 62:235-258.

Reilly, S.B., A.D. Gottscho, J.M. Garwood, and W. B. Jennings. 2010. Phylogenetic analysis of common garter snake (*Thamnophis sirtalis*) stomach contents detects cryptic range of a secretive salamander (*Ensatina eschscholtzii oregonensis*). Herpetological Conservation and Biology 5: 395-402.

Reilly SB, and D.B. Wake. 2019. Taxonomic revision of black salamanders of the *Aneides flavipunctatus* complex (Caudata: Plethodontidae) PeerJ 7:e7370.

Ritland, R.M. 1955. Studies on the post-cranial morphology of *Ascaphus truei*. Journal of Morphology 97:215-282.

Roelants, K. and F. Bossuyt. 2005. Archaeobatrachian paraphyly and Pangaean diversification of crown-group frogs. Systematic Biology 54:111-126.

Stebbins, R. C. 2003. *A field guide to western reptiles and amphibians*. 3rd edition. Houghton Mifflin, Boston, Massachusetts.

Reptiles of the Klamath Mountains

Chris Feldman and Justin Garwood

Introduction

Similar to amphibians, reptiles are ectotherms whose lives are tied to the weather and dictated by climate. Their seasonal and daily activity patterns shift dramatically based on air temperature, and in the Klamath Mountains, the annual cycle includes a long hibernation period between fall and spring, spring emergence and breeding, and summer or even early fall reproduction. Like amphibians, reptiles are small and secretive, generally living cryptic lives out of sight, so finding one is often a surprising and serendipitous event. Although far fewer species occur in the Klamath Mountains relative to the warmer Great Basin and arid southwest, there are plenty of striking reptiles in our region. Explorers are often wonderstruck when they come across a California mountain kingsnake or a large group of northwestern pond turtles quietly basking on a floating log. Many reptiles (namely snakes) also bring out innate fears in people—even though few snakes in the U.S. are actually dangerous. In fact, the Klamath Mountains hold only one venomous species that might cause serious injury, the western rattlesnake.

Globally, reptiles are the most species rich group of tetrapods—the lineage of vertebrates that stem from a fishy ancestor and evolved lungs and four limbs over 380 million years ago (Ma), and now includes amphibians, reptiles, birds, and mammals (see page 289). In fact, reptiles and mammals share a common ancestor (they are sister lineages), and birds are a type of dinosaurian reptile whose closest living relatives are crocodilians and turtles! Even removing birds (~10,550 species) from the reptile count, there are more kinds of non-avian reptiles than any other land vertebrates (11,570 species) (Uetz et al. 2021). Excluding birds, 61 percent of reptiles are lizards, 34 percent are snakes, 3 percent are turtles, and the remaining 2 percent includes amphisbaenians (worm lizards), crocodilians, and rhynchocephalians (tuatara) (Uetz et al. 2021).

This chapter explores "traditional" reptile diversity (exclusive of birds, which get their own sub-chapter), ecological roles, and unique morphologies, life histories, and behaviors across the diverse mosaic of habitats found throughout the Klamath Mountains.

Diversity and Biogeographic Patterns

Twenty species of native reptiles were documented in the Klamath Mountains by the end of the 20th century (Fitch 1936, Bury 1970, Bury and Pearl 1999). Since that time, the cryptic and regionally restricted forest sharp-tailed snake was described as a new species (Feldman and Hoyer 2010). In addition, multiple isolated records of the striped racer have turned up in the eastern Klamath Mountains in Shasta and Trinity counties. These additions bring the current Klamath reptile richness to 22 extant native species from 8 families (Table 1). In contrast, only 20 reptiles occur in the entire half of Oregon west of the Cascades and north of the Klamath Mountains, while only 12 species are represented in western Washington (St. John 2002).

Table 1. Reptiles of the Klamath Mountains Geological Province, their habitats, and biogeographic associations.

Common Name	Scientific Name	COMMON HABITATS[1]	BIOGEOGRAPHY[2]
SNAKES (SQUAMATA)			
Boas	**Boidae**		
▸ northern rubber boa	*Charina bottae*	MC, CH, CO, SA, DL, RO	Widespread
Colubrid Snakes	**Colubridae**		
North American racer	*Coluber constrictor*	CO, MG, CM, CH, SA	Widespread
striped racer	*Masticophis lateralis*	CM, CO, RO, MG	Southern
striped whipsnake	*Masticophis taeniatus*	CM, RO, SJ, CO	Southern
forest sharp-tailed snake	*Contia longicauda*	CH, RI, DL, RO	Regional
common sharp-tailed snake	*Contia tenuis*	CH, RI, DL, RO	Southern
ring-necked snake	*Diadophis punctatus*	CH, RI, DL, RO	Widespread
California kingsnake	*Lampropeltis californiae*	MG, CM, CO, DL, RI	Southern
California mountain kingsnake	*Lampropeltis zonata*	CO, MC, CM, RO, DL	Southern
gopher snake	*Pituophis catenifer*	MG, CO, MC, DL, SA	Widespread
▸ aquatic garter snake	*Thamnophis atratus*	ST, LP, SF, RI	Southern
▸ mountain garter snake	*Thamnophis elegans elegans*	MG, CO, LP, RE	Widespread
▸ coast garter snake	*Thamnophis elegans terrestris*	MG, CH, RI	Regional
▸ northwestern garter snake	*Thamnophis ordinoides*	RI, CH, MG	Northern
▸ common garter snake	*Thamnophis sirtalis*	LP, RI, MG, CO, SF	Widespread
Vipers	**Viperidae**		
▸ Western Rattlesnake	*Crotalus oreganus*	CO, MC, CH, RO, CM, SA	Widespread
LIZARDS (SQUAMATA)			
Alligator Lizards	**Anguidae**		
▸ northwestern alligator lizard	*Elgaria coerulea principis*	CO, MC, CH, LP, MG	Northern
▸ Shasta alligator lizard	*Elgaria coerulea shastensis*	CO, MC, CH, LP, MG, SF, MC	Regional
southern alligator lizard	*Elgaria multicarinata*	CO, MG, MC, RI, RO, CH, DL	Southern
Spiny Lizards[3]	**Phrynosomatidae**		
common sagebrush lizard	*Sceloporus graciosus*	MC, CO, RO, SA, SJ, RI	Widespread
western fence lizard	*Sceloporus occidentalis*	CO, MC, CM, RO, MG, SJ, RI	Widespread
Skinks	**Scincidae**		
western skink	*Plestiodon skiltonianus*	CO, DM, MC, RI, DL	Southern
Whiptails	**Teiidae**		
western whiptail	*Aspidoscelis tigris*	RI, WS, UG	Southern
TURTLES (TESTUDINES)			
Pond Turtles	**Emydidae**		
northwestern pond turtle	*Actinemys [Emys] marmorata*	ST, LP, RI, RE, CH	Southern
red-eared slider	*Trachemys scripta*	RE, ST, LP	Introduced

1. HABITATS—Common habitats for each species are listed in order of importance.
Aquatic: Streams and Rivers (ST), Lakes and Ponds (LP), Reservoir (RE), Washes/ Sand (WS)
Moist Forest: Riparian (RI), Mixed Conifer/ Hardwood (CH)
Dry Forest: Mixed Conifer (MC), Mixed Conifer/ Oak Woodland (CO), Subalpine Conifer (SA)
Chaparral and Shrublands: *Ceanothus*/ Manzanita (CM), Sagebrush/ Juniper (SJ)
Rock and Burrows: Rock Outcrops/ Cliffs/ Talus (RO), Underground Burrows (UG), Downed Logs (DL)
Meadows: Meadow/ Grassland (MG), Subalpine Fen (SF)

2. BIOGEOGRAPHY—Range wide distribution relative to Klamath Mountain Province (KMP). Regional: Portion of range in KMP and regionally endemic; Northern: Most of range north of KMP; Southern: Most of range south of KMP; Widespread: Range broadly spans north and south of KMP; Introduced: Exotic species.

3. An early museum specimen of the Blainville's horned lizard (*Phrynosoma blainvillii*) was apparently collected in the Klamath Mountains in "Kennett" CA, a historic Shasta County town now flooded by Shasta Reservoir (Stebbins 2003). If correct, the Shasta population would have been a far northern disjunction from the nearest extant populations, which occur in the foothills around Chico and Oroville (Butte County, CA). Although this single specimen has generated a persisting biogeographic mystery, horned lizards are presumed to be absent from the Klamath Mountains today.

▸ Live Birth (viviparous); all others are egg layers (oviparous)

Reptiles

Reptile diversity and distributions in the Klamath Mountains are largely defined by five ecological communities: open grasslands in valleys, mixed conifer and oak woodlands historically maintained by frequent fires, cooler conifer forests with open patches, aquatic and riparian habitats, and the narrow arid eastern slopes of the range largely occurring in a rain shadow (Figure 2a in Climate Chapter). Within these mosaics are important microhabitats that local reptile species need and which further refine species distributions. Access to particular soils, rocks, sun exposure, water sources, or plant associations is required for nesting, hibernating, basking, foraging, or social interactions. Such abiotic and biotic conditions help determine where reptile species persist in the region (discussed for each taxon in species accounts). For example, rattlesnakes in the Klamath Mountains largely hibernate in south-facing talus slopes with relatively gentle inclines.

Because reptiles require environmental heat to warm their bodies, there is a general trend of decreasing species richness with increasing elevation and latitude. Indeed, we see these same global patterns in the Klamath Mountains, with a host of species at lower and middle elevations, but only a few cold-adapted reptile species, such as the northern rubber boa and northern alligator lizard, at the highest sites. Likewise, reptile diversity in the Pacific Northwest (PNW) generally decreases from south to north. The Klamath Mountains are part of the southwestern end of the PNW and hold the greatest reptile diversity in the PNW, west of the Cascades. Moving east of the Cascade Ranges brings warmer and more arid conditions that create the Sagebrush Steppe of the northern Great Basin which supports a different reptile assemblage.

Beyond the ecological factors, unique histories of dispersal, colonization and fragmentation contribute to diversity. These patterns are well studied regionally for amphibians but haven't been investigated as intensively in reptiles—though many species have been examined at broader scales, allowing us to at least consider the origins of Klamath reptiles. Of the 22 extant species in the Klamath Mountains, only the forest sharp-tailed snake could be considered a "near endemic". There are also two subspecies with a large portion of their ranges within the region, the Shasta alligator lizard and mountain garter snake. All the other reptiles are either broadly distributed south of the Klamath Mountains or are widespread across western North America (Table 1). While genetic analyses are required to test regional origin and history, work to date suggests most species have either dispersed northward along the Coast Ranges, likely during Holocene warming in the past 10,000 years (Feldman and Spicer 2006) or dispersed westward from the Cascade Ranges and Sierra Nevada. Reptiles that appear to have moved from the south include the ring-necked snake (Feldman and Spicer 2006), common garter snake (Hague et al. 2017) and western fence lizard (Bouzid et al. 2021). The northern alligator lizard (Lavin et al. 2018) and terrestrial garter snake (Hallas et al. 2021) likely arrived from the east using suitable habitat corridors. Only two reptile taxa occur largely north of the Klamath Mountains, and thus either originated in the region and spread north or dispersed southward into the Klamath (Table 1). Future genetic work may uncover unique patterns of diversity and endemic lineages while highlighting regions that function as genetic corridors or barriers.

Order Testudines – Turtles

Turtles are among the most distinctive and recognizable animals in the world. No other vertebrate has such a unique body plan—with robust limbs, large heads, and short, stout trunks fully encased in bones overlaid with tough keratinized scales. This ancient reptile clade extends back to the late Triassic Period (~220 mya) and is a testament to the turtle mold, with ~360 species recognized today (Rhodin et al. 2017). Turtles are found just about everywhere on Earth, including all continents except Antarctica, many remote islands, and all world's oceans. These charismatic reptiles are symbols of longevity, wisdom, persistence, and good health.

Family Emydidae – Pond Turtles

The New World pond turtles (family Emydidae), with ~50 species, are an ecologically important group of semi-aquatic freshwater turtles, with a few estuarine and terrestrial species (Ernst and Lovich 2009). The center of emydid origins and diversity is in the eastern US, with only a few emydids found out west. One of these, the northwestern pond turtle (formerly *Clemmys marmorata* and now either *Emys* or *Actinemys* depending on the authority) is the only native turtle in the Klamath Mountains (Table 1). They are a true western endemic, occurring west of the Sierra-Cascade crest from California north to Washington's Puget Sound (Stebbins 2003). They extend further inland east of the Klamath Mountains where the upper Klamath and Pit rivers have cut through the southern Cascade Range (St. John 2002). They are relatively small (length up to 8.5 in, 22 cm), with flat, tan-brown, olive, and black shells with dark flecking or lines. They are easily distinguished from the non-native red-eared slider—which have been introduced to isolated regional habitats by humans—that have obvious yellow stripes on the shell and along the head and neck.

Despite their name, northwestern pond turtles in the Klamath Mountains tend to be more strongly associated with riverine habitats than ponds or lakes. Rivers and creeks bisect every corner of the range, and pond turtles are distributed in sunny, medium to low-gradient reaches with large boulder-lined pools with slower runs containing vegetated undercut banks. These same reaches usually give rise to broad, elevated floodplain benches that are used by turtles for nesting and overwintering. Isolated turtle populations are found in natural silt-lined ponds

A western pond turtle adult (left) and hatchling (right) captured by Justin Garwood.

The Long Secret Lives of Northwestern Pond Turtles

Most wild animals in the Klamath Range have short lifespans. For example, mountain lions only live to ~13 years. Black bears rarely make it three decades with an average lifespan of ~20 years. Turtles, on the other hand, are world-renowned for their impressive longevity—with some captive land tortoises surpassing 175 years! One difficulty in aging turtles is that few studies have encompassed a lifetime, especially in the wild. Young turtles develop conspicuous annual growth rings (annuli) on their shells like annual tree growth rings. For Northwestern pond turtles, these annuli are reliable for the first 10 to 15 years depending on the individual and habitat (e.g., muddy vs abrasive rocky substrates). Beyond this timeframe, the annuli wear down and become indistinguishable leaving only a smoothly polished plastron.

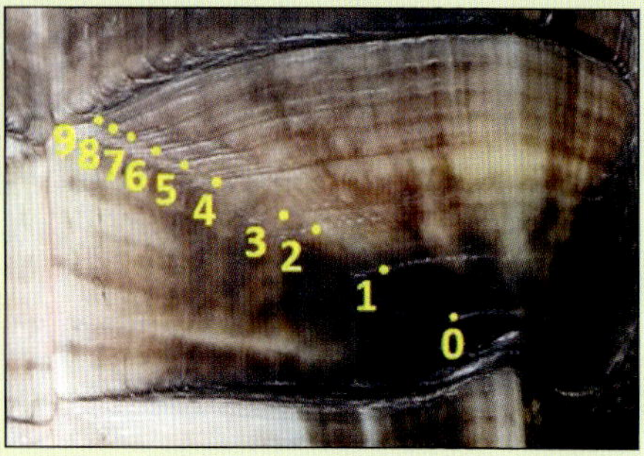

Pond turtle plastron showing annual growth rings (annuli).

Bruce Bury has been monitoring Northwestern pond turtles (*Emys marmorata*) for over 50 years in a tributary of the South Fork Trinity River. This has allowed him to estimate the longevity of the only turtle species in the Klamath Range. He began his study in 1968 while earning his PhD at UC Berkeley, marking around 500 pond turtles over three years by filing unique triangular notch combinations along their shell margins. He has continued marking turtles ever since, recapturing a number of his original marked animals and marking new ones during each visit.

Bruce holding an adult pond turtle (#50) in 2018—when it was 55 years old—next to a hatchling with Don Ashton looking on. Photo by Jamie Bettaso.

Fifty years after the inception of the study, Bruce and his biologist daughter Dr. Gwen Bury, led a 2018 expedition with other Klamath turtle biologists. This team surveyed the same stream section and found numerous marked turtles. They identified nearly 20 turtles that were marked >30 years earlier, and 10 turtles marked between 40 and 50 years earlier (Bury et al. 2019). A five-year old turtle was marked in 1968—the same year as the Apollo 8 Moon mission—making her 55 years old in 2018! Despite this remarkable record, we still do not know the longevity of pond turtles in the Klamath Mountains. The 55 year old female did not show much sign of aging and was gravid (carrying eggs) at the time of capture (Bury et al. 2019). This lack of long-term turtle longevity data demonstrates that conservation planning should reach decades into the future.

and lakes upslope of rivers, and the two habitats are often connected by migrating or dispersing individuals. This is especially evident in montane natural lake and pond districts found along the western flank of the Klamath Mountains. There are also few upper montane records perched above major drainages in the Yolla Bolly Mountains (4,900 ft; 1,494 m) and the Trinity Alps (4,930 ft, 1,503 m), but these cases are few due to habitat rarity and isolation. Northwestern pond turtles have also colonized reservoirs and and homestead ponds regionally.

Pond turtles are usually seen when basking and use different basking modes depending air and water temperatures. Aerial basking allows for turtles to gain heat from the sun and hot rocks or logs while above the water, while aquatic basking takes advantage of warm water locations in ponds and streams. They are alert while aerial basking and thus seldom found directly on shorelines but preferring partially submerged logs or rocks which offer clear views of potential foes and a quick escape. With excellent color vision and hearing, a wary disposition, and cryptic appearance, pond turtles often slip into water unnoticed—many places have far more turtles than are actually observed. Aquatic basking turtles usually keep their bodies just below the water surface while craning their heads above water.

Pond turtles are generalist omnivores that forage on annelids, insects, crustaceans, amphibians, fish, and dead animals (carrion). They will also eat plant material such as pond lily inflorescences, willow and alder catkins, and even filamentous algae (Bury et al. 2012). Predators are largely dictated by a turtle's size and whether the turtle is in water or on land. River otters are aquatic predators of adult turtles while herons and egrets, introduced bullfrogs, and introduced centrarchid fishes take hatchlings (Holland 1994). Black bears and raccoons commonly destroy nests and eat turtle eggs, so hatchling survival varies greatly. However, turtles are extremely long-lived (see longevity sidebar) allowing for sustained population persistence across low and high recruitment years—at least within intact habitats lacking invasive predators.

There are many fascinating elements to the intricate annual lifecycle of northwestern pond turtles. They are highly aquatic from March to September and sometimes overwinter in permanent ponds and lakes by burrowing into the mud (Ultsch 2006). For those living in rivers and temporary ponds, overwintering turtles usually migrate upslope in the fall (Ultsch 2006). Here they spend the winter buried in the duff layer and return to the water in the spring. Courtship and copulation take place in the water and females develop up to 13 eggs (Bury et al. 2012). Nesting takes place on land between June and July when the female finds (or returns) to an isolated nesting bench with southern exposure, short vegetation, and compact soils. She will often void her bladder to soften the soil as she excavates a nest chamber. She then lays her elliptical-shaped eggs, plugs the chamber opening with soil, and returns to the water for the summer. The eggs incubate in the warm soils (which determines the sex of the turtle—see reptile reproduction sidebar) throughout the summer and hatch in the fall. Hatchlings remain in the sealed nest chamber throughout their first winter. Miniature pond turtles, the size of a quarter, dig themselves out of the nest in the spring and disperse to water. The spatial separation of these distinct but necessary habitats reveals the complex relationship our turtles have with the landscape.

Order Squamata – Lizards and Snakes

Squamata (meaning "scaly") includes lizards, snakes, and a little-known group of fossorial beasts called worm lizards. With approximately 11,182 species, squamates are by far the most species rich reptile group, containing ~96 percent of living reptile diversity (Uetz et al. 2021). In fact, squamates are the most diverse group of tetrapods, with even more species than birds. Over 70 anatomical traits diagnose squamates (Gauthier et al. 1988), but perhaps the most obvious and unusual are hemipenes (singular is hemipenis). These are paired copulatory organs that develop from out-pockets of the hind-wall of the vent (cloaca) and are stored inside-out at the base of tail. Another interesting squamate trait is the pattern of ecdysis, or skin molting. Rather than constantly shedding little bits of skin like mammals or scales as in other reptiles, the entire outer layer of the skin is shed at once. Because the outer layer of skin is composed of dead cells, a growing squamate must literally bust out or wriggle free of its old threads to fit into its more well-fitting new clothes that have developed under the old skin.

Lizards

Lizards are a diverse (>7000 species placed in 41–46 families) and often abundant and conspicuous component of most ecological communities (Pianka and Vitt 2003). They are found just about everywhere on Earth except Antarctica. From sea level to above tree line, from the tropics to all the world's deserts, to north of the arctic circle and to remote oceanic islands, lizards display a remarkable range of ecomorphological diversity. Most are terrestrial generalists while some specialize living in rocky outcrops, are arboreal, or even burrowing. Given this diversity, lizards are difficult to generalize, but most are opportunistic insectivores eating whatever they can find and fit in their mouths. In fact, lizards are known for their

Common sagebrush lizard. Photo by Spencer Riffle.

Reptiles

role as invertebrate regulators, which has cascading effects in ecological communities (Pianka and Vitt 2003). They also serve as an abundant prey to larger lizards, snakes, corvids, birds of prey, and mammals.

One neat lizard feature is the trick many use to avoid predators, which is the ability to shed the tail when attacked, termed caudal autotomy (Clause and Capaldi 2006). Lizards have conscious control over when to release the tail. In most lizard species, the muscles in the fractured tail fire automatically, twitching the tail violently and distracting the predator (for up to 30 minutes in some species) while the lizard makes its hasty getaway. Most lizards can then regenerate the tail by growing a rod of cartilage from the fracture plane. However, this is energetically expensive and the new tail is not as long or flexible as the original because it lacks the series of small bones that permit articulation. Thus, the decision to retain or release is not a simple one, must occur in the blink of an eye, and might be the difference between life and death.

The Klamath Mountains hold six extant lizard species from four families (Table 1). Despite this low diversity, our lizards occupy a vast range of habitats, from the lowest river canyons to the highest subalpine meadows.

Family Phrynosomatidae – Spiny Lizards

Phrynosomatids are a North American lineage of 165 species that range from Panama to southern Canada, with greatest diversity in the southwest US and northeast Mexico. These diurnal lizards are abundant and conspicuous components of ecological communities, and come in a variety of forms. The spiny lizards (genus *Sceloporus*) are the most successful lineage of this family (110 species), and the Klamath Mountains contain two species: the western fence lizard and the sagebrush lizard. Another phrynosomatid (Blainville's horned lizard) may have once occurred in the region, based on a single specimen collected in an area now under the Shasta Reservoir (see footnotes section in Table 1).

Spiny lizards are the most commonly encountered and familiar lizards in our region and across most of the west. They are known as "blue-belly" lizards because of the distinct blue patches on their throats and along the chest and abdomen. They are also easily recognized by densely overlapping keeled dorsal scales, giving them a prickly dragon-like appearance. With a relatively large head, distinct neck, stocky frame, and tail just slightly longer than body they have a classic lizard look. With long powerful limbs and long toes and claws, they are excellent sprinters but also exceptional climbers. Because they are capable of moderate color shifts by shuttling melanin (the pigment responsible for dark coloration), dorsal coloration varies substantially based on location, temperature regimes, time of day, social status, and even mood.

Our two species are quite similar in appearance, but the western fence lizard is larger (total size up to 8 in, 20 cm) than the sagebrush lizard (6 in, 15 cm) (Espinoza 2009, Ryan 2009). In addition, fence lizards have larger but fewer dorsal scales, while sagebrush lizards have many more dorsal scales that are smaller and less prickly, and more granular shaped scales on their thighs, giving them a smoother overall appearance. Both species tend to be shades of browns or greys and may

Saurian Superheroes

Ticks often attach to lizards on parts of the body that are difficult for lizards to scratch and remove them. Photo by Justin Weist.

If there is a lesson to be learned from 2020 (and hopefully we've learned many), it's that our encroachment into wild places and exploitation of wildlife can potentially have dire "spillover" consequences. Zoonoses are diseases that jump from animals to humans, and Lyme disease is among the most well-known zoonotic diseases in the US, afflicting an estimated 300,000 people each year. The disease causes a range of ailments, from fever and fatigue to permanent facial palsy, arthritis, heart complications and a host of other muscular and nervous issues. And though Lyme disease is now treatable if caught early, there is still no cure.

The disease is caused by spirochete (corkscrew-shaped) bacteria of the genus *Borrelia* that make their homes in western black-legged ticks (*Ixodes pacificus*) in the West and deer ticks (*Ixodes scapularis*) out East. These ticks are not especially picky feeders and will take a blood meal from just about any terrestrial wildlife they encounter, spreading the bacteria from one wildlife host to the next. Certain wildlife species, notably rodents and birds, offer particularly good refuge to the spirochete and serve as important bacteria reservoirs, keeping *Borrelia* prevalent in an area. Humans accidentally pick up offending ticks by walking through tall grass or brushing alongside shrubs where ticks are waiting to grab a ride—that cotton sock or denim jean feels about the same as a hairy deer leg or fuzzy rabbit ear to a tick. If the tick makes it to your soft flesh, then it might transmit the spirochete from its gut to its salivary glands, and then to you (note: this takes roughly 36 hours). If you're in the northeastern US, then your chances of getting the Lyme disease-causing bacteria are relatively high. But if you're hiking through the Klamath Mountains, or elsewhere in the west, then you should have little fear because our native lizards, especially western fence lizards and southern alligator lizards, have natural immune defenses to the bacteria.

When infected ticks bite these reptiles and take a blood meal, they also suck up the antibodies in lizard blood that actually kill off the disease-causing bacteria. Thus, lizards essentially clear ticks of the *Borrelia* infection. In addition, lizards tend to be more abundant than other wildlife hosts of ticks, further diluting the pool of infected ticks in areas where lizards are common. Thanks to a superior immune system, and healthy population sizes, lizards actually help remove Lyme disease from landscapes. One of us recently counted 23 ticks on a single fence lizard captured along the McCloud River! There are no reports of Lyme disease in the counties that contain the Klamath Mountains. So the next time you're tramping about in the Klamath Mountains, and you see a lizard with a tick on its neck, be sure to thank that saurian superhero for making your outdoor experience a little more carefree.

Western fence lizard. Photo by Spencer Riffle.

possess faint stripes that run from the eye to the tail, with nearby dark wedges or chevrons that interrupt the stripe.

The western fence lizard and sagebrush lizard are the most widespread and abundant lizards in the Klamath Mountains. They are habitat generalists but are most often found in open manzanita and buckbrush chaparral, rocky grasslands, oak savannah and woodlands, open pine forests, rocky canyons, open riparian zones, and the sage-steppe and open juniper woodlands of the eastern Klamath. They require sunlight penetration and boulders, rockpiles, talus slopes, logs, stumps, woodpiles, fences, etc., for basking and surveying. However, the two species seldom overlap. The western fence lizard occupies hotter and more arid sites, while the sagebrush lizard occurs in cooler and damper microclimates at higher elevations [up to 7,100 ft (2,164 m)].

Both species are visually oriented sit-and-wait predators with excellent focal abilities and color vision (as tetrachromats, they see a wider spectrum of colors than we can perceive). They remain perched on their lookout and then dart out to nab prey using a fleshy, sticky tongue. They consume mainly insects such as flies, termites, grasshoppers, moths, butterflies but even beetles, noxious ants, venomous bees, wasps, spiders, and scorpions (Stebbins and McGinnis 2012).

Spiny lizards are strongly diurnal and heliothermic, requiring time basking in the sun to warm up and maintain optimal body temperature. In addition, they are social and territorial, using visual displays, scent cues, and even physical combat to maintain social order and communicate with potential mates (Stebbins and McGinnis 2012). As such, they are conspicuous—seemingly always about, perched on rocks, moving under shrubs, or performing pushup or head-bob displays to defend turf. Males hold larger territories encompassing smaller home ranges of several females. Both sexes possess blue ventral markings which are more vibrant and conspicuous in males—which expand their dewlaps to show off throat color and puff up chests to expose blue bellies. They can also develop green, turquoise, or blue dorsal markings. Males also use wax-secreting pores along the inner thighs used to scent mark territory. If push-up displays and scents don't settle turf wars, then male combat ensues—it can be as vicious as a professional wrestling match.

This diurnal activity puts a target on spiny lizards, and they fall prey to a host of predators. Because alligator lizards can function at cooler temperatures, they prey on vulnerable spiny lizards—especially when below their thermal optimum. The athletic western whiptail lizard will also take small spiny lizards by simply outrunning them in open sandy habitats. Racers and whipsnakes are the fastest moving snakes in the Klamath Mountains and are highly specialized at visually hunting lizards, especially spiny lizards (Stebbins 2003). Both species lay eggs in loose soil in the late spring and early summer with the smaller sagebrush lizard having a much lower fecundity (3–7 eggs) compared to the larger western fence lizard (3–17) (St. John 2002). Longevity is thought to be 3–5 years (Tinkle et al. 1993).

Family Anguidae – Alligator Lizards

Lizards of the family Anguidae occupy cool and moist environments in temperate and tropical areas of the Northern Hemisphere. The 130 species are elongate with relatively short limbs or no limbs at all (limblessness has evolved more than once in the group). They are covered in scales that are embedded with bone (called osteoderms) providing a tough armor coating. Our region contains alligator lizards (genus *Elgaria*), a well-known and charismatic group (seven species) endemic to western North America. Alligator lizards get their names from a superficial resemblance to alligators—tough dorsal scales, long but thickset body and tail, short but stout limbs, and broad triangular head. And it's not just looks that are reminiscent of miniature gators, but their fierce aspect, bold manner, and vicious bite—beware! The dorsal scales of alligator lizards are relatively smooth but keeled along the center and surprisingly hard, while the ventral scales are glossy and smooth, allowing more efficient movements over terrain. Dorsal colors vary greatly, from dull yellow and brown to olive and slate grey.

The Klamath Mountains host two species, the southern alligator lizard and the northern alligator lizard. These, the largest bodied regional lizards, can be difficult to distinguish. The southern alligator lizard is typically longer (body size up to 8 in, 20 cm) than the northern alligator lizard (up to 5.5 in, 14 cm), with a relatively longer tail that can be up to twice the length of the body, while the tail of the northern species rarely exceeds one and half times the body. The southern species has bright yellow eyes and distinctive dark crossbands along back, while the northern species has duller tan or bronze eyes and a mottled back.

To complicate identification more, two subspecies of northern alligator lizards occur here (Table 1). The Shasta alligator lizard is the larger of the two (body size up to 5.5 in, 14 cm) with a more robust head and a stunning variable appearance—sometimes with slate grey heads and yellow-green dorsum that may have dark black mottling (Fitch 1935, Smith 1946). This form has a regional distribution centered around the Klamath Mountains. The northwestern alligator lizard is found at the far northwest edge of Klamath Mountains, and tends to be smaller (rarely larger than 4 in, 10 cm) and usually a shade of brown (Fitch 1935, Smith 1946).

Alligator lizards can operate at lower temperatures and dwell in more humid spots than other regional lizards (Stebbins and McGinnis 2012). This is likely

Shasta alligator lizard (top, photo by Alan St. John) and southern alligator lizard (bottom, photo by Spencer Riffle).

why they have the broadest distribution and longest active periods among lizards in the Klamath Mountains. This physiological advantage allows them to occupy a broader range of cooler habitats including mature forests, wetland edges, and riparian zones—they can even swim quite well! Generally, they are secretive and prefer habitats with an abundance of cover like decaying logs, leaf litter, talus, and boulders. The two species usually segregate spatially, with southern alligator lizards preferring habitat below 4,000 ft (1,219 m) in grasslands, chaparral, oak woodlands, open mixed conifer, and also riparian zones with dappled light. The northern alligator lizard tolerates even cooler conditions, and is found at elevations and habitats that are too cold for its southern cousin—up to up to 7,400 ft (2,225 m). They are encountered along high-elevation lakes, wetlands, meadows, and riparian zones, as well as mixed conifer-hardwood forests and mature conifer forests. Not surprisingly, this species has the northern-most distribution of any lizard in the Pacific Northwest—extending well into southern British Columbia (Stebbins 2003).

Alligator lizards are impressive, agile predators that tackle any small animal they can overpower and consume. They stalk terrestrial prey with slow forward movements using a keen sense of smell. Thanks to a partially forked tongue, they can track prey through litter or into burrows. Good climbers, they will forage in shrubs and low trees with the aid of a prehensile tail that wraps around twigs and branches. A diverse diet of Earthworms, slugs, and arthropods that even includes dangerous prey such as centipedes, scorpions and spiders, including black widows are taken (Stebbins 2003, Thill et al. 2022). Alligator lizards will also prey on small lizards, newborn rodents, bird eggs, and even bird hatchlings (St. John 2002).

Alligator lizards are prey for vertebrates including snakes, mammals, birds of prey, and corvids. However, these irascible lizards don't go down without a fight. They will use any number of anti-predator tricks to avoid being eaten, depending on the predator and situation (Fitch 1935). Most evade predators with clever escapes, such as slithering into deep grass like a snake with legs tucked alongside the body, diving into water and remaining submerged, or even climbing into shrubs. When cornered, they might feign death or instead puff-up their bodies, wave their tails, and gape and hiss in an angry display of bluster. If seized they fight aggressively by biting with surprising force, twisting about in alligator-like fashion, and smearing feces, urine, and a foul liquid from anal scent glands. Some will even bite their own tails to form a loop too large for small predators to swallow! They also drop their tails if needed, which can spasm for tens of minutes.

Alligator lizard courtship involves the male holding the neck of the female in his jaws. Because they are locked in the same place for many hours or even a day, courtship is a common sight in the spring and early summer. Southern alligator lizards lay 8–14 eggs underground and the female, like in other anguids, stays with the eggs though what level of care she provides and for how long remains unknown. In contrast, northern alligator lizards give birth to fully formed young. This reproductive strategy is thought to allow mothers to more carefully control embryo development through their own thermoregulatory behaviors. It is postulated that species that have live birth can thus occupy cooler habitats (see reptile reproduction sidebar). Young take 2–3 years to reach sexual maturity, and then likely live for up to a decade or more (Fitch 1935).

Family Scincidae – Skinks

By some measures, skinks are the most successful group of saurians, with over 1700 species and the widest distribution of all lizard families, occurring on every continent (except Antarctica) and most oceanic islands (Utez et al. 2021). Although the Klamath Mountains host just a single species, its sleek and striking appearance is a rewarding sight.

The western skink is a small (7.5 in, 19 cm) cylindrical lizard with glossy, rounded scales. The head is small, covered by large scales, holding small, dark eyes (Stebbins 2003). The limbs are reduced and appear too small for the elongate body. Sharp contrasting black, creamy yellow, and copper stripes extend down the length of the body where they fade into a rounded radiant blue tail. As they age, the vibrant blue tail gradually fades to a more greyish or coppery tone.

Western skinks are found in open, sunny habitats throughout the Klamath Mountains in elevations below 5,000 ft (1,524 m). They are common in grasslands, oak savannah with scattered rocks and rotting logs, mixed oak-conifer woodlands, and even open gallery pine forests. They also frequent the edges of riparian zones, from rivers to meadows, as long as there is enough cover–like logs, rocks, and deep leaf litter. Western skinks are even found in serpentine barrens and along south facing roadcuts with piles of loose cobbles and flat boulders. They seem to select cover objects and burrows that maintain a more humid envelope compared to the surrounding landscape.

Western skink. Photo by Jamie Bettaso.

Though western skinks are active by day and common, they are seldom seen out in the open. This has to do with their small size, secretive nature, and tendency to live and forage under cover—inside rotting logs, below boulders, and through leaf litter. Thus, they are most often seen when moving a log or flipping a stone. Keep watch for a flash of that bright tail scurrying from leaf pile to rodent burrow.

Western skinks prey on beetles, crickets, spiders, and isopods using a keen sense of smell that involves frequent tongue flicking (using a partially forked tongue) (Stebbins and McGinnis 2012). Predation involves cat-like stalking and pouncing (Nussbaum et al. 1983). Due to their secretive nature, the primary predators of this species are unknown but likely include alligator lizards, snakes, small mammals, and ground dwelling birds. To avoid predators, they scurry under or into cover, using their smooth scales, rounded bodies, and reduced limbs to push into grass, litter or soft soil and vanish. If seized, they writhe violently, trusting their slick (osteoderm reinforced) scales will permit escape. They are also quick to part with their vivid blue tails which rapidly twitch and distract predators. Western skinks are oviparous and the female will dig into moist soil and lay a clutch of 2–10 eggs. As in many skinks, the mother tends her eggs and guards the hatchlings briefly before dispersal (St. John 2002). Longevity isn't known, but related species reach 5–7 years.

Family Teiidae – Whiptails

The teiids are a New World lizard family, ranging from Chile and Argentina into the northern US and across the West Indies. The 170 species are diurnal, alert, and active animals that occupy a range of habitats. A successful genus, *Aspidoscelis*, occurs in the US. The roughly 50 species are key components of southwestern deserts and other arid landscapes. In the Klamath Mountains, the western whiptail lizard is the only species.

This is the longest lizard here (up to 13 in, 33 cm) with a long, dark tapering tail that is two or even three times longer than the body. The whiptail's back is covered in tiny granular scales while the belly is protected by 8 rows of large, square scales. Unique black blotches, long golden-brown stripes down the back, and black and white marbling along the sides of the abdomen are distinct among our regional

Western whiptail. Photo by Chad Marks-Fife.

lizards. The relatively narrow head has symmetrical scales and a spear-shaped snout for probing litter and soil in search of prey. Large muscular hind limbs with long toes make whiptails the fastest lizard in the region. Males tend to be slightly larger with stouter forelimbs and darker ventral coloration.

The western whiptail is the rarest lizard in the Klamath Mountains and little is known regarding regional status or distribution. Here they prefer flat open areas lacking dense grasses but with thinly dispersed shrubs used for cover including sandy or gravelly river bars, dry washes, and rocky benches. They are occasionally found in open chaparral along the arid foothills near Shasta and Whiskeytown reservoirs. Early collections by Grinnell and Camp (1917) and by Henry Fitch in 1934 found them near the confluence of the Pit and McCloud rivers that were later flooded by Shasta Reservoir (Museum of Vertebrate Zoology records, UC Berkeley). These observations are not surprising given documentation in nearby lowlands of the Sacramento River watershed and throughout California's Central Valley. Remarkably, western whiptails have been found further west at isolated locations along a 39-mile stretch of the Trinity River from Lewiston in 2021 (iNaturalist obs.[2021]), Douglas City (Bury 1970), and earlier in 1955 around Big Bar (Slater Museum of Natural History). They have also been found in the past decade near Junction City and Big Bar along flat open sandy river benches with southern exposure (Jamie Bettaso and Don Ashton, pers. comm.). The flood regime of the Trinity River has been dramatically reduced since it was dammed, causing riparian vegetation to encroach into the floodplain potentially reducing historic whiptail lizard habitats in this region.

The western whiptail is most likely encountered in mornings but may stay active later into the day compared to other species because of a high tolerance for heat. However, during the hottest periods of the summer, they commonly become inactive underground (aestivate) and later reappear after peak temperatures decrease (Storm and Leonard 1995). These alert and energetic animals are constantly on the go, roaming from shrub to shrub, looking about with quick, jerky movements, and digging in leaf litter and soil. Prey are detected by sight, sound, and scent, with the aid of a partially forked tongue. They eat spiders, mites, scorpions, centipedes, grasshoppers, butterflies, termites, beetles, and even other lizards (Stebbins 2003). They produce up to four eggs in underground burrows. Longevity is likely 3–7 years (Petermann et al. 2017).

Serpentes – Snakes

Snakes are the premier limbless vertebrates. While body elongation and limblessness has occurred many times in vertebrates, from eels and caecilians to worm lizards and legless lizards, snakes are by far the most successful of evolution's many experiments with this interesting body plan (Greene 1997). With nearly 4,000 species (in 25–31 families), snakes represent over ⅓ of living reptile diversity and occur almost everywhere on the globe (absent only from Antarctica and some islands) (Uetz et al. 2021). And though snakes might appear to be just simple tubes with scales, they actually exhibit a surprising degree of specialization and body size variation.

Snakes are actually highly derived lizards, so in an evolutionary sense, are just a special kind of lizard (in the same way that we are just a type of Old World primate). However, snakes differ from other lizards in so many key respects that they have always been treated as separate. To achieve limblessness and body elongation, snakes have turned off the genes that build limbs in other tetrapods and turned up the genes that add elements to the backbone. Thus, the long snaky form emerges from adding pre-caudal (trunk) vertebrae—up to 300–400, more than any other vertebrate animal (Greene 1997). Other unique traits include no eyelids—snake eyes are always open, and covered by a clear scale (called the spectacle)—and no external ears, though snakes can still detect low frequency sounds, especially ground-borne vibrations (Lillywhite 2014).

All snakes are predators and most are carnivores. Both snakes and lizards are limited to prey they can fit in their mouths because they cannot chew or use claws to tear chunks. Snakes, unlike lizards, have flexible skulls and lower jaws specialized for accommodating larger prey. The functional morphology of the snake skull is an evolutionary marvel that allows snakes to ingest bulky prey that can be larger than the snake's head or torso, or even weigh more than the snake itself (Greene 1997)! Snake vision is less acute than lizard, and they rely strongly on scent to find prey. With a marvelous forked tongue to "taste" the air, water, and substrate, it can

Northern rubber boa. Photo by Spencer Riffle.

pick up different concentrations of odor on the left and right sides, allowing them to smell in stereo and to track prey with incredible efficiency (Lillywhite 2014). It's no wonder that many biologists consider snakes among the ultimate predators.

The Klamath Mountains contain 15 different snake species from three families (Table 1). These animals range from diminutive slug eaters of the leaf litter, to graceful garter snakes that frequent the same swimming holes as you, to pit vipers that are key in regulating local rodent populations.

Family Boidae – Boas

The word "boa" invokes images of a powerful serpent whose heavy coils droop from tropical tree branches, ready to constrict hapless prey in the jungle. As such, people are often surprised to learn that the Klamath Mountains are home to one of three boa species found in the United States. Our very own, the northern rubber boa, is rather short and stout (<32 in, 83 cm) and less imposing than its tropical cousins (Stebbins 2003). In fact, they are among most gentle and docile of snakes.

The rubber boa is uniformly brown on top ranging from tan to darker olive-brown or gray-brown, with a lighter underside. Fine, glossy-smooth scales and loose skin cover the body, creating a rubbery appearance. Their eyes are small, and generally yellow, with distinct cat-like vertical pupils. They also have short blunt tails, shaped like a head, terminating in larger scales. These combined features make this snake one of the most unique in the Klamath Mountains. They are also among the most widespread snakes in the region, occurring in a wide variety of habitats and elevations, even up to 7,500 ft (2,286 m) where they frequent grasslands, meadows, riparian edges and rocky hillsides across mixed evergreen, and montane forests. Despite this geographic and ecological breadth, they are rarely encountered (or noticed) because of their cryptic coloration, secretive and sluggish nature, and crepuscular and nocturnal habits. Aside from finding rubber boas under logs, rocks or other cover, most boa observations are made around dusk or during night hikes (using a headlamp) when they are out in the open, even on cooler nights that dip into the mid-50s (13°C) and far too cold for the other regional snakes.

Rubber boas use their strong bodies and small head with countersunk jaws to burrow into soft or loose soil, downed decaying wood, rock outcrops, and rodent burrows. They are also good climbers and have been found several feet up in trees, presumably in pursuit of prey. They kill their prey through constriction—wrapping body coils tightly around their victim until the heart is stopped. Primary prey are mice and nestling mice but also birds and their nestlings, lizards, snakes, and even salamanders (Storm and Leonard 1995, Rodriguez-Robles et al. 1999). Because rubber boas are slow they have adopted an unusual antipredator strategy. When confronted by a foe, they will curl the body over the head and then waive the blunt tail above the body, as a decoy head—if the predator bites the decoy, it will get a mouth full of foul musk and feces (Stebbins and McGinnis 2012). This ruse works well, judging by the number of scars seen on the thick tails of older rubber boas.

Rubber boas are slow growing and late to mature, taking roughly six years for males and eight years for females to become reproductive, and able to live up to 35 years old (Hoyer 1974 and pers. comm.). Like almost all boids, they are viviparous, giving live birth up to eight young in the summer (Hoyer 1974, Stebbins 2003).

Family Colubridae – Colubrid snakes

The Colubridae represents the most species rich snake family in the world with over 2030 species, accounting for 52 percent of all snake diversity (Utez et al. 2021). Thirteen of the 22 reptile species (59 percent) found in the Klamath Mountains are colubrids (Table 1). From gopher snakes stretching to nearly two meters in length, down to diminutive sharp-tailed snakes smaller than a pencil, this group is full of unique wonders. Species occurring in the Klamath Mountains can be divided into 5 general groups: garter snakes, gopher snakes, kingsnakes, racers and whipsnakes, and small woodland snakes.

Garter Snakes (*Thamnophis*)

Garter snakes are among the most common and ubiquitous snakes in the US. The 35 species range across nearly all North America, from southeast Alaska and the lower Northwest Territories in Canada (further north than any other reptile) down to Costa Rica and Panama. These familiar animals are popular in the pet trade and in research labs because of their docile nature, ease of care, and ability to habituate to captivity. Because they are diurnal and active, and generally semiaquatic, they are frequently encountered in the same places we prefer to visit, recreate in, and exploit including lakes, ponds, streams, and picturesque meadows.

There are four species in the Klamath Mountains: the aquatic garter snake, terrestrial garter snake, common garter snake, and northwestern garter snake. Note our area also contains two distinct subspecies of the terrestrial garter, the coast garter snake and mountain garter snake that are morphologically and ecologically distinct and may be incipient species (Hallas et al. 2021). All garter snakes are relatively slender and small to moderate in size, with adults ranging from 12–40 in

Clockwise from top left: Aquatic garter snake (photo by Alan St. John), coast garter snake, common garter snake, northwestern garter snake. Photos by Spencer Riffle.

Reptiles

(30–102 cm). Garter snakes are instantly recognizable by a bold and bright dorsal stripe running down the middle of the back, from just behind the head to above the tail, and a similar lateral stripe along each side of the body which demarcates the dorsal and ventral scales. Though sometimes muted or faded in some populations of aquatic and northwestern garter snakes, stripes are usually obvious. The bold, midbody stripe is not found on other regional snakes.

Local garter snakes can be difficult to distinguish and animals in hand are needed for examination of subtle traits. To add to the difficulty, there is variation across and within populations and some forms will even hybridize—making it difficult for even the experts to correctly identify the species. Refer to the ecological differences described below or a field guide (such as St. John 2002, Stebbins 2003). Posting photos to an informed community such as iNaturalist is also a great way to get feedback and learn the species. Refer to the plates here to get you started.

Klamath Mountain species are found in slightly different habitats, suggesting some degree of niche partitioning to avoid competition with one another, though some overlap does occur. The aquatic garter snake is aptly named, as it is rarely more than a few meters from water. It is the only snake in our region that exploits lotic habitats, favoring rivers, creeks, and streams with rocky substrates, woody debris, and riparian vegetation. It is also found at ponds and lakes from low to high elevations. Here it meets the common garter snake, which is found in the same ponds and lakes with natural vegetation and shallow banks. But common garters are also found in wet meadows, fens, and marshy habitats. The two forms of terrestrial garter snake also have distinct habitat preferences, with the coastal form found in coastal fields and dry meadows while the mountain form is found at ponds, lakes, reservoirs, as well as the wet seeps leading to and from such water bodies. The northwestern garter snake prefers meadows, as well as natural and man-made forest patches with sufficient grass and other low-lying vegetation. It is interesting to note that coastal garter snake and northwestern garter prefer similar habitats—they are not associated with standing water but inhabit places with moisture in the form of coastal fog. These are the only lineages restricted to the outer Coast Ranges and western maritime slope of the Klamath Mountains. The northwestern garter snake has a limited distribution in the region, largely found west of the Cascade Range in Washington and Oregon, but then squeezed along the coast from near Bridgeview, OR to Trinidad, CA, the terminus of the species range. All the other species can be found throughout the entire range of the Klamath Mountains (see range map).

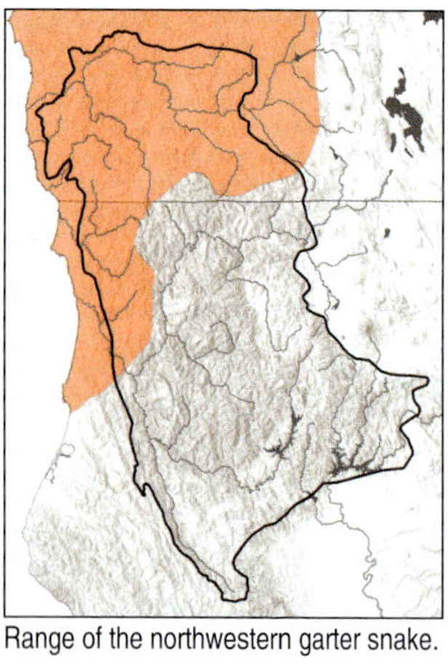
Range of the northwestern garter snake.

Garter snakes often occur locally in great abundance and thus play a key role in regulating prey populations (both amphibian and fish) and in transferring nutrients between aquatic and terrestrial systems (Rossman et al. 1996). All species are active foragers, using relatively good eyesight to spot movement or the outline of prey, coupled with a keen sense of smell to stay on the hunt (the forked tongue works even under water). These animals often work the edge of a pond, then dart in to grab a calling frog, or probe the margins of a log to find a salamander hidden underneath. Of these species, the aquatic garter snake is the most capable aquatic hunter, taking both invertebrate and vertebrate prey, ranging from leeches and tadpoles to newts and fish (Rossman et al. 1996). The species has an impressive ability to forage in the cold streams of the Klamath Mountains, diving for long periods of time and surfacing with larval coastal giant salamanders or trout. The aquatic garter snake will even use their tongues (a colorful red and black) as little lures to catch fish—that's right, some garter snakes can fly fish with a forked tongue (Welsh and Lind 2000). The terrestrial and common garter snake are known for having perhaps the broadest diets of all North American snakes (Rossman et al. 1996). They will eat just about anything including carrion, Earthworms, banana slugs, bullfrogs, toxic toads, poisonous newts, prickly sculpin, spiny fence lizards, fledgling meadowlarks, deer mice, and voles. In fact, both species may even employ venom (using modified salivary glands at the rear of the mouth) to help immobilize prey (though they lack fangs and must clamp on prey to let venom trickle into wounds), and some larger terrestrial garters have been seen trying to "constrict" small mammals by pressing them against burrow walls or throwing sloppy coils around their fuzzy victims (Vest 1981, de Queiroz and Groen 2001). The northwestern garter snake differs from its three larger and often more aquatic cousins in focusing on small terrestrial prey—chiefly slugs, worms, and amphibians.

Garter snakes are prey to a wide array of predators owing to their snackable size but also because they are out by day, foraging in and around open aquatic habitats. Predators include fish, bullfrogs, other snakes as well as mesocarnivores like racoons, skunks, felids, and canids. Predatory birds, especially corvids and also large wading birds such as herons and egrets also prey on them (Rossman et al. 1996). Because of the large selection of predators, garter snakes are always on the lookout, ready to dive into water, flee into thickets, or hide in nearby burrows. Natural selection has sculpted quickness, and the stripe along the back doesn't provide predators with a fixed reference point—making it difficult for prey to judge where to grasp the snake. Interestingly, those that are more botched than striped tend to freeze and rely on crypsis, rather than bolt (Brodie 1993). If cornered, garter snakes may put up a blustery display, coiling the body, flattening and widening the head, and striking. Garter snakes rarely bite but larger aquatic garter snakes do. All snakes also possess paired musk glands located at the base of the tail (called anal scent glands) that they use in defense by expelling a stinky liquid. Though many snakes are hesitant to use the musk glands, garter snakes have a quick trigger and an especially foul musk.

Mating occurs early in the season, typically as animals emerge from hibernation in early to mid-spring. Females are capable of storing the sperm from multiple male partners (for multiple years) to improve the genetic variation of each litter. All garter snakes give birth to live young, typically by late summer or early fall. The number of offspring produced varies and is related to female body size. In our area, most females produce litters of 6–10 neonates, but as few as 1–3 or as many as 20–30 are possible. Mark-recapture work suggests that most species can live for 12–15 years in the wild, and possibly longer (Fitch 1999).

The Wild Reproductive Biology of Reptiles

The reptiles of the Klamath Mountains display much of the remarkable variation in reproductive biology that characterizes non-avian reptiles as a whole. For example, there are dozens of asexually reproducing squamates (lizards and snakes) from across the globe. That's right, no daddies needed. Some lizard and snake mommies are capable of producing exact clones of themselves—a feat no birds or mammals appear capable of.

In our region, both the aquatic and terrestrial garter snakes are known to rarely reproduce asexually, if females have no mating opportunities for many years. Several North American rattlesnakes can perform this same trick, so it's a good bet that our Pacific rattlesnake females can too. And though our western whiptail lizard is an ordinary bisexual species, it has over a dozen unusual relatives in the southwest—species of whiptails that are unisexual. These all female whiptail species formed by hybridization events between bisexual species, creating hybrids that are now capable of generating identical female clones.

On the topic of males and females, even the mode of sex determination varies in non-avian reptiles. In birds and mammals, the sex of the developing embryo is determined by genes that initiate gonadal development. This genotypic sex determination (GSD) results in female birds when embryos have ZW sex chromosomes (males are ZZ) or male mammals when embryos have XY chromosomes (females are XX). Reptiles show both of these patterns. For example, the phrynosomatid lizards, such as our western fence lizard, have XY males just like us mammals, while all snakes have ZW females like birds. Other reptiles have odd variations of those sex chromosomes, such as ZZW, ZWW, or even ZZWW.

And perhaps most bizarre, some reptiles have no sex chromosomes at all! Instead, sex determination is environmental, and results from the temperatures that embryos experience during a critical developmental period. This temperature-dependent sex determination (TSD) is seen across the reptile tree, including many lizard groups, tuatara, all crocodilians, and most turtle lineages. For example, in our western pond turtle, low temperatures result in clutches of male turtles, while higher temperatures result in females, and intermediate temperatures produce clutches of mixed sex of offspring.

Following mating, sex determination, and fetal development, mothers need to bring their offspring into the world. All birds do this by laying eggs (oviparity), while all mammals except the monotremes give birth (viviparity). Here again, non-avian reptiles surprise us by exhibiting both reproductive modes. While all crocodilians, turtles, and tuatara are oviparous, roughly 20 percent of all squamate species are viviparous. In fact, viviparity has evolved independently in lizards and snakes over 120 times, more so than in any other tetrapod group. There are lineages that are entirely viviparous (e.g., night lizards), lineages that contain species that are either oviparous or viviparous (e.g., phrynosomatid and anguid lizards, boid and colubrid snakes), and there is even a lizard species where females can reproduce either way (even in the same litter!).

A female western skink protecting her clutch of eggs along the Chetco River. Photo by Alan St. John.

Squamates also show a great range in the degree to which developing fetuses are connected and nourished by the mother, from relatively simple egg retention in utero, where the fetus is largely provisioned by yolk reserves, to forms of placentation where the developing fetus forms an intricate connection to the mother's uterus, every bit as complex and sophisticated as in placental mammals, and rivaling that of marsupial mammals. In some of these species (e.g. some scincid lizards), reptile mothers have prolonged pregnancies and give birth to only one or two well-developed young that are so large (e.g., >30 percent the size of mother) that most mammal moms would cringe at the thought of delivery.

Why viviparity has evolved so many times in squamates is a topic of great interest. Herpetologists hypothesize that viviparity is an adaptation to cold conditions, allowing squamates to occupy habitats and parts of the globe that might be too cold for most oviparous reptiles. Under this model, moms can actively thermoregulate while pregnant, keeping their developing young at warmer and more stable temperatures than eggs would experience in a nest at the same site. Indeed, the proportion of viviparous lizard and snake species increases with both elevation and latitude. For example, 6 of the 15 snake species (40 percent) of the Klamath region are viviparous, and of those, only the live-bearing species can be found occupying alpine habitats along the highest peaks in the Klamath Mountains. Similarly, the region's two alligator lizard species appear to segregate by habitat and elevation, with the oviparous southern alligator lizard (*Elgaria multicarinata*) leaving cooler coniferous forests and meadows to its viviparous northern alligator lizard (*E. coerulea*) cousin. Thus, aspects of reproductive biology may be partly responsible for the ecogeographic patterns seen in Klamath Mountain reptiles.

Gopher Snakes (*Pituophis*)

Gopher snakes are the largest-bodied and among the longest serpents in the Klamath Mountains, regionally reaching up to 72 in (183 cm), but capable of even greater sizes (100 in, >2.5 m!) in the southwest deserts (Stebbins 2003). They are cream to yellow-tan with extensive brown to black blotching across the body—providing excellent camouflage. Their dorsal scales are well-keeled but the surface is still smooth. The ventrum is usually cream or yellow and may have dark spots from the dorsum. The head is relatively large and triangular, with a conspicuous dark stripe that runs through the eyes and down to the posterior margin of the upper jaw.

Gopher snakes are diurnal but sometimes crepuscular or nocturnal when warm enough. They are among the most frequently observed snakes in the Klamath Mountains and are particularly common in lowlands with an abundance of burrowing rodents—especially grasslands, chaparral, oak savannah, and dry open conifer woodlands. They have even been seen in the Trinity Alps subalpine to 6,600 ft (2,011 m) on south facing slopes with pine and huckleberry oak. Gopher snakes are also common, welcome inhabitants around dwellings and farms where they serve as rodent control agents.

Gopher snakes are powerful constrictors that take a range of prey including lizards, snakes, birds, and mammals (St. John 2002, Stebbins 2003). However, they prefer small mammals like voles, mice, chipmunks, gophers, ground squirrels, and rabbits (Rodríguez-Robles 2002). They probe open rodent burrows, brush, and along the edges of rock outcrops using their heads as spades and moving aside loose soil. They are skilled climbers and will ascend shrubs and trees to feed on bird eggs and young birds. Predators include raptors, especially large hawks and eagles, California kingsnakes, and mesocarnivores. To avoid predation, they mimic rattlesnakes. They already resemble rattlesnakes in stout body and similar coloration but gopher snakes really complete the mimicry when cornered—wrapping their body and recoiling their head in the same striking pose used by rattlesnakes. Gopher snakes even puff up their bodies and widen their heads (by flaring out the quadrate bones) to more closely resemble vipers, and then rapidly vibrate their tails against dry vegetation to make a rattling sound. If more bluster is needed, they will add forceful hissing noises, the loudest of any North American snake, and then strike, often with the mouth closed, in a final attempt to bluff predators (Ernst and Ernst 2003). Sadly, this impressive display works almost too well—humans confuse harmless gopher snakes with venomous rattlesnakes and kill them indiscriminately. Gopher snakes are egg layers, and those in our area produce up to nine eggs per year (Nussbaum et al. 1983). These snakes take several years to reach sexual maturity and can live up to three decades (Snider and Bowler 1992).

Juvenile gopher snake. Photo by Spencer Riffle.

Racers and Whipsnakes (*Coluber* and *Masticophis*)

Racers and whipsnakes are long slender snakes with narrow necks and large eyes set on well-defined heads. Large, smooth (unkeeled) dorsal scales give them a sleek appearance. The Klamath Mountains contain three species: the North American racer, striped racer, and striped whipsnake. Most snake babies closely resemble adults, but not those of North American racers. The young are born with dark-brown blotches along the body, set on a lighter cream or tan background (quite similar to the gopher snake), and the blotches gradually fade as young mature and develop their solid adult coloration The North American racer is the only species lacking stripes, and is usually tan or olive on top with a yellowish belly. This snake is also the smallest among the three, rarely exceeding 36 in (91 cm). In contrast, the striped racer is much longer at up to 60 in (152 cm), while the striped whipsnake reaches an impressive 72 in (183 cm) (Stebbins 2003), sharing the title of longest snake in the Klamath Mountains with the gopher snake. While all three species are harmless, they can be quite aggressive and are well-known for biting if (mis) handled—a bad play on the genus name among herpetologists is to call them "Nasty-cophis" because of their ornery disposition.

The North American racer has the broadest distribution among the three Klamath species. They occur throughout the entire Klamath range from valley floors to subalpine meadows to 6,500 ft (1,982 m), indicating a broad tolerance of environmental conditions. They prefer open areas including grasslands, chaparral, edge habitats, and open riparian zones such as dry meadows. In contrast, the striped racer and striped whipsnake are among rarest reptiles in the region. Striped racers enter the range from the southeast with records restricted to open chaparral-lined valleys and wide river canyon foothills in the upper Sacramento River and Trinity River watersheds below 3,000 ft (914 m) (Bury 1970). Outside of the Klamath they are broadly distributed in the California foothills surrounding the Central Valley and south along the California coast into Baja (Stebbins 2003). The striped whipsnake is even more rare than the striped racer in the Klamath Mountains, with less than a dozen records in Oregon and even fewer in California (Fitch 1936, Reilly et al. 2011, VertNet, iNaturalist). Based on these few records, striped whipsnakes are restricted to the lower eastern flanks of the Klamath Mountains north of Mount Shasta and in isolated lowlands below 3,000 ft (914 m) including Etna

North American racer. Photo by Spencer Riffle.

Striped racer. Photo by Spencer Riffle.

Striped whipsnake. Photo by Alan St. John.

Valley, Hilt, Applegate Valley, Illinois Valley, and potentially the Rogue Valley. In these areas they use rocky slopes with open mixed conifer and hardwoods with a brushy understory (St. John 2002). They are broadly distributed outside of the Klamath throughout the Great Basin to the southeast and Columbia River basin to the North, where they are found in shrub-steppe habitat and other desert communities.

You might catch a glimpse of a racer or whipsnake scurrying across a road or path, or perhaps deliberately hunting in the grass or chaparral, watchful with head elevated above the body to spy prey. These active foragers have impressive stamina and are constantly on the move. Unlike many snakes they use keen eyesight to detect the shape or movements of prey. Racers and whipsnakes then use speed and athleticism to intercept their quarry. In fact, these snakes are the fastest in North America, and among the fastest in the world, having been clocked at sprints of up to 10 mph! They are also adept climbers and will hunt in shrubs and chaparral—especially the striped racer and striped whipsnake. All three species feed on amphibians, lizards (especially spiny lizards), snakes, birds and their eggs, and small mammals (Stebbins 2003). Unusual among most North American snakes, the young will prey on insects, especially grasshoppers and caterpillars. The North American racer has the broadest diet among the three species, while the striped racer and striped whipsnake tend specialize on other reptile species, eating lizards and other snakes—including rattlesnakes (they are resistant to the venom)! None of these snakes possess venom or constrict prey (despite the scientific name of the North American racer) but instead simply overpower and eat smaller prey. As active diurnal animals, these are vulnerable to a wide range of predators and

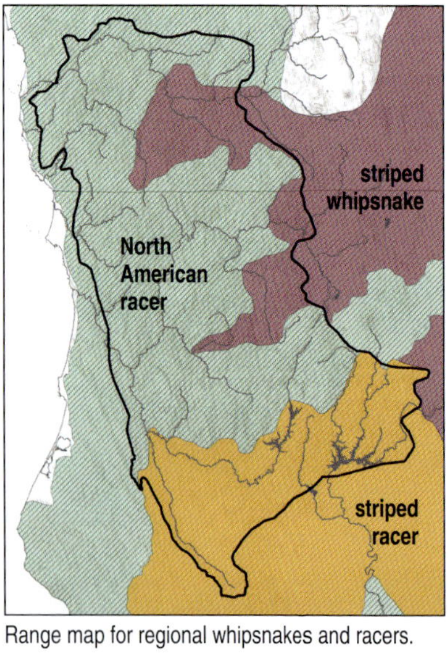

Range map for regional whipsnakes and racers.

Reptiles

are known prey of other snakes, mesocarnivores, and especially raptors (Ernst and Ernst 2003). All three species are egg layers, producing clutches of up to 12 eggs each year (Stebbins 2003). Racers and whipsnakes are thought to live 5–10 years in the wild (Brown and Parker 1984).

Kingsnakes (*Lampropeltis*)

The Klamath Mountains have two species of kingsnakes—the California kingsnake and the California mountain kingsnake. They are among the most beautiful snakes in the region with smooth and glossy dorsal scales adding luster to striking bands of sharply contrasting colors. The California kingsnake has dark brown or black bands offset by cream or white bands, while the California mountain kingsnake has the same colors but the dark bands are interrupted by red. These bold markings are thought to disrupt the outline of the body to visual predators and prey, especially at dawn and dusk when shadows are long and these snakes are active. The pattern is also thought to disorient predators because the bands blur together with rapid movement (called flicker fusion) so that a fleeing snake appears unicolor (Pough 1976). But when the snake freezes, it becomes multicolor and cryptic again, confusing the predator which is now searching for a moving unicolor target. Both kingsnake species rarely exceed 40 inches (102 cm) in length in the Klamath Mountains, but the typical California kingsnake is usually larger and stouter than the California mountain kingsnake.

Although the broad ranges of the two kingsnakes encompass most of the Klamath Mountains, both species have unique habitat preferences that restrict their regional distributions. The California kingsnake is mostly found in open habitats below 2,500 ft (914 m), preferring river floodplains, valley farmlands, chaparral, and lower oak and pine woodlands with rocky or woody debris cover. In contrast, the mountain kingsnake is mainly associated with montane oak and pine dominated forests on slopes perched above mountain stream gorges with an abundance of surface cover objects such as rotting logs, large boulders, stable talus, and leaf litter. They are found as low as 900 ft (274 m) in the rugged Smith River canyon in Gasquet, California, and likely reach the upper montane zone to 5,000 ft (1,524 m) in more inland areas. Both species find their northern range limits just north of the Klamath Mountains and appear to be uncommon in Oregon (Reilly et al. 2011).

California kingsnake (left) and California mountain kingsnake (right). Photos by Spencer Riffle.

Both kingsnakes are crepuscular and most active at dawn and again at dusk—shifting to nocturnal habits during the hottest parts of summer. They are incredibly strong constrictors that can kill large prey, including snakes their own size. Both species are generalist predators, preying on lizards and other snakes, reptile eggs, birds, and rodents. California kingsnakes have a broader diet that also includes frogs and are famed for their ability to kill and consume large rattlesnakes to whose venom they are imperious (Wiseman et al. 2019). Predators of kingsnakes include raptors and mesocarnivores (Ernst and Ernst 2003). Both species are egg layers, with the larger California kingsnake producing a larger annual clutch of 6–12 eggs while the smaller mountain kingsnake producing 3–9 eggs annually (Stebbins 2003). These are long-lived animals, known to reach 25–30+ years (Snider and Bowler 1992).

Small Woodland Snakes (*Contia* and *Diadophis*)

Not all the snakes of the Klamath Mountains are big and bold—there are some tiny gems, seldom glimpsed, but quietly living out of sight in their own hidden world. The common sharp-tailed snake, forest sharp-tailed snake, and ring-necked snake are three such treasures and are the smallest and least seen regional ophidians. Because they are so small, only seasonally active, and secretive by nature, they dwell in interstitial places humans can only ponder—inside logs, under bark, in leaf duff, and between rocky rubble.

From top to bottom: Forest sharp-tailed snake, common sharp-tailed snake, ring-necked snake. Photos by Spencer Riffle.

Sharp-tailed snakes are diminutive, usually just 8–12 inches long (20–30 cm). The forest species is slightly larger, occasionally attaining lengths of 18 inches (45 cm) and with a longer tail (and more caudal scales), while the common sharp-tail rarely reaches 16 inches (40 cm) and possesses a shorter tail (Feldman and Hoyer 2010). Both species possess a blunt head set on an indistinct neck with relatively thick body and a tail that ends in a sharp point. The function of the tail-spine remains unknown, but may function in feeding or burrowing, providing extra purchase when anchored into substrates (Stebbins and McGinnis 2012, R. Hoyer, pers. comm.). The smooth dorsal scales are usually a light red to rusty brown to grey, and there may be two faint lines running alongside the body. The common sharp-tailed snake tends to be redder than the forest species. The ventrum is cream to grey, with distinct dark crossbands at the front of each ventral scute. These dark bands are thinner in the forest sharp-tail, which also have a paler underside, especially the tail (Feldman and Hoyer 2010).

The ring-necked snake is a slightly longer and thinner snake, with adults reaching 8–24 in, (25–65 cm), and even larger in the southwest (Stebbins 2003). This is an unmistakable animal, with a darkened head followed by a striking orange neckband (1–2 scales thick), a light green, olive, slate grey or brownish body, and then a yellow-orange to orange-red underside. Black flecking may suffuse the bright ventral colors, especially at the anterior, which grade from yellow-orange into deeper shades of orange and red moving towards the posterior, such that the underside of the tail may be an arresting scarlet.

The two sharp-tailed snake species occupy different parts of the Klamath Mountains and different habitats. The common sharp-tailed snake favors open habitats at lower elevation with plenty of cover, particularly open meadows, grassland, chaparral, oak savannah and mixed oak and conifer woodland. The forest sharp-tailed snake prefers cooler habitats in mixed oak-conifer forests and at edges or patch openings in mature conifer forests. The two species can be found together where oak and conifer mix. However, their ranges seldom overlap.

Sharp-tailed snakes are cold tolerant and chiefly nocturnal; they can sometimes be seen moving about early on cool mornings or at dusk. These snakes are active early in spring (by March or April) but then hide below ground again as warm weather approaches, often vanishing by early summer (Cook 1960). They re-emerge in early or mid-fall, for a second pulse of activity before a winter hibernation. Ring-necked snakes have similar diel and seasonal patterns, most seen at twilight in spring and early summer. However, once they disappear for the season they do not resurface.

Sharp-tailed snakes are thought to be slug specialists, and have surprisingly long, recurved teeth perfect for grasping slippery mollusk prey (Stebbins and McGinnis 2012). They also take Earthworms and small salamanders. Ring-necked snakes in our area take salamanders as well, but also prey on small frogs and prefer little lizards and other snakes, even taking sharp-tailed snakes and small garter snakes they immobilize with venom delivered from small fangs held at the rear of the mouth (O'Donnell et al. 2007, Wiseman et al. 2007)). Predators are many, including any small or medium sized predator that encounter these slow and innocuous snakes, such as alligator lizards, kingsnakes, shrews, weasels, raccoons, jays, and domestic

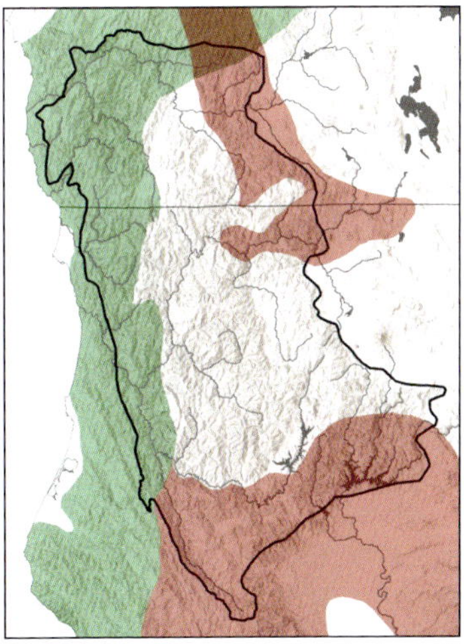

Regional range map for forest sharp-tail (green) and the common sharp-tail (maroon).

cats (Ernst and Ernst 2003). When threatened, sharp-tailed snakes hide their head inside tight coils and remain in a stiff ball, or flip the body over so the ventrum is exposed, which bears resemblance to toxic millipedes. Ring-necked snakes will tuck the head beneath some loose curls while flipping most of the body over to expose the bright ventral colors. At the same time, they twist the tail into a tight spiral (resembling to a corkscrew or thimble), and may wave the tail about, expelling a nasty musk (Greene 1997)—one of the most pungent of western snakes. All three forest gems are oviparous, laying small clutches of relatively large oblong eggs in summer that hatch in the fall. Sharp-tailed snakes generally lay 1–5 eggs and ring-necked snakes 1–10, with the largest females laying the most eggs (up to 18 in some big southwestern forms). Longevity in sharp-tailed snakes is known to exceed 6 years in the wild (Govindarajulu et al. 2011; R. Hoyer pers. comm.) and is likely longer, especially considering that ring-necked snakes can reach 15–18 years in the wild (Fitch 1999).

Family Viperidae – Vipers

Rattlesnakes (genus *Crotalus*) are a unique and diverse group (~50 species) of pit vipers that originated in southwestern North America (likely Mexico) and now reach as far north as Canada and as far south as Argentina. We have just one representative in the Klamath Mountains, the western rattlesnake, the only venomous snake that is potentially dangerous to humans. Though often confused for gopher snakes, rattlers are quite distinctive. Their heads are wide, triangular, and distinct from the neck. They have large eyes with vertical pupils and paired forward-facing heat sensory pit organs between the nostrils and eyes (and larger than nostrils). They are also wide-bodied snakes, though rarely exceeding a length of 40 in (102 cm) regionally (Nussbaum et al. 1983). Finally, they have short and stout tails that terminate in an obvious rattle which is formed by a series of hollow interlocking scale segments. The body color usually matches the dominate substrate and is thus remarkably variable among populations and individuals. Rattlesnake scales are strongly keeled and overlapping, creating a rough texture and muted appearance—different from the glossy sheen of other snakes.

Western rattlesnakes are broadly distributed across much of the western United States where they occupy a wide range of ecological communities. The species is found throughout the Klamath Mountains across all elevations to 7,500 ft (2,286 m) in grassland, chaparral, oak woodland, and open coniferous woodland. They prefer areas with abundant small mammals, especially those with abundant burrows, open rocky outcrops, or large woody debris. Western rattlesnakes are active by day in the cool spring and fall, but then shift to crepuscular

A western rattlesnake consuming a Douglas squirrel in the Scott Mountains. Photo by Colin Harris.

Reptiles

and nocturnal habits during the summer as nights warm. They are mostly sit-and-wait predators and use their keen senses to find a small mammal den or runway, and then lie in wait, sometimes for days, to ambush prey (Stebbins and McGinnis 2012). Using their sophisticated heat-sensing pit organs (sensitive to the temperatures of endothermic prey even 15 m away) they intercept game in total darkness (Ebert and Westhoff 2006). With a quick strike, venom is squeezed from muscular glands at the rear of the head into two specialized hollow teeth (fangs) that fold out when the mouth is open. The snake then quickly releases its target to avoid injury and the venom begins its work. Rattlesnake venom is a complex cocktail of peptides, small proteins, and large enzymes designed to induce hypotension (so prey pass out), promote venom diffusion through cells (and reduce coagulation), and destroy blood, muscle, and other tissues (Mackessy 2010). The venom also includes a homing molecule that acts like a beacon—releasing an odor that rattlers quickly locate. In addition, venom injected inside the prey begins to digest game from the inside out, speeding up the digestion process. Thus, the benefits of venom are vast, allowing rattlesnakes to subdue and consume large and dangerous prey that would otherwise be too difficult to catch and consume (Greene 1997). Indeed, rattlesnakes are well known for eating prey that actually weigh more than they do! The menu is full of small mammals, including a diverse suite of rodents, such as deer mice, chipmunks, squirrels, and even rabbits (Nussbaum et al. 1983). When young, they tend to feed on lizards, producing venom specific to reptile prey, before switching to rodents as they mature (Mackessy et al. 2003).

Despite their venomous punch, rattlesnakes fall prey to their own predators, which include birds of prey, such as great horned owls and red-tailed hawks, and canids including foxes and coyotes (Ernst and Ernst 2003). California kingsnakes and striped whipsnakes are resistant to rattlesnake venom and will occasionally prey on western rattlesnakes (Stebbins 2003). Nevertheless, the bite of a rattlesnake is a good deterrent to predation, and rattlesnakes are renowned for their imposing defensive display—body coiled, head and neck elevated in an S-shaped curve ready to strike, and tail buzzing its “don’t tread on me” anthem.

Rattlesnakes give live birth to 5–12 young once a year, or every other, as conditions allow (Klauber 1972). Rattlesnake moms are good parents. Just a few decades ago herpetologists discovered that after giving birth, the litter stays with the mom for 1–2 weeks, until they have had their first shed and can then see better and use their heat-pits effectively. She stays with them, and even shuttles them in and out of shelter, and will corral them and protect them from any predators (Greene et al. 2002). In fact, newer work shows that females will congregate in rookeries, and will even baby sit for each other while other moms are out trying to get water or a meal (E. Taylor, pers. comm.). Rattlesnakes are long-lived, though a common misperception is that a segment is added to the rattle each year, so that one can count the buttons and determine a snake’s age (like counting tree rings). It is true that a new button is added to the rattle with each skin shed, so the rattle does grow over time as the snake grows. However, a rattlesnake could molt several times in a season, or a rattle might easily break off, so the length of the rattle has little bearing on the age of the snake. Using more scientific approaches, we know that rattlesnakes live 15–30 years and possibly 50–60 years (Snider and Bowler 1992).

Acknowledgements

Many thanks to Kevin Wiseman, Jamie Bettaso, and Jim Agee for reviewing the chapter. Jamie Bettaso, Don Ashton, Bruce Bury, and Richard Hoyer provided many details on Klamath reptiles that enriched this chapter. Spencer Riffle and Alan St. John provided most of the exceptional photographs used throughout the chapter. California Department of Fish and Wildlife, VertNet, HerpMapper, and iNaturalist provided useful datasets for developing range maps. Special thanks to Jamie Allen and family for documenting a western range extension of common sharp-tail snakes into Seiad Valley, CA. Lastly, to Jackson Shedd for discussions regarding the enigmatic Blainville horned lizard.

Literature Cited and Suggested Reading

Bouzid, N.M., J.W. Archie, R.A. Anderson, J.A. Grummer, and A.D. Leaché. 2021. Evidence for ephemeral ring species formation during the diversification history of Western Fence Lizards (Sceloporus occidentalis). Molecular Ecology.

Brodie III, E.D. 1993. Consistency of individual differences in anti-predator behaviour and colour pattern in the garter snake, *Thamnophis ordinoides*. *Animal Behaviour* 45:851-861

Brown, W. S. and W.S. Parker. 1984. Growth, reproduction and demography of the racer, *Coluber constrictor mormon*, in northern Utah. In: *Vertebrate ecology and systematics*. University of Kansas Museum of Natural History, Special Publication 10:13-40.

Bury, R. B. 1970. A biogeographic analysis of the herpetofauna of Trinity County, California. *Journal of Herpetology* 4:165-178.

Bury, R.B. and C.A. Pearl. 1999. Klamath-Siskiyou herpetofauna: biogeographic patterns and conservation strategies. *Natural Areas Journal* 19:341-350.

Bury, R.B., D.R. Ashton, H.H. Welsh Jr, D.A. Reese, and D.J. Germano. 2012. Synopsis of biology. Pages 9-19 *In* Bury, R. B., H.H Welsh Jr, D. J. Germano, D.J. Ashton, editors. Western pond turtle: biology, sampling techniques, inventory and monitoring, conservation, and management. Olympia, Washington: Society for Northwestern Vertebrate Biology. *Northwest Fauna* 7.

Bury, R., G. Bury, D. Ashton, J. Bettaso, and D. Germano. 2019. Longevity of the Western Pond Turtle (*Actinemys marmorata*) based on studies over 50 years. Abstracts from the 2019 annual meeting of the Society of Northwestern Vertebrate Biology, Grand Mound, WA. Northwestern Naturalist 100:132-164.

Clause, A. R. and E.A. Capaldi. 2006. Caudal autotomy and regeneration in lizards. *Journal of Experimental Zoology Part A: Comparative Experimental Biology* 305:965-973.

Cook, S.F. 1960. On the occurrence and life history of *Contia tenuis*. *Herpetologica* 16:163-173.

de Queiroz, A. and R.R. Groen. 2001. The inconsistent and inefficient constricting behavior of Colorado western terrestrial garter snakes, *Thamnophis elegans*. *Journal of Herpetology* 35:450-460.

Ebert, J. and G. Westhoff. 2006. Behavioural examination of the infrared sensitivity of rattlesnakes (*Crotalus atrox*). *Journal of Comparative Physiology A* 192:941-947.

Ernst, C.H. and E.M. Ernst. 2003. *Snakes of the United States and Canada*. Smithsonian Books, Washington DC.

Ernst, C.H. and J.E. Lovich. 2009. *Turtles of the United States and Canada*. John Hopkins University Press, Baltimore, MD.

Espinoza, R.E. 2009. Western Fence Lizard *Sceloporus occidentalis* Baird and Girard, 1852. In: *Lizards of the American Southwest*. Rio Nuevo Publishers, Tucson, AZ. 234-237.

Feldman, C.R. and G.S. Spicer. 2006. Comparative phylogeography of woodland reptiles in California: repeated patterns of cladogenesis and population expansion. *Molecular Ecology* 15:2201-2222.

Feldman, C.R. and R.F. Hoyer. 2010. A new species of snake in the genus *Contia* (Squamata: Colubridae) from California and Oregon. *Copiea* 2010:254-267.

Fitch, H.S. 1935. Natural history of the alligator lizards. *Transactions of the Academy of Science of Saint Louis* 29:1-38.

Fitch, H.S. 1936. Amphibians and reptiles of the Rogue River Basin. *American Midland Naturalist* 20:381-424.

Fitch, H.S. 1999. A Kansas snake community: composition and changes over 50 years. Krieger Publishing, Malabar FL.

Gauthier, J., R. Estes, and K. de Queiroz. 1988. A phylogenetic analysis of Lepidosauromorpha. In: *Phylogenetic relationships of the lizard families*. Stanford University Press, Stanford, CA. 15-98.

Govindarajulu, P.,et al. 2011. Relevance of life-history parameter estimation to conservation listing: case of the sharp-tailed snake (*Contia tenuis*). Journal of Herpetology 45:300-307.

Greene, H.W. 1997. *Snakes: the evolution of mystery in nature*. U. of California Press, Berkeley, CA.

Greene, H.W., May, P.G., Hardy Sr, D.L., Sciturro, J.M. and T.M. Farrell. 2002. Parental behavior by vipers. In: *Biology of the Vipers*. Eagle Mountain Publishing, Eagle Mountain UT. 179-205.

Grinnell, J. and C.L. Camp. 1917. A distributional list of amphibians and reptiles of California. University of California Publications in *Zoology* 17:127-208.

Hague, M.T., C.R. Feldman, E.D. Brodie Jr, and E.D. Brodie III. 2017. Convergent adaptation to dangerous prey proceeds through the same first-step mutation in the garter snake *Thamnophis sirtalis*. Evolution 71:1504-1518.

Hallas, J.M., T.L. Parchman, and C.R. Feldman. 2021. The influence of history, geography and environment on patterns of diversification in the western terrestrial garter snake. *Journal of Biogeography*.

Holland, D.C. 1994. The western pond turtle: habitat and history. Final Report. for U.S. Department of Energy, Bonneville Power Administration, Portland, Oregon.

Hoyer, R.F. 1974. Description of a rubber boa (*Charina bottae*) population from western Oregon. Herpetologica, 275-283.

Klauber, L.M. 1972. *Rattlesnakes: their habits, life histories, and influence of mankind*. University of California Press, Berkeley CA.

Lavin, B.L., Wogan, G.O., McGuire, J.A, and C.R Feldman. 2018. Phylogeography of the northern alligator lizard (Squamata, Anguidae): hidden diversity in a western endemic. *Zoologica Scripta* 47:462-476.

Lillywhite, H.B. 2014. *How snakes work: structure, function and behavior of the world's snakes*. Oxford University Press, New York, NY.

Mackessy, S.P. 2010. Evolutionary trends in venom composition in the western rattlesnakes (*Crotalus viridis sensu lato*): toxicity vs. tenderizers. *Toxicon* 55:1463-1474.

Mackessy, S.P., K. Williams, and K.G. Ashton. 2003. Ontogenetic variation in venom composition and diet of *Crotalus oreganus concolor*: a case of venom paedomorphosis? *Copeia*, 2003:769-782.

Nussbaum, R.A., E.D. Brodie, Jr., and R.M. Storm. 1983. Amphibians and Reptiles of the Pacific Northwest. University Press of Idaho, Moscow, ID, USA.

O'Donnell, R.P., K. Staniland, and R.T. Mason. 2007. Experimental evidence that oral secretions of northwestern ring-necked snakes (*Diadophis punctatus occidentalis*) are toxic to their prey. *Toxicon,* 50:810-815.

Petermann, H., N.M Koch, and J.A. Gauthier. 2017. Osteohistology and sequence of suture fusion reveal complex environmentally influenced growth in the teiid lizard *Aspidoscelis tigris*—implications for fossil squamates. Palaeogeography, Palaeoclimatology, Palaeoecology, 475: 12-22.

Pianka, E.R. and L.J. Vitt. 2003. *Lizards: windows to the evolution of diversity*. University of California Press, Berkeley, CA.

Pough, F.H. 1976. Multiple cryptic effects of crossbanded and ringed patterns of snakes. *Copeia,* 1976:834-836.

Reilly, J., E. Meyers, D. Clayton, and R. Nauman. 2011. Snake surveys in Jackson, Josephine and southern Douglas counties, Oregon. Bureau of Land Management. 19p.

Rhodin, A.G.J., J.B. Iverson, R. Bour, U. Fritz, A. Georges, H.B. Shaffer, and P.P. van Dijk. 2017. Turtles of the World: Annotated Checklist and Atlas of Taxonomy, Synonymy, Distribution, and Conservation Status (8th Ed.). *Chelonian Research Monographs* 7:1-292.

Rodríguez-Robles, J.A. 2002. Feeding ecology of North American gopher snakes (*Pituophis catenifer*, Colubridae). *Biological Journal of the Linnean Society* 77:165-183.

Rodríguez-Robles, J.A., C.J. Bell, and H.W. Greene, 1999. Gape size and evolution of diet in snakes: feeding ecology of erycine boas. *Journal of Zoology* 248:49-58.

Rossman, D.A., N.B. Ford, and R.A. Seigel. 1996. *The garter snakes: evolution and ecology*. University of Oklahoma Press, Norman OK.

Ryan, M.J. 2009. Common Sagebrush Lizard *Sceloporus graciosus* Baird and Girard, 1852. *In* Lizards of the American Southwest. Rio Nuevo Publishers, Tucson, AZ. 218-221.

Smith, H.M. 1946. *Handbook of Lizards: Lizards of the United States and of Canada*. Comstock Publishing Company, Ithaca, NY.

Snider, A.T. and J.K. Bowler. 1992. Longevity of reptiles and amphibians in North American collections (2nd Ed). *Herpetological Circular* 21.

St. John, A. 2002. *Reptiles of the Pacific Northwest: California to Alaska, Rockies to the coast*. Lone Pine Publishing, Renton WA.

Stebbins, R.C. 2003. *A field guide to western reptiles and amphibians* (3rd Ed). Houghton Mifflin, Boston MA.

Stebbins, R.C. and S.M. McGinnis. 2012. *Field guide to amphibians and reptiles of California*: (revised edition). University of California Press, Berkeley CA.

Storm, R.M. and W.P. Leonard. 1995. *Reptiles of Washington and Oregon*. Seattle Audubon Society, Seattle WA.

Thill, V., H. Moniz, M. Teglas, M. Wasley, and C. Feldman. 2022. Preying dangerously: Black widow spider venom resistance in sympatric lizards. R. Soc. Open Sci. 2022, 9, 221012.

Tinkle, D.W., A.E. Dunham, and J.D. Congdon. 1993. Life history and demographic variation in the lizard *Sceloporus graciosus*: a long-term study. *Ecology*, 74:2413-2429.

Uetz, P., P. Freed, R. Aguilar. and J Hošek (eds.). 2021. *The reptile database*, http://www.reptile-database.org, accessed June 2021.

Ultsch, G. 2006. The ecology of overwintering among turtles: Where turtles overwinter and its consequences. *Biological Reviews* 81:339-367.

Vest, D.K. 1981. The toxic Duvernoy's secretion of the wandering garter snake, *Thamnophis elegans vagrans*. *Toxicon* 19:831-839

Welsh Jr, H.H. and A.J. Lind. 2000. Evidence of lingual-luring by an aquatic snake. *Journal of Herpetology* 34:67-74.

Wiseman, K.D., K.R. Marlow, J.S. Minton, and A.E. Herman. 2007. Natural history notes: *Thamnophis couchii* (Sierra garter snake). Predation. Herpetol. Rev. 38 (3):345.

Wiseman, K.D., H.W. Greene, M.S. Koo, and D.J. Long. 2019. Feeding ecology of a generalist predator, the California kingsnake (*Lampropeltis californiae*): why rare prey matter. *Herpetological Conservation and Biology* 14:1-30.

Birds of the Klamath Mountains

Greg Gray, Michael Kauffmann, Dan Barton, and Tiana Williams-Clausen

Introduction

Birds, members of Class Aves, are the most speciose group of four-limbed vertebrates (Superclass Tetrapoda). While taxonomists recognize approximately 10,000 species worldwide, recent applications of evolutionary species concepts, and our improving understanding of cryptic diversity, has led some to estimate the true number of bird species is much higher. This suggests we have yet to fully describe the diversity of birds worldwide.

Likewise, we are still uncovering the evolutionary origins of birds and flight. The relationship of birds to dinosaurs is well-accepted, but our understanding of exactly how extant birds relate to the richly preserved, feathered, and sometimes flighted dinosaur fossils discovered in recent decades is somewhat lacking. Yet appreciate birds we do! Birds play a pivotal role in the origins and contemporary practices of indigenous and colonial cultures around the world—most likely due to birds being the most easily watched and accessible components of vertebrate faunas. Thus, birds play a vital role in understanding countless stories, symbols, words, and metaphors within and among cultures (see Forest for the Birds sidebar). Birds also hold a unique position in the evolving culture and history of field naturalists—given their accessibility, cultural significance, and beauty—the hobby and history of "birding" has propelled extensive contributions of amateur naturalists to their study, and participation by both amateur and professional naturalists in their cultural celebration, scientific study, and conservation.

A Crossroads in Avian Biogeography

The Klamath Mountains are an intersection of species richness and avian biogeography, connecting the bird communities of the humid coastal rainforests with the inland bird communities of the arid West and the cool climate of the high mountains. Just as the region's flora and fauna are strongly influenced by a combination of sharp and soft gradients in temperature, elevation, seasonality, aridity, soil types, and geology—and varying degrees of isolation—so too are its birds. Bordered by the Coast Ranges, the Cascades, the Sierra Nevada, and the Central Valley, the Klamath Mountains share similarities with these inland provinces but are also directly affected by the coastal climate. This results in a unique melding of bird assemblages that includes species from these adjoining bioregions.

Trail et al. (1997) highlighted this crossroads concept, compiling a list of 37 breeding species that reached a range limit in the Klamath Mountains: 18 reaching a western limit, 3 northern and eastern limits, 8 a northern limit, and 7 a southern limit. This pattern of range convergence underscores the habitat variability the region provides, resulting in a species mix that includes rufous-crowned sparrows (*Aimophila ruficeps*) from dry chaparral lowlands, evening grosbeaks (*Coccothraustes vespertinus*) from northern coniferous forests, black-billed magpies (*Pica hudsonia*) from Great Basin shrub-steppe, and chestnut-backed chickadees (*Poecile rufescens*) from humid coastal forests.

Evening grosbeak (*Coccothraustes vespertinus*) by Greg Gray.

Allen's hummingbird (*Selasphorus sasin*) by Rob Fowler.

Most of the bird species seen on a trip through the Klamath Mountains are also found in other areas of the western United States, but seldom do so many occur in close regional proximity or in sympatry (range overlap). This results in counter intuitive species assemblages, in which the coastal taxon of the Steller's jay (*Cyanocitta stelleria frontalis*), a subspecies associated with humid coastal forests, overlaps with the white-headed woodpecker (*Leuconotopicus albolarvatus*), a species of higher, drier pine forests (Anderson and Grinnell 1903). The convergence of geologic provinces and habitat types is further illustrated by the sympatry of the black-capped chickadee (*Parus gambeli*), mountain chickadee (*Parus rufescens*), and chestnut-backed chickadee (*Poecile rufescens*)—one of the few regions in North America where this occurs.

The crossroads concept is underscored by two of our smallest migratory visitors, the Allen's hummingbird (*Selasphorus sasins*) and the rufous hummingbird (*S. rufus*), which occupy a 190-mile-long hybrid zone of overlap from southern Oregon into Humboldt County, California, centering on the coast and western edge of the Klamath Mountains. The rufous hummingbird breeds from southern Oregon to Alaska, while Allen's hummingbird breeds along a narrow strip of the California coast up to the Oregon border. Myers et al. (2019) examined behavioral and morphological data between the two and determined a small contact zone along southern coastal Oregon and the Klamath Mountains, extending north into the breeding range of rufous and south into the breeding range of Allen's. As if these two *Selasphorus* spp. are not difficult enough to positively identify, the hybrids make this task much more challenging!

Klamath Birds: By the Numbers

The Klamath Mountains avifauna includes approximately 244 expected native birds, 156 of which are confirmed or probable as breeding within the region (Table 1; Trail et al. 1997, Harris 2006, eBird obs.). Local indigenous knowledge and historical documentation indicate that the region supported additional species within the near history (see the California condor [*Gymnogyps californianus*] sidebar for one such example). Of the extant breeding species, 97 (62%) are in the Order Passeriformes (passerines; songbirds). The remaining 59 breeding species are distributed among 14 orders, with Caprimulgiformes (nighthawks, poorwills, swifts, hummingbirds; 10 species), Piciformes (woodpeckers; 10 species), Strigiformes (owls; 9 species), and Accipitriformes (hawks and eagles; 8 species) holding the bulk of the rest. Most of these species are migratory in some capacity, a theme that we explore throughout this chapter.

The remaining 88 species visit the Klamath Mountains during their nonbreeding season or as stopover habitat during migration. Almost all the Charadriiformes (shorebirds and gulls) are rare and short-term visitors, stopping in larger waterbodies or valleys (e.g., Lewiston Reservoir, Lake Selmac; Grants Pass vicinity from eBird obs.) during migration to rest and refuel before continuing their journeys. The only breeding shorebirds in the region are the ubiquitous killdeer (*Charadrius vociferus*) and spotted sandpiper (*Actitis macularius*), both of which are present year-round and widespread along rivers and lakeshores. Passerine birds like the

Table 1. Summary of 51 native bird families and 244 expected bird species of the Klamath Mountains Geomorphic Province. Note that year-round presence does not imply that the species is a resident, but rather that individuals of a species are present year-round. Many year-round species are migratory and more abundant during breeding or nonbreeding seasons.

Family	# of genera	Breeding		Non-breeding season or in migration	Total # of species
		Summer only	Year-round		
Anatidae (ducks, geese)	13	0	4	26	30
Odontophoridae (quail)	2	0	2	0	2
Phasianidae (grouse)	2	0	2	0	2
Podicipedidae (grebes)	3	0	1	5	6
Columbidae (pigeons, doves)	2	0	2	0	2
Caprimulgidae (nighthawks, poorwills)	2	2	0	0	2
Apodidae (swifts)	3	3	0	0	3
Trochilidae (hummingbirds)	3	4	1	0	5
Rallidae (rails)	3	0	3	0	3
Gruidae (cranes)	1	0	0	1	1
Recurvirostridae (stilts)	1	0	0	1	1
Charadriidae (plovers)	1	0	1	1	2
Scolopacidae (sandpipers, etc.)	7	0	1	14	15
Laridae (gulls, terns)	4	0	0	9	9

Family	# of genera	Breeding		non-breeding season or in migration	Total # of species
		Summer only	Year-round		
Gaviidae (loons)	2	0	0	2	2
Phalacrocoracidae (cormorants)	1	0	0	1	1
Ardeidae (herons, egrets)	5	0	2	4	6
Threskiornithidae (ibis)	1	0	0	1	1
Cathartidae (vultures)	1	0	1	0	1
Pandionidae (osprey)	1	0	1	0	1
Accipitridae (kites, hawks, eagles)	6	0	7	4	11
Tytonidae (barn owl)	1	0	1	0	1
Strigidae (owls)	7	1	7	1	9
Alcedinidae (kingfishers)	1	0	1	0	1
Picidae (woodpeckers)	5	0	10	2	12
Falconidae (falcons)	1	0	3	1	4
Tyrannidae (flycatchers)	5	8	1	2	11
Vireonidae (vireos)	1	2	1	0	3
Laniidae (shrikes)	1	0	0	2	2
Corvidae (jays, crows, ravens)	6	0	7	0	7
Paridae (chickadees, titmouse)	2	0	4	0	4
Alaudidae (larks)	1	0	0	1	1
Hirundinidae (swallows)	6	7	0	0	7
Aegithalidae (bushtit)	1	0	1	0	1
Sylviidae (wrentit)	1	0	1	0	1
Regulidae (kinglets)	2	0	1	1	2
Sittidae (nuthatches)	1	0	3	0	3
Certhiidae (creeper)	1	0	1	0	1
Polioptilidae (gnatcatchers)	1	1	0	0	1
Troglodytidae (wrens)	5	1	5	0	6
Cinclidae (dipper)	1	0	1	0	1
Mimidae (thrashers, mockingbirds)	2	0	1	1	2
Turdidae (bluebirds, thrushes)	5	1	6	0	7
Bombycilliadae (waxwings)	1	0	1	0	1
Motacillidae (pipits)	1	0	0	1	1
Fringillidae (finches)	6	1	8	2	11
Passerellidae (sparrows, junco, etc.)	12	6	11	2	19
Icteriidae (chat)	1	1	0	0	1
Icteridae (blackbirds, meadowlarks)	5	1	2	2	5
Parulidae (warblers)	4	7	2	1	10
Cardinalidae (tanagers, buntings)	3	3	0	0	3
TOTALS	**154**	**49**	**107**	**88**	**244**

sooty fox sparrow subspecies complex (*Passerella iliaca* [*unalaschensis* Group]), sooty fox sparrow) and ruby-crowned kinglet (*Corthylio calendula*) arrive from the north in large numbers during the nonbreeding season, wintering in lower elevation woodlands of the Klamath Mountains or adjacent coastal plain.

The vast majority of the Anseriformes (geese and waterfowl) are present only during the nonbreeding season and, like the shorebirds, most of the sightings occur on larger waterbodies or valleys. In general, waterfowl are not abundant or widespread in the Klamath Mountains, although the Scott Valley floodplain likely hosted much greater numbers pre-colonization.

While breeding season habitat for migratory birds garners the bulk of the attention amongst Nearctic naturalists, the critical role that nonbreeding (wintering) habitat and migration stopover habitat play in the avian life cycle cannot be understated. As detailed in the following section, the Klamath Mountains play a pivotal role in all three phases for numerous species.

Bird Migration: To and Through the Klamath Mountains

Any exploration of Klamath Mountain bird communities would be incomplete without an overview of one of the defining features of the class Aves: migration. Bird migration is a large-scale, synchronized movement that has evolved over millions of years, in which species travel both short and long distance to take advantage of emerging food resources (budding plants, nectar, insects, seeds) and abundant nesting locations. The common pattern in North America involves flying north to breed in the temperate or arctic summer and returning to the wintering grounds in warmer regions to the south—often in tropical latitudes. These movements are cyclical, predictable, and often accompanied by hormonal signals developed in response to environmental cues (e.g., day length, changing temperatures). Long-distance migrants comprise over half of North America's breeding birds.

Likewise, these species comprise a significant component of breeding birds found in the Klamath Mountains. From a hermit warbler (*Setophaga occidentalis*) singing atop a white fir in the Trinity Alps to a northern rough-winged swallow (*Stelgidopteryx serripennis*) circling low over the Rogue River, these migrants have journeyed thousands of miles north to breed in the Klamath Mountains—starting as a trickle in March and culminating in peak migration by the end of May.

The Klamath Mountains, with a wide diversity of habitat types across an elevational gradient, serve as breeding habitat, migration stopover habitat, and wintering habitat for numerous long-distance migrants as well as resident species. Here are the predominant bird migration terms as they relate to species inhabiting the Klamath Mountains:

- **Neotropical Migrants**: long-distance migrants that breed in Canada and the United States during our summer and spend our winter in Mexico, Central America, South America, and the Caribbean Islands. Approximately 54% of bird species that breed in North America, comprising billions of birds, migrate to the Neotropics for 6-8 months (Albert et al. 2020) before returning to breed. The majority are songbirds but there are also many shorebirds, some raptors, and a few waterfowl species.

- **Short-Distance Migrants**: birds that migrate only a short distance from lower to higher elevations (elevational migration), or birds that typically do not migrate south of the United States.
- **Residents**: birds that generally do not migrate and are able to obtain sufficient food resources year-round. These species have evolved strategies to survive in the colder winter climates, taking advantage of the winter exodus of the Neotropical migrants to exploit the more limited cold-season food resources.
- **Irruptive Movements**: often used in association with finches that breed in the northern boreal forests of Canada and journey south in search of better seed production in forests at lower latitudes of the United States. These irruptive species are following the food, ranging widely in huge numbers, sometimes wintering at these lower latitudes.

Just as the Klamath Mountains are a crossroads for species, the region is also an international intersection for migrants. As the Neotropical migrants depart for their southern wintering ranges, other migrants like the Townsend's warbler (*Setophaga townsendi*) and ruby-crowned kinglet use the Klamath Mountains as their southern wintering habitat after nesting at more northerly latitudes. The lower elevations of the Klamath Mountains provide a mild overwintering climate for these migrant species from October onward.

The breeding and nonbreeding habitat of the Klamath Mountains is illustrated by the divergent migration strategies of two subspecies groups of fox sparrow: the sooty fox sparrow and the thick-billed fox sparrow (*Passerella iliaca [megarhyncha* Group]). The sooty fox sparrow breeds along the coast of Canada and Alaska and migrates south for the winter, spending its nonbreeding season in large numbers at low elevations along the Pacific coast (Fraser et al. 2018), including the Klamath Mountains. In contrast, the thick-billed fox sparrow group breeds in montane chaparral habitats in the Sierra Nevada, Klamath Mountains and southern Cascades, migrating south of the Klamath Mountains for the winter. The sooty fox

What's in a name?

The practice of adopting standardized names of American birds is one that is deeply rooted in colonialism and settler practices. Western science has largely disregarded indigeneity and the cultural significance of bird species to all native cultures. Because many of the historical figures, after whom birds are currently named, are becoming known for other actions and beliefs beyond their scientific accomplishments, eponymous names are currently under substantial scrutiny. Notably John Kirk Townsend, for whom Townsend's warbler and Townsend's solitaire are currently named, oversaw the abduction of the remains of indigenous peoples of the West. The removal of his eponymous name is appropriate and in the early stages of happening. Many scientists today suggest entirely ending the practice of eponymous naming of birds is appropriate. We begrudgingly use the standardized English names of birds throughout but look to new and less deeply colonial naming (and re-naming) practices of birds in the future.

sparrow arrives as the thick-billed fox sparrow departs, wintering a few aerial miles from where the thick-billed fox sparrow spent its summer (Zink and Weckstein 2003).

During migration, most species alternate long migratory flights with stopover periods, inhabiting food-rich patches in which they can rest and refuel before continuing their journeys. These stopover habitats, and the food resources they provide for weary migrants, are a critical link in the avian annual cycle and are increasingly recognized as contributing to a significant portion of annual mortality (Albert et al. 2020). Poor conditions in one phase of the annual cycle (breeding, nonbreeding, migration, and stopover habitat) can have carryover effects into the next phase, resulting in reduced breeding success (Norris et al. 2004) and undoubtedly has an impact on survival.

For many species migration is not an all-or-nothing strategy. Many short-distance migrants, like the golden-crowned kinglet (*Regulus satrapa*), have individuals and populations that reside year-round in a particular region, while other individuals may fly north to the boreal forests to breed in the summer (Swanson et al. 2020). These species can express more than one of these migratory strategies within different portions of their geographic or elevational distributions. Further, many species show a "leap-frog" pattern of migration in which individuals that breed at moderate latitudes migrate the shortest distances, whereas the high-latitude breeders often winter furthest south (Cox 1985). This plasticity in migration and breeding strategies provides a measure of population-level resilience to changes in habitat, with the population having built-in flexibility to adapt to a changing environment.

Bird Distributions and Communities

As we saw in the Plant Communities chapter, it is difficult to simplify the complex geology and topography of the Klamath Mountains into general habitat types and communities. We will highlight a few notable bird communities based on a specific habitat type but, to provide context, we first divide the mountains into four general elevational zones (Figure 1): (1) a varied and diverse lower- to upper-montane zone of riparian corridors; (2) oak woodlands; (3) mixed-evergreen forests of conifers and hardwoods; and (4) an upper montane and subalpine zone of the high elevation conifer forests. In the subalpine zone, the conifers are widely spaced and patchy with an open understory of herbaceous vegetation, wet meadows, and shrubs.

Since typical geographic barriers seen in less mobile fauna are non-existent for most birds, it is important to note that bird assemblages are varied and overlapping. Species may occupy one or more of these elevational zones, keying in on structural attributes (e.g., tree canopy cover, a deciduous subcanopy, forest seral stage, shrub layer, presence of water) that may be present in different plant communities in each of these zones. Riparian habitats span the mountains from the bottom to the top, and certain species, like the Wilson's warbler (*Cardellina pusilla*) and black-headed grosbeak (*Pheucticus melanocephalus*), may be found in a mixed-conifer evergreen riparian corridor along the Trinity River or at 6,000 feet elevation (1,829 m) in a simple willow-dominated riparian patch along a subalpine lake in the Trinity Alps.

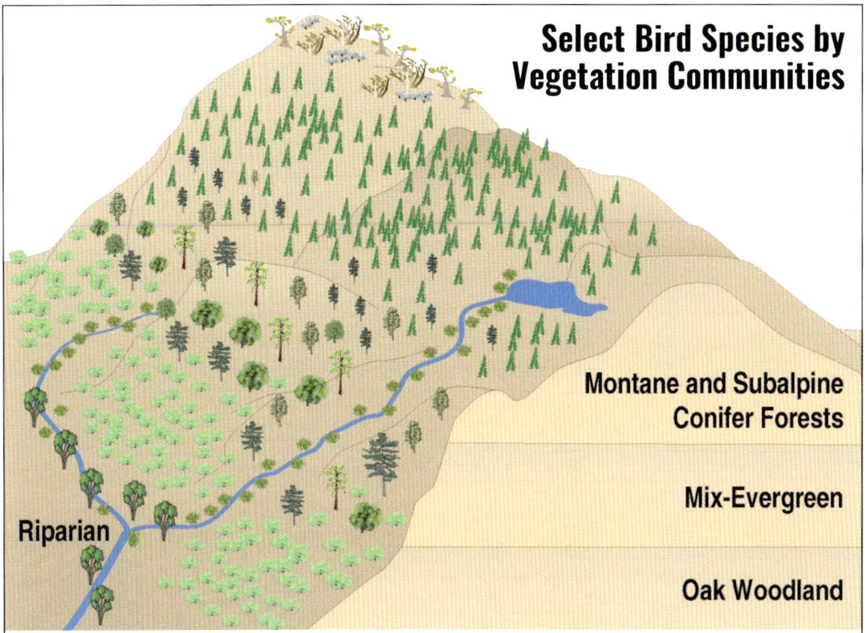

Select Riparian Birds
downy woodpecker, Pacific-slope flycatcher, western wood-pewee, warbling vireo, northern-rough winged swallow, lesser goldfinch, song sparrow, yellow-breasted chat, Bullock's oriole, common yellowthroat, yellow warbler, Wilson's warbler, black-headed grosbeak

Select Oak Woodland Birds (see list in Figure 2, page 395)

Select Mixed-Evergreen Birds
ruffed grouse, pileated woodpecker, red-breasted sapsucker, chestnut-backed chickadee, bushtit, Pacific wren, Cassin's vireo, MacGillivray's warbler, varied thrush, black-throated gray warbler, Nashville warbler, purple finch, western tanager

Select Montane Conifer Birds
mountain quail, sooty grouse, Calliope's hummingbird, northern goshawk, flammulated owl, white-headed woodpecker, olive-sided flycatcher, Clark's nutcracker, mountain chickadee, golden-crowned kinglet, red-breasted nuthatch, mountain bluebird, Townsend's solitaire, Cassin's finch, Lincoln's sparrow, green-tailed towhee, yellow-rumped warbler, hermit warbler, fox sparrow

Figure 1. Select Klamath Mountain birds by habitat.

The structural diversity of habitats has significant influence on bird species diversity and distributions. The proportion of tree canopy cover to open understory, shrub layer, edge habitat and forest patch size, and forest seral stage all dictate where birds live and feed. Take the chipping sparrow (*Spizella passerine*) which is often associated with both oak woodlands and the subalpine zone—two distinctly different forest communities located at different elevations. Both these habitats share the commonality of open, expansive understories with minimal tree cover.

Vegetation structure varies over short distances and so do bird communities. Both communities shift rapidly in response to disturbance but usually evolve more slowly as a response to cues born from gradual changes in plant communities. Birds can also shape the plant communities in which they occur (see Clark's nutcracker sidebar).

Because these complex vegetation communities blend and abut it is easy to see the bird diversity that is nurtured within. A walk in a low-elevation, cool and damp north-facing bend in the Marble Mountains reveals a Pacific-slope flycatcher (*Empidonax difficilis*) as it sallies for insects from the branches of a bigleaf maple. A Swainson's thrush (*Catharus ustulatus*) sings unseen from a clump of small Douglas-fir in the same riparian corridor. A quarter-mile down the path and around another bend, a wrentit (*Chamaea fasciata*) sings from thick manzanita and ceanothus shrubs in a 10-year-old high-severity burn, while its mate weaves together a cup-shaped nest somewhere in thick chaparral. Nearby, in a fire-killed snag looming above the newly emerging chaparral, beetle larvae chew cellulose while a northern flicker (*Colaptes auratus*) hammers away at a new nest cavity. The beetles will soon forage for the woodpeckers. Further upslope, in the higher elevations, a vast mélange of wildflowers, sedges, grasses, and willows emerge from a montane meadowland dotted with the liquid songs of the ground-nesting Lincoln's sparrow (*Melospiza lincolnii*).

What follows is an exploration of these habitats, beginning in the lower elevation and moving upward. We dive in a bit deeper by highlighting riparian, oak woodlands, chaparral, montane conifer forests, and mountain meadows—presenting a broad overview of where particular bird species are likely found.

Riparian Birds

As we explore in Chapter 5, water is a key abiotic factor contributing to beauty and biodiversity. Watersheds connect the Klamath Mountains, providing a narrow, moist corridor though which species move and remain connected to the productivity that the presence of water provides. Along these riparian systems, a variety of obligate riparian birds are found year-round or arrive as migrants in summer. Riparian corridors range across all elevations, from broad, multi-structured river canyons to shrubby alder- and willow-dominated thickets in high mountain wet meadows.

Lazuli bunting (*Passerina amoena*) by Kyle Keegan.

Riparian habitats cover less than one percent of the landscape in western North America yet they are used by more breeding bird species than any other habitat in North America (Knopf et al. 1988). Skagen et al. (2005) estimated that 50% of breeding birds in the western United States breed exclusively in riparian habitats. Riparian habitats support a disproportionately large number of bird species and greater densities of birds than other forested habitats. Studies have demonstrated the importance of these insect-rich havens

Forests for the Birds

Birds have an important role in the Traditional Ecological Knowledge (TEK) for the First Peoples of the Klamath Mountains. Species like violet-green swallows and mourning doves are phenological indicators, relaying subtle changes in weather and habitat conditions. The California condor and pileated woodpecker play integral roles in World Renewal dances—ceremonies that bring balance to the world through the gift of their feathers in dancers' regalia. Other species, like varied thrushes and mountain quail are important food sources. Biodiversity is inherently important to all cultures and birds are, in this measurable way, important members of the landscape and indicators of ecosystem health.

Pileated woodpecker (Iktakatákaheen) is used in Karuk Brush Dance regalia. Photo from the Lake Family Collection, courtesy of Frank Lake.

To reinvigorate collaborative stewardship in the Klamath Mountains, and bring the land back to a place of ecosystem balance, First Peoples are working with western scientists and other land managers to reintroduce fire. Birds are moving to the forefront of this data-driven decision making. After over 100 years of suppression, cultural practitioners are sharing their knowledge of birds to aid in restoration efforts. For example, by looking at the timing of bird nesting, fledgling molt, and peak fall forage, prescribed burning is being planned more effectively and efficiently. Looking toward upland communities, TEK surrounding seasonal or elevational migrants becomes important as well (Karuk Tribe 2019). Today many grasslands and meadows of the high elevations are disappearing due to lack of fire (Lake 2013). Grasslands have been historically significant for many species of birds and also broad-leaved herbs, native annual and perennial grasses, insects, mammals, reptiles, and amphibians—all of which contribute to regional biodiversity but also cultural and environmental health.

Man in a Brush Dance costume holding a dance knife. Photo from the Robert's Collection, Cal Poly Humboldt, RS78.

as stopover habitat for a wide variety of migrants during their seasonal migrations. A September day along a Trinity River riparian corridor can yield a bonanza of long-distance migrants making their way south—including mixed-species flocks of dozens of individuals, comprised of yellow warblers (*Setophaga petechia*), willow flycatchers (*Empidonax traillii*), lazuli buntings (*Passerina amoena*), Cassin's vireo (*Vireo cassinii*) and more—frenetically replenishing their exhausted food reserves in preparation for the next southbound flight.

The low-elevation riverine riparian corridors support some of our most iconic piscivorous (fish-eating) birds—common merganser (*Mergus merganser*) and belted kingfisher (*Megaceryle alcyon*). Common mergansers, waterbirds that often rely on pileated woodpecker cavities for nesting, quietly float down rivers, often close to shallow shorelines, where they dive for fish, aquatic invertebrates, and frogs with their thin sharp bills. They are visual pursuit predators, diving from above and sometimes corralling fish schools with the help of other mergansers.

In contrast, the belted kingfisher, the lone representative of its order Coraciiformes in the region (and throughout most of the Unites States), always hunts alone. It is an aerial predator that typically perches over open water before plunge-diving for fish. Territorial and solitary, individuals are often seen chasing one another away from their preferred fishing holes, calling out with their loud and repeated rattle calls. Belted kingfishers, similar to mergansers, nest in cavities, but uniquely in burrows they excavate in vertical banks. As top predators in aquatic food chains, both the merganser and the kingfisher have been studied as bio-indicators of environmental toxins, especially mercury contamination and pesticides. As piscivores with different predatory techniques, they illustrate the evolution of birds that depend upon fish and aquatic habitat for survival.

A year-round resident with a propensity for elevational movement in the summer is the American dipper, the only true aquatic songbird in North America. It is widely known for hunting in fast moving (lotic) streams by probing along the bottom and catching invertebrates. However, a study by Garwood et al. (2009) found that dippers also take advantage of high elevation lentic environments in the Klamath Mountains. They surveyed 16 high elevation lakes in the Trinity Alps and found dippers at 10 of them, always in areas with steep, rocky, littoral zones. At these sites they found that dippers fed on caddisfly larvae and even amphibian larvae. Because American dippers are often considered obligate stream foragers, these findings suggest they are also dependent, seasonally, on areas that were not previously known to be ecologically important. While nest sites were never found, juveniles were seen feeding with adults. The team surmises that fledglings follow parents to these high-elevation sites later in the summer due to the abundant food sources. This is yet another reminder of how much more there is to learn about all species across the Klamath Mountains.

Pacific slope flycatchers (*Empidonax difficilis*) arrive in the summer to breed in wet and shady mixed-evergreen woodlands throughout the lower elevations of the Klamath Mountains. Here they sally-about hunting a variety of invertebrates including spiders, bees, flies, and moths. While often difficult to spot, males are heard throughout the day, and sometimes even at night during breeding season, when they call with a distinct, up-slurred peeWEET to mark their territory and find a mate.

Pacific slope flycatcher (*Empidonax difficilis*) by Greg Gray.

Other summer-arriving insect lovers of the lower elevation, broad river valleys are the yellow-breasted chat (*Icteria virens*) and Bullock's oriole (*Icterus bullockii*). Yellow-breasted chats prefer dense shrubbery in which they build nests and feed on a variety of invertebrates including beetles, bees, mayflies, moths, and caterpillars. Just above the dense shrubbery, in taller cottonwoods, Bullock's orioles are found. While orioles also breed in oak woodlands and mixed-evergreen forests, riparian corridors are frequented by oriole families throughout the summer as they feast on invertebrates, summer nectar, and late summer berries.

As songbirds forage for insects and kingfishers for fish, the sharp-shinned hawk (*Accipiter striatus*), our smallest hawk, forages for other birds. The sharp-shinned hawk is one of three sympatric accipiters—also the Cooper's hawk (*Accipiter cooperii*) and northern goshawk (*Accipiter gentilis*). All three have short, rounded wings and long narrow tails adapted for powerful acceleration and maneuverability through trees as they chase prey. The sharp-shinned hawk hunts smaller songbirds almost exclusively, and its not uncommon to see a passerine flock "explode" in all directions as this small accipiter dives in for a potential meal. While not dependent on riparian corridors, they frequently hunt them due to the density of songbirds that forage in these zones.

From lowland river valleys to high elevation lakeshores, the dance and call of the spotted sandpiper (*Actitis macularius*) is hard to miss. As the most widespread shorebird in North America, they make their journey from the subtropics or coasts of North American beaches up to the rivers and lakes of the northern latitudes to feast on the influx of invertebrates during the summer. Nests are built on the ground, near water, where they scrape out a depression and line it with grasses and small woody material. Notably territorial, spotted sandpiper females serially mate with multiple males, which provide most of the parental care in a mating system known as resource-defense polyandry (Oring and Lank 1982). Watch where you step next time you are walking around a subalpine lake and look for a spotted sandpiper—most likely an incubating male—as it leads you away from its nest.

Oak woodlands

In the Klamath Mountains, deciduous oak woodlands are biodiversity islands in a largely coniferous landscape. Located in the lower montane zone, oak woodlands are an iconic habitat that holds tremendous floristic, avian, and cultural value. The oak woodlands of the Klamath Mountains are among the most biodiverse terrestrial habitats in North America, supporting hundreds of species. Approximately 26 bird species are highly associated with oaks of the region (Altman and Stephens 2012), often relying on this unique community to complete one or more phases of their life history cycle (Figure 2).

Because of the occurrence of multiple oak tree species within these habitats, the communities take on a variety of characteristics. These include tall, well-spaced stands of Oregon white oak (*Quercus garryana*) along river benches, dense forests of tanoak (*Notholithocarpus densiflorus*) above those benches, California black oak (*Q. kelloggii*) mixed with low-elevation chaparral on the eastern slopes of the range, or Brewer's oak (*Q. garryana* var. *breweri*) and huckleberry oak (*Q. vaccinifolia*) in montane chaparral. These keystone habitats, and the structural and floristic diversity they offer, serve as seasonal pathways for many migrating species including warblers, flycatchers, and hummingbirds. Long-distance migrants like the Townsend's warbler and the hermit warbler stop to forage in the insect-rich oak woodlands on their way to coniferous breeding zones (Block et al. 1992).

The widespread persistence of many oak communities was overseen by the managed fire regimes of the First Peoples. This was interrupted and greatly altered with European settlement. Frequent, low-intensity fire favors oaks because conifer saplings, such as Douglas-fir, are excluded from the system through the reduction of understory competition when fire moves through. The absence of fire allows for widespread invasion of oak habitats by conifer species, eventually shading out and killing the oak trees, resulting in a conversion to conifer forest. Indigenous Tribes implement fires not only to exclude conifers but also to favor the health and vigor of oak trees and acorns—ensuring that insect infestation of acorn mast are minimized through periodic fire.

The open canopies of oak woodlands ensure that more light reaches the understory and increases plant diversity, which in turn lead to more invertebrates and pollinator species (moths, butterflies, beetles, bumblebees) and the vertebrates that prey upon them—notably the birds. Furthermore, the open canopies and widely-spaced oak stands ensure that the grassland habitat thrives. Western meadowlarks (*Sturnella neglecta*) require large tracts of open grassland throughout their full life cycle and are absent in areas where oak-prairie habitat has

Blue-gray gnatcatcher (*Polioptila caerulea*) by Greg Gray.

Birds

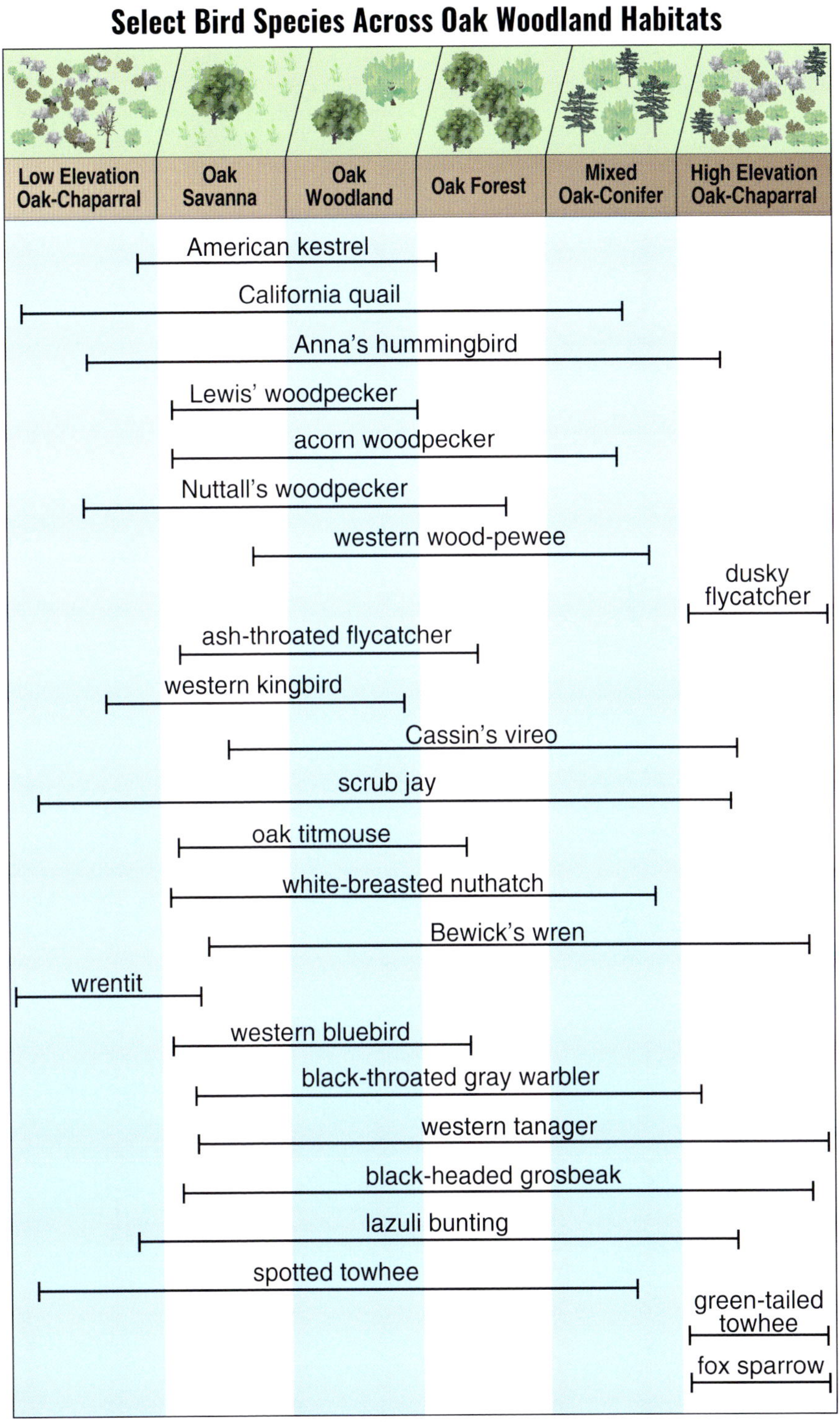

Figure 2. Adapted from Altman and Stephens (2012).

been converted to conifer forest. Fire is a essential driver in maintaining this habitat.

The most important element for birds in oak woodlands are mature, large-diameter trees that produce a complex structure including cavities for nesting and branch structure for roosting. Oaks support significantly more cavities than conifers because they are more adept at compartmentalizing injuries or insect damage. This ensures oaks survive but also offers foraging opportunities for cavity excavators such as acorn woodpeckers (*Melanerpes formicivorus*) which eat the many invertebrates that live in the compartmentalized decay cavities and store bounties of oak acorn mast in neat rows in their storage (granary) trees. Once excavated, species like ash-throated flycatchers, at their northern range extension, and black-capped chickadees, at their southern range extension, use these pockets for nesting. Stretching beyond the cavities are the mushroom-like canopies of old oaks which offer protection and nesting surfaces for blue-gray gnatcatchers (*Polioptila caerulea*) and black-throated gray warblers (*Setophaga nigrescens*), as well as perching opportunities for larger birds like turkey vultures (*Cathartes aura*).

Oak titmouse (*Baeolophus inornatus*) by Jeff Todoroff.

With this complexity in structure also comes diversity of life. On these live trees, epiphytic plants (see sidebar page 180) and invertebrates thrive. These smaller organisms provide foraging opportunities through increased food-web complexity for species like house wrens (*Troglodytes aedon*) and Hutton's vireos (*Vireo huttoni*). Berries of the mistletoe (*Phorodendron* spp.) provide a winter food source in the form of fruits for western bluebirds (*Sialia mexicana*) among others. So important are these epiphytic keystones to birds, that wintering western bluebirds live in extended family groups that defend territories based on mistletoe abundance, relying on this little plant for food and thermal warmth in the leaf-scarce canopies of winter (Kraaijeveld and Dickinson 2001).

Unlike other oak woodland communities of the western United States, shrubs in the understory are a more dominant habitat feature in the Klamath Mountains. Here, poison oak (*Toxicodendron diversilobum*), manzanita (*Arctostaphylos* spp.), and ocean-spray (*Holodiscus discolor*) add more structure and diversity to the oak woodlands. Many Klamath Mountain species are usually not as strongly associated with oak woodlands across the remainder of their respective ranges but use oak woodlands here because of an increased shrub understory. These include species like Bewick's wren (*Thryomanes bewickii*) and spotted towhee (*Pipilo maculatus*), which forage for insects and berries below the shrubs.

And then there are acorns. Acorns have high caloric value and are an important food source for many vertebrates. Acorn woodpeckers live in complex family groups of breeding and cooperative non-breeders that maintain and defend their acorn harvest and granary trees throughout the year; they store acorns communally and raise young together. In the fall and winter, after the acorns have ripened, band-tailed pigeons (*Patagioenas fasciata*) and Lewis's woodpeckers (*Melanerpes*

Woodpeckers of the Klamath Mountains

Woodpeckers (family Picidae) are keystone species and indicators of regional forest bird diversity. Their foraging and nesting activities have a positive influence on the abundance and richness of other forest birds. While they forage, they remove the bark of dead and dying trees, exposing what is underneath for other birds. When they create nesting cavities, these are sometimes later utilized by other species—known as secondary cavity nesters—including northern saw-whet owls (*Aegolius acadicus*), swifts, bluebirds, wrens, nuthatches, and more. Because of their importance to ecosystem diversity, woodpeckers guide forest management decisions based on their influence on the overall community dynamics and health (Drever et al. 2008). Woodpeckers are also early colonizers of newly burned forests across the region. In the Klamath Mountains, there are 12 species in 5 genera, which is an indicator in itself of regional biotic diversity.

Here we touch on black-backed woodpeckers (*Picoides arcticus*) and Lewis's woodpeckers (*Melanerpes lewis*), two of the region's rarest regional woodpeckers. Black-backed woodpeckers are a common species in select areas of the Cascades, just east of the Klamath Mountains, but are a rare sighting in the region. Although the species does occur in unburned forest, it is most common in recently burned areas. A hopeful outcome of the large-scale burning of the Klamath Mountains is that these fires continue to create novel habitats for more black-backed woodpeckers. Lewis's woodpeckers have an interesting life cycle in the Klamath Mountains. In summer, they breed almost exclusively in the eastern part of the range including in the Scott Valley and at the eastern fringes near Yreka, California. In winter, occasional migrants arrive in the oak woodlands in the western parts of the range, including the Hoopa Valley, where they forage for acorns. This is surely one of our most unforgettable woodpeckers for color (pink and green), size (crow-like), and foraging techniques (flycatcher and gleaner). As is the case with many good things, Lewis's woodpeckers are in decline due to the loss of extensive areas of oak savannah dominated by large trees—which is the required habitat for this cavity-nesting bird.

Woodpeckers from top to bottom: downy (*Picoides pubescens*), hairy (*Leuconotopicus villosus*), pileated (*Dryocopus pileatus*), white-headed (*Picoides albolarvatus*), Lewis's (*Melanerpes lewis*).

Birds

lewis) use them as a critical food source when other resources are scarce. California scrub jays (*Aphelocoma californica*), acorn woodpeckers, and oak titmice (*Baeolophus inornatus*) cache acorns and serve as seed dispersers. To put the acorn woodpecker's role as dispersal vector in perspective: Koenig et al. (2008) estimated that an individual acorn woodpecker harvests 325 acorns per year, and with a population size of 7.3 million individuals in California, this species harvests over 2.4 billion acorns per year in California! While over half are consumed, many are abandoned in their caches. Thus, this species clearly plays a pivotal role in oak demography.

Band-tailed pigeon (*Patagioenas fasciata*) by Jeff Todoroff.

Chaparral Birds

Chaparral is one of the most distinctive environments within the California Floristic Province (CFP). It is characterized by drought-tolerant woody shrubs like manzanitas, scrub oaks, and California lilac (*Ceanothus* spp.) that are shaped by fire, dry summers, and winter rain. The Klamath Mountains are near the northern extent of the CFP, and have less chaparral cover than mountains further south. Still, this plant community type covers most of the eastern foothills, many south-facing mountain slopes, and upward into the montane conifer belt and subalpine on the same aspect (see discussion page 209). Although many species may travel through chaparral, only a few breed or reside year-round in this tough, fire-prone environment.

While most of the montane chaparral birds in the Klamath Mountains are migratory, the lower elevation chaparral of the eastern Klamath Mountains hosts year-round residents. The range of the wrentit nearly defines that of coastal scrub and interior chaparral from Baja California, Mexico north to the Columbia River. From this preferred habitat they do not wander, except into oak woodlands and mixed-evergreen forests in the northern part of their range. In fact, studies suggest they rarely disperse much more than 1,000 feet (305 m) from where they were born (Baker et al. 1995). Wrentits are at a taxonomic crossroads, so crossed that their family status perplexed ornithologists for a century. *Chamaea* is a monotypic genus and they are the sole representative of their taxonomic family in the Americas—with the rest of their family primarily occurring in Southeast Asia and Africa. Curious and inquisitive, keep an ear cocked for a loud, accelerating bouncing-ball trill emitted from a nearby entanglement of shrubs. Wrentits form lifelong pair bonds and both sexes build the nest and incubate the eggs, an unusual trait in North American songbirds (Geupel and Ballard 2020).

A breeding male's loud, pure, and descending QUEE-ar, echoing between steep-walled rocks, is an unforgettable character of the mountain quail (*Oreortyx pictus*). Though not immediately apparent, if you spend any amount of time in the Klamath Mountains you will easily hear, but rarely see, one. Mountain quail use a wide variety of habitats but are generally found foraging in dense understory. Here they frequent thickets of manzanita, huckleberry, elderberry (*Sambucus* sp.), and California lilac—montane chaparral at its finest!

Green-tailed towhee (*Pipilo chlorurus*) by Kyle Keegan.

The green-tailed towhee, its deep olive feathers edged in yellow-green and its head crowned with rufous, is difficult to see and a reward when spotted. It is a large, migratory sparrow that breeds in montane chaparral across the Klamath Mountains. They spend summers scratching at leaf litter for berries and invertebrates, only occasionally emerging to perch on a green-leaf manzanita (*Arctostaphylos patula*) or outcropping rock to offer a querulous call. Green-tailed towhees are only distantly related to the larger spotted towhees with which they often co-occur. Instead, they are more related to similar-looking tropical Emberizid sparrows limited in distribution to Central and South America—and with which they neighbor in the winters, after their long-distance migrations south. In the summers, in the Klamath Mountains, green-tailed towhees are often neighbors with the aforementioned thick-billed fox sparrows. Fox sparrows are easier to hear than see because they are constantly scratching through the leaf-litter, foraging for invertebrates.

Montane Conifer Forests

Climbing higher into the Klamath Mountains, passing above 5,000 feet (1,524 m) in elevation, the forests shift from open, mixed-evergreen woodlands with madrones, oaks, and pines to dark evergreen forests with true firs (*Abies* spp.) and mountain hemlocks (*Tsuga mertensiana*) higher still. This shift is due to winter snow, which limits a plant's ability to survive—conifers are the best in dealing with both winter cold and summer heat. With evergreen conifers as the dominant cover, these landscapes are more reminiscent of north of the 45 parallel than sub-tropical zones to the south. Here, birds flourish in a landscape that requires either seasonal arrival or hearty, year-round adaptability. Bird density is highest in the Klamath Mountain conifer forests during the summer months when Neotropical migrants arrive to take advantage of plant productivity and the insects and seeds that follow—all creating high-calorie forage for a variety of migratory species. This productivity occurs mostly adjacent to the dense conifer forests, as is discussed in greater detail in the meadow section, but many species find shelter and some forage in the thick canopies of evergreen conifers.

Numerous thrushes (Turdidae) inhabit the montane conifer forests of the Klamath Mountains. The most widespread, in both numbers and habitat types,

Made for each other

Whitebark pine (*Pinus albicaulis*) is set apart from other 5-needle pines in that its cone does not open at maturity and its seed is "wingless." This is a characteristic of stone pines subsection (Cembrae) within the larger five-needle subgenus *Strobus*. Whitebark pine is the only stone pine in North America and is solely dependent on Clark's nutcrackers (*Nucifraga columbiana*) for seed distribution and future seedling recruitment. The birds open the cone, collect the seeds, and cache them. Over time, this has developed into a keystone mutualistic relationship. Whitebark regeneration is, therefore, often manifested as seedling clumps when young and collections of multiple trees at maturity due to the unused caches left by the birds.

Clark's nutcrackers and whitebark pine in the Trinity Alps. Photo by Justin Garwood.

A study in Arizona found that the combined caching ability for a flock of 150 birds was several thousand seeds per acre, totaling nearly 4 million seeds per flock, with the combined nutritional value equivalent to 650 pounds of fat (Lanner 1996). Around 20 percent of these caches go unused and then germinate in the following years. During whitebark pine surveys in the Trinity Alps Wilderness, several unretrieved caches were found in the form of whitebark pine saplings and seedlings on isolated mountain tops including Packer's Peak and Seven Up Peak (Kauffmann 2013).

From Mount Shasta westward into the Klamath Mountains, whitebark pines are found in only four regions (Trinity Alps, Scott Mountains, Salmon Mountains, and Marble Mountains) with several mountain tops in those regions having fewer than 50 trees. Because of the scarcity of whitebark pine and the fact that seed production does not occur annually, nutcrackers in the Klamath Mountains also consume foxtail (*P. balfouriana*) and western white pine (*P. monticola*) seeds—among others. It is rare to hear the nutcracker's robotic squawk and meet the whitebark pine by road in the Klamath Mountains—you have to climb a mountain.

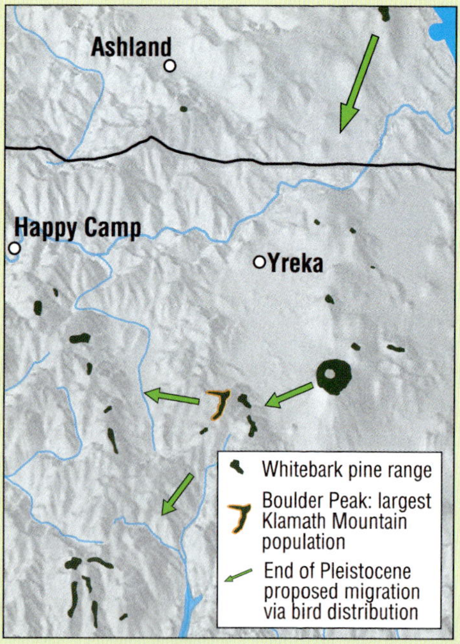

Birds

is the American robin (*Turdus migratorius*). This species is found from low elevation rivers to mountain tops across its range. Its relative, the Swainson's thrush, a Neotropical migrant that visits the region from Central and South America, has perhaps the most striking song of any Klamath Mountain bird. Listen for its ethereal, spiraling song emanating from a deciduous understory in a closed canopy conifer forest, particularly during the crepuscular hours of dawn and dusk. Territorial males often engage in singing duels, with songs increasing in volume and frequency, sometimes accompanied by a physical altercation and a chase through the woods.

The Swainson's thrush is most common in the western Klamath but extends inland, into drier habitat, where wet microsites persist late into summer. They consume insects in the summer and fruits in the fall, where they are one of the latest Neotropical migrants to leave—often remaining in their breeding habitat until October. Be careful not to confuse them with their close cousin, the hermit thrush (*Catharus guttatus*), a short-distance migrant that occupies a broad range of forest habitat from 500–4,000 feet (152–1,220 m). The hermit thrush is the only member of its genus (*Catharsus*) to winter in the continental United States (typically at lower elevations).

At the tops of conifer crowns, or on an ancient snag of a long-dead conifer, flycatchers are a stunning member of the regional montane conifer forests in summer. Perhaps the most noticeable is the olive-sided flycatcher (*Contopus cooperi*). Compared to other pewees (*Contopus* spp.) this is by far the largest—more than twice the size of a western wood pewee (*C. sordidulus*). The distinct "quick-three-beers" whistle will draw your eyes to this tuxedoed beauty. Of all flycatchers that breed in the United States, the olive-sided has the longest migration—with some travelling over 7,000 miles (11,265 km) each spring and fall. Hammond's flycatcher (*Empidonax hammondii*) are also mature montane conifer forest specialists. Their preferred habitat can overlap with dusky flycatchers (*Empidonax oberholseri*) in the highest elevations and in montane chaparral at the edge of conifer forests. They perch, and then quickly fly outward and upward, to catch a variety or prey including butterflies, moths, flies, and beetles. Listen for their somewhat high-pitched "pip" calls coming as a chorus as dusk settles in. Dusky flycatchers and yellow-rumped warblers (*Setophaga coronata*) make a similar but lower-tenored "whit" call—which Hammond's flycatchers do not make. This can aid in differentiating the two flycatcher species. Morphologically the tips of the primary feathers are shorter and stubbier in dusky flycatchers and pointier, longer, and narrower in Hammond's—but even experienced bird watchers have trouble separating the two with eyes alone. The distinct calls and habitat preferences are the best clues as to which species you are observing.

Olive-sided flycatcher (*Contopus cooperi*) by Jeff Todoroff.

Year-round residents of the conifer forests are a hearty bunch. To survive in the higher elevations, bird species must shift food sources throughout the seasons—especially in winter when food is most scarce. Many of the birds that survive in these environments turn to bark gleaning in the winter. A study in the northern Sierra Nevada, El Dorado County, found that incense-cedar (*Calocedrus decurrens*) was the most common winter foraging substrate for red-breasted nuthatches, brown creepers, golden-crowned kinglets (*Regulus satrapa*), and ruby-crowned kinglets. These species exploit incense-cedar scales (*Xylococculus macrocarpa*), and insects that live under the loose, flakey bark of small saplings or on the branches of larger trees. While these birds also foraged on the bark furrows of other species including Douglas-fir (*Pseudotsuga menziesii*), sugar pine (*Pinus lambertiana*), and white fir (*Abies concolor*) all three showed a preference for incense-cedar—presumably due to the scale as an additional food source (Morrison et al. 1989).

Mountain chickadee (*Poecile gambeli*) by Rob Fowler.

Perhaps most iconic of the high elevation songbirds is the indomitable mountain chickadee (*Poecile gambeli*), a species that forgoes migration, braving the snowclad winters in search of conifer seeds and insects and surviving partly on seed caches acquired during months with more plentiful food (Pravosudov and Clayton 2001). In winter months, the "chicka-dee-dee" call can alert you to a wintertime flock of mountain chickadees, nuthatches, brown creepers, downy woodpeckers, and more taking advantage of the food availability vacated in the wake of the Neotropical migrant exodus.

One of the most conspicuous and intrepid birds in winter conifer forest birds of the Klamath Mountains is the Canada jay (*Perisoreus canadensis*). They are opportunists with the ability to cache winter stores by producing special saliva that molds their fall harvest into sticky, preserved blobs they conspicuously store behind bark, under lichen, and in conifer needles—all above the height of the predicted winter snowline (Dow 1965). This is a true boreal species with populations as far north as the Beaufort Sea in northern Alaska. Here in the Klamath Mountains, they are at the southern extent of their coastal range with an extension south into Mendocino Country through the redwood belt.

The finches (Fringillidae), a family of passerines with large, conical bills adapted for cracking and eating seeds and nuts, are iconic birds of the conifer forests. There are 17 North American finches, including red crossbills, goldfinches, evening grosbeaks, pine siskins, and Cassin's finches (*Haemorhous cassinii*) and their stocky bodies and large jaw muscles bely their colorful plumages. Eleven species of finches occur in the Klamath Mountains, most of which breed of here. Two non-breeding species, the gray-crowned rosy finch (*Leucosticte tephrocotis*) and pine grosbeak (*Pinicola enucleator*), are extremely rare irruptive winter visitors; both species having been observed only a handful of times atop Mount Ashland, and the grosbeak once in Josephine County. The real value of the Klamath Mountains to finches lies in the winter.

Two-thirds of North America's total number of finches breed in the Canadian boreal forests of the northern latitudes, underscoring the importance of these vast forests as reservoirs for breeding songbirds. Finches are highly dependent on the boreal forests for conifer seed production, and during times of large-scale collapse of seed production of boreal pine, spruce, and fir forests, huge numbers fly thousands of miles south in search of better seed availability—a large-scale movement known as an irruption. Several species, including red crossbills and pine siskins, will visit the Klamath Mountains in large numbers in the winter during these irruptive movements. Keep an eye out for their huge flocks (up to 200 individuals in pine siskins) at lower elevations during the winter.

Another irruptive visitor and occasional breeder is the evening grosbeak, a large yellow finch adorned with a massive greenish-yellow bill used for crushing conifer seeds that are too large for pine siskins and smaller birds. This unmistakable and seldom-seen songbird makes up for its simple song with an elaborate dance. A male grosbeak, attempting to attract a prospective mate, raises his head and tail while drooping and vibrating his wings, all the while swiveling back and forth. In response, the female may acknowledge his dancing prowess with a bow, initiating back-and-forth bows of appreciation and courtship. Sadly, this bird's overall continental population is down 90%, a startling decline for this enigmatic wanderer (Rosenberg et al. 2019).

On the talus slopes of the higher mountains, where rock-strewn rubble adorns the base of the ridges and slopes, listen for the loud trills of a rock wren (*Salpinctes obsoletus*) as it bounces and bobs from boulder to boulder. While this rocky denizen is found at all elevations in appropriate rocky terrain, these higher talus slopes are a reliable viewing location for an attentive, sure-footed bird-watcher. The rock wren is exceptional in avian nest-building lore, as the nest is built upon a foundation of hundreds of small, flat stones within a crevice of larger boulders typically with a constructed stone ramp or walkway leading out. Less than 0.3 percent of bird species worldwide (remember, there are over 10,000 species) carry stones as part of their nest architecture. A study of rock wren nests in Colorado postulates that

Rock Wren (*Salpinctes obsoletus*) by Christopher Lindsey.

Prey-go-neesh Returns

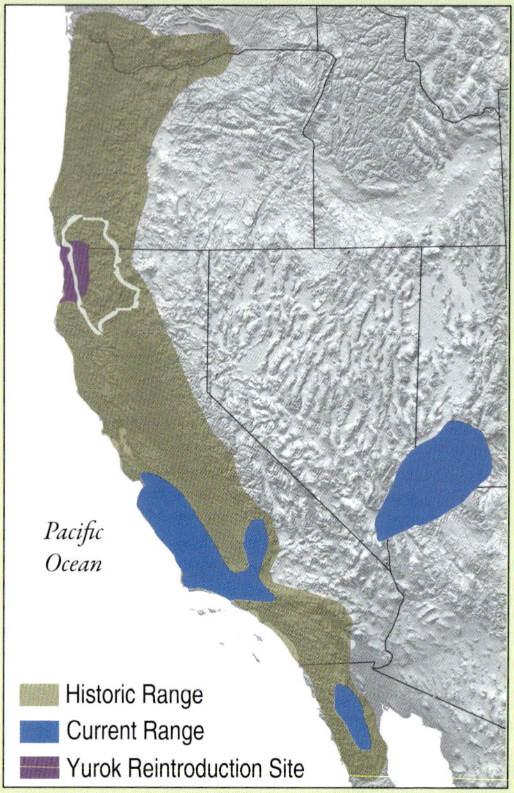

The Yurok people have served as stewards of their homeland since time immemorial. However, Yurok do not consider themselves to be the first people in this place, but rather inheritors of a world already established by the spirits, including those tied to the wildlife who persist within our ecological community today. Rather than viewing wildlife species as less highly derived than humans, a common Western conception, they are recognized as teachers—both wiser and more knowledgeable than humans. One such spirit is Prey-go-neesh, or California condor (*Gymnogyps californianus*).

Condors fill a valuable ecological niche as obligate scavengers, providing critical resource breakdown and recycling through consumption of carcasses across a variety of interior and coastal ecoregions. Having a wingspan of as much as 9.5 feet (2.9 m), California condors are the largest land-based bird in North America, and once ranged from Baja California, Mexico to the Columbia River. With the influx of Euro-American colonizers, who introduced new threats and heavily impacted condors' critical environmental resources, that range and their population decreased dramatically. Some of the most significant impacts were introduction of lead as an environmental contaminant, decimation of megafauna populations they relied on for food, razing of old-growth forest and nesting habitat, fire suppression, and planting of tree farms in key prairie linkages that served as foraging grounds. Condors were locally extirpated by the end of the 19th century, and teetered on the edge of extinction with only 22 individuals remaining in the world by the mid-1980s. Thankfully, the conservation community recognized the importance of this keystone species. Intense recovery efforts were made, including establishment of a captive breeding program and several heavily managed but successful wild-release areas in the southern half of their range. However, initial restoration efforts did not include the northern half of condors' historical range, limiting their recovery potential and leaving large swaths of the North American west coast with a longing for their return home.

Prey-go-neesh serve as a focal point for wildlife and habitat restoration for the Yurok people because of their traditional ties to world renewal. Yurok and surrounding Tribes consider themselves to be world renewal people, whose reason for being is to keep the world in balance. Considered spiritually powerful and fundamentally a good-hearted being, Prey-go-neesh was one of the first people of the world, and helped establish and carry out Yurok world renewal ceremonies. These are considered high ceremonies, and prey-go-neesh continues to contribute today through use of their feathers in regalia and as the carrier of ceremonial dancers' messages as they do their part to bring the world back into balance.

A panel of Yurok elders specifically tasked with prioritizing natural and cultural resource restoration needs identified condor as the most important terrestrial species to return to Yurok ancestral territory. To this end, the Yurok Tribe Wildlife Department was established in 2008, with condor as its flagship species. With condor serving as a beacon of holistic restoration, the Wildlife Department has since grown significantly, expanding into other threatened and endangered species monitoring and management, cultural species management, youth and community outreach and education, and habitat restoration. Condors' restoration to Yurok lands in far northern California should be considered the launching point for expansion throughout the Pacific Northwest and reclamation of condors' historical range. Restoring condor to its ancestral home in Yurok country exemplifies not only the Yurok people's traditional role as environmental stewards and world renewal people, but also the Tribe's status as a sovereign nation with an obligation to and capacity for significant contributions to conservation.

— **Tiana Williams-Claussen**, Director of the Yurok Tribe Wildlife Department

Weighing over 20 pounds with a wingspan as much as 9.5 feet, the California condor, or Prey-go-neesh, is North America's largest bird. The Yurok Tribe is working to reintroduce them to northwest California. Photo by Chris West.

this energetically-costly nest structure may serve to alert the incubating female to a potential predator through acoustic or vibrational signals transferred across the stone walkway (Warning and Benedict 2015).

An uncommon yet widespread predator inhabiting large tracts of mature forests is the northern goshawk, the largest of the three North American accipiters. The goshawk hunts a variety of prey, including tree squirrels, grouse, corvids, woodpeckers, and larger songbirds. They are territorial and defensive of their nests and will strike at potential intruders (including humans) who get too close. My (Greg's) first goshawk sighting was an unforgettable experience: while backpacking to the southernmost stand of Pacific silver fir (*Abies amabilis*) in North America, located in the Marble Mountain Wilderness between Diamond and Hancock lakes. My first glimpse of these regionally-rare firs was accompanied by an equally rare sighting—a goshawk perched in one of them. It quietly eyed me askance for the next five minutes. Lucky for me it decided I wasn't worth an intruder strike!

Mountain Meadows

Intermixed with the vast coniferous forests of the region are pockets of non-coniferous vegetation that includes shrubs and herbs, culminating in sweeping meadow systems in the subalpine community of the higher montane zone. These pockets host grasses, sedges, and forbs in lower, wetter regions and transition into a variety of broad-leaved shrubs, like Sitka alder (*Alnus viridis*) and wild plums (*Prunus* spp.) moving upslope—where water is less available. The presence of herbs and broad-leaved shrubs and trees has an important influence on food and cover resources for wildlife in adjacent conifer forests (Hagar 2007). Because we focus on chaparral separately, here we highlight select birds that primarily rely on meadow systems.

As small floodplain and riparian zones, montane wet meadows are biodiversity hotspots that retain and slowly release water into the late summer—maintaining stream flow and groundwater levels near the surface in an increasingly drying landscape. Wolfe et al. (2019) refer to them as moisture refugia and, as the breeding season progresses, they become progressively more important to bird species reliant on arthropod food resources to successfully complete nesting and post-breeding molt. As ecotonal transition areas, they not only support the riparian-associated birds but also forest-dwelling birds that use these areas to forage, collect nesting

Nashville warbler (*Leiothlypis ruficapilla*) by Dan Barton.

MacGillivray's warbler (*Geothlypis tolmiei*) by Barton.

material, visit with recently fledged young, and disperse to post-breeding (we explore this concept further in the Migration Pathways section). Researchers in the Sierra Nevada concluded that almost every bird species that breeds or migrates through the region uses montane meadows at some point during their annual cycle (Siegel and DeSante 1999).

Migratory warblers explore various microhabitats in and around mountain meadow systems. Nashville warblers (*Leiothlypis ruficapilla*) breed in shrubbery adjacent to meadows and lakes. Watch for them in willow thickets and alder groves where they move among branches, picking the underside of leaves for insects. MacGillivray's warblers are associated with shrubs and deciduous vegetation at the edge of more dense conifer forests, often along small headwater creeks. Wilson's warblers prefer willow, alder, and other shrubs near streams. The loud, warbling song of the aptly-named warbling vireo (*Vireo gilvus*) can often be heard from stands of black cottonwoods (*Populus trichocarpa*) along the riparian belt of these meadows. A mountain meadow complex along the edge of an alpine lake in the Trinity Alps can host all these species. Keep an eye out for their yellow hues and emphatic songs as they bounce in and out of willow pockets in search of food and nesting material.

A variety of breeding sparrows are found in mountain meadow systems, highlighted—due to its rarity—by the vesper sparrow (*Pooecetes gramineus*). Listen for the vesper sparrow's sweet slurs and trills at twilight in meadow complexes around the Oregon border and northward. Poker Flat, at the edge of the Siskiyou Wilderness, has historically been a dependable place to find them. The meadows around Mount Ashland are another good location for the species. They are ground-dwellers, preferring to explore the world through dense grasses and shrubs. Vesper sparrows are all but absent (or yet documented) in the southern and eastern Klamath Mountains but are more typical in similar habitats to the north.

Birds of the Night

Birds that enjoy the night life are quite different than diurnal birds. These nocturnal and crepuscular species vary by having camouflaged plumage, large eyes, and an enhanced sense of hearing. Often, if spotted in their daytime roost, they will remain motionless. These species have chosen this nocturnal lifestyle to pursue alternative food sources available at night. The most obvious of which are small mammals like deer mice (*Peromyscus* spp.) but others pursue insects or flowers.

Common poorwills (*Phalaenoptilus nuttallii*) are a summer migrant into the more arid, shrubby, and open areas in the high-country of Klamath Mountains. While more common in the intermountain west and locally in the eastern parts of the range, they have been infrequently documented within sight of the Pacific Ocean on the Siskiyou Crest of California. Because they are nocturnal insectivores of the high country, if conditions are harsh they can enter a state of torpor to save energy until things improve. When conditions are right, common poorwills hunt from dusk to dawn from flats on ridgelines or mountain summits. They prefer moths, grasshoppers, and beetles which they consume through their unusually large mouths. Seldom seen but often heard, listen for their gentle and repetitive "poowWILLup" song emanating from the nighttime darkness.

There are 10 owl species that occur in the Klamath Mountains, ranging from small to large and rare to widespread. From the common barn owl (*Tyto alba*) to the uncommon long-eared (*Asio otus*) and flammulated owls (*Psiloscops flammeolus*) it is never short of exhilarating to hear or see an owl. I (Michael) once slept on a ridgeline in the southern Klamath Mountains and had a flammulated owl calling after sunset, a northern saw-whet owl (*Aegolius acadicus*) calling in the middle of the night, and a northern pygmy-owl (*Glaucidium gnoma*) calling as the sun rose the next morning. There is something about owling that, when successful, is never forgotten.

The flammulated owl and northern saw-whet owl, while superficially similar in that they are small owls that live in forests, have evolved two very different migration strategies. The tiny reddish-brown flammulated owl is a range-restricted, Neotropical migrant that breeds in pine forests of the western United States and British Columbia and winters in a narrow range of central Mexico south to the highlands of Guatemala and El Salvador (Stock et al. 2006). The northern saw-whet owl, on the other hand, is a wide-spread short-distance migrant that winters in North America, occupying a wide range of forested habitats and elevations. Why the difference in migration strategies for these two small owls? The flammulated owl's diet consists almost entirely of insects, including moths, beetles, and crickets, whereas the saw-whet owl is primarily carnivorous, feeding on mice and other small mammals. The low insect abundance in higher latitudes in the winter months means the flammulated owl must migrate to southern latitudes to avoid the dramatic reduction in its prey items. Since small mammals are present year-round in the higher latitudes, the saw-whet's diet allows it to forgo a long-distance trek.

Barred owl (*Strix varia*) by Greg Gray.

Two species of large cryptic owls from the same genus, the Federally threatened northern spotted owl (*Strix occidentalis caurina*) and the barred owl (*Strix varia*), are at a crossroads, literally and figuratively. The northern spotted owl of the Pacific Northwest prefers older-growth, humid coastal forests. This forest type has been degraded and fragmented due to aggressive forestry practices, resulting in range-wide population decline. Concurrently, in the twentieth century, the barred owl—formerly absent from the Pacific Northwest—has expanded its range westward through the northern boreal forests and then south into northern California because of its tolerance for forest fragmentation and second-growth forests. This has brought their once separate ranges into contact and the more aggressive barred owl displaces (and sometimes hybridizes with) the spotted owl, further threatening this already declining species.

The larger barred owl has had negative impacts on spotted owls by altering behavior, excluding habitat, and increasing regional extinction events. Several approaches to "help" spotted owls have ensued. The most effective and controversial has been the experimental removal of barred owls from some spotted owl habitat. A pilot study in the Hoopa Valley, in the western Klamath Mountains of California, demonstrated that removal of barred owls, in combination with conservation of suitable forest conditions, can slow or even reverse the rate of population decline in spotted owls (Diller et al. 2016). Research is continuing with the hopes of saving the northern spotted owl.

Migration Pathways

Be it riparian corridors, mid-elevation oak-grassland habitats, or high mountain meadows, migrants follow the phenology of flowering plants—and the insect communities they support—as they stop over during migration, stay and nest, or stay for the winter.

One such migrant, the rufous hummingbird, performs a 4,000-mile (6,437 km) elliptical migration from its wintering habitat in Central America to its breeding grounds in the Pacific Northwest (primarily north of the Klamath Mountains). The spring migration is initiated by males, followed several weeks later by females, as they follow the coastal belt northward. In March and April, look for these rufous gems as they range over the region in search of flowers to fuel their migration. During post-breeding southbound migration, males primarily head down the Rocky Mountains, while females and juveniles cover a broader swatch, including the high mountain meadows of the Klamath Mountains (Gillespie et al. 2020). These birds take advantage of the late blooms of the high mountain meadow flowers like columbine, larkspur, and penstemon as they rest and recharge on their southward migration.

Montgomerie and Gass (1981) studied rufous hummingbird use of a high mountain meadow system in the Trinity Alps in July and August. They found that during southbound migration hummingbirds foraged almost exclusively at the flowers of giant paintbrush (*Castilleja miniate*) and western columbine (*Aquilegia formosa*). Here, they establish temporary territories for a few days before moving on (presumedly to another meadow further south). The meadow habitats were saturated with hummingbirds more than 95% of the time, highlighting the importance of these meadow systems as migration stopover habitat. In addition, it highlights the evolution of its migration patterns to be synchronous with the phenology of the flowering plants occupying these higher meadows.

While some Klamath Mountain migrants undertake arduous long-distance treks, others, known as elevational migrants, take a more casual approach to accessing food resources during different seasons. The dark-eyed junco (*Junco hyemalis*), a mid- to high-elevation mountain breeder, forms large post-breeding flocks that mainly feed on insects. They slowly move down in elevation as the days grow shorter and colder while their winter diet switches mostly to seeds. These large flocks eventually settle out in low-elevation woodlands or along the coast, taking advantage of the relatively mild coastal climate to overwinter before beginning their upslope breeding migration the next spring.

In a recent study of migrants in the greater Klamath River Basin, Wiegardt et al. (2017) revealed that some long-distance migrants perform short-distance inland or upslope movements to preferred locations where adults molt body and flight feathers prior to initiating their migrations. This is known as elevational molt migration. In the study, orange-crowned warblers (*Oreothlypsis celata*) moved higher in elevation post-breeding and Audubon's yellow-rumped warblers (*Setophaga coronate auduboni*) moved further inland to molt after breeding. Orange-crowned warblers bred at sites that averaged 1,600 feet (488 m) in elevation, whereas they molted at sites that were 4,900 feet (1,494 m) in elevation. These pre-migration upslope movements demonstrate that species take advantage of differences in food availability and that a diversity of structure and habitats by elevation is crucial to migrants as they increase their energy reserves prior to molt and migration. Each step of their migratory journey, from the wintering range, to migration stopover habitat, to breeding range, to pre-molt habitat, is a crucial component of the avian annual cycle.

Conclusion

The avifauna of the Klamath Mountains is rich in abundance and diversity, representing a unique mix of species due to the region's location. This diversity arises from local habitat structure and the abundance of life nurtured by its primary producers—the plants. The birds, in turn, support and shape these plant communities through pollen transfer and seed dispersal, ensuring botanical gene flow and resilience in the face of ever-changing environments.

Whether it's a Swainson's thrush that has arrived after a month-long trek from Panama, a newly hatched rufous hummingbird attempting to make its first perilous migration to central Mexico, or a wrentit that hasn't ranged outside of a 1,000-foot radius its entire life, the Klamath Mountains host some 244 avian species—each with its own life history and ecological role.

Most of these avian species are understudied and more breeding season—and nest location—records are needed for the rare or more cryptic species. We encourage you to use eBird to document what you see and where you see it. Upload a recorded call or picture (if you are lucky enough to capture that) to eBird or iNaturalist. As community scientists, we can all contribute to our understanding of this region's avifauna through documenting the common and rare birds of the region.

Acknowledgements

Thanks to Frank Fogarty, Jim Agee, Mark Bailey, and C.J. Ralph for reviewing the chapter and providing thoughtful insight and suggestions. Photographic contributions from Dawn Blake, Rob Fowler, Jeff Todoroff, Christopher West, Christopher Lindsey, Kyle Keegan, Justin Garwood, and Frank Lake are greatly appreciated.

Literature Cited and Suggested Reading

Albert, S., J. Wolfe, J. Kellerman, T. Sherry, B. Stutchbury, N. Bayly, and A. Ruiz-Sánchez. 2020. Habitat ecology of Nearctic–Neotropical migratory landbirds on the nonbreeding grounds. *Ornithological Applications*, 122, Issue 4:1-18.

Altman, B. and J. Stephens. 2012. Land Managers Guide to Bird Habitat and Populations in Oak Ecosystems of the Pacific Northwest. American Bird Conservancy and Klamath Bird Observatory. 82 pp.

Anderson, M. and J. Grinnell. 1903. Birds of the Siskiyou Mountains, California: A problem in their distribution. Proc Acad Nat Sci Pa. 55:4-15.

Baker M., N. Nur, and G. Geupel. 1995. Correcting biased estimates of dispersal and survival due to limited study area: theory and an application using wrentits. *The Condor.* 1;97(3):663-74.

Block, W., J. Ganey, K. Severson, and M. Morrison. 1992. Use of oaks by Neotropical migratory birds in the Southwest. Ecology and Management of Oak and Associated Woodlands. US Department of Agriculture, Forest Service General Technical Report RM-218. pp.65-70.

Cox, G. 1985. The evolution of avian migration systems between temperate and tropical regions of the New World. *The American Naturalist* 126(4): 451-74.

Diller, L., K. Hamm, D. Early, D. Lamphear, K. Dugger, C. Yackulic, C. Schwarz, P. Carlson, and T. McDonald. 2016. Demographic response of northern spotted owls to barred owl removal. *Jour. Wild. Mgmt.* 80: 691-707.

Dow, Douglas D. 1965. The role of saliva in food storage by the Gray Jay. *The Auk* 82: 139-154.

Drever, Mark, K. Aitken, A. Norris, and K. Martin. 2008. Woodpeckers as reliable indicators of bird richness, forest health and harvest. *Biological Conservation* 141. 624-634.

eBird, Cornell Lab of Ornithology, Ithaca, New York. Available: http://www.ebird.org.

Fraser, K., A. Roberto-Charron, B. Cousens, M. Simmons, A. Nightingale, A. Shave, R. Cormier, and D. Humple. 2018. Classic pattern of leapfrog migration in Sooty Fox Sparrow (*Passerella iliaca unalaschcensis*) is not supported by direct migration tracking of individual birds, The Auk 135:572-582.

Garwood, J., K. Pope, R. Bourque, and M. Larson. 2009. High mountain lakes provide a seasonal niche for migrant American Dippers. *Wilson Journal of Ornithology* 121:600-609.

Geupel, G., and G. Ballard. 2020. Wrentit (*Chamaea fasciata*), version 1.0. In Birds of the World (P. G. Rodewald, Editor). Cornell Lab of Ornithology, Ithaca, NY, USA. https://doi.org/10.2173/bow.wrenti.01

Gillespie, C., S. Contreras-Martinez, C. Bishop, and J. Alexander. 2020. State of the Rufous Hummingbird science and conservation. Western Hummingbird Partnership, Boulder, CO.

Hagar, J. 2007. Wildlife species associated with non-coniferous vegetation in Pacific Northwest conifer forests: A review. Forest Ecology and Management, Volume 246, Issue 1.

Harris, S. 2006. *Northwestern California Birds*, 3rd edition. California: Living Gold Press.

Karuk Tribe Department of Natural Resources. 2016. Karuk Tribe Climate Change Vulnerability Assessment.

Kauffmann, M. 2013. *Conifers of the Pacific Slope.* Backcountry Press, Kneeland, CA.

Koenig, W., J. McEntee, and E. Walters. 2008. Acorn harvesting by acorn woodpeckers: annual variation and comparison with genetic estimates. *Evolutionary Ecology Research* 10: 811-822.

Knopf, F., R. Johnson, T. Rich, F. Samson, and R. Szaro. 1988. Conservation of Riparian Ecosystems in the United States. *The Wilson Bulletin* 100, no. 2; 272–84.

Kraaijeveld, K. and J. Dickinson. 2001. Family-based winter territoriality in western bluebirds: the structure and dynamics of winter groups. *Anim. Behavior* 61:109-117.

Lake, F.K. 2013. Trails, fires, and tribulations: Tribal resource management and research issues in northern California. Occasion: *Interdisciplinary Studies in the Humanities*, 5, pp.1-22.

Lanner, R. 1996. *Made for Each Other: A Symbiosis of Birds and Pines.* Oxford University Press, USA.

Montgomerie, R. and C.L. Gass. 1981. Energy limitation of hummingbird populations in tropical and temperate communities. *Oecologia* 50: 162-165.

Morrison, Michael L., D.L. Dahlsten, S. M. Tait, R.C. Heald, K. A. Milne, D. L. Rowney, 1989. Bird foraging on incense-cedar and incense-cedar scale during winter in California. Res. Pap. PSW-RP-195. Berkeley, CA: U.S. Department of Agriculture, Forest Service, Pacific Southwest Forest and Range Experiment Station. 16 p.

Myers, B., D. Rankin, K. Burns, and C. Clark. 2019. Behavioral and morphological evidence of an Allen's × Rufous hummingbird (*Selasphorus sasin × S. rufus*) hybrid zone in southern Oregon and northern California. *The Auk*, Volume 136, Issue 4, 1 October 2019, ukz049.

Norris, D., P. Marra, T. Kyser, T. Sherry, and L. Ratcliffe. 2004. Tropical winter habitat limits reproductive success on the temperate breeding grounds in a migratory bird. Proceedings of the Royal Society of London. Series B: Biological Sciences 271(1534):59-64.

Oring L. and D. Lank. 1982. Sexual selection, arrival times, philopatry and site fidelity in the polyandrous spotted sandpiper. *Behavioral Ecology and Sociobiology.* Jun;10(3):185-91.

Pravosudov V., and N. Clayton. 2001. Effects of demanding foraging conditions on cache retrieval accuracy in food-caching mountain chickadees (*Poecile gambeli*). Proceedings of the Royal Society of London. 268(1465): 363-8.

Rosenberg, K., A. Dokter, P. Blancher, J. Sauer, A. Smith, P. Smith, J. Stanton, A. Panjabi, L. Helft, M. Parr, and P. Marra. 2019. Decline of the North American avifauna. Science, 366 (6461), pp.120-124.

Siegel, R., and D. DeSante. 1999. The draft avian conservation plan for the Sierra Nevada bioregion: Conservation priorities and strategies for safeguarding Sierra bird populations. Institute for Bird Populations report.

Skagen, S., R. Hazlewood, and M. Scott. 2005. The Importance and Future Condition of Western Riparian Ecosystems as Migratory Bird Habitat. USDA Forest Service Gen. Tech. Rep. PSW-GTR-191.

Stock, S., P. Heglund, G. Kaltenecker, J. Carlisle, L. Leppert. 2006. Comparative ecology of the flammulated owl and the Northern saw-whet owl during fall migration. Journal of Raptor Research, 40(2):120-129

Swanson, D., J. Ingold, and R. Galati. 2020. Golden-crowned Kinglet (*Regulus satrapa*), version 1.0. In Birds of the World (A. F. Poole, Editor). Cornell Lab of Ornithology, Ithaca, NY, USA

Trail, P., R. Cooper, and D. Vroman. 1997. The breeding birds of the Klamath/Siskiyou region. Proceedings of the First Conference on Siskiyou Ecology. Cave Junction, Oregon: Siskiyou Project and The Nature Conservancy.

Warning, N. and L. Benedict. 2015. Paving the way: Multifunctional nest architecture of the Rock Wren. The Auk, Volume 132, Issue 1, Pages 288–299.

Wiegardt, A., J. Wolfe, C. Ralph, J. Stephens, and J. Alexander. 2017. Postbreeding elevational movements of western songbirds in Northern California and Southern Oregon. Ecology and Evolution 19: 7750-7764.

Wolfe, J., J. Alexander, J. Stephens, and C. Ralph. 2019. A novel approach to understanding bird communities using informed diversity estimates at local and regional scales in northern California and southern Oregon. *Ecology and Evolution*. 9: 4431– 4442.

Zink, R. and J. Weckstein. 2003. Recent Evolutionary History of the Fox Sparrows (Genus: *Passerella*). Auk. 120. 522-527.

Mammals of the Klamath Mountains

Karen Reiss, Michael Kauffmann, and Chris Feldman

Introduction

Upon returning from time in nature, trip tales inevitably include animal encounters. We get particularly excited over mammal sightings because these are relatively rare—and usually fleeting—with the notable exception of squirrels and chipmunks in campgrounds, parks, or yards. Sightings are rare because most mammals are nocturnal. Birds and mammals are alone among animals in being endothermic homeotherms: they derive energy from food to keep their temperature high and stable. Provided adequate food to fuel metabolism, mammals and birds thrive in climates, seasons, and times of day not as accessible to other animals. Birds typically dominate daytime hours, while most mammals are active at night. This is, perhaps, correlated with birds' visual acuity while mammals have keen olfaction. Diurnal mammals are typically ungulates (hoofed mammals like deer and elk) and all sciurids (squirrels including chipmunks) except for flying squirrels. Other mammals seen in daylight are crepuscular—active at dawn, dusk, and also at night.

Endothermy also explains mammalian responses to food shortages and changing food supplies. Seasonal bird migration is well known, but some mammals also seasonally migrate. In the Klamath Mountains, elk are the most conspicuous migrators and follow forage abundance. Many mammals vary their diet opportunistically, including "carnivores" that may be quite frugivorous (fruit-eating) or insectivorous (insect-eating) at times, and rodents that are typically granivorous (seed-eating) but also consume hypogeous or below-ground fungi like truffles. Another mammalian response to dwindling calories is hibernation or torpor. True hibernators achieve a drop in metabolic rate that affects body temperature as well as respiratory and heart rate. Slowing down the system enables mammals to wait out food shortage, usually in winter, but requires a significant caloric reserve to "jump start" the system during arousal. True hibernation only occurs in mammals small enough to store sufficient calories either externally in the form of caches (some squirrels) or internally in the form of thermogenic fat (most notably, bats, but also some squirrels). The Klamath region's climate is temperate so true hibernation is relatively rare. However, torpor, a mild depression of body temperature and associated metabolic variables, is regionally common. Torpor occurs for short duration—like a chipmunk on a cold morning might appear "dead" but warm them up in a pocket and they'll get lively relatively quickly. Bears use torpor as well, for they are simply too large to store enough food to jump-start after drastic decreases in body temperature. On the other hand, they are diet generalists and opportunistic feeders and easily find alternative foods as seasons change.

Among other defining characteristics that we share with all mammals are hair and mammary glands. Humans are poor examples of hairy beasts but most mammals are well-covered. Hair is especially malleable and serves many adaptive functions. It certainly insulates with variable density and length but is densest in aquatic mammals and denser in winter than summer in mammals inhabiting temperate climates. Mammals are often identified by their hairs, but color can vary dramatically within species and even within individuals based on geography, seasonality and age (Zimova et al. 2014). Gloger's rule predicts that endotherms will be darker the closer they are to the equator, possibly because darker pigments are more resistant to the bacterial decomposition facilitated by humid tropical environments, or because darker pigments offer more protection against increased UV radiation near the equator. Nevertheless, there are many examples of species at the same latitudes whose color varies with habitat. In the shadowed forests of the Klamath Mountains, darker pelages are more common than they are in sunnier inland habitats where individuals of the same species may be lighter (e.g., chipmunks; Sutton and Patterson 2000). Evolution has sculpted mammalian pelage to blend in with the environment such that regionally we see a limited range of dusky Earth tones including gray to brown to black with occasional splashes of white. All white winter forms of higher latitudes are mostly absent here, even at higher elevations (e.g., short-tailed weasels), but seasonal growth and molting of more insulating undercoats are more common. While there are occasionally subtle color differences between juveniles and adults, and more rarely, between females and males, mammalian pelage is, frankly, much less colorful and dramatic than bird plumage. This goes hand in hand with the majority of mammals being less visually inclined and having fairly limited color perception compared with birds.

Because mammals are mostly nocturnal, well camouflaged, and frisky, they are rarely seen. This means that we must rely more on detective work to infer their presence. Studying scat (fecal remains) and tracks has always been key to detecting presence and interpreting patterns of mammalian diversity. Recently, there has been a huge increase in the scientific use of camera traps, outfitted with infrared for night-time use, to ascertain distribution and abundance of mammals (see eMammal.org). Sometimes the only evidence is toothmarks in camping gear!

In this chapter, we will take a tour of the mammals of the Klamath Mountains (Table 1). Many of these charismatic critters can travel long distances and thus, along with birds, their natural histories often extend to other mountain ranges of western North America. For these reasons, mammal conservation issues are often broader than a single region: some bats may migrate to far-flung wintering habitats and other taxa may travel narrow forested migration corridors that connect the Klamath Mountains to other nearby regions such as the Cascade Range. At the same time, Klamath region mammals are relatively understudied compared to those in closer proximity to populous areas of the West. The modern genomic revolution in systematic biology is leading to the discovery of cryptic species and subspecies (see marten and flying squirrel sidebars, and chipmunk text below). Unique stories are emerging that are intertwined with the region's historical geological and climatic complexity and there are undoubtedly correlated conservation implications. What follows are the tales of both the ubiquitous and the rare at a crossroads of diverse mammalian habitats.

Order Rodentia - Rodents

Rodents, which include mice, rats, squirrels, chipmunks, porcupines, woodrats, gophers, and beavers (but not rabbits and hares) are characterized by a pair of continuously growing incisors in the upper and lower jaws, flanked by a wide toothless space called a diastema—useful for moving and masticating vegetation

Table 1: Native Mammals of the Klamath Mountains ‡No recent record

Rodents (Rodentia)

Family Aplodontiidae
mountain beaver *(Aplodontia rufa)*
Family Castoridae
American beaver *(Castor canadensis)*
Family Erethizontidae
porcupine *(Erethizon dorsatum)*
Family Geomyidae
Botta's pocket gopher *(Thomomys bottae)*
Mazama pocket gopher *(Thomomys mazama)*
Family Heteromyidae
California kangaroo rat *(Dipodomys californicus)*
Great Basin pocket mouse (*Pergonathus parvus*)
Family Cricetidae
brush mouse (*Peromyscus boylii*)
deer mouse *(Peromyscus maniculatus)*
pinyon mouse (*Peromyscus truei*)
bushy-tailed woodrat *(Neotoma cinerea)*
dusky-footed woodrat *(Neotoma fuscipes)*
western harvest mouse *(Reithrodontomys megalotis)*
white-footed vole (*Arborimus albipes*)
red tree vole *(Arborimus longicaudus)*
Sonoma tree vole *(Arborimus pomo)*
western red-backed vole *(Clethrionomys californicus)*
California vole *(Microtus californicus)*
creeping vole *(Microtus oregoni)*
Family Zapodidae
western jumping mouse *(Zapus princeps)*
Pacific jumping mouse *(Zapus trinotatus)*
Family Sciuridae
golden-mantled ground squirrel *(Callospermophilus lateralis)*
California ground squirrel *(Otospermophilus beecheyi)*
Douglas squirrel *(Tamiasciurus douglasii)*
gray squirrel (*Sciurus griseus)*
shadow chipmunk *(Neotamias senex)*
yellow-pine chipmunk *(Neotamias amoenus)*
Siskiyou chipmunk *(Neotamias siskiyou)*
Townsend's chipmunk *(Neotamias townsendii)*
Humboldt's flying squirrel *(Glaucomys oregonensis)*

Hares and Rabbits (Lagomorpha)

Family Leporidae
Oregon snowshoe hare (*Lepus americanus klamathensis*)
black-tailed jackrabbit (*Lepus californicus*)
brush rabbit (*Sylvilagus bachmani)*

Ungulates (Artiodactyla)

Family Cervidae
Roosevelt elk *(Cervus canadensis roosevelti)*
Columbian black-tailed deer (*Odocoileus hemionus columbianus*)
Family Antilocapridae
‡pronghorn *(Antilocapra americana)*

Insectivores (Eulipotyphla)

Family Soricidae
marsh shrew (*Sorex bendirii*)
Pacific shrew *(Sorex pacificus)*
water shrew (*Sorex palustris)*
fog shrew *(Sorex sonomae)*
Trowbridge's shrew *(Sorex trowbridgii)*
vagrant shrew (*Sorex vagrans)*

Insectivores (Eulipotyphla) continued

Family Talpidae
broad-footed mole *(Scapanus latimanus)*
coast mole *(Scapanus orarius)*
Townsend's mole *(Scapanus townsendi)*
shrew-mole *(Neurotrichus gibbsii)*

Bats (Chiroptera)

Family Molossidae
western mastiff bat (*Eumops perotis)*⊙
Brazilian free-tailed bat (*Tadarida brasiliensis)*
Family Vespertilionidae
pallid bat (*Antrozous pallidus)*
Townsend's big-eared bat (*Corynorhinus townsendii)*
big brown bat (*Eptesicus fuscus)*
spotted bat (*Euderma maculatum)*
silver-haired bat (*Lasionycteris noctivagans)*
red bat (*Lasiurus borealis)*
hoary bat (*Lasiurus cinereus)*
California myotis (*Myotis californicus)*
small-footed myotis (*Myotis ciliolabrum)*▪
long-eared myotis (*Myotis evotis)*
little brown myotis (*Myotis lucifugus)*
fringed myotis (*Myotis thysanodes)*
long-legged myotis (*Myotis volans)*
Yuma myotis (*Myotis yumanensis)*
western pipistrelle (*Pipistrellus hesperus)*

Carnivores (Carnivora)

Family Canidae
common gray fox (*Urocyon cinereoargenteus*)
‡Sierra Nevada red fox *(Vulpes vulpes necator)*
coyote (*Canis latrans*)
wolf *(Canis lupus)*
Family Felidae
mountain lion *(Puma concolor)*
bobcat (*Lynx rufus)*
Family Mustelidae
ermine (*Mustela erminea)*
long-tailed weasel (*Mustela frenata)*
river otter (*Lontra canadensis)*
Pacific marten (*Martes caurina)*
fisher (*Pekania pennant)*
American mink (*Neovison vison*)
American badger (*Taxidea taxus)*
‡wolverine (*Gulo gulo*)
Family Mephitidae
western spotted skunk (*Spilogale gracilis)*
striped skunk (*Mephitis mephitis)*
Family Procyonidae
ringtail cat (*Bassariscus astutus*)
common raccoon (*Procyon lotor)*
Family Ursidae
black bear (*Ursus americanus*)
‡grizzly bear (*Ursus arctos*)
Family Otariidae
California sea lion *(Zalophus californianus)*
Family Phocidae
harbor seal *(Phoca vitulina)*

⊙ One audio recording by Pierson and Rainey 2007
▪ Audio recorded near Klamath Mountains by Pierson and Rainey 2007

around the mouth. With nearly 2,600 mammal species, some 40 percent are in the Order Rodentia making it the most speciose and diverse mammalian order by far (followed by the bats). Rodents are terrestrial, arboreal, fossorial, semiaquatic, or even urban, and so are found everywhere. And though some rodents are insectivorous or fungivorous, most are herbivorous or granivorous. Their ecological importance, due to their sheer numbers in some habitats and remarkable species richness across ecological communities cannot be overstated. Their foraging can dramatically affect the abundance of the (mostly) plants that they feed on. They serve as important seed dispersers through caches that may not be fully retrieved and seed or spore transport on fur or in feces. And many burrow, whether to create shelter or store food, engineering the soil of their habitats. All these behaviors have consequences for plant community composition. They also form the primary prey base for carnivores (as well as predatory birds and reptiles), again, due to their numbers and ubiquitous presence in all habitats.

Family Aplodontidae - mountain beavers

The mountain beaver, boomer, or sewellel is an interesting Pacific Northwestern endemic. It is the last remaining member of a long-lasting family, one that split from the remainder of the squirrel clade in the mid-Eocene ~40 million years ago (Piaggio et al. 2013). They retain several less derived features including a jaw muscle arrangement that produces less bite force than other rodents, and kidneys that cannot concentrate urine especially well which, perhaps, restricts them to moist climates and habitats. Their cheek teeth are constantly growing (a feature shared with voles) which helps them withstand abrasion and erosion from even the toughest plant foods.

Despite their name, mountain beaver are neither beaver nor exclusively found in the mountains. They range from sea-level to the montane conifer belt—up to 6,100 feet (1,860 m) in the Trinity Alps—and reach peak densities in mid-seral forests with complex shrub understories that have abundant herbaceous forage. They live singly in burrows that consist of a network of tunnels built into deep moist soils, often on slopes adjacent to small streams or marshy meadows with emergent forbs. The burrow entrances often hold clumps of green vegetation used for both a food cache and for nesting material.

This wet hillslope in the high elevations of the Siskiyou Wilderness is home to a mountain beaver. Photo by Michael Kauffmann. Inset is a mountain beaver at its tunnel system entrance. Photo by Ivan Yates.

Family Castoridae - beavers (see The Hidden Lives of Beavers sidebar in Water Chapter on pages 134-135)

Family Erethizontidae - porcupines

The North American porcupine in an unmistakable animal, protected by a shield of spiny quills, some of which are barbed and detachable (but not a projectile). Predators that have gambled badly on tackling porcupines have been found dead, with quills embedded in their mouths or paws. The North American porcupine occurs throughout much of North America where it is a habitat generalist that consumes a wide variety of plants. This broad distribution is partly possible due to an ability to tolerate harsh climates and low-quality diets like bark and conifer needles (Coltrane 2012). Compared to other rodents, porcupines have an exceptional tolerance for cold, which is even more remarkable when one considers that New World porcupines evolved in tropical climates, and most porcupine species currently inhabit tropical regions. Porcupines maintain an energy balance across the year by foraging on broadleaf trees, fruits, and shrubs during the spring and summer and switching to tree bark and needles in the winter.

California appears to be the southwestern range limit where it is now designated a Species of Greatest Conservation Need. Population decline most likely started with poisoning in a misguided effort to save trees so they could be logged. Porcupine populations in the Orleans/Somes Bar area in the late 1960's and early 1970's were locally abundant when the Forest Service nailed strychnine-laced salt baits to trees. The State of California was still paying a $5.00 bounty on porcupine noses in the early 1980's. Further declines can be attributed to forest regeneration and a subsequent decline in preferred habitat after the logging boom of the 1970s also contributed. Most of the reports over the past 100 years come from around the periphery of the range including reports in coastal dune forests and along the western edges of the Cascades. However, porcupine remains have been found, recently, in the subalpine zone of the Trinity Alps at ~6,300 feet (1,920 m) (J. Garwood, unpublished data). Also, an increase in porcupine sightings in the mid Klamath River region is a hopeful sign. The Karuk, Yurok, Hupa, and Chilula Tribes have cultural uses for porcupine quills (Driver 1939) indicating long presence in the region (see sidebar on the next page).

Family Geomyidae - pocket gophers

Pocket gophers are small fossorial or burrowing mammals with about 35 species that are all endemic to North America. Pocket gophers are so named because they have fur-lined cheek pouches, or pockets. Their bodies have typical features associated with fossorial life such as long-clawed forefeet, tiny eyes, and short tails, but they are, in fact, chisel-tooth diggers. Their long incisors are used to excavate, and they have furry flaps that close behind their teeth to protect against soil ingestion giving them a characteristic "buck-tooth" look. The cheek pouches are used to safely store food—virtually any part of a plant—that they collect until they return to storage chambers in their complex burrow systems. The entrances to their tunnels are usually framed by a crescent-shaped ridge of tilled soil, and the entrance itself may be plugged with loose dirt. These tunnel systems are also used by other vertebrates including snakes, salamanders, and small mammals. Despite their reputation as garden pests, in the wild, gopher soil excavation has positive effects on soil

Cultural Uses Of Porcupine

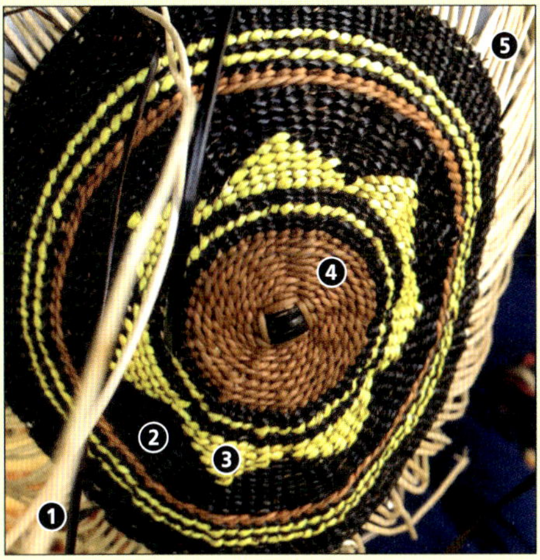

Porcupine quills are the only non-plant material used in the basket weaving of northern California Native People—though sometimes shell beads and feathers are used for adornment. Because porcupine quills are difficult to procure and require a great degree of skill to weave, these baskets are highly valued. Historically, quills were harvested by tossing an elk or deerskin blanket over the porcupine and then extracting the quills from the blanket after the unharmed animal had been released. Today, road kill serves as the primary source for quills, mainly from the Scott and Shasta Valley region.

This ceremonial Medicine Hat, by Lisa Morehead-Hillman, is in the early stages of weaving. Lisa is a Karuk cultural practitioner who is one of only a few still creating such works. Porcupine quills play an important role in the creation of both ceremonial hats and baskets for many Tribes across the region. This ceremonial hat serves as a potent reminder of the intimate relationship that continues to persist between indigenous people and the Klamath Mountains.

Reference the numbers in the image above with the description provided below for an example of this relationship.

1. The preferred foundation materials (sticks) consist of hazel (*Corylus cornuta*), willow (*Salix* spp.), or ceanothus (*Ceanothus* spp.). Hazel is harvested from low to mid elevation patches which are managed using frequent low intensity surface fire. Willow is harvested along river banks and riparian areas from plants which have been pruned, coppiced or treated with medium intensity surface fire. Ceanothus is harvested from low to mid elevation patches managed with frequent low intensity surface fire or by pruning and coppicing of individual plants. This material is especially prized for fine basketry work due to its long, fine sticks and lack of taper.
2. Black overlay material used in the basketry of Klamath River Tribes is prepared for use by separating out the black from the red portions of the maidenhair fern stem. Maidenhair fern (*Adiantum aleuticum*), is also known by the common English term(s) five finger fern or black fern.
3. Porcupine quills, another overlay material, are dyed yellow using wolf lichen (*Letharia* spp.).
4. The red/orange overlay material is prepared for use by dying the two strong inner fibers of the woodwardia fern (*Woodwardia fimbriata*) stem. A small section from the cambium layer of the white alder (*Alnus rhombifolia*) is crushed and soaked in water with the woodwardia fibers.
5. Another common overlay material, although not found in this basket, is beargrass (*Xerophyllum tenax*). Beargrass is harvested from recently burned plots under open canopy in well-spaced mixed forests at higher elevations.

structure and chemistry—the tilling oxygenates the soil, and their scat and urine add nitrogen. The combination of these physical and chemical effects with their dietary choices affect local plant community structure.

Botta's pocket gopher and the Mazama pocket gopher are two species in the Klamath Mountains with overlapping distributions though the latter may prefer more open habitats. Botta's pocket gopher is widespread throughout the southwest from Texas to California and it reaches its northern range limit in the Klamath Mountains. Across this range, they are found in virtually all friable soils up to the subalpine zone. Mazama pocket gophers are native to the Pacific Northwest from Washington, where they are considered threatened (Stinson 2020), south to the Klamath Mountains where they reach their southern range limit. They are one of the smallest of all gophers and named after Mount Mazama, the volcano that formed Crater Lake—which is where the species was first described.

Family Heteromyidae - kangaroo rats and pocket mice

The Family Heteromyidae includes kangaroo rats, kangaroo mice, and pocket mice. They are a diverse radiation found mainly in southwestern North America, especially in xeric grasslands and deserts. All share with the pocket gophers the presence of fur-lined external cheek pouches in which they can store food for transport to their below-ground caches. In the relatively humid Klamath region, only two species are found. The large and mostly bipedal California kangaroo rats enjoy open country in the eastern Klamath, becoming more abundant on the Modoc Plateau and in the Central Valley. The smaller and more quadrupedal Great Basin pocket mice range further west into the region but are only found in low elevation grasslands.

Family Cricetidae – New World mice and rats

One of the most species rich and ecologically significant rodent families is the Cricetidae. The Subfamily Cricetinae includes the New World mice and rats, including woodrats, while the Subfamily Arvicolinae has a circumboreal distribution and includes voles, lemmings and muskrat. The Klamath Mountains do contain several important cricetine and arvicoline species which, while small, nocturnal, and shy, are nevertheless abundant and a critical source of food for all carnivorous vertebrates (including Pacific giant salamanders).

Klamath cricetines include deer mice, woodrats, and harvest mice, the first two of these being the most conspicuous and important ecologically. The deer mouse is arguably the most abundant, widespread, and well-studied mammal in North America as it is simply found everywhere and often in great abundance. It is not uncommon to see a deer mouse peek out looking for scraps in the light of a campfire or to see nibble marks on personal items left out overnight. The slightly smaller brush mouse may be found in Klamath region chaparral and other brushy scrub. The pinyon mouse is nominally a denizen of pinyon pine woodlands but is found in other lower elevation coniferous forests and can be distinguished from other mice by their much larger "Mickey Mouse" ears and a longer tail. The smallest mouse of the region is the tiny western harvest mouse which is restricted to grassy areas.

Two species of woodrat (or packrat) are found in the Klamath mountains. The dusky-footed woodrat is usually at lower elevations, almost always associated with oaks. The bushy-tailed woodrat is found with rocky outcrops in montane forests, often associated with pinyon pine and juniper. Woodrats are four times the size of a deer mouse and represent more calories in a predation event making them the desired food of many predators. They are charming, with the peculiar habit of building large, above-ground dens called middens which are constructed from sticks and twigs, small stones and pebbles, and other assorted items. Middens provide unique refuges for smaller animals including invertebrates, amphibians, and reptiles. Middens appear to be passed down along matrilines and are known to be used for decades. So the large stick nest you're looking at under the oak might actually be 80 years old, and has been used by 30 generations of woodrats—each making its own home-repairs and modifications. The rainy regional climate takes a toll on most of these woodrat houses. While most don't persist for eons, Samwel Cave has a midden that dates back 18,000 years (see sidebar on the next page)—a longevity more common in the dry desert southwest. Well-preserved middens are an incredibly valuable source of paleoecological data for scientists interested in the effects of climate change on plant and animal communities.

Arvicoline rodents have rounder faces and smaller, rounder, well-furred ears—giving them a cuteness advantage over cricetine mice. Native arvicolines in the Klamath include tree voles, red-backed vole, and meadow voles. Red tree vole, Sonoma tree vole, and white-footed vole are the only truly arboreal arvicolines in the world—their genus, *Arborimus*, means "tree-mouse." The first two of these species are ecologically similar, being among the few mammals that live in and feed primarily on needles and twigs of conifers. They are geographically separated by the Klamath River (Blois and Arbogast 2006), a boundary that likely has broad phylogeographic significance. Red tree voles live in western Oregon and northwestern California north of the Klamath River and Sonoma tree voles live in coastal California from the Klamath River south to southern Sonoma County (Forsman et al. 2016). Both prefer old-growth forest or at least large Douglas-firs (*Pseudotsuga menziesii*), their preferred food. Moisture from rain and fog is obtained from the needles along with carbohydrates and the needle's central resin ducts that remain after feeding are used as an antimicrobial nesting material. Intricate nests are built on branches near the main tree trunk, and sometimes in cavities and hollow up to 250 feet off the ground. The Sonoma tree vole is a Species of Special Concern due to loss of habitat from timber harvest, fire, and habitat conversion (Chinnici et al. 2012).

The white-footed vole prefers alder-rich riparian habitats in the western Klamath Mountains and is the least arboreal and least associated with conifers of the three tree voles. This species is also believed to be in decline due to logging but all tree voles are challenging to study due to their disinterest in baited traps. The western red-backed vole is terrestrial and restricted to older mesic forests with abundant hypogeous fungi. Yes, they detect truffles by scent and consume them, but more broadly, they are important spore dispersers for the mycorrhizal fungi associated with conifers in these habitats. The meadow voles (*Microtus* sp.) are typically herbivorous and found in grassy areas where they form surface runways covered with thatch.

The Data are in the Strata

Jessica Blois (left) and Uma Ramakrishnan (right) excavate a woodrat midden in Samwel Cave. Photo courtesy of Jessica Blois.

Samwel and Potter Creek caves of the eastern Klamath Mountains are rich with animal fossils. The regional limestone first came to the attention of scientists in the late 1800s with the discovery of late Triassic ammonites. By 1901, John C. Merriam and Annie Alexander began documenting the caves, spending the next nine years unEarthing late Triassic marine reptiles and Pleistocene fossils. Merriam and Alexander's collections are stored in the University of California Museum of Paleontology in Berkeley where they continue to be used to refine our understanding of the California landscape in the Pleistocene and how biotic communities responded as climates shifted in the Holocene.

In the mid-2000s, Jessica Blois followed Merriam and Alexander by crawling into Samwel Cave. With hard-hat, headlamp, and coveralls she and her team excavated woodrat middens with fossil deposits. By analyzing bone collagen with radiocarbon dating, their findings go back 18,000 years. Along with placing timestamps in these stratified midden layers the researchers also identified many of the fossil fragments to the species level and could interpret shifts in small mammal populations over that time.

When the Pleistocene ended around 11,700 years ago, biotic communities were altered by the interplay of a climate shift from colder and dryer to warmer

This photo, circa 1904, shows non-human bones from the Pleistocene wrapped and ready to be hoisted out of Samwel Cave. Image courtesy of University of California Museum of Paleontology, Berkeley, California.

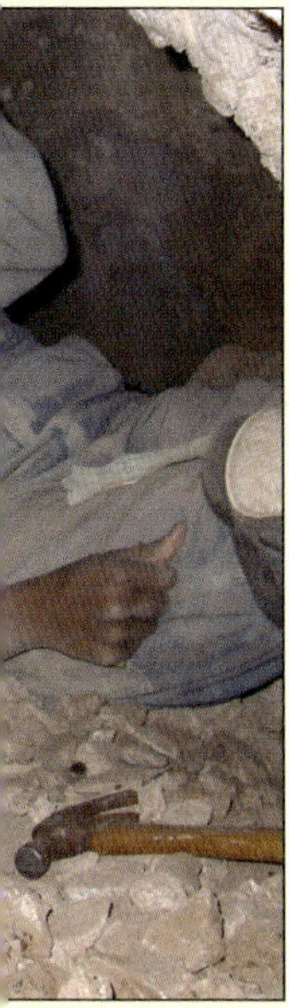

and wetter. When this happened, species shifted their geographic distributions. Data from the middens showed that during the Pleistocene the small mammal community remained relatively stable in the fossil record. At the beginning of the Holocene some small mammal species were dominant while others became scarce. Some species like the Mazama pocket gopher and mountain beaver have been extirpated while chipmunks, ground squirrels, and voles composed a smaller proportion than they had previously (Blois et al. 2010).

These shifts occurred in two waves. The first occurred between 15,000 and 14,000 years ago when there was a shift in small mammal species abundance (deer mice and Botta's pocket gopher became more abundant while the Mazama pocket gopher declined) but no species disappeared completely. An extinction pulse happened between 11,000 and 7,500 years ago when some species disappeared from the midden records and others experienced population reductions. Climate change and increasing densities of *Homo sapiens* both played a role in these extinctions (Broughton et al. 2018). Today, the pocket gophers and mountain beavers which once inhabited the lower elevations around Samwel Cave live exclusively in higher elevation, cooler areas. Other species like deer mice have had marked population increases in the Holocene because these mice are generalists who have a keen ability to pioneer new habitat. These findings are interesting on many levels, including implications for the future. As our climate continues to rapidly shift will this drive more restructuring of small mammal communities, cause a loss of biodiversity like the extinction pulse documented earlier in the Holocene, or cause a rising dominance of native opportunists?

Table 2: Select Mammalian fossil deposits in Samwel and Potter Creek Caves, Klamath Mountains. Based on Feranec et al. 2007 and Feranec 2009. Symbols: †, extinct; ‡ recent regional extirpation, extant

Artiodactyla (deer, sheep, etc.)
†*Euceratherium collinum* (shrub-ox)
‡*Antilocapra americana* (pronghorn)
†*Oreamnos americanus* (mtn. goat)
‡*Ovis* sp. (sheep)

Carnivora (cats, dogs, etc.)
†*Arctodus pristinus* (short-faced bear)
†*Canis dirus* (dire wolf)
‡*Canis lupus* (gray wolf)
Lutra canadensis (river otter)
†*Panthera leo atrox* (American lion)
Spilogale gracilis (spotted skunk)
‡*Ursus arctos* (grizzly bear)

Chiroptera (bats)
Antrozous pallidus (pallid bat)
†*Desmodus stocki* (vampire bat)

Insectivora (shrews and moles)
Scapanus latimanus (broad-footed mole)

Lagamorpha (rabbits)
Lepus americanus (snowshoe hare)
Lepus californicus (jackrabbit)
‡*Sylvilagus auduboni* (desert cottontail)
Sylvilagus bachmani (brush rabbit)

Perissodactyla (horses)
†*Equus occidentalis* (western horse)

Primates
Homo sapiens (human)

Rodentia (rodents)
‡*Aplodontia rufa* (mountain beaver)
Castor canadensis (American beaver)
Erethizon dorsatum (porcupine)
Peromyscus maniculatus (deer mouse)
Thomomys bottae (Botta's pocket gopher)
†*Thomomys microdon* (pocket gopher)

Proboscidea (elephants)
†*Mammut americanum* (American mastodon)
†*Mammuthus primigenius* (woolly mammoth)

Xenarthra (anteaters, tree sloths)
†*Nothrotheriops shastensis* (ground sloth)

Family Zapodidae – jumping mice

Pacific and western jumping mice are convergent with heteromyids in morphology—large hind legs, long tails, and short forelegs, all correlated with saltatory or hopping behavior —but the jumping mice are usually found in wet meadows and riparian habitats. They are rarely seen, but a sometimes greenish cast to their backs and their hopping behavior lends them to occasionally being mistaken for frogs.

Family Sciuridae – ground squirrels, tree squirrels, flying squirrels, chipmunks

Squirrels round out our rodent discussion and include ground squirrels, tree squirrels, chipmunks, and flying squirrels. All but the flying squirrel are diurnal and so squirrels are important large-sized prey for diurnal predators and are more often seen by humans than perhaps any other mammal.

The ground squirrels include the lower elevation California ground squirrel and the montane golden-mantled ground squirrel. The California ground squirrel is a grizzled black-and-white with a darker patch on its back. The golden-mantled ground squirrel has conspicuous black and white side stripes and a golden head (but no face stripes, distinguishing it from chipmunks). It is especially conspicuous during its active foraging because, as a higher elevation hibernator, it must put on a good layer of fat before winter. Both ground squirrels den in burrows that also provide habitat for reptiles, amphibians, and smaller mammals.

Yellow-bellied marmot (*Marmota flaviventris*), really an enormous ground squirrel, are absent from the Klamath Mountains. Reports place them just east of Ashland, Oregon in the southern Cascades and they appear in the fossil record from the late Pleistocene in Samwel Cave (Feranec et al. 2007) but no other records exist for the Klamath Mountains despite the apparent presence of suitable high elevation habitat. If they ever were there, they could have been eliminated by predation, competition, or warming climates that constricted suitable habitat to a point where populations were not viable. Alternatively, post-Pleistocene warming may have prevented dispersal from nearby populations across intervening lowlands to take up residence in the Klamath Mountains.

The tree squirrels include Douglas squirrel and the gray squirrel. Douglas squirrel is the smaller of the two and are usually heard long before being seen because their vigorous and scolding chitter is one of the loudest sounds of the forest. It is a conifer seed specialist from the coast to interior montane and subalpine region and frequently drops cones from trees (another characteristic sound) and leave a tell-tale trailside pile of denuded cone scales. The gray squirrel is larger, and all gray, with an especially bushy gray tail. It prefers low-elevation mixed forests including oak woodlands and acorns are its preferred food source. It is often busy collecting and scatter-hoarding acorns, a behavior that benefits oak dispersal because despite their excellent spatial memory some caches ultimately remain uneaten. Gray squirrels also create empty caches to deceive on-lookers, both mammals and birds, an interesting strategy that consumes energy but must ultimately pay off in deterring cache thieves.

Chipmunks are distinguished from golden-mantled ground squirrels by black and white facial stripes in addition to the stripes on their sides and back (see images and map page 424). They are broadly granivorous and serve as important seed-dispersers. Forest dwelling chipmunks are known to eat both above- and below-ground fungi, disperse fungal spores, and play a role in supporting conifer symbioses with mycorrhizae. The small yellow-pine chipmunk has a similar range and habitat to the golden-mantled ground squirrel, inhabiting the interior mountains of the montane West including the most xeric forests in the higher elevations of the western Klamath Mountains.

Throughout the more mesic forests of the Pacific Northwest resides the "Townsend's group" of chipmunk—all about twice the size of the yellow pine chipmunk and generally less brightly colored. The group is distinguished from other chipmunks by their large size and pure white tips on the tail hairs. They were once considered a single far-ranging species but are now divided into four species distinguishable only by genital bone or baculum anatomy, three of which are in our region. North of the Rogue River, from the northernmost portion of the Klamath Mountains northwards through the coastal forests of Oregon and Washington, resides the Townsend's chipmunk. From the Rogue River south to the Klamath River resides the Siskiyou chipmunk. South of the Klamath River and north of the Eel River, across Humboldt, Siskiyou, and Trinity counties, lives the shadow (or Allen's) chipmunk which also ranges in the interior northwards into the southern Oregon Cascades and southwards into the Sierra Nevada of California—with an isolated population in the Warner Mountains. The fourth species of the group, the yellow-cheeked chipmunk (*Neotamias ochrogenys)*, lies west of the Eel River in the coastal areas and the King Range, and so is out of our region of interest.

The Rogue, Klamath, and Eel are three big rivers with vast drainages and their role creating and maintaining these species remain unclear. No chipmunk could or would traverse the wide gravel bars and/or deep flowing water near their mouths. Yet the situation changes upstream where these rivers and their tributaries are fed by rushing creeks coursing through dense forests and are crisscrossed with unlikely barriers like mossy logs. Also, since the headwaters of the Rogue and Klamath both arise from the slopes of what was Mt. Mazama, these three species are contiguous and may be interbreeding or even incompletely diverged in the forest surrounding Crater Lake. Genomic studies to clearly define species identity throughout these river drainages are currently underway.

The Humboldt's flying squirrel lives in the coastal lowland forests, but rarely seen. Photo Nick Kerhoulas.

Our last sciurid, flying squirrels, don't actually fly with true powered flapping flight. Instead, they glide, or fall with style (as Buzz Lightyear from *The Toy Story* would say). They run up the bole of a tree and then leap, wing membranes outstretched, until—thwack—they clasp the trunk of the next tree, which they then scamper up and repeat. The wing membranes are, in fact, a flap of fur covered skin that extends from wrist to ankle, and their flattened tail aids in steering. Flying squirrels are the only nocturnal squirrels in North America, and

Klamath Mountain Chipmunks

IMAGE NOTES: *T. amoenus* is the smallest of the Klamath chipmunks. The photo of *T. siskiyou* is of an inland morph—note that while brighter than coastal populations, it still has the diffuse stripes characteristic of the Townsend's group.

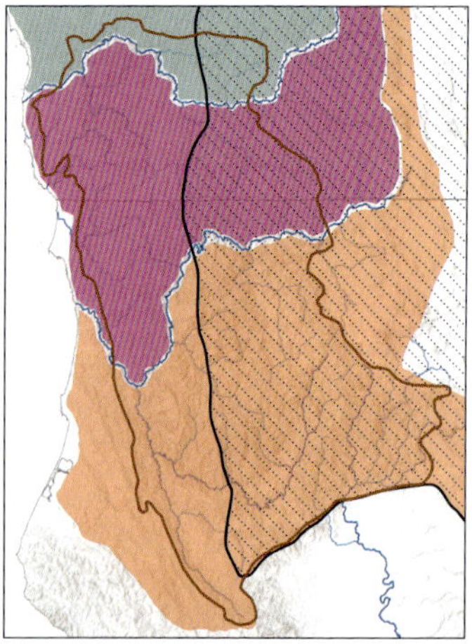

as such, are prey for those nocturnal carnivores that catch them. Not surprisingly, owls are efficient flying squirrel predators. These squirrels eat fungi, and move relatively large distances through the forest, playing a significant role in fungal spore dispersal. Until recently, mammalogists recognized two species of flying squirrels in North America but in 2017 a third was added with the formal description of the Humboldt flying squirrel, *Glaucomys oregonensis* (Arbogast et al. 2017). Though slightly smaller and darker, this cryptic species was once considered to be a northern flying squirrel (*G. sabrinus*) but genetic studies and, again, genital bone anatomy, forced a revision. The more widespread northern flying squirrels inhabit boreal coniferous forests of the northeastern U.S., Canada, and Alaska as well as relictual habitat in the higher elevations of the Appalachian and Rocky Mountains. The Humboldt flying squirrel replaces it in Pacific Slope coastal forests from southern California sky islands northwards through the Sierra Nevada, Klamath Mountains, Coast Ranges, and Cascades—always in conifer and mixed-conifer forests.

Order Lagomorpha – Rabbits and Hares

Lagomorphs are herbivores and, given their relatively large size compared to most rodents, are a coveted prey for mid-sized and larger carnivores. Of the three Klamath Mountain lagomorphs, all are in the Family Leporidae and include one rabbit and two hares. Rabbits are generally smaller, with shorter ears, ambling locomotion, and helpless altricial young. Hares are larger with longer ears, especially long hind limbs, speedy bounding locomotion, and capable precocial young. Not surprisingly, then, rabbits are shyer and prefer to hide from their enemies in dense brush while hares are speedy runners in meadows and prairies.

The brush rabbit requires the dense shrubs of young forest understory or other thick brushy habitat like chaparral. The blacktailed jackrabbit is found from sea level to subalpine across western North America, and in the Klamath Mountains, approaches the northern extent of its range. They are generally found in lower elevations along the outer reaches of the region where they inhabit the xeric, open habitats to the south and east in California. Snowshoe hares are found in diverse habitats across northern North America with their southernmost limits in high elevation areas of the Cascades, Rockies, and Appalachians. The Oregon snowshoe hare in the Klamath Mountains is at the southern edge of its western range. Its differences from the blacktail jackrabbit are all traits that serve snowshoe hare well in winter: smaller ears that lose less heat, larger furrier feet that sink less into the snow, and white winter pelage that provides camouflage against snowy backgrounds.

Oregon snowshoe hare (*Lepus americanus klamathensis*). Photo by Ivan Yates.

Why the charismatic lagomorph, the American pika (*Ochotona princeps*) is absent from the region despite apparently suitable habitat and their presence nearby is another question for biogeographers. Pikas are talus denizens that are present in the highest elevations

Mammals

of the Sierra Nevada, Warner Mountains, and Oregon Cascades but there is no evidence, historical or fossil, of their presence in the Klamath Mountains. Possible explanations for their absence parallel those for other high elevation absentees like the yellow-bellied marmot.

Order Artiodactyla – Even-toed Ungulates

Artiodactyls are the even-toed hoofed mammals. Even-toed means they use either two or four toes to support their weight on the ground, and each of these toes is surrounded by a structure called a hoof, or unguis, equivalent to a human fingernail. This foot structure is what causes the characteristic two-toed track of their cloven hooves. This is in contrast to the odd-toed Perissodactyla with their single hoof print like horses (or three-pronged in tapirs and rhinoceros).

Family Cervidae – deer and elk

Among artiodactyls, cervids are characterized by their bony antlers, usually only in males, that are shed annually. Associated with this sexual dimorphism in head ornamentation often are interesting social behaviors (e.g., male-male competitions) and mating systems (e.g., polygynous systems in which one male is dominant and copulates with many females). Cervids are typically browsers, large-bodied, frequently live in groups, and forage and defecate over large areas—so their foraging behavior has pronounced effects on vegetative communities. Cervids are hefty prey for carnivores that can take them down, whether pack-hunting canids or the more adept solitary predators like mountain lion.

Today, there are only two native hoofed mammals in the Klamath Mountains—Columbian black-tailed deer and Roosevelt elk—each a subspecies of a diverse species complex spread across the west. Elk are considerably larger than deer and have a conspicuous broad white rump compared with deer that have white only on the underside of their otherwise black tail. Elk are also more likely to be in large numbers, whether "harems" of females and their calves or "bachelor herds", while deer are almost always alone or in small family groups. Columbian black-tailed deer are ubiquitous from the river valleys to the upper montane belt, favoring the forest edges for both food and habitat availability.

Today, several herds of elk are found in the Klamath mountains and surrounding lowland areas but it hasn't always been so. While they once roamed across northwest California they declined by the mid-1800s due to increased hunting, active extermination campaigns by European settlers, and the introduction of cattle. The Klamath Mountain population was near extinction by the mid-1900s, and north coast populations were down to an estimated 2,000 individuals in 1965. Reintroduction efforts were undertaken as a collaborative between California Fish and Game, the Karuk Department of Natural Resources, and the Klamath National Forest. Now thriving populations in the Klamath Mountains move from high-elevations in summer to river valleys in the winter, always preferring open meadows with adjacent scrub or forest for additional browse and cover. In late summer, mature males enter the rut, a period of hormone-driven stereotyped competition for females that includes loud vocalizing called bugling and antler clashes. The winner then controls the harem of females through the fall breeding season. Never approach elk any time of year.

Family Bovidae – bison, bighorn sheep, and domesticated species (cows)

There are currently no native bovids in the Klamath Mountains, though there are Pleistocene sheep records from Samwel and Potter Creek caves, and bighorn sheep (*Ovis canadensis*) records from the flanks of Mount Shasta in 1808 (Grinnell 1937), and historically, bison may have approached the eastern edge of the region (Grinnell 1937). Domesticated bovids, on the other hand, are quite common throughout the Klamath Mountains beginning in mid-July when they are brought into some high country meadow and lake complexes to escape valley heat.

Family Antilocapridae – pronghorn antelope

Pronghorn are beautiful tawny-and-white ungulates and the sole species in this uniquely North American family. They are named for their two-pronged "pronghorn," found in both males and females, that consists of a permanent conical bony core and a black outer keratinous sheath that is shed. Pronghorn tie for the record of the fastest terrestrial mammal on the planet, 60 mph (the other being the cheetah, but the cheetah has less endurance). In northern California, pronghorn are mainly on the Modoc Plateau, reaching as far west as eastern Siskiyou County. While they don't reach the Klamath Mountains today there is a Karuk word, yupinsiih, meaning antelope, and early European reports place them between Little Shasta—suggesting historical presence in the area. Early Europeans report that they were "most plentiful" between Little Shasta Valley and Gazelle (Grinnell 1937).

Order Eulipotyphla – shrews and moles

Shrews and moles are small mainly insectivorous mammals. All are small with elongated and sharply pointed cusps on their teeth which are well suited for penetrating the chitinous exoskeletons of their insect prey. Venomous saliva is not uncommon, enabling these small mammals to subdue relatively large and potentially dangerous prey like centipedes, scorpions, and beetles.

Family Soricidae - shrews

Most shrews are tiny and voracious. They forage day and night, sometimes taking small vertebrates, worms, and even nuts and seeds—in addition to insects—driven by a constant need for metabolic energy to replace thermal loss from the high relative surface area of their bodies. Five species of shrews are found in the Klamath Mountains, all in the genus *Sorex*. These are the smallest mammals regionally. It is not uncommon to have multiple shrew species in the same area, but studies examining mechanisms that reduce competition between these sympatric species (i.e., niche partitioning) are lacking in the Klamath Mountains.

The Pacific shrew is native to western Oregon and northwest California in damp areas along creeks and downed trees. The fog shrew has a similar range to the Pacific shrew but extends further south toward San Francisco in chaparral and coastal conifer forests. American water shrew has a much broader range across the temperate mountains of North America. Rarely found far from water, they are adapted to forays into water with hind feet that contain specialized hairs that act as webbing, and pelage that traps air bubbles to keep the body warm and provide buoyancy. In the Klamath Mountains, water shrews are commonly found in subalpine fens with ponds for hunting amphibian larvae, as is the larger marsh shrew. Trowbridge's shrew is a specialist along the Pacific Slope where they prefer forested areas. The vagrant shrew is broadly present in open woodlands and meadows near water west of the continental divide. This species sometimes uses echolocation for orientation, a sensory modality that may be present in the others as well.

Family Talpidae - moles and shrew moles

Moles are only slightly larger than shrews and are specialized for fossorial life with correlated stout skulls, muscular forearms, and broad front feet. They are humeral-rotation diggers that essentially do the breaststroke through the soil. They excavate extensive burrow systems, and then patrol their burrows for unsuspecting invertebrate prey. The openings to their burrows differ from those of gophers in being round mounds that may or may not have an apparent opening, whole gopher mounds are semicircular. Unlike shrews, moles have a relatively low metabolic rate for their body size, thought to be an adaptation for the low oxygen levels of subterranean life.

Talpids of the Klamath Mountains include three species of fossorial moles and one shrew-mole. The moles are in the genus *Scapanus* and all three are specific to western North America with overlapping ranges in the Klamath. The broad-footed mole is found throughout from Baja California, Mexico north into southern Oregon, preferring moist, rich soils with scant vegetation (Verts and Carroway 2001). Pacific mole is at its southern limit in the Klamath Mountains and ranges north into British Columbia, preferring moist forests and open areas where they primarily hunt Earthworms but are capable of surfacing for scavenging and juvenile dispersal. Townsend's mole, one of North America's largest moles, is exclusively found along the Pacific Coast in the Klamath Mountains and extending north and inland to the west slopes of the Cascades up to southwestern British Columbia. Townsend's moles prefer meadows, floodplains, and open fir forests where they build permanent tunnels.

The American shrew-mole is a Pacific Northwest endemic found in damp forests from California to British Columbia. Phylogenetically a mole, it behaves somewhat like a shrew. Its metabolic rate is in between the depressed rate of moles and the elevated rate of shrews. Their foraging style is less fossorial than most moles, but they dig more than most shrews. Most curiously, many naturalists boast at least one dead shrew-mole in their freezer, usually found stiff and in the middle of a trail with no apparent damage. This phenomenon has never been explicitly investigated but may reflect a low thermal tolerance in an animal with a fossorial ancestry but a propensity for above-ground excursions.

Order Chiroptera - bats

Bats are the second most diverse group of mammals, after rodents, with approximately 1,430 species worldwide. They are usually insectivorous but some are carnivorous (eating fish and frogs), nectarivorous (nectar-feeders), or sanguivorous (i.e., vampire bats). Worldwide abundance and dietary diversity bespeaks their extraordinary ecological importance across diverse habitats. Because they fly, bats effectively transport nutrients from one habitat to another. Because they use riparian corridors as flyways, they connect aquatic and terrestrial food webs. Insectivorous species play crucial roles in controlling flying insect populations, particularly pro-

This collection of photos by Ivan Yates was taken in 2018 during a bat inventory at Oregon Caves National Monument and Preserve.
From top to bottom: Fringed myotis (*Myotis thysanodes*), California myotis (*Myotis californicus*), Townsend's big-eared bat (*Corynorhinus.townsendii*), and Townsend's big-eared bat hibernation roosts.

nounced in areas inhabited by colonial species where an entire colony (ranging from dozens to millions of bats) will embark on nightly foraging flights.

Through a combination of mist-netting and acoustic monitoring, eighteen species have been detected in the Klamath Mountains, including twelve known to breed locally (Weller and Stricker 2012; Long and Weller 2019). Many migrate seasonally over long distances, but some species appear to be year-round forest residents (Gellman and Zielinski 1996). In addition to roosting in natural structures like tree cavities, rock crevices, and caves, they will also use human-created structures like old cabins, bridges, and mine shafts that are abundant in the Klamath Mountains. Chances of seeing wild bats are best during dusk, especially at slow moving river or creek stretches or at lakes and ponds. From spring through fall, our local bats emerge nightly to forage on insects, and almost always take a quick drink before hunting. Members of two bat families are found in the Klamath Mountains: the free-tailed bats (Family Molossidae) and the evening bats (Family Vespertilionidae).

Family Molossidae - free-tailed bats

Free-tailed bat species in North America prefer the subtropics but are migratory in summer. Of molossids, only the Brazilian free-tailed bat is regularly detected in the Klamath Mountains. Local free-tailed bats have migrated to the northern extent of their range in southwest Oregon, Nevada, and Utah, and return to Baja California in the winter. While regarded as one of the most abundant mammals in North America (along with deer mice) it is a Species of Special Concern due, in large part, to its tendency to roost in extremely large numbers making their roosting populations vulnerable to human influence. Another molossid, the western mastiff bat, has been detected regionally with audio (Pierson et al. 2007, Long and Weller 2019).

Family Vespertilionidae - evening bats

Evening bats (Vespertilionidae) are the most widely distributed bats: over 500 species are found worldwide including members of the diverse genus *Myotis*. They range across a variety of habitats on every continent except Antarctica. Most are insectivores but some eat fish, birds, or amphibians. In a Klamath Mountain study (Pierson and Rainey 2007) the five most frequently captured species of all regional bats were Yuma myotis, silver-haired bat, little brown myotis, big brown bat, and California myotis. Less common were western pipistrelle, long-legged myotis, hoary bat, small-footed myotis, and the Brazilian free-tailed bat discussed above. Twelve of the 16 vespertilionid species found were confirmed as having local reproduction (Pierson et al. 2007). Klamath Mountain bat species composition and relative abundance appear quite different from other California mountain ranges. Particularly striking is the relative abundance of silver-haired bats and relative rarity of free-tailed bats in the Trinity Mountains. (Pierson et al. 2007, Long 2019).

Order Carnivora - Carnivores

Most of our predators are in this order but the name is a bit misleading because many of the species do not eat meat exclusively, and rather, incorporate plants into their diet when needed and available. The meat-eating carnivorans of the Klamath Mountains range in size from the short-tailed weasels, about the size of an elongated toilet paper tube, to mountain lions. While most are dependent on the abundant rodent and lagomorph prey base, the largest feed on ungulates. The most omnivorous of the carnivorans are raccoons and bears, so it is not surprising that these are the most common marauders of human food caches. Because the order contains the top predators in the trophic web they play a dynamic role in the distribution and abundance of all other terrestrial taxa, especially mammals.

Family Canidae - coyotes and foxes

Canids have the typical carnivoran dentition but behind the carnassial teeth they also have broad molars that can be used to crush bone, access marrow, and also allow opportunistic feeding on vegetation. They tend to be social with close family ties during pup rearing and/or cooperative hunting behaviors. The history of canid diversity in the Klamath Mountains is complicated by historical extirpations of once widespread species, introductions of captive species, immigrations, and species that may be locally abundant but are rare across the region.

Gray fox are common throughout North and South America and the most common canid in the Klamath Mountains. They favor a broad range of habitats and are the most arboreal of canids so it is not impossible to see them climbing a tree (or a fence). Red foxes first colonized North America from Asia around 200,000 years ago or earlier and the Sierra Nevada subspecies was once found in the Klamath Mountains. Red foxes suffered harshly at the hand of European colonizers, and later, escapees from fox farms became naturalized and recolonized areas where wild foxes were extinct. The Grinnell surveys (Grinnell 1937) reported the Sierra Nevada subspecies from Mt. Shasta, forests around Lassen Peak, and the upper elevations of the Sierra Nevada from Tulare to Sierra Counties, but there are relatively few recent reports from the Sierra and none from Mount Shasta or the Klamath Mountains.

Coyote are found throughout North America and are commonly associated with more open habitats but began expansion into North American forests around 1900—perhaps as a result of increasing forest fragmentation and consequent edge-effects (Hody 2018). In the Klamath Mountains they range from uncommon to locally abundant. Large packs are seen in some areas including in the Yolla Bolly-Middle Eel Wilderness—likely due to this area's proximity to the Coast Range and Central Valley foothills where coyotes are more common. In other areas there may be cryptic individuals lurking in dark forests and in high elevation glaciated cirques where their nightly chatter will either amuse or raise the hair on your neck.

Gray wolves are a historical mystery in California wherein only two specimens were taken in the state by early Euro-American explorers—one near Lassen and the other in the Mojave (Grinnell 1937). Based on the pelt from Lassen, the wolves that were here were most likely *Canis lupus nubilus* from the Great Plains and not timber wolves (*C. l. gigas*) from the north. If timber wolves did occur in the Klamath they most likely inhabited the western slopes where they hunted Roosevelt elk (Grinnell 1937). Cultural evidence of Klamath wolves include Tribal regalia of the Karuk, including ikxâavnamichtaahkoo (white wolf) and the Hupa and Yurok, who all use wolf headdresses in their dance regalia. A pelt, part of a Karuk family's regalia, was taken near what is now Orleans an estimated 250 years ago. It was a member of a large pack who had surrounded a group of women and girls who were camped in their fall acorn gathering grounds. This individual was taken by one of the men sent to rescue the group (Leaf Hillman, personal communication).

The story of wolves in the Klamath Mountains may be taking another turn. OR-7, a wolf born in northeast Oregon wandered over 1,000 miles and became the first wolf documented in California since 1924. In 2014, this male settled to breed in the Oregon Cascades near the California border—only 40 miles or so from the eastern edge of the Klamath Mountains. In 2018 OR-44 was tracked into the Klamath Mountains crossing the Marble Mountains, the Klamath River, and then into the Siskiyous. OR-103 has been spending some time in the eastern Klamath Mountains in Siskiyou County (USFWS pers. comm.). Animosity toward wolves continues today in the same way white settlers fueled their extermination in the first place. We can only hope that this negative paradigm will shift toward acceptance.

Family Felidae - mountain lions and bobcats

Felids have shorter faces and tooth rows than canids, reflecting the absence of crushing molars behind the carnassial teeth and making them obligate flesh eaters. Felids are also distinguished by their power and agility. If one is lucky enough to see a mountain lion a first impression is often of rippling muscle under a short coat of fur. They have manual dexterity owing to a nimble ability to rotate their forearm from a paw-facing-down position to a paw-facing-up position, serving them equally well when extending a grasping paw towards the hind end of pursued prey, handling their subdued victims, or climbing trees. They also differ from canids in being solitary, generally hunting alone. It's worthwhile to be alert to their surprisingly large, clawless tracks, and their furry, flat-ended scat.

Mountain lion (or cougar) and bobcat are the Klamath Mountain's native cats. Mountain lion are significantly larger than bobcat, have relatively even coloration, and a long black-tipped tail. They are found in a diversity of habitats, but males especially require large and uninterrupted (300–700 km^2) home ranges to thrive. Like most cats, adults are solitary, but females will roam with their offspring for over a year. In the Klamath, black-tailed deer and elk are preferred foods, and carcasses will be cached under leaves and woody debris to provide multiple days' worth of meals. There is some evidence from the region that black bear arousing early due to warming winters may pilfer these caches and necessitate greater energy expenditure on the part of mountain lions to keep themselves fed. Bobcat, larger than a burly house cat but substantially smaller than mountain lion, have short "bob-tails" and grizzled or even spotty coats. They are found broadly across the U.S. but are replaced by the larger lynx (*Lynx canadensis*) to the north in Canada. These mesocarnivores can subsist on small to medium-sized prey including rodents and lagomorphs but they are capable of taking prey as large as deer.

Family Mustelidae - weasels, otters, martens, mink, fishers, badgers, and wolverines

Fisher (*Pekania pennanti*) in the interior lowland forests along the Klamath River. Photo by Will Harling.

Mustelids are surprisingly diverse, have some of the most compelling natural histories of all the Klamath Mountain carnivores—and are among the most rarely seen. When spotted, one's first impression is often of a rather elongated animal with short limbs and conspicuous backbone flexibility. There are at least eight species that call the region home, two of which are primarily aquatic. The weasels are the smallest, followed by marten and mink, then river otter and fishers (not likely to be confused with one another) and badgers (uncommon), and finally, the largest mustelid, the unlikely wolverine. American and Pacific martens are discussed in a sidebar (page 434-435) and what follows is an overview of the other regional mustelids.

There are two weasels in the genus *Mustela* in the Klamath Mountains. The short-tailed weasel or ermine has a circumboreal distribution, reaches its southernmost range in northwest California, and is rare. The larger long-tailed weasel is widely found south of the 50th parallel all the way to northern South America in a variety of habitats. Both weasels use their elongated bodies to advantage—hunting burrowing rodents. Fisher is a late-successional forest-dweller that ranges from Canada to the northern United States. Northwest California and southwest Oregon are a western stronghold with isolated populations in the southern Sierra Nevada that were recently listed as endangered under the Federal Endangered Species Act. Fisher depend on forests with dense canopies and large diameter trees (live or dead; conifers or hardwoods) that contain cavities in which to den. In suitable habitat, they are found at any elevation and are usually solitary. These mesocarnivores feed on a wide variety of small mammals, birds, reptiles, insects, plants, fruit, and even carrion, and they are one of the few predators that specializes hunting and eating porcupine. Fisher (and marten) are unusual among mustelids in having retractable claws, which affects their tracks.

River otter are large aquatic weasels with a stout body significantly larger than a domestic cat and a thick and slightly flattened tail. They live in the northern half of California north into Canada where they inhabit small streams to large rivers, marshes, and lakes, and near-coast animals will venture into the ocean, confusing onlookers who mistake them for the much larger and more unevenly distributed sea otter (*Enhydra lutris*). River otter are widespread in the Klamath Mountains where they serve as the top predator in aquatic ecosystems. They most commonly prey on fishes and crayfish, but also take insects, freshwater mussels, amphibians, garter snakes, and even turtles. Surprisingly, river otter are documented in high elevation glacial lakes in the Klamath Mountains up to 6,900 feet (2,103 m) where they forage on introduced trout and native amphibians (Garwood et al. 2013). It is unlikely these historically fishless lakes always supported river otters, given their insatiable appetites. Rather, fish introductions may have expanded otter distributions into these subalpine lakes.

American mink have a similar distribution to river otter in both California and across North America where they, too, inhabit streams and marshes from sea level to around 7,000 feet (2,100 m). They shouldn't be confused with river otter because they are significantly smaller, slighter than a domestic cat, with a conspicuous white chin patch. In the western Klamath Mountains they are common along the major rivers and streams and in the eastern Klamath Mountains they are sometimes seen swimming in reservoirs. Like river otter, they feed primarily on aquatic vertebrates.

American badger range across California where they are typically found in open territory with friable soils and are generally absent from the dense forests of northwest California and southwest Oregon. There are occasional historical and current records of badgers in the Klamath Mountains but they are more common in grasslands to the east and south. A 1933 report puts badgers near Bully Choop Mountain in the southeastern Klamath Mountains, Grinnell spotted badgers near South Yolla Bolly (Grinnell 1937), and a 1946 census from the Six Rivers National Forest counts 60 badgers killed from the region of the southern Trinity River but none from Gasquet or Orleans (Connors 1998). There are recent reports of badger near Big Bar on the Trinity River (2004 Cal Poly Humboldt Vertebrate Museum) and around Lewiston (2020, Kris Burton pers. comm). Where found, badgers are capable long-clawed scratch-diggers that use what is essentially a running motion to excavate extensive burrow systems that have well-maintained hemispherical openings surrounded by churned soil that lead deeply underground. They will feed on available rodents as well as snakes and ground nesting birds.

There is evidence wolverines were present in the Klamath Mountains before EuroAmerican colonization but today there is a mystique and a lot of wishful thinking

Martens of the Klamath Mountains

Martens are solitary carnivores in the weasel family (Mustelidae). The American marten (*Martes americana*) was, until recently, the only marten species recognized in North America—described as living within the temperate and boreal coniferous forests from the Atlantic and Pacific coast. However, the Pacific marten (*Martes caurina*) was split from the American marten based on genetic and morphological differences (Dawson and Cook 2012). The range of the Pacific marten is largely in montane and coastal coniferous forests west of the Rocky Mountain crest, while the range of the American marten occurs to the north and east of the Rocky Mountain crest. The genetic split between these two species of martens in North America most likely originated from the persistence of marten populations in two disjunct glacial refugia during the Pleistocene.

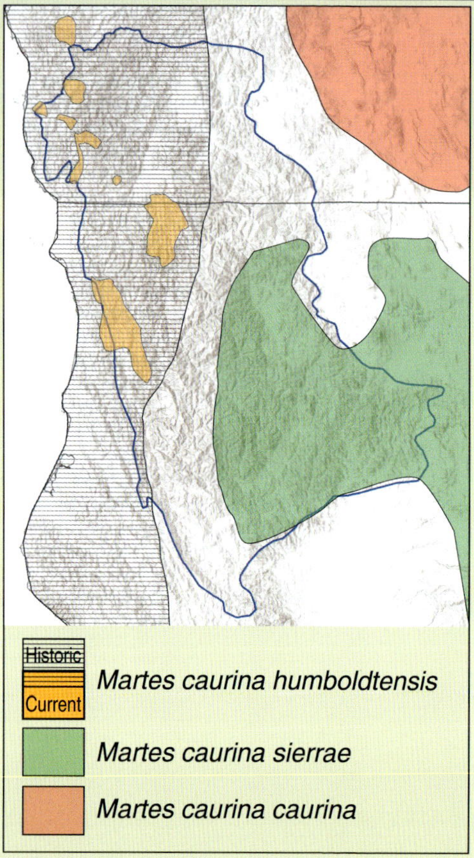

Range of martens in the Klamath Mountain Region. Based on Slauson et al. 2019.

Both North American marten species have long and narrow bodies (about as long as your forearm to the tips of your fingers), overall brown pelage with distinctive throat coloration varying from orange to yellow to cream, large and distinctly triangular ears, and a bushy tail that is proportionally equivalent to about 75 percent of the body length. Two subspecies of the Pacific marten live in the Klamath Mountains (see range map) where ssp. *humboldtensis* prefers low-elevation late-successional coastal forest with dense shrub cover and ssp. *sierrae* prefers high-elevation true fir forests that receive large amount of snow (Slauson et al. 2019). Compared to the Sierra marten, the Humboldt marten is darker, with a richer golden tone to the underfur and has a smaller throat patch that is more cream colored than orange and yellow—though pelage coloration is variable and change with wear.

The Humboldt Marten (*Martes caurina* ssp. *humboltensis*)

The Humboldt marten was historically found throughout the mesic, fog-influenced forests from northwestern Sonoma County in California northward to the Columbia River in Oregon. Detections of Humboldt martens today are in three forest types including moist Douglas-fir (*Pseudotsuga menziesii*), moist forest on serpentine soils, and dune forests on coastal terraces with shore pine (*Pinus contorta* subsp. *contorta*). Extensive shrub layers dominated by shade-tolerant ericaceous species are a consistent understory feature between all three habitats. The occurrence of the Humboldt marten in low elevation forests that receive little snowfall appears unique compared to other subspecies in the Pacific Northwest which all occur primarily in high elevation forests that receive large amounts of snow.

Because of their valuable pelts, martens were trapped in the coastal forests of California and Oregon beginning in the late 1800s. Due to declining harvest numbers, the trapping was closed indefinitely in California in 1946 but the species continued to decline due to the unregulated logging of late-successional forest across northwest California. Oregon recently banned trapping, though only for ssp. *humboldtensis* in the western part of the state. In 1996, the Humboldt marten was believed extinct in California, but a single population was rediscovered on the Six Rivers National Forest in Del Norte County, California (USFWS 2018). In response, it was listed as endangered under the California Endangered Species Act in 2018. Thanks to recent extensive camera-trap surveys in California and Oregon, we know the Humboldt marten has been eliminated from 93 percent of its historical range. We also know that the Klamath Mountains are the last remaining stronghold for the subspecies—deep in the Siskiyou Mountains (Gamblin 2019). Most recent marten detections in California have occurred in largely unmanaged moist Douglas-fir-associated forest types, where individuals select large patches (>80 ha) of late-successional forest as habitat.

The Sierra Marten (*Martes caurina* ssp. *sierrae*)

In the eastern Klamath Mountains, including the higher elevations of the Salmon and Trinity Mountains—eastward into the southern Cascades and Sierra Nevada—Sierra martens are found in mature, dense conifer forests or mixed conifer-hardwood forests. They prefer landscapes with a high percentage of canopy cover, a large amounts of coarse woody debris on the forest floor, and tree cavities or boulder fields. These inland high-elevation forests provide prey, protection, den sites, large snags, and downed logs that are utilized for denning and rest.

LEFT: A family of Humboldt martens in Prairie Creek Redwoods National and State parks. Photo by Keith Slauson. RIGHT: A Sierra marten in the subalpine forest of the Trinity Alps captured in late spring by Ryan Bourque.

surrounding these mustelids. The Karuk have distinct words for wolf and wolverine suggesting that both were present historically. A documented report of one killed in 1893 near Carberry Ranch in Shasta County, just east of the McCloud-Pit River area of the Klamath Mountains (Merriam 1899), also places them in the region. More recently, a single wolverine was documented in the Sierra Nevada which is thought to have originated in Idaho or Montana, having dispersed some 650 miles (1,000 kilometers) (Moriarity et al. 2009). They are notorious wanderers, so it is likely that wolverines have stepped foot on Klamath soils in the last century.

Family Mephitidae – striped and spotted skunks

Striped skunk and western spotted skunk are the two skunks found throughout western North America. Both species prefer disturbed areas, forest edge, or dry meadows—including unforested rocky and brushy areas. The smaller spotted skunks are shyer, more forest associated, and less likely to be seen. Both species range from sea level to around 4,000 feet (1,300 m) and nest in burrows left behind by other digging mammals of similar size. Both prey on small arthropods, ground-nesting bird eggs, rodents, and small reptiles. Both are masters of self-defense owing to pungent secretions emitted from anal glands. If you or your dog has an unfortunate encounter, the secret recipe—developed by organic chemist Dr. Bill Wood at Cal Poly Humboldt—combines one quart hydrogen peroxide, ¼ cup baking soda, and a teaspoon of dish detergent...leave on for five minutes before rinsing!

Family Procyonidae - raccoon and ringtail

Raccoon and ringtail are both found in the Klamath Mountains, though of course, raccoons are found almost everywhere in North America. These adaptable opportunists are as likely to be seen night-prowling along streams as foraging through trash cans in urban areas. In the wild, they are found across all elevations but typically associated with water, whether streamside riparian zones or in high elevation wet meadows and lake basins. Raccoons are known for "dousing" behavior, the splashing and patting of food under water. This behavior has been variously and, to date, unsatisfactorily explained as either fastidiousness, an attempt to trigger sensitive nerve endings, or a conditioned response related to foraging for aquatic prey such as crayfish. Regardless of the reason, this behavior is consistent with myriad examples of their acute manual dexterity—they can unzip your backpacks and untie your bear-bagged food sacs!

Ringtails are smaller than raccoons and named for their striking lemur-like black- and white-striped tail. They have a penchant for taking up residence in human constructions, whether cabins or mine shafts. The Klamath Mountains are the northern extent of their range and they are forest-dwellers here, while in the southwestern United States and south to southern Mexico are found along rocky cliffs. Agile climbing is facilitated by an ability to rotate their hindfeet in three distinct foot and ankle joints, so called hind-foot reversal. Raccoons and ringtails are both technically omnivores, but ringtail teeth are much more carnivore-like and ringtail diets consequently consist of more wild-caught prey. Because they are relatively small carnivores, they are prey for great horned owls, red tail hawks, bobcat, and even raccoons.

Family Ursidae - bears

Ursids are the largest carnivorans but their dentition attests to broad omnivory, with long tooth rows terminating in stout and relatively flat-topped molars. Grizzly bears, the most carnivorous of bears, once roamed across the entirety of western North America including the Klamath Mountains and were the top predator. In the Klamath Mountains they were most common along low-elevation river valleys where they hunted salmon. Merriam (1899) reported that while formerly abundant in the Shasta region grizzlies were rare at the time of his book's publication. The last recorded grizzlies killed in northwest California appear to have been in 1868 on the Mattole River in coastal Humboldt County (Grinnell 1937) and in 1910 along Swift Creek in Trinity County (Agee 2007). Grizzly bear features prominently in many traditional Karuk stories, including origin stories, communal songs to protect the family while digging potatoes, or picking berries. Grizzly bear was the impetus for designing doors on traditional Karuk houses and sweathouses.

Black bear, which can be black, brown, or reddish-gray, are the only bears here today and are numerous across the Klamath Mountains, in the Coast Range north of San Francisco, and throughout the Sierra Nevada and Cascades—they are now the largest terrestrial carnivore due to the extirpation of the grizzly. The Hoopa Valley Reservation, which spans the Trinity River and includes portions of the western Trinity Alps, reports the highest black bear density in the country, lamentable because some bears here girdle conifers to obtain sugar-rich sapwood, killing trees in the process. While this behavior is deemed bad for trees by foresters, it does increase suitable habitat for some cavity-nesting birds (Mendia et al. 2019) and thins dense second-growth forests. Black bear-preferred regional habitat is in the montane conifer belt in the range of 3,500–7,000 feet (1,000–2,200 m) but they will explore lower elevations, particularly in the coastal region and inland river valleys.

Black bear are quite omnivorous, their diet following regional phenological patterns that include the emergence of catkins, grasses, sedges, and young animals in late spring, leaves and berries in summer, and followed by acorns and other nuts in the fall. Insects are taken throughout the year, including raids on the nests of high-energy subterranean yellowjackets and ants. Storing fat for winter is critical. While some individuals may migrate to lower and more temperate elevations, others may slumber with their winter metabolism supported by stored fat. Black bears have been less problematic for visitors in the Klamath Mountains—where they typically avoid humans—than other more heavily visited areas in California like Yosemite National Park. However, this is changing. Bears are habituating to the bad wilderness ethics of increased numbers of mountain visitors, particularly in the Trinity Alps Wilderness, and have begun raiding camps and breaking into cars. Properly storing food on your next backpacking trip can save a bear from learning bad habits, and quite possibly, save a bear.

Pinnipeds: Family Phocidae and Otariidae - seals and sea lions

The pinnipeds are literally the "fin-footed" carnivores. Mammalian systematists have deliberated whether they are each other's closest relatives (embraced in a Family Pinnipedia) or descended from separate lineages that similarly and convergently evolved aquatic forms (most notably, seals from the Mustelidae and sea lions from the Ursidae), but there has never been any doubt that they are carnivorans on some level. The two groups differ in their degree of aquatic specialization, with sea lions being less specialized than seals. While both have limbs that are modified as fins, sea lions can turn their hind legs at the hip for some semblance of four-footed beach walking with their bellies off the sand. Seals cannot and so are relegated to inchworm-like rolling blubber locomotion. Sea lions also retain their ancestors' ear flaps (though tiny) while seals have none, and sea lions are at minimum twice the size of a seal and never have the spots that might be found on seals.

Despite the fact that this is a book about the Klamath Mountains, rivers are accessible corridors for these aquatic mammals. California sea lions and harbor seals are oceanic but both swim up rivers to hunt anadromous fishes. They love fish and can be found upriver seasonally in pursuit of migrating adult salmon, Pacific lamprey, and steelhead, in the Rogue, Chetco, Smith, and Klamath rivers when flows are sufficient for their passage in fall and winter. Later, they switch foraging pursuits to the ocean-going juveniles. Seeing one of these marine mammals gliding through a river in a forested gorge many miles from the Pacific Ocean is surreal a but it really does happen!

Acknowledgements

Many thanks to all who contributed and reviewed this chapter. Leaf and Lisa Hillman provided a keen review, multiple checkpoints for culturally appropriate content, and deeper historical understandings. Michael Parker and Jim Agee reviewed the entire chapter for content and accuracy. Jenny Lynn Hutchinson and Keith Slauson helped with the marten sidebar. Jessica Blois provided feedback on the cave mammals sidebar. Thanks to the photographers including Keith Slauson, Ryan Bourque, Will Harling, Ken Chamberlain, Phil Georgakakos, Ivan Yates, and Nick Kerhoulas.

Literature Cited and Suggested Reading

Agee, J.K. 2007. *Stewards Fork, A Sustainable Future for The Klamath Mountains.* University of California Press.

Appel, C.L., et al. 2017. Distribution of the North American Porcupine (*Erethizon dorsatum*) in Northern California. Open Educational Resources. 3.

Arbogast, B. et al. 2017. Genetic data reveal a cryptic species of New World flying squirrel: *Glaucomys oregonensis. Journal of Mammalogy* 98:1027-1041.

Blois, J.L. and B. S. Arbogast. 2006. Conservation Genetics of the Sonoma Tree Vole (*Arborimus pomo*) Based on Mitochondrial and Amplified Fragment Length Polymorphism Markers." *Journal of Mammalogy* 87, no. 5 (October 17, 2006): 950–60.

Blois, J., J. McGuire, and E. Hadly. 2010. Small mammal diversity loss in response to late-Pleistocene climatic change. *Nature* 465:771-774.

Broughton, J.M., Weitzel, E.M. 2018. Population reconstructions for humans and megafauna suggest mixed causes for North American Pleistocene extinctions. *Natural Communities* 9, 5441.

Carraway, L. and B.J. Verts. 1993. *Aplodontia rufa*. Mammal Species. *The American Society of Mammalogists* 432:1-10.

Chinnici, S.J., D. Bigger, and E. Johnson. 2012. Sonoma tree vole habitat on managed redwood and Douglas-fir forestlands in north coastal California. In: Standiford, Richard B. et al. Proceedings of coast redwood forests in a changing California. Gen. Tech. Albany, CA: Pacific Southwest Research Station, Forest Service, U.S. Department of Agriculture. pp. 389-397

Coltrane, J.A. 2012. Redefining the North American Porcupine (*Erethizon dorsatum*) as a facul-tative specialist herbivore. *Northwestern Naturalist* 93:187-193.

Conners, P.A. 1998. A History of the Six Rivers National Forest. Commemorating the First 50 Years. USDA Forest Service Pacific Southwest Region. SRNF. Eureka, California.

Dawson, N.G. and J.A. Cook. 2012. Behind the genes: diversification of North American martens (*Martes americana* and *M. caurina*). In: Aubry, K.B.[et al.], eds. *Biology and conservation of martens, sables, and fishers: a new synthesis.* Ithaca, NY: Cornell University Press: 23–38.

Driver, H.E. 1939. Anthropological Records 1:6, Culture Element Distribution in Northwest California. University of California Press.

Feranec, R. 2009. Implications of Radiocarbon Dates from Potter Creek Cave, Shasta County, California, USA. *Radiocarbon*, 51:931-936.

Feranec, R.S., et al. 2007. Radiocarbon dates from the Pleistocene fossil deposits of Samwel Cave, Shasta County, California, USA. *Radiocarbon* 49:117-21

Forsman, E.D., et al. 2016. Tree voles: an evaluation of their distribution and habitat relation- ships based on recent and historical studies, habitat models, and vegetation change. Gen. Tech. Rep. PNW-GTR-948. Portland, OR: U.S. Department of Agriculture, Forest Service, Pacific Northwest Research Station. 119p.

Gamblin, H.E. 2019. Distribution and habitat use of a recently discovered population of Humboldt martens in California. HSU theses and projects. 338.

Garwood, J., et al. 2013. Use of historically fishless high-mountain lakes and streams by neartic river otters (*Lontra canadensis*) in California. Northwestern Naturalist 94:51-66.

Gellman, S.T. and W.J. Zielinski. 1996. Use by Bats of Old-Growth Redwood Hollows on the North Coast of California. *Journal of Mammalogy* 77, no. 1.

Grinnell J., J. Dixon, and L. Linsdale. 1937. *Fur bearing mammals of California*, vol 1 and 2. Berkeley: University of California Press.

Hody, J.W, and R. Kays. 2018. Mapping the expansion of coyotes (*Canis latrans*) across North and Central America. ZooKeys, 759 81-97.

Long, C. and T. Weller. 2019. Six Rivers National Forest NABat Stationary Detector Sites 2018 Results. Batcon.org. Bat Conservation International.

Mammal Diversity Database. (2020). Mammal Diversity Database (Version 1.31) [Data set]. Zenodo. http://doi.org/10.5281/zenodo.4139818

Melquist, W.E., P. J. Polechla, JR., and D. Toweill. 2003. River Otter (*Lontra canadensis*). In: Feldhamer GA, et al. editors. *Wild mammals of North America: Biology, management, and conservation.* Baltimore, MD: Johns Hopkins University Press. p 708–734

Mendia, S.M., M.D. Johnson, and J.M. Higley. 2019. Ecosystem Services and Disservices of Bear Foraging on Managed Timberlands. *Ecosphere* 10, no. 7.

Merriam, C.H. 1899. *Results of a biological survey on Mount Shasta, California.* U.S Department of Agriculture.

Moriarty, K.M. et al. 2009. Wolverine Confirmation in California after Nearly a Century: Native or Long-Distance Immigrant? *Northwest Science* 83, no. 2 April: 154–62.

Piaggio, A. et al. 2013 Molecular phylogeny of an ancient rodent family (Aplodontiidae), *Journal of Mammalogy*, Volume 94, Issue 3, Pages 529–543.

Pierson, E.D., W.E. Rainey. 2007. Bat Distribution in the Forested Regions of Northwestern California. Prepared for the California Department of Fish and Game.

Schwartz, M. et al. 2020. Pliocene–Early Pleistocene Geological Events Structure Pacific Martens (*Martes caurina*). *Journal of Heredity.* 1–13. doi:10.1093/jhered/esaa005

Slauson, K.M. et al. 2019. A conservation assessment and strategy for the Humboldt marten in California and Oregon. Gen. Tech. Rep. PSW-GTR-260. Arcata, CA: U.S. Department of Agriculture, Forest Service, Pacific Southwest Research Station. 124 p.

Soule, M.E. et al. Reconstructed Dynamics of Rapid Extinctions of Chaparral-Requiring Birds in Urban Habitat Islands. *Conservation Biology* 2, no. 1 (March 1988): 75–92.

Sutton, D. and B.D. Patterson. 2000. Geographic Variation of the Western Chipmunks *Tamias Senex* and *T. Siskiyou*, with Two New Subspecies from California. Journal of Mammalogy 81:299-316.

Stinson, D.W. 2020. Mazama Pocket Gopher Recovery Plan and Periodic Status Review, 110.

U.S. Fish and Wildlife Service. 2018. Species status assessment report for the coastal marten (*Martes caurina*), Version 2.0. July 2018. Arcata, CA.

Verts, B.J. and L. Carraway. 2001. *Scapanus latimanus.* Mammalian Species. 666:1-7. Wilkins, K. 1989. *Tadarida brasiliensis.* Mammalian Species. 331:1-10.

Weller, T.J. and C.A. Stricker. 2012. Northern California Redwood Forests Provide Important Seasonal Habitat for Migrant Bats. In: Standiford, Richard B.et al. Tech. Coords. Proceedings of Coast Redwood Forests in a Changing California: A Symposium for Scientists and Managers. Gen. Tech. Rep. PSW-GTR-238. Albany, CA: Pacific Southwest Research Station, Forest Service, U.S. Department of Agriculture. Pp. 447-457 238 (2012): 447–57.

Zimova, M., et al. 2018. Function and Underlying Mechanisms of Seasonal Colour Moulting in Mammals and Birds: What Keeps Them Changing in a Warming World?: Function and Basis of Seasonal Coat Colour Moults. Biological Reviews 93, no. 3 (August 2018): 1478–98.

Change and Stewardship

"In all the years I have spent standing or sitting on the banks of this river, I have learned this: the more knowledge I have, the greater becomes the mystery of what holds that knowledge together, this reticulated miracle called an ecosystem." —**Barry Lopez**, *The Naturalist*

The Significance of the Klamath Mountains

This book celebrates the natural history of the Klamath Mountains through stories of diversity and resilience over deep time. Shaped by geology, these mountains form an ancient jigsaw puzzle and topographic mosaic dissected by big-shouldered river canyons and sharp ridgelines that create localized climatic gradients. Within the geomorphic province, the rocks are much older than in surrounding regions. This dichotomy has allowed many distinct evolutionary lineages of plants and animals to adapt, survive, and sometimes speciate where elsewhere they became extirpated long ago.

As we have seen, the diversity of life here is vast and includes snails, butterflies, amphibians (Figure 1), conifers, and so much more. The bottom line is this roughly 12,000 square mile (31,080 km^2) mountain range is home to one of the most biodiverse temperate mountain ranges on Earth. Even to this day the Klamath Mountains continue to surprise us with cryptic diversity and unique endemic species hiding in plain sight. More recent discoveries include rare obligate serpentine soil plants, terrestrial salamanders, cave dwelling ice crawlers, and even a new wingless stonefly genus (*Dracoleuctra siskiyou*) found in a single high-elevation spring (Lee and Baumann 2022). The Klamath Mountains also have a rich cultural history where numerous Native American Tribes manage and tend natural resources. Life has been interconnected here for millennia.

To showcase the intricate natural history of the Klamath Mountains in a single volume, we had to omit discussions on the many foundational conservation and societal issues the region is facing. Several of these topics are very complex and deserve their own treatments. There are many sources that have explored the region's most pressing conservation issues including *Salmon and Acorns Feed the People* by Kari Marie Norsgaard (2020), the Karuk Climate Adaptation Plan (2019), *Steward's Fork* by Jim Agee (2007), and *Northwest California* by John Sawyer (2006). Many also highlight the Klamath Mountains as a critical biological reserve for temperate North America (Wagner 1997, Noss et al. 1999, Dellasala et al. 1999, Strittholt et al. 2001, Wallace 2003, Olson et al. 2012, Sarr et al. 2015). Most of these discuss the resiliency of Klamath Mountains despite the region being exposed to a legacy of landscape, ecological, and cultural destruction that started with the gold rush over 170 years ago. The Klamath Mountains were proposed as a World Heritage Site and UNESCO Biosphere Reserve over 25 years ago (Vance-Borland et al. 1995) but the certification process requires significant planning and support from the public and our elected officials. Perhaps this campaign can be rekindled with renewed enthusiasm as the Klamath Mountains are an essential biodiversity reserve in the face of climate change.

The Klamath Mountains have not escaped significant losses of biodiversity. Grizzly bears (*Ursus arctos*), California condors (*Gymnogyps californianus*), and bull trout (*Salvelinus confluentus*) are a few that have disappeared in the past century and it's difficult to grasp the ecological and cultural impacts these extirpations have had. Some species can return with carefully planned reintroductions. The imminent return of the California condor, initiated by the Yurok Tribe, will be a test-case for biological resiliency. Others, such as the Franklin's bumble bee (*Bombus franklini*) last seen in 2006, or the *Nebria praedicta* ice beetles described in 2008 on the rapidly disappearing Grizzly Glacier, are perhaps much more challenging conservation issues due to the habitat decline, endemism, and rapid climate change. Despite all this, the isolation and ruggedness of the Klamath Mountains has buffered some of the damage and development enacted in other ecologically and culturally important landscapes around the world.

If you are reading to this point in the book you clearly love the Klamath Mountains—and thus know your responsibility as a steward. Our hope is this book deepens our connection to and understanding of the Klamath Mountains but to

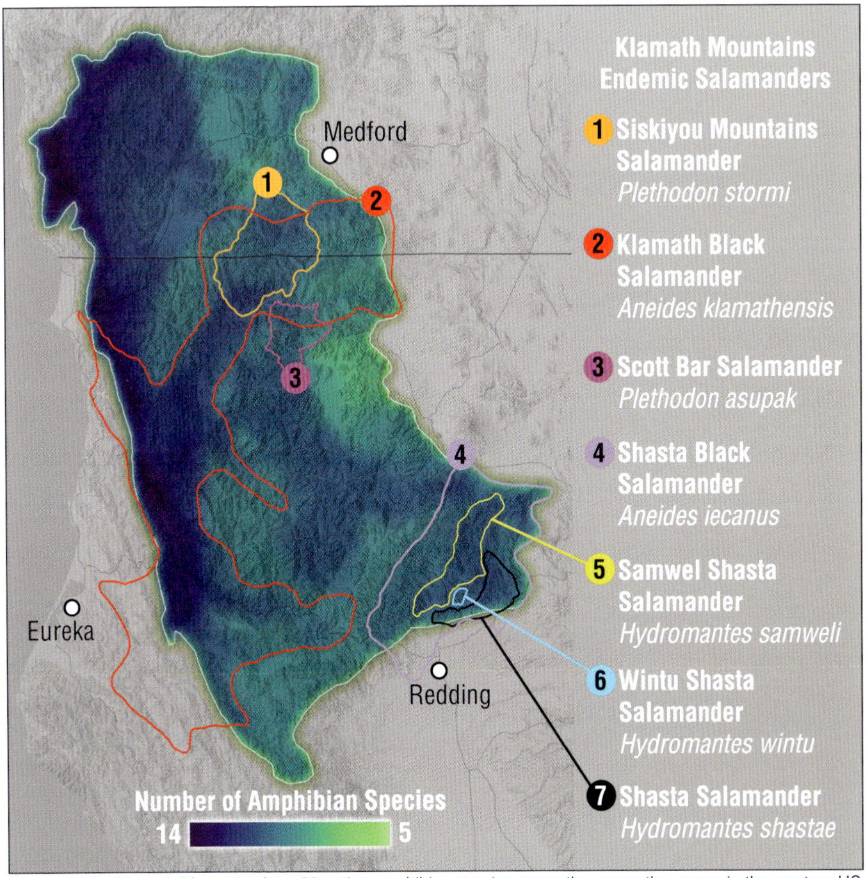

Figure 1. The Klamath Mountains host 26 native amphibian species, more than any other range in the western US. This heat map displays amphibian species richness across the region along with seven regionally endemic terrestrial salamanders, from the family Plethodontidae. Species richness analysis and map by Eric Samson, data by Justin Garwood.

also highlight just how much more there is to learn. In this, our final chapter, we briefly summarize some major threats to the region and its biodiversity but also look to the future and see what stewards are doing to preserve, promote, and protect the Klamath Mountains for future generations.

A Changing Climate Impacts Everything

Climate change is the dominant force affecting human wellbeing and all ecosystems on the planet (Ostberg et al. 2018). Shifts in seasonal and annual weather patterns are increasing stress on ecosystems, biodiversity, communities, and the health of people. For the North Coast region of California, which includes most of the Klamath Mountains, average annual maximum temperatures are likely to increase by 5–9 °F through the end of the 21st century based on an average of ten priority climate models under the "business-as-usual" emissions scenario (Grantham et al. 2018). Interior regions will experience the greatest degree of warming. Projections specifically for the Klamath region indicate temperature increases of up to 3.5–4.5 °F in the next 50 years (as compared to the period of 1971–1999; Garfin et al. 2014). Projected maximum summer temperatures are expected to exceed 105 °F over the same period. Winter season temperatures are also expected to increase by greater than 5–7 °F by mid-century and 8–11 °F by the end of the 21st century.

Projections of precipitation are more variable, but annual precipitation for the Klamath Mountains and the North Coast region are generally projected to be within or near the historical range of variability for the 21st century. However, there is an expectation of increased variability. Precipitation is also expected to be delivered in more intense storms and within a shorter wet season. The North

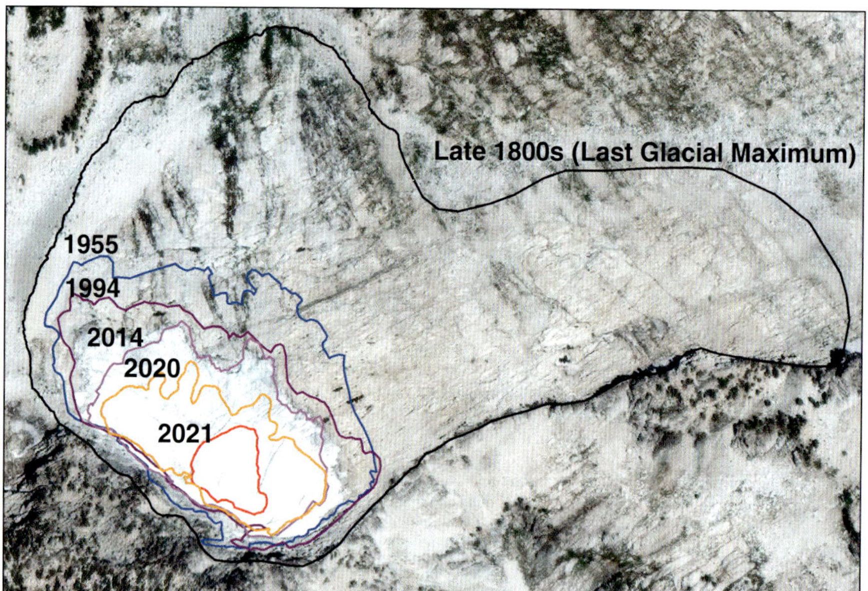

Figure 2. Digitized outlines of Grizzly Glacier, Trinity Alps, California, from the late 1800s through 2021. Recent accelerated glacier retreat, caused by increasing temperatures and persistent drought, will likely eliminate this last remaining Little Ice Age glacier in the Klamath Mountains over the next few years. Adapted from Garwood et al. (2020) and Garwood; unpublished data (2022). Base satellite image from July 2014.

Figure 3. Two images of the headwaters of Grizzly Creek at Grizzly Meadows. Notice the shift in tree density in the upper meadow and decreased snowpack on the north slopes of Thompson Peak.

TOP PHOTO from July 23rd, 1911 by Annie M. Alexander and Louise Kellogg used with permission from Museum of Vertebrate Zoology, UC Berkeley.

BOTTOM PHOTO from nearly the same location on 108 years later on September 20th, 2019 by Justin Garwood.

Coast region is already experiencing the most intense storms in the state in terms of three-day maximum precipitation (Swain et al. 2018). A shorter wet season (i.e., earlier spring drying and later onset of fall rains) are resulting in prolonged dry seasons and reduced soil moisture conditions, even when annual precipitation totals remain relatively consistent. Future predictions include a higher likelihood of extreme wet years and extreme dry years (drought), and the "average" rainfall year will be less common. The potential for a dramatic increase in the frequency of wet and dry year extremes has been termed "precipitation whiplash" and a new climate regime for the region (Swain et al. 2018).

It is likely less precipitation will fall as snow and total snowpack will be a small fraction of its historical average (Figures 2 and 3). One study (Flint et al. 2013) predicts that the April 1 snow cover will decline to 11 percent and April 1 snow water equivalent will decline from 10.3 inches (1951–1980) on average to 1 inch by end of the 21st century. Thus, with current and projected increases in temperature, shorter wet seasons, and loss of snowpack regardless of changes in total annual precipitation, collectively these lead to increases in climatic water deficits and enhanced drought magnitude and duration (Diffenbaugh et al. 2015; Cayan et al. 2018). Climatic water deficits that may increase by 10–19 percent mid-century and 16–32 percent by the end of the century translate to lower soil water availability and increasing stress experienced by vegetation and ecosystems (Micheli et al. 2018). Yet the Klamath Mountain landscape is topographically complex and is situated near the Pacific Ocean. These factors produce many pockets of favorable temperature and precipitation regimes that stand to buffer against some of the most extreme effects of climate change imposed on the region leading to key refugia for biodiversity.

The Effects of Atypical Wildfire

Wildfires will continue to be a major disturbance in the region and changes in patterns of fire behavior are among the most urgent and pressing dimensions of climate change in the Klamath Mountains (Figure 4). Future wildfire projections suggest a longer fire season, an increase in wildfire frequency, and an expansion of the area susceptible to fire. Also, increasingly dry winters combined with nearly a century of fire suppression are changing the frequency and severity of wildfire and these drastically alter available habitat and food at all elevations. As we have discussed in the First Peoples, Climate, Fire, and Plant Communities chapters, Klamath Mountain vegetation has been shaped by annual lightning strike fires and managed for thousands of years with purposeful fires ignited by Native Peoples to maintain important resources. These fires have traditionally been low intensity. Today, with suppression and extraction practices, most forests in the Klamath Mountains are suffering from a lack of fire which increases intensity when fire does occur. Coupled with changes in warming, low soil moisture, wind, insects, diseased forests, and excessive fuel loads, we are seeing a severe increase in fire frequency, length of fire season, elevation at which fires occur, and size of fires.

Invasive Species and Diseases

The arrival of new pests and pathogens further exacerbate the effects of a changing climate on native species in the Klamath Mountains. Invasive plants such as star

Figure 4. The drastic shift in forest composition from 2004 (top) and 2019 (bottom) as seen from English Peak, Marble Mountain Wilderness. The 2014 White's Fire burned at a high intensity through a dense, fire-prone forest. Top photo by Jay Smith bottom photo by Michael Kauffmann.

thistle (*Centaurea* sp.), Himalayan blackberry (*Rubus armeniacus*), scotch broom (*Cytisus scoparius*), and several grasses are well established within the Klamath Mountains. In addition to crowding out native species, invasive plants exacerbate fire behavior by increasing fuel load between trees or shrubs and increasing flame lengths. Invasive plant pathogens are also an issue and include species like sudden oak death (*Phytophthora ramorum*), Port Orford-cedar root rot (*P. lateralis*), and white pine blister rust (*Cronartium ribicola*) which are leading to the decline and eventual death of a variety of tree species with effects including a decrease in biodiversity and an increase in fuel accumulation. Some native pathogens have also become more lethal to host populations as environmental conditions imposed by altered habitat and climate change benefit disease outbreaks. Dams have altered flow, impacted water quality, and have starved the Klamath River of sediments. These, along with increasing water temperatures from climate change, have increased the frequency and severity of fish die-offs for both adult and juvenile salmonids, especially Chinook salmon. Major dam removals and accompanied river restoration are on the horizon and will hopefully reduce the frequency and severity of these fish disease epidemics.

The introduction of non-native animals is also posing threats to native wildlife, especially in aquatic environments where several predacious aquatic invasive species are shifting ecosystem dynamics. The most notable aquatic invasive species for the Klamath Mountains are non-native trout populations planted over the last century for sportfishing across hundreds of high-elevation lakes. These predatory fishes wreak havoc and displace native, defenseless aquatic breeding amphibians into shallower fishless habitats that are more prone to drying (Welsh et al. 2006, Ryan et al. 2014). In recent years, managers have attempted to balance fishery management with protecting native amphibian communities by eliminating fish stocking in critically important habitats. However, active fish removal in key waters is likely still needed. The American bullfrog, a highly invasive and voracious predator, has become established in reservoirs, large rivers, and ponds (Fuller et al. 2011). Bullfrogs are also being introduced and are invading into illegally constructed ponds used primarily for cannabis (*Cannabis sativa*) farm irrigation. These artificial habitats have proliferated across the region and will likely facilitate the dispersal of bullfrog populations while reducing biodiversity. The best way to reduce the regional spread of bullfrogs is through education and to allow artificial ponds to dry down annually to break reproduction cycles as larvae usually take over a year to metamorphose into a frog.

In terrestrial habitats, recent arrivals of aggressive competitors, such as the barred owl (*Strix varia* see Bird chapter) are displacing native fauna. Invasive wild turkeys (*Meleagris gallopavo*) and wild pigs (*Sus scrofa*) are establishing populations in the periphery of the Klamath Mountains, mainly in regions with grasslands and mixed oak-conifer woodlands. These invasive game species are opportunistic omnivores and cause major ground disturbance leading to erosion and loss of forest duff from indiscriminately rooting and foraging for any plant, animal, or fungus they can find. Licensed hunters can pursue these invasive pigs and turkeys following regulations established by California and Oregon fish and wildlife agencies.

Threats to Biodiversity

Habitat loss is the greatest threat to biodiversity across the world. It takes a variety of forms including ecosystem conversion from mega-fires, logging and mining caused biodiversity declines, wetland destruction, cattle grazing impacts on public lands, water withdrawals for agriculture, and even decreased snowpack that affects stream flow and temperature in aquatic systems. Habitat loss is often linked to climate change and can also result from more aggressive invasive species taking hold in a new area and out-competing or preying on native species. Klamath Mountain ecosystems have a relationship with various natural disturbance processes such as lightning-caused and culturally-practiced fires, floods, drought, landslides, and disease. In other cases, animals that were more abundant in the past, such as beavers and their persistent stream damming habits, or large herds of elk disturbing the vegetation through browsing, generated natural disturbance. Today, many of these natural disturbance regimes are out of balance, with some now being too destructive while others are marginalized, threatening the health of the greater ecosystem. Humans are also directly impacting fragile habitats by careless recreation activities, such as 4-wheeling through montane wetlands, while also loving certain destinations to death as wilderness visitation has grown exponentially in the past few decades.

The Klamath River basin has only two core wild relict runs of spring Chinook salmon—in the Salmon and South Fork Trinity rivers. Removing Klamath dams and restoring access to the Upper Klamath basin will hopefully reestablish former habitats and reduce the likelihood of extinction for imperiled Klamath River spring Chinook. To buffer against decreasing summer stream flows and increasing summer temperatures, cold-water sources and refugia in the Klamath Mountains need to be protected and restored wherever possible. There are also many opportunities for restoring habitat lower in watersheds by re-connecting cold-water sources in floodplains which have been extensively altered by dams, historic mining, beaver declines, sediment, roads, and development. Reconnecting these cool waters to warmer streams provides local refuges for aquatic life during summer heatwaves.

The Klamath Mountains are a region with ancient human habitation but only relatively recent resource-extractive industry. Clearcut logging, cannabis cultivation, and commercial fishing have broad impacts on biodiversity through habitat loss, pesticide use, surface water diversions, erosion and sedimentation, and continued harvesting of fish from declining stocks. Overfishing affects all stream trophic levels, fish population viability, and prey availability for several carnivores including minks, otters, pinnipeds, and black bears. Ill-advised logging practices affect water quality and habitat availability for the most devoted forest dwellers. Despite their potential for impacts, currently, fishing and clearcut logging are regulated through quotas and timber harvest rules instilled by local, state, and federal regulations representing substantial improvements from historic methods. At the time of this writing, Oregon, long known for its minimal protections on riparian zones, is on the verge of approving a landmark overhaul of its forest practice rules after successful negotiations between timber companies and conservation groups.

Cannabis cultivation in the Klamath Mountains has had far-reaching ecological impacts from the "green rush" boom and bust over the past 20 years, one of which is the conversion of mid-elevation montane forest to landscapes supporting cannabis cultivation which include poorly maintained roads that bleed sediment into creeks and flat-topped greenhouse terraces with consequential loss of habitat (Figure 5). The accumulated debris and chemicals at many of these operations is

Figure 5. Forest clearing and headwater stream diversions by illegal cannabis operations have been commonplace in many regions of the Klamath Mountains over the past few decades. Submitted photo.

disheartening. Recent state legalization and permitting mandates require cultivators to comply with state and local environmental laws to mitigate for potential significant impacts on water, habitat, and wildlife. Yet this only applies to permitted operations, and an overwhelming number of illegal unpermitted operations remain intact throughout this remote and rugged region.

Arguably the most debilitating anthropogenic influence from cannabis cultivation in the Klamath Mountains is widespread poisoning at all trophic levels from numerous toxic pesticides and fertilizers used by illegal cannabis operations. For example, Northern California fisher, as well as other carnivores including mountain lions and predatory birds such as northern spotted owls have been negatively impacted by the bioaccumulation of anticoagulant rodenticides associated from cannabis cultivation within the Klamath Mountains (Gabriel et al. 2012, Gabriel et al. 2018, Franklin et al. 2018). Rodenticides are widely used at illicit cannabis cultivation sites and are not legal for use in many of these settings. Due to their use and the amounts of rodenticides applied, they can create a graveyard of toxic carrion but also have the potential to leach into groundwaters. Several studies have reported values upwards of 85percent sampled animals in forested portion of the Klamath Mountains were exposed to one or more different rodenticide chemicals (Gabriel et al. 2015). Trespass cartel-style operations on our public lands are the most troubling because extremely toxic chemicals that are banned in the US, in addition to rodenticides have been used indiscriminately. For example, Carbofuran, an insecticide that is extremely toxic to all animals in very small amounts, including humans (Thompson et al. 2017) has been documented repeatedly at cannabis sites in the Klamath Mountains. The largescale cleanup and stabilization of countless illegal operations will take decades to remediate at the current course.

Stewardship Across the Klamath Mountains

Despite many challenges, there is so much good news when it comes to the future of the Klamath Mountains. In response to rapid change, passionate community members have formed grass-roots organizations with collaborative goals to restore the land. These people envision healthy watersheds, abundant fish and wildlife populations, as well as best practices for land stewardship. The vision for relationships with the land are increasingly embracing Traditional Ecological Knowledge (TEK) with a focus on restoring human responsibility to enhance these relationships. TEK is deep knowledge of a place, painstakingly discovered by Native Peoples who have adapted to, and interpreted, the landscape over thousands of years.

As we have explored at various levels throughout this book, land stewardship practices have been interrupted by genocide, settlement by European immigrants, and the infiltration of non-Native land management practices—exemplified by fire exclusion. By reconnecting communities with open communication around land-management practices, relationships between Federal and State governments, Tribes, resource conservation districts, watershed councils, fire safe councils, timber companies, and others in the community are growing. Groups like the Rogue Riverkeepers, Klamath-Siskiyou Wildlands Center, Smith River Alliance, Mid-Klamath Watershed Council, Hayfork Watershed Research and Training Center,

Figure 6. Prescribed burning as part of the Klamath River TREX program. Photo by Scott Harding.

Salmon River Restoration Council, and the Scott River Watershed Council are healing the landscape at a community level through partnerships, education, government restoration grant programs, a variety of philanthropic groups and private donors, and countless volunteer hours. Many are partnering with Tribes to learn about the TEK that has been passed down for generations while also forging new approaches to complex problems from legacy issues such as river mining tailings, overstocked forests, and invasive species.

Collaborative work is reshaping the land for the better. At the forefront is the incorporation of prescribed fire using traditional cultural burning practices to achieve a wide range of results. These include increasing landscape heterogeneity while promoting culturally important vegetation characteristics, such as food resources. They are also working to improve forest resilience near residential areas by mechanically reducing fuel loads and restoring fire processes that have been lost due to fire exclusion. Through these collaborative efforts, communities are seeing not just the economic value that is achieved but also the landscape health that promotes the interdependence and interconnection of all species.

Leading these efforts is the Prescribed Fire Training Exchange (TREX, Figure 6). This cooperative provides training at the local level and offers fire practitioners a holistic perspective on what rural communities really need. In essence, it promotes the needs of diverse entities, including Tribes, private landowners, Federal and State agencies, contractors, academics and international partners—while incorporating local community values and issues to build the right kinds of capacity in the right places. Some of these goals (adapted from Margo Robbins' PowerPoint at www.mkwc.org) include:

1. To develop the capacity to reintroduce fire safely and effectively
2. Provide participants with experience and training in conducting controlled burns
3. To provide fire-dependent basket materials for basket weavers (hazel and bear grass)
4. Prairie restoration (bring the elk home)

5. Improve acorn production and other traditional food sources
6. Help local landowners achieve a safer environment for their homes

In addition to bringing back fire, many other efforts are happening on the ground to heal the land. From fisheries restoration to American beaver (*Castor canadensis*) restoration, groups are enlisting members of their communities to implement grassroot actions to slowly begin to wipe the historic scars of logging, mining, and other extractive practices to bring back an equilibrium that has been missing since pre-contact. To learn more about these efforts, visit WWW.KLAMATHMOUNTAINS.ORG to find examples of past and future projects.

Conclusion

Climate change is real and happening on such a large scale that it is overwhelming. Ultimately, climate change is the product of unsustainable Western land management practices and the rise of political and economic systems. All these actions were initiated by a few for the economic benefit of only a few more. The mental health dimensions of climate change are many. For our region they include stress and anxiety related to food insecurity and wildfires coupled with the impacts of ecosystem decline and species loss. Ultimately, climate adaptation is about restoring human responsibilities and appropriate relationships to the natural world and in the Klamath Mountains this is happening—and for that we should all be thankful.

Literature Cited and Suggested Reading

Agee, J.K. 2007. Steward's Fork: a sustainable future for the Klamath Mountains. University of California Press, Berkeley, CA.

Bartholomew, J.L., M.J. Whipple, D.G. Stevens, and J.L. Fryer, J.L. 1997. The life cycle of *Ceratomyxa shasta*, a myxosporean parasite of salmonids, requires a freshwater polychaete as an alternate host. The Journal of Parasitology 83:859-868.

Cayan, D.R., D.W. Pierce, and J.F. Kalansky. 2018. Climate, drought, and sea-level rise scenarios for the Fourth California Climate Assessment. California's Fourth Climate Change Assessment, California Energy Commission. Report number: CCCA4-CEC-2018-006.

DellaSala, D., S. Reid, T. Frest, J. Strittholt, and D. Olson. 1999. A global perspective on the biodiversity of the Klamath-Siskiyou ecoregion. Natural Areas Journal 19:300-319.

Diffenbaugh, N.S, D.L. Swain, and D. Touma. 2015. Anthropogenic warming has increased drought risk in California. Proceedings of the National Academy of Sciences 112:3931-3936.

Flint, L.E., A.L. Flint, J.H. Thorne, and R. Boynton. 2013. Fine-scale hydrologic modeling for regional landscape applications: the California Basin Characterization Model development and performance. Ecological Processes 2:25.

Franklin, A.B., Carlson, P.C., Rex, A. et al. 2018. Grass is not always greener: rodenticide exposure of a threatened species near marijuana growing operations. BMC Research Notes 11:94.

Fuller, T., K. Pope, D. Ashton, and H. Welsh, Jr. 2011. Linking the distribution of an invasive amphibian (*Rana catesbeiana*) to habitat conditions in a managed river system in northern California. Restoration Ecology 19:204-213.

Gabriel, M.W., L.W. Woods, R. Poppenga, R.A. Sweitzer, C. Thompson, S.M. Matthews, J.M. Higley, et al. 2012. Anticoagulant rodenticides on our public and community lands: spatial distribution of exposure and poisoning of a rare forest carnivore. PLoS One7: e40163.

Gabriel, M.W., L.W. Woods, G.M. Wengert, N. Stephenson, J.M. Higley, et al. 2015. Patterns of natural and human-caused mortality factors of a rare forest carnivore, the fisher (*Pekania pennanti*) in California. PLoS One 10(11): e0140640.

Gabriel, M.W., L.V. Diller, J.P. Dumbacher, G.M. Wengert, J.M. Higley, et al. 2018. Exposure to rodenticides in Northern Spotted and Barred Owls on remote forest lands in northwestern California: evidence of food web contamination. Avian Conservation and Ecology 13(1):2.

Garfin, G. et al. 2014: Ch. 20: Southwest. Climate Change Impacts in the United States: The Third National Climate Assessment, J. M. Melillo, Terese (T.C.) Richmond, and G. W. Yohe, Eds., U.S. Global Change Research Program, 462-486. doi:10.7930/J08G8HMN.

Garwood, J., A. Fountain, K. Lindke, M. van Hattem, and H. Basagic. 2020. 20th century retreat and recent drought accelerated extinction of mountain glaciers and perennial snowfields in the Trinity Alps, California, USA. Northwest Science 94:44-61.

Grantham, T.E.W., D.M. Carlisle, G.J. McCabe, and J.K. Howard. 2018. Sensitivity of streamflow to climate change in California. Climatic Change. https://doi.org/10.1007/s10584-018-2244-9.Karuk Tribe. 2019. Karuk Climate Adaptation Plan. 232p.

Karuk Climate Adaptation Plan. 2019.>>>Not sure if this is the plan or the 2016 assessment?

Lee, J.J., and R.W. Baumann. 2022. Dracoleuctra, a new wingless stonefly genus from southern Oregon, U.S.A. (*Plecoptera: Leuctridae*). Zootaxa 5115 (2):258-266.

Micheli, E., C. Dodge, L. Flint, and T. Comendant. 2018. Climate and natural resource analyses and planning for the North Coast Resource Partnership: a technical memorandum summarizing data products. Dwight Center for Conservation Science at Pepperwood, Santa Rosa, CA.

Norsgaard, K.M. 2020. Salmon and acorns feed our people: colonialism, nature, and social action. Rutgers University Press. New Brunswick, NJ.

Noss, R., J. Strittholt, K. Vance-Borland, C. Carroll, and P. Frost. 1999. A conservation plan for the Klamath-Siskiyou ecoregion. Natural Areas Journal 19:392-411.

Olson, D., D. DellaSala , R. Noss , J. Strittholt , J. Kass , M. Koopman , and T. Allnutt. 2012. Climate change refugia for biodiversity in the Klamath-Siskiyou ecoregion. Natural Areas Journal 32:65-74.

Ostberg, S., J. Schewe, K. Childers, and K. Frieler. 2018. Changes in crop yields and their variability at different levels of global warming, Earth System Dynamics 9:479-496.

Ryan, M., W. Palen, M. Adams, and R. Rochefort. 2014. Amphibians in the climate vice: loss and restoration of resilience of montane wetland ecosystems in the western US. Frontiers in Ecology and the Environment 12:232-240.

Sarr, D., A. Duff, E. Dinger, S. Shafer, M. Wing, N. Seavy, and J. Alexander. 2015. Comparing Ecoregional Classifications for Natural Areas Management in the Klamath Region, USA. Natural Areas Journal 35: 360-377.

Sawyer, J.O. 2006. Northwest California: a natural history. University of California Press, Berkeley, CA.

Strittholt, J., and D. Dellasala. 2001. Importance of roadless areas in biodiversity conservation in forested ecosystems: Case study of the Klamath-Siskiyou ecoregions of the United States. Conservation Biology 15:1742-1754.

Swain, D.L., B. Langenbrunner, J.D. Neelin, and A. Hall. 2018. Increasing precipitation volatility in twenty-first century California. Nature Climate Change 8:427-433.

Thompson, C.M., M.W. Gabriel, and K.L. Purcell. 2017. An ever-changing ecological battlefield: marijuana cultivation and toxicant use in western forests. The Wildlife Professional May/June: 42-46.

Vance-Borland, K.R., R.F. Noss, J. Strittholt, P. Frost, C. Carroll, and R. Nawa. 1995. A biodiversity conservation plan for the Klamath/Siskiyou region. Wild Earth 5(4):52-59.

Wagner, D.H. 1997. Klamath-Siskiyou region, California and Oregon, U.S.A. Pages 74-76 in S. D. Davis, V. H. Heywood, O. Herrera-MacBryde, J. Villa-Lobos, and A. C. Hamilton, editors. Centres of plant diversity. Volume 3. The Americas. World Wildlife Fund for Nature and World Conservation Union, Information Press, Oxford, United Kingdom.

Wallace, D. 2003. The Klamath Knot, 20th anniversary edition. University of California Press, Berkeley, CA.

Welsh Jr., H.H., K. Pope, and D. Boiano. 2006. Sub-alpine amphibian distributions related to species palatability to non-native salmonids in the Klamath Mountains of northern California. Diversity and Distributions 12:298-309.

Acknowledgements

Thanks to Jim Agee, Rosemary Sherriff, Mourad Gabriel, and Michael van Hattem for reviewing and improving this chapter. Many more contributed through their research and love of the Klamath Mountains.

Index

Note: Page numbers in *italics* or with *fig.* indicate photos or other illustrations. Page numbers with *t* indicate tables. Page numbers in **bold** indicate key discussions which usually include photos or other illustrations.

E

F

G

H

L

M

N

P

S

T

U

X

Y

Z

The Klamath Mountains
Wilderness Areas and Roads

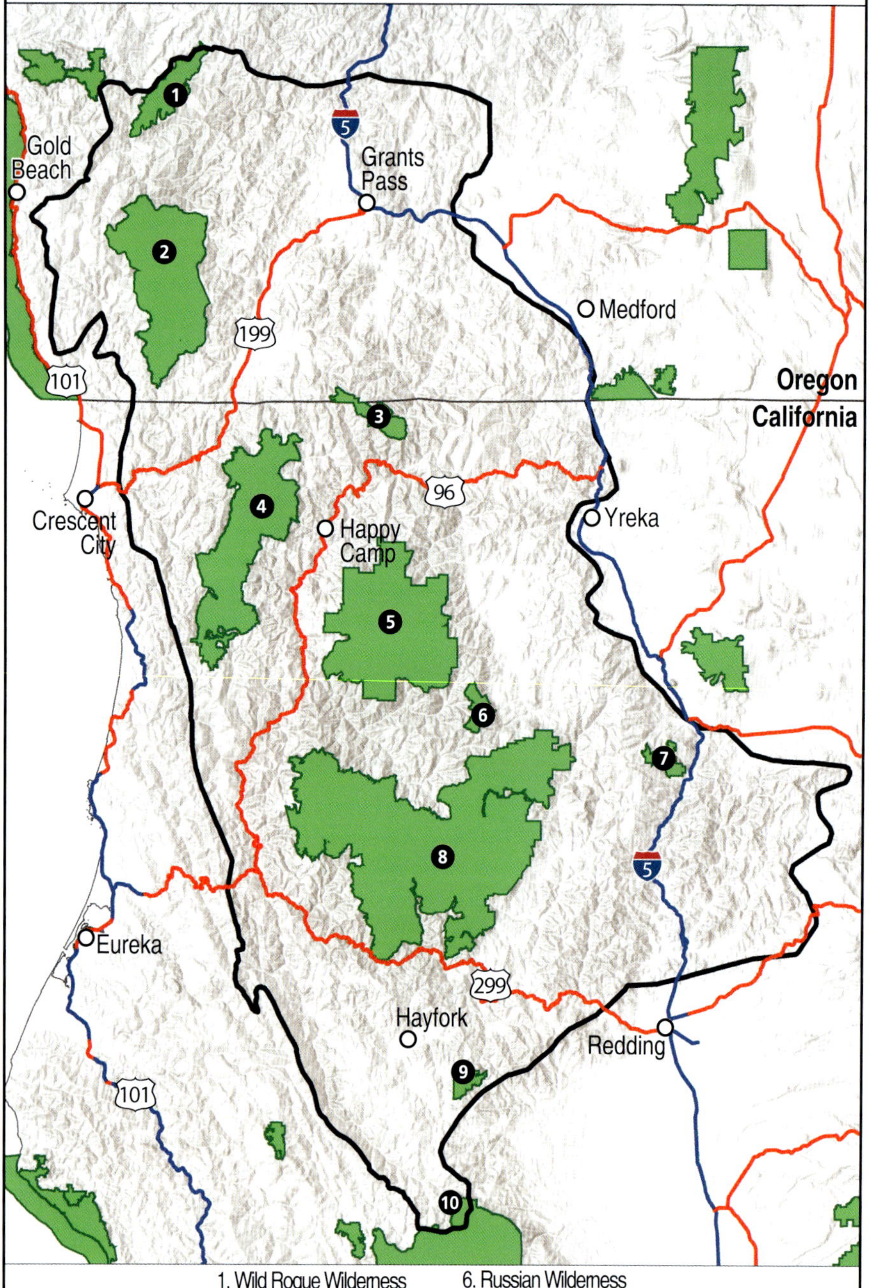

1. Wild Rogue Wilderness
2. Kalmiopsis Wilderness
3. Red Buttes Wilderness
4. Siskiyou Wilderness
5. Marble Mountain Wilderness
6. Russian Wilderness
7. Castle Crags Wilderness
8. Trinity Alps Wilderness
9. Chanchelulla Wilderness
10. Yolla Bolly-Middle Eel Wilderness

Wild Rogue Wilderness by Michael Dotson.

Kalmiopsis Wilderness by Mark Bailey.

Red Butte Wilderness by Michael Kauffmann.

Siskiyou Wilderness by Michael Kauffmann.

Marble Mountain Wilderness by Kauffmann.

Russian Wilderness by M. Kauffmann.

Trinity Alps Wilderness by M. Kauffmann.

Yolla Bolly-Middle Eel Wilderness by MK.

Editors and Authors

MICHAEL KAUFFMANN grew up in the eastern deciduous forests of Virginia and would stare out the window in the winter, hooting for barred owls and wonder...why are some of those trees still green? He nurtured his evergreen conifer fascination in the Appalachian Mountains with a biology degree from Virginia Tech. Upon graduation, found a job teaching outdoor education in California. Sans family, alone in the West, he began to hang out with giant sequoias and foxtail pines on weekends. The high country conifers of southern California entertained him for many years until the temperate rainforest eventually called him north to Humboldt County where he earned a biology Master's degree from Cal Poly Humboldt for his first book *Conifer Country*. He now lives in Kneeland, California, with his wife, Allison, and his sons, Sylas and Asa. He has also authored *Conifers of the Pacific Slope* and co-author of *Field Guide to Manzanitas* and *California Desert Plants*. Besides writing, he works as a devoted Kindergarten through College teacher, is the founder and president of the Bigfoot Trail Alliance, and a research plant ecologist. ***Climate, Cryptogams, Invertebrates, Plant Communities, Birds,*** and ***Mammals.***

JUSTIN GARWOOD has a deep love for the Klamath Mountains. He was raised on the edge of the Trinity Alps in Lewiston, California and after a stint in carpentry, pursued his education at Cal Poly Humboldt where he received a BS in Fisheries Biology and an MS in Wildlife Management and Conservation. He is currently an Environmental Scientist for the California Department of Fish and Wildlife with a focus on fisheries and herpetology in subalpine wetlands and lakes of Northern California. He also studied salmonid populations in Humboldt Bay watersheds and Redwood Creek before becoming the Smith River aquatic biologist where he developed and implemented a basin-wide salmon, steelhead, and coastal cutthroat trout monitoring program for 10 years. One of his long-term research projects is studying population trends and demography of a restored Cascades Frog population in the Trinity Alps for the past 20 years. As an amateur, he also monitors snowfield and modern glacier declines in the Klamath Mountains as the region faces a rapid climate transition. He lives above the Mad River in Humboldt County with his wife Rebecca and his children, Nova and Zane. ***Climate, Water, Invertebrates, Fish, Amphibians,*** and ***Reptiles.***

Authors

James K. Agee is professor emeritus of forest ecology in the School of Environmental and Forest Sciences at the University of Washington, Seattle. He has a fondness for the Klamath as he has been visiting the region since he was a youngster. ***Western Science Arrives, Editing.***

Mark Bailey is a long-time science instructor who has taught multiple topics over a nearly 40-year career. His favorite subjects include Geology, Astronomy, Biology, Prehistory, and Classical Mediterranean History. ***Geology.***

Dan Barton is an Associate Professor at Cal Poly Humboldt and a keen appreciator of birds. ***Birds.***

Dawn Blake is a member of the Hoopa Valley Tribe and has spent nearly 20 years as a wildlife biologist. She is currently the Forestry Director for the Yurok Tribe and is concerned with uniting forest management with culture. ***Klamath as Teacher.***

Tom Carlberg has been a cryptogamic botanist for many years, working for the Forest Service, private contractors, and non-profit organizations. His ongoing interest is mapping the range and distribution of lichen species across California. ***Cryptogams (lichens).***

Rod Crawford is a professional arachnologist who has been curating the spider collection at the Burke Museum in Seattle for the past 50 years. In the past, he has also worked with cave biology, and in 1992-93 performed a cave invertebrate survey at Oregon Caves National Monument. ***Invertebrates (spiders), Editing.***

Brian Dykstra is a botanist, farmer, and founder of the Native Bee Society. ***Invertebrates (Hymenoptera).***

Julie Evens has worked as a vegetation ecologist and botanist for over 20 years. Since 2001 she has served as is the Vegetation Program Director for the California Native Plant Society. ***Plant Communities.***

Chris Feldman is a western herpetologist who studies the evolution and ecology of local reptiles and amphibians. He is a Professor in the Department of Biology at the University of Nevada, Reno (UNR), and also serves as the Co-Director and Curator of Vertebrates in the UNR Museum of Natural History where he studies the evolution and ecology of local reptiles and amphibians (and sometimes mammals). ***Reptiles, Mammals.***

Greg Gray is a wildlife biologist with the U.S. Fish and Wildlife Service, where he works with private landowners and partners to restore oak woodlands, streams, and wetlands. When he is not found birding around Humboldt Bay with his dog Lucy, he is backpacking the Trinity Alps in search of birds and alpine lakes. ***Birds.***

Paul C. Hammond is an expert in the taxonomy and biology of butterflies and moths. He serves as a curator of Lepidoptera in the Oregon State Arthropod Collection at Oregon State University in Corvallis. ***Invertebrates (butterflies).***

Chris Lee is a forest pathologist with Calfire. His work focuses on a variety of pests and pathogens affecting the forests of the North Coast and Klamath Mountains of California. ***Forest Pests.***

Matt Johnson is a Professor at Cal Poly Humboldt and fascinated by all aspects of wildlife Habitat Ecology—from the theoretical to the practical. ***Klamath as Teacher.***

Jeff Kane is an Associate Professor of Fire Ecology and Fuels Management and Director of the Wildland Fire Lab and Fire Resilience Institute at Cal Poly Humboldt. He has authored over 50 journal articles that broadly focus on providing information to better manage and restore fire-prone ecosystems in an era of rapid change. ***Fire.***

Lucy Kerhoulas is a professor in the Forestry and Wildland Resources Department at Cal Poly Humboldt. She studies many aspects of western forests, a few of which include tree physiology, canopy epiphytes, and forest drought responses. ***Cryptogams.***

Julie Kierstead is a retired federal botanocrat who still loves botanizing the Klamath Mountains. Occasionally she finds something new and publishes it. ***Plant Communities.***

Andrew Kinziger is a Professor in the Department of Fisheries Biology at Cal Poly Humboldt where he has taught in the area of fisheries management and mentored M.S. level graduate students since 2003. ***Fishes.***

Frank Lake works for the US Forest Service. His research involves wildland fire effects, traditional ecological knowledge, climate change, and ethno-ecology with an emphasis on cultural management and fire ecology of forest, shrub, grassland and riparian environments in the Klamath Mountain bioregion. ***First Peoples.***

Jordan Mayor is interested in ecosystem scale patterns in above and below ground nutrient cycling, the biogeography of microbial functions, and the preservation of open space. Sometimes this takes him to created wetlands, alpine treelines, and remote tropical rainforests. ***Cryptogams (fungi).***

Len Mazur graduated from Cal Poly Humboldt in 2019 with a degree in Ecology and Restoration and a minor in Botany. He works in natural resource management as a field botanist and prescribed fire technician in Sonoma County, California. ***Invertebrates (mollusks).***

Michael Murray earned his masters from Cal Poly Humboldt studying subalpine meadows of the Marble Mountains and then a Ph.D. in fire history and health of whitebark pine at the University of Idaho. After six years as the Terrestrial Ecologist at Crater Lake National Park, he gained his current occupation with the British Columbia Ministry of Forests. ***Plant Communities.***

Keith Parker is a Senior Biologist with the Yurok Tribe of Northern California, co-managing the lower 44-miles of the Klamath River. Keith received both his undergraduate and graduate degree at Cal Poly Humboldt on a National Science Foundation Graduate Research Fellowship and a Robert and Patricia Switzer Foundation environmental fellowship. ***Klamath as Teacher.***

Eric Peterson is a lifelong biologist who grew up within the Klamath region and developed an intense interest in lichens. With a Ph.D. in plant ecology, Eric has also worked in vegetation ecology and river restoration. ***Cryptogams (lichens).***

Tim Palmer is the award-winning author of 29 books about rivers, conservation, and adventure travel. He is a photographer of America's natural landscapes and a dynamic public speaker with inspiring slide shows about a range of environmental topics. ***Water.***

Charley Reed is a Karuk, Yurok and Hupa person of the Klamath-Trinity watershed in northern California. His work explores his inherent responsibility to manage and protect the spring Chinook salmon and other more-than-human relatives. ***Klamath as Teacher.***

Karen Reiss is a Biology Professor at College of the Redwoods where she teaches a myriad of zoology courses, trains California Naturalists, and curates the CR Natural History Collection. She received her Ph.D. in Zoology from Cornell University and has researched topics as wide-ranging as mammalian intestinal physiology, ant-eating mammal feeding, and most recently, Pacific Northwestern chipmunk biogeography and speciation. ***Mammals.***

Dana Ross is an independent entomologist who specializes in the conservation of rare, threatened, and endangered butterflies in addition to performing Lepidoptera inventories at important ecological sites throughout the Pacific Northwest. He lives in Corvallis, Oregon, where he serves as a Curatorial Associate at the Oregon State Arthropod Collection. ***Invertebrates (butterflies).***

Rosemary Sherriff is a professor in the department of Geography, Environment, and Spatial Analysis at Cal Poly Humboldt. ***Climate.***

David Rains Wallace has published more than twenty books on conservation and natural history, including *The Monkey's Bridge* (a 1997 New York Times Notable Book) and *The Klamath Knot* (1984 Burroughs Medal). He has written articles for the National Geographic Society, The Nature Conservancy, the Sierra Club, and other groups. ***Prologue.***

Tiana Williams-Claussen is a Yurok Tribal member, born and raised in northwestern California. A Harvard graduate and Director of the Yurok Tribe Wildlife Department, she combines Yurok traditional ecological knowledge and the tools of Western science to do her part to advance holistic restoration within Yurok ancestral territory. ***Birds.***

Dana York is a botany research associate at California Academy of Sciences. He received his received his M.S. in Botany from California State University, Fresno, and his B.S. in Forest/Natural Resource Management from Cal Poly Humboldt. He is the author of *The Illustrated Flora of Sequoia and Kings Canyon National Parks*. ***Cryptogams.***

Special Thanks

I (Michael) would like to thank Allison Poklemba for supporting this decade-long journey. She helped with all aspects including covering kid duty, editing and critiquing text and illustrations, photography, designing the book cover, and cultivating a community of Klamath Mountains enthusiasts during our 2021 webinar series and beyond. My kids, Sylas and Asa, have helped me see the Klamath Mountains with new eyes throughout the time this book was taking shape. I look forward to many more years of exploration with them. I would also like to thank my parents for allowing me to explore, ask questions, and seek out answers—they helped me develop and refine this scientific approach to life. Justin Garwood was the best co-editor I could have hoped for. His diligence and persistence made this a much better offering than I could have created on my own. A project like this could have tested a friendship but I'm happy to report ours is headed toward old-growth. Copy editing (a weakness of mine) was provided by many including Jim Agee (amazing contributions throughout the process of creating the book), Matt Ritter (Preface), Rod Crawford, and Dylan Neubauer. Lisa and Leaf Hillman helped me recognize and refine a western science perspective and relearn it from the perspective of Native Peoples—for this I will always be grateful. There are so many more—including all the co-authors—thank you all for your love and support.

I (Justin) would like to thank Rebecca Garwood for her unwavering dedication and support throughout this project—it simply could not have happened without you. To my kids Nova and Zane—thank you for giving me confidence in the future, bringing me writing tea, and making me a kid again while joining me on many crazy aquatic and terrestrial research missions for this project. Thanks to my parents Paul and Lorna Garwood, who allowed my siblings and I to "get lost" so we could find the magic of the Klamath Mountains. I owe my co-editor and good friend Michael Kauffmann special gratitude for trusting me to help lead this project. Michael is a creative force, and he inspires me to do my very best. He also offered an open mind that allowed me to discover and unlock my own creativity leading to many "trust me" assertions that miraculously worked out. Backcountry Press (i.e., Michael and Allison) is a rare gem dedicated to the promotion of accurate natural histories—where quality, beauty, and compassion pour into everything they create. Thank you for allowing this project to slowly manifest resulting in an exceptionally unique and thorough book. Ken Jarvela, Michael Zontos, and Spencer Riffle—you are truly inspiring artists and there is a spot for you around my campfire anytime. Thank you for adding amazing art and photos in the spirit of many vintage natural history books we all grew up with. I have spilled all my knowledge into this project, but it is only a fraction of what others have collectively brought. To all the authors—thank you for joining this adventure without reservation and purely for the love of the Klamath Mountains. To Jim Agee and all the many others who provided subject-matter and copy edits—your detailed reviews and honest suggestions greatly enhanced this book. Last, to all the naturalists—this book is a starting place so please keep teaching me and I will try to do the same.